AF616830

THE BIOLOGY OF TUMORS

PEZCOLLER FOUNDATION SYMPOSIA

Recent volumes published by Plenum Press:

Volume 4 • CELL ADHESION MOLECULES: Cellular Recognition Mechanisms
Edited by Martin E. Hemler and Enrico Mihich

Volume 5 • APOPTOSIS
Edited by Enrico Mihich and Robert T. Schimke

Volume 6 • NORMAL AND MALIGNANT HEMATOPOIESIS: New Advances
Edited by Enrico Mihich and Donald Metcalf

Volume 7 • CANCER GENES: Functional Aspects
Edited by Enrico Mihich and David Housman

Volume 8 • GENOMIC INSTABILITY AND IMMORTALITY IN CANCER
Edited by Enrico Mihich and Leland Hartwell

Volume 9 • THE BIOLOGY OF TUMORS
Edited by Enrico Mihich and Carlo Croce

A Continuation Order Plan is available for this series. A continuation order will bring delivery of each new volume immediately uponn publication. Volumes are billed only upon actual shipment. For further information please contact the publisher.

THE BIOLOGY OF TUMORS

Edited by

Enrico Mihich
Roswell Park Cancer Institute
Buffalo, New York

and

Carlo Croce
Kimmel Cancer Institute
Philadelphia, Pennsylvania

PLENUM PRESS • NEW YORK AND LONDON

Library of Congress Cataloging-in-Publication Data

The biology of tumors / edited by Enrico Mihich and Carlo Croce.
p. cm. -- (Pezcoller Foundation symposia ; v. 9)
"Proceedings of the Ninth Annual Pezcoller Symposium on the Biology of Tumors, held June 4-7, 1997 in Rovereto, Italy"--t.p. verso.
Includes bibliographical references and index.
ISBN 0-306-45932-9
1. Carcinogenesis--Congresses. 2. Cancer cells--Congresses. 3. Carcinogenesis--Molecular aspects--Congresses. 4. Cancer--Genetic aspects--Congresses. I. Mihich, Enrico. II. Croce, Carlo. III. Pezcoller Symposium on the Biology of Tumors (1997 : 7, 19Rovereto, Trento, Italy) IV. Series.
[DNLM: 1. Neoplasms--genetics congresses. 2. Neoplasms--physiopathology congresses. 3. Molecular Biology congresses. 4. Cell Transformation, Neoplastic--genetics congresses. 5. Gene Expression Regulation, Neoplastic congresses. W1PE995 v.9 1998]
RC268.5.B59 1998
616.99'407--dc21
DNLM/DLC
for Library of Congress 98-39857
CIP

Proceedings of the Ninth Annual Pezcoller Symposium on the Biology of Tumors, held June 4 – 7, 1997, in Rovereto, Italy

ISBN 0-306-45932-9

A Division of Plenum Publishing Corporation
233 Spring Street, New York, N.Y. 10013

http://www.plenum.com

10 9 8 7 6 5 4 3 2 1

Printed in the United States of America

THE PEZCOLLER FOUNDATION

The Pezcoller Foundation was created in 1979 by Professor Alessio Pezcoller (1896–1993) who was the chief surgeon of the S. Chiara Hospital in Trento from 1937 to 1966 and donated a substantial portion of his estate to support its activities. The Foundation also benefits from the cooperation of the Bank Cassa di Risparmio di Trento e Rovereto.

The main goal of this non-profit foundation is to provide and recognize scientific progress on life-threatening diseases, currently focusing on cancer. Towards this goal, the Pezcoller Foundation awards the Pezcoller Prize, every two years in recognition of highly meritorious contributions to medical research. It also sponsors a series of annual symposia promoting interactions among scientists working at the cutting edge of basic oncological sciences. The award selection process is managed by the European School of Oncology in Milan, Italy, with the aid of an international committee of experts chaired by Professor U. Veronesi. The symposia are held in the Trentino Region of Northern Italy and their scientific focus is selected by Enrico Mihich with the collaboration of an international standing symposia committee. A program committee determines the content of each symposium.

The first symposium focused on *Drug Resistance: Mechanisms and Reversal* (E. Mihich, Chairman, 1989); the second on *The Therapeutic Implications of the Molecular Biology of Breast Cancer* (M.E. Lippman and E. Mihich, Co-Chairmen, 1990); the third on *Tumor Suppressor Genes* (D.M. Livingston and E. Mihich, Co-Chairmen, 1991); the fourth on *Cell Adhesion Molecules: Cellular Recognition Mechanisms* (M.E. Hemler and E. Mihich, Co-Chairmen, 1992); the fifth on *Apoptosis* (E. Mihich and R.T. Schimke, Co-Chairmen, 1993); the sixth on *Normal and Malignant Hematopoiesis: New Advances* (E. Mihich and D. Metcalf, Co-Chairmen, 1994); the seventh on *Cancer Genes: Functional Aspects* (E. Mihich and D. Housman, Co-Chairmen, 1995); the eighth on *Genomic Instability and Immortality in Cancer* (E. Mihich and L. Hartwell, Co-Chairmen, 1996). The tenth symposium (1998) will focus on *The Genetics of Cancer Susceptibility* (R. Klausner, E. Mihich and L. Strong, Co-Chairmen).

PREFACE

The Ninth Annual Pezcoller Symposium entitled "The Biology of Tumors" was held in Rovereto, Italy, June 4–7, 1997. It focused on the genetic mechanisms underlying heterogeneity of tumor cell populations and tumor cell differentiation, on interactions between tumor cells and cells of host defenses, and the mechanisms of angiogenesis.

With presentations at the cutting edge of progress and stimulating discussions, this symposium addressed issues related to phenomena concerned with cell regulation and cell interactions as determined by activated genes through the appropriate and timely mediation of gene products. Important methodologies that would allow scientists to measure differentially genes and gene products and thus validate many of the mechanisms of control currently proposed were considered, as were the molecular basis of tumor recognition by the immune system, interactions between cells and molecular mechanisms of cell regulation as they are affected by or implemented through these interactions. The molecular and cellular mechanisms of tumor vascularization were also discussed. It was recognized that angiogenesis provides a potential site of therapeutic intervention and this makes it even more important to understand the mechanisms underlying it.

We wish to thank the participants in the symposium for their substantial contributions and their participation in the spirited discussions that followed. We would also like to thank Drs. James Allison, Thierry Boon, Giulio Draetta, Douglas Hanahan, Rakesh Jain and David Livingston, for their essential input as members of the Program Committee, and Ms. A. Toscani for her invaluable assistance. The aid of the Bank Cassa di Risparmio di Trento e Rovereto and the Municipal, Provincial, and Regional Administrations in supporting this symposium through the Pezcoller Foundation is also acknowledged with deep appreciation. Finally, we wish to thank the staff of Plenum Publishing Corporation for their efficient cooperation in the production of these proceedings.

Enrico Mihich
Carlo Croce

CONTENTS

1

THE ROLE FOR *ink*4a IN MELANOMA PATHOGENESIS

One Gene, Two Products, Multiple Pathways

Jason Pomerantz,[1] Nicole Schreiber-Agus,[1] Nanette Liegeois,[1] Alice Tam,[1] Kenneth P. Olive,[1] Ronald A. DePinho,[1,*] and Lynda Chin[1,2]

[1]Department of Microbiology and Immunology
[2]Division of Dermatology, Department of Medicine
Albert Einstein College of Medicine
1300 Morris Park Avenue, Bronx, New York 10461

1. INTRODUCTION: *ink4a, RAS,* AND THE GENETICS OF MALIGNANT MELANOMA

Malignant melanoma is a disease with high metastatic potential and poor clinical response to current therapeutic measures[1]. It represents a significant health crisis given its high rate of increase in incidence; by year 2000, one in 76 Americans will develop melanoma[2]. Although the molecular pathogenesis of this disease is poorly understood, predisposition to melanoma appears to have a strong genetic component. Tumor surveys and kindred analyses have uncovered several potential chromosomal "hot spots" including frequent loss of 6q and 10q, non-random karyotypic alterations of chromosome 1, and 9p21-associated deletion/mutation[1]. The latter appears to be the most compelling etiological link to melanoma in that cytogenetic, linkage and molecular analyses have documented a high incidence of 9p21 germline and somatic mutations in both familial and sporadic melanomas[3–6].

In mouse and man, the 9p21 locus has the capacity to encode at least three potent growth inhibitors (Figure 1). It contains the closely linked *ink4a* and *ink4b* genes that encode the highly related G1 cyclin-dependent kinase inhibitors, $p16^{INK4a}$ and $p15^{INK4b}$, respectively[7–10]. The *ink4a* gene, through alternative first exon usage and reading frames, also encodes for a second, distinct, growth inhibitor protein $p19^{ARF}$ (ARF for alternative reading

* To whom correspondence should be addressed: tel 718-430-2822; fax 718-430-8972; email depinho@aecom.yu.edu

The Biology of Tumors, edited by Mihich and Croce
Plenum Press, New York, 1998.

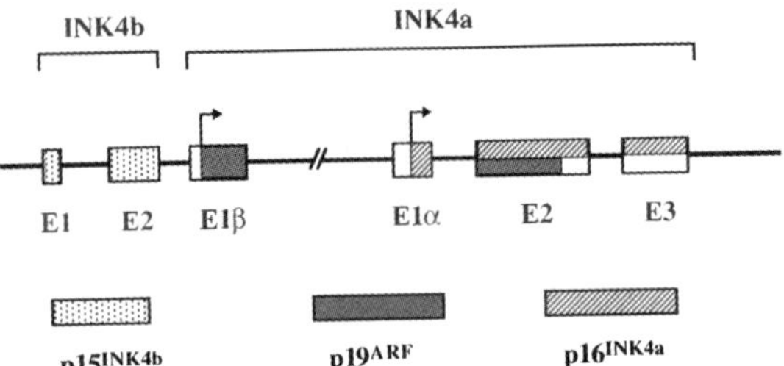

Figure 1. Schematic diagram of the 9p21 locus. *Ink4a* and *ink4b* encode the highly related cyclin-dependent kinase inhibitors, p16^{ink4a} and p15^{ink4b}, respectively. Through alternative first exon usage, the *ink4a* gene encodes another cell cycle inhibitor termed p19^{ARF} (alternative reading frame).

frame)[11]. This complex genomic organization coupled with the common occurrence of large homozygous 9p21 co-deletions[3–6] hampered initial efforts to designate *ink4a* versus *ink4b* as the 9p21 melanoma susceptibility gene, and to delineate p16^{INK4a} versus p19^{ARF} contribution to *ink4a* tumor suppressor function. Large homozygous deletion of 9p21 represents an unusual loss-of-function profile for a tumor suppressor locus, which classically presents with inactivating point mutations on the remaining wild type allele after sustaining deletion of one allele[12,13]. As such, this presentation raises the possibility that more than one 9p21 gene product may be targeted for inactivation in tumorigenesis. While initial supposition favored the concept of *ink4a* and *ink4b* elimination, observations of germline mutations which exclusively targeted the *ink4a* gene in melanoma susceptible individuals[14,15] indicated that *ink4a*, as opposed to *ink4b*, is the principal target. For *ink4a,* a clear anti-neoplastic role for p16^{INK4a} is supported by the existence of germline mutations that exclusively target the p16^{INK4a} reading frame[16]. However, this observation does not rule out a potential cooperating role for p19^{ARF} as a tumor suppressor. In fact, the frequent occurrence of 9p21 deletions/mutations which dually affect both p16^{INK4a} and p19^{ARF} coding sequences has led us to speculate that both of these products contribute to *ink4a*-mediated tumor suppression.

In addition to loss of 9p21 tumor suppressor activity, the Receptor Tyrosine Kinase (RTK)-RAS-MAPK pathway is thought to play a prominent role in the stepwise phenotypic progression from a normal melanocytic phenotype to metastatic malignant melanoma. Studies in transgenic mice[17,18] and in a fish melanoma model[19] have shown that over-expression and/or activation of RTKs in melanocytes result in melanoma. Although activating *RAS* mutations have been observed in melanoma cells, a clear causal role for activated *RAS* in melanocyte transformation has yet to be demonstrated experimentally. Stable transfection of activated H-*RAS* in cultured mouse[20,21] and human[22] melanocytes generated fully transformed melanoma cells capable of anchorage independent growth and tumorigenicity in nude mice. Some mutational analyses revealed a high frequency of N-*RAS* mutations in primary melanomas[23]. Moreover, several groups[24–26] have correlated mutational profiles with tumor stages of non-cultured melanoma samples and reported a higher frequency of *RAS* mutations in metastatic and recurrent tumors, suggesting a role for *RAS* activation in disease progression rather than initiation. On the other hand, Albino et al. reported a significantly higher frequency of activating *RAS* mutations in cultured melanoma cell lines (24%) than in non-cultured melanomas (5–6%)[27] raising the possibility that *RAS* mutations may be a consequence of the inherent genomic instability of transformed cells. Furthermore, a transgenic mouse model in which activated H-*RAS* was overexpressed in melanocytes generated melanocytic hyperplasia but not a melanoma phenotype[28]. In short, a causal role for activating *RAS* mutations in melanoma development has remained a point of controversy.

Below, we review a series of recent studies from our laboratory in which we have attempted to verify the role of *ink4a* in tumor suppression as well as to understand its functional interrelationship to *RAS* activation in melanoma development *in vivo*. To accomplish these objectives, an *ink4a* knockout mouse model was constructed and the impact of loss of *ink4a* gene function was examined in the context of melanocyte-directed mutant H-*RAS* transgene expression. Using this model and cell culture-based transformation assays, a number of issues have been addressed including the tumor suppressor role of *ink4a in vivo*, the relative contribution of $p19^{ARF}$ and $p16^{INK4a}$ to *ink4a* tumor suppressor activity, and the significance of *ink4b* loss in 9p21-associated cancers.

2. RESULTS AND DISCUSSION

2.1. The *ink4a* Gene in Tumor Suppression

2.1.1. Lessons Learned from the ink4a KO Mice. In an effort to verify directly the role of *ink4a* in oncogenesis, we have previously generated a germline *ink4a* null allele that eliminates both $p16^{INK4a}$ and $p19^{ARF}$[29]. Genotype analysis of live offspring from heterozygous intercrosses revealed that all three genotypes were present in the expected Mendelian ratio. Disruption of the *ink4a* gene was verified in homozygotes on the genomic, transcript and protein levels. Moreover, compensatory changes in the expression of other *ink4* family members was not evident from an analysis of CDK4 complexes in co-immunoprecipitation studies. Homozygous null *ink4a* mice were found to be viable and fertile, albeit with some decline in reproductive capacity after 4 to 6 months of life. With progression through post-natal life, these mice exhibit advanced extramedullary hematopoiesis presenting with marked infiltration and enlargement of the spleen, liver and lung. The basis for this progressive hematopoietic disorder is not understood at present but does not result from peripheral anemia.

By far, the most prominent phenotypic manifestation of *ink4a*-deficiency is the marked predisposition to tumor formation. These mice developed spontaneous tumors at an early age; 69% of observed animals developed histologically confirmed tumors with an average latency of 7 months. In addition, *ink4a*-deficient mice were highly susceptible to tumor induction by carcinogens with 50% of them succumbing to tumors by 2 months of age. The tumor spectrum in the *ink4a* knockout mice included predominantly malignant fibrosarcoma and B cell lymphoma (Table 1). The complete absence of malignant melanoma was an unanticipated outcome given the prominent representation of this cancer type in humans harboring germline *ink4a* mutations[8]. Species differences in tumor spectra have been observed for other tumor suppressors such as *Rb*[30]. The genetic basis for these differences is not understood, but may relate to variability in the expression or activity of cell type-specific cancer susceptibility modifiers and/or micro-environmental differences. Assuming that cooperation with additional pro-oncogenic modifiers could facilitate the development of malignant melanoma in *ink4a*-deficient mice, we assessed the cooperative effects of *RAS* activation and *ink4a* loss in melanocyte transformation *in vivo*. The selection of *RAS* was based upon the common involvement of the RTK-RAS-MAPK pathway, the occurrence of *RAS* mutation in human melanoma, and the enhanced transforming activity of activated H-*RAS* in *ink4a*-deficient fibroblasts[29].

2.1.2. Building a Mouse Model for Malignant Melanoma. Transgenic mice were generated in which activated H-*RAS* was overexpressed specifically in melanocytes. Melanocyte-specific transgene expression was achieved using the mouse tyrosinase gene promoter and a strong melanocyte-specific enhancer located far upstream of the tyrosinase

Table 1. Spectrum of histologically-confirmed tumor types arising in mice homozygous or heterozygous for the *ink4a* null allele

Tumor Histology	*ink4a* +/−	*ink4a* −/−
fibrosarcoma	50%	45%
lymphoma	25%	33%
Squamous cell carcinoma	25%	9%
sarcoma	none	6%
angiosarcoma	none	3%
liposarcoma	none	3%
malignant melanoma	none	none

promoter region[31]. Several tyr-*RAS* mice spontaneously developed cutaneous and, less commonly, ocular tumors. These tumors presented either as amelanotic dermal nodules with marked telangiectasia or exophthalmos resulting from enlarging retro-orbital mass. The common cutaneous sites of tumor formation include the torso, pinna of the ears, tail and perineum (Fig. 2). Although these tumors were highly invasive locally, metastatic disease was not detected in any tumor-bearing mice.

Histologically, these tumors were composed of spindle cells with prominent epithelioid features and varying degrees of melanization. They exhibited strong positive immunoreactivity to a melanocyte-specific marker, TRP1[32]. The ocular tumors appeared to emerge from the pigmented retinal epithelium. In early stages of tumorigenesis, these ocu-

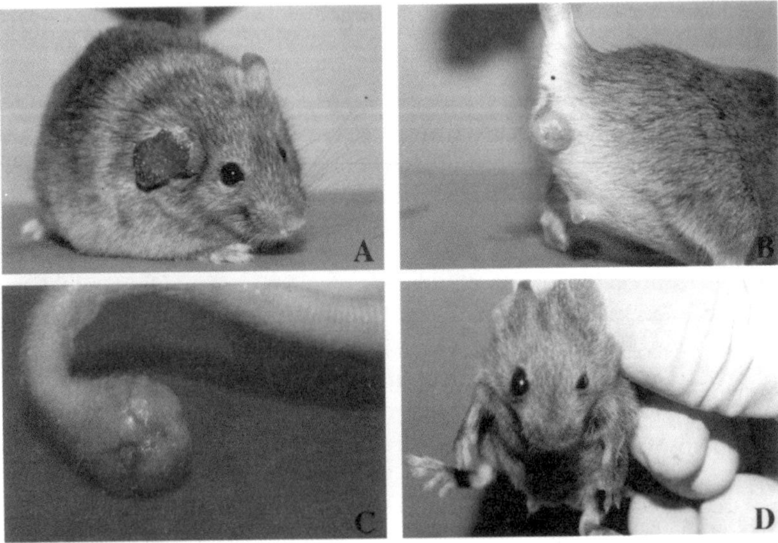

Figure 2. Tyrosinase-*Ras* transgenic mice develop multiple primary melanomas at cutaneous (A,B,C) and ocular (D) sites.

lar neoplasms were heavily pigmented and exhibited strong S100 immunoreactivity. Subsequently the tumors underwent a distinct morphological transition characterized by loss of pigmentation and S100 immunoreactivity, a finding consistent with the occurrence of additional genetic lesions beyond *RAS* activation and loss of *ink4a* function.

2.1.3. Role of ink4a in Melanoma Development. As discussed above, tumor-prone homozygous null *ink4a* mice do not develop melanomas. Thus, a number of studies were conducted to assess the consequences of *RAS* activation and *ink4a*-deficiency on melanocyte growth *in vivo*. First, we analyzed Tyr-*RAS* transgenic mice harboring one null allele for *ink4a* to determine whether the remaining wildtype allele is eliminated during melanoma development, and if eliminated, whether the *ink4b* gene is consistently co-deleted as well. In all cases, tumors arising in the tyr-*RAS ink4a*+/– mice sustained a deletion encompassing the wildtype *ink4a* gene with maintenance of the knockout allele. On the contrary, only a subset of these tumors eliminated the neighboring *ink4b* gene sequences; and more significantly, no tumor sustained deletion of *ink4b* without deletion of *ink4a*. Thus, one interpretation of the frequent co-deletion of *ink4a* and *ink4b* in this subset of mouse tumors, as well as in published studies of human cell lines and clinical tumor samples[3,4,6,33], may be that the loss of *ink4b* results from *ink4a* deletional events which randomly extend to *ink4b* rather than from a biological requirement for genesis of this cancer type. Support for the innocent bystander scenario is derived from the demonstration that *ink4b* sequences remain intact in *tyr-RAS* tumors arising in mice homozygous null for *ink4a*, i.e., melanomas in which deletion of *ink4a* is not genetically required. Together, these data suggest that *ink4a* is the principal target for tumor-associated chromosomal loss and that the occasional elimination of *ink4b* likely reflects its close proximity to the *ink4a* gene.

The second and more definitive study of a role for *ink4a* in melanoma was a comparison of the incidence of melanoma development in tyr-*RAS* mice in the presence or absence of *ink4a* deficiency. During an observation period of 6.5 months, none of the *tyr-RAS ink4a*+/+ mice developed grossly apparent tumors or signs of ill health. In the *tyr-RAS ink4a* 94Symbol"-/94Symbol"- cohort, 65% of the mice developed histologically verified melanomas (Table 2a). Among these tumors, cutaneous melanomas predominated (Table 2b), and no metastatic lesion was observed. The onset of melanoma development is 2 months, and that for non-melanocytic tumors is over 4 months. Thus, *tyr-RAS* mice are particularly susceptible to the development of melanomas in the absence of *ink4a* gene function, a finding which directly implicates *ink4a* in melanoma suppression *in vivo*.

2.2. Role for the *RAS* Pathway in Melanoma

Although activating *RAS* mutations have an established role in the genesis of other cancers, the mouse model reported here provided strong support for its oncogenic role in malignant melanoma as well. In this model, the development of multiple, locally invasive primary melanomas without evidence of metastatic spread clearly indicates that additional genetic events beyond *RAS* activation and *ink4a*-deficiency are required for progression to metastatic disease. This outcome stands in contrast to previous studies which suggested a role for *RAS* activation in more advanced stages of melanoma (i.e. progression rather than initiation), particularly in promoting a metastatic phenotype (see above).

The initiator role served by *RAS* is undoubtedly a necessary molecular step in our model since *ink4a*-deficiency alone does not generate melanoma. However, the prolonged tumor latency in the absence of *ink4a*-deficiency suggests that *RAS* activation is not by itself a potent inducer of melanoma. The weak oncogenic activity of mutant H-*RAS* in

Table 2. A) Tyrosinase-*Ras* mice are particularly susceptible to melanoma development in the absence of *ink4a*. B) Tyrosinase-*Ras* melanomas originate primarily at cutaneous sites and less commonly from the retinal pigmented epithelium

A

Tyr-RAS Tumors	*ink4a* +/+ (n=25)	*ink4a* −/− (n=31)
melanoma	none	20
fibrosarcoma	none	2
melanoma and fibrosarcoma	none	1
Deaths of unknown cause	none	6

B

Tumor Sites	# Tumors
Tail	6
Back	7
Ear	3
Abdomen	1
Perineum	1
Eye	3

melanocytes *in vivo* was also evident in the previous transgenic study where melanocyte-specific expression of mutant *H-RAS* resulted only in melanocytic hyperplasia without melanoma development[28]. This phenotypic difference may relate to our use of the far up-stream tyrosinase enhancer element[31], our utilization of germline *ink4a* mutations, and variability in genetic background. On another level, the modest oncogenic actions of *RAS* in melanocyte transformation may be in accord with recent studies demonstrating that overexpression of activated H-*RAS* in primary fibroblasts induces a G1 arrest and premature cellular senescence and that H-RAS-induced mitogenesis or oncogenesis requires an accompanying immortalizing event such as *ink4a* or *p53*-deficiency[34]. This requirement for antecedent or concomitant immortalization events to elicit *RAS*-induced transformation of cultured cells matches well with the synergistic actions of *RAS* activation and *ink4a* deficiency in melanocyte transformation observed in this study.

2.3. *ink4a*: One Gene, Two Proteins, Multiple Pathways

2.3.1. Evidence Supporting an Anti-Oncogenic Role for p19^{ARF}. The knockout mouse studies (above) have verified that an intact *ink4a* gene plays an integral role in cellular growth control, functions in pathways linked to cellular senescence, and guards against the development of cancer *in vivo*[29]. At the same time, however, the simultaneous disruption of two distinct ORFs precluded definitive assessment of whether the loss of p16^{Ink4a}, p19^{ARF}, or both, is responsible for the increased susceptibility to neoplastic transformation. In human cancers and tumor cell lines, the vast majority of *ink4a* gene mutations/deletions affect both p16^{Ink4a} and p19^{ARF} coding sequences because they involve the shared exon 2 region[16,35–38].

As an initial assessment of its function, we examined the capacity of p19^{ARF} to suppress malignant transformation of primary cells. Using the highly quantitative rat embryo fibroblast cooperation assay, we have demonstrated potent p19^{ARF}-induced suppression of primary cell transformation by Myc/RAS and E1a/RAS, as well as anti-neoplastic synergy with p16^{Ink4a}. Although little is known about its mechanism of action, p19^{ARF} transcript levels are observed to be up-regulated in cell lines in which either p53 is inactivated through genetic mutation, mdm2 is overexpressed, or temperature-sensitive SV40 T-ag is induced[11]. These observations, coupled with the fact that p19^{ARF}, like p53, is purported to act in G1 and G2M, raised the possibility of a functional connection between p19^{ARF} and the p53 pathway. A potential mechanistic link also was supported by the observations that SV40 Large T antigen transformation was refractory to the inhibitory effects of p19^{ARF} and that dominant-negative p53 mutants significantly attenuated the anti-oncogenic effects of p19^{ARF}. Correspondingly, the anti-transformation potential of p19^{ARF} is markedly reduced in p53-deficient MEFs. Together, these results suggest that p19^{ARF} functions in a p53-dependent pathway. As such, the frequent involvement of *ink4a* in cancer pathogenesis may represent a strategic genetic route for efficient transformation in that tumor suppressor activities of proteins in both the Rb and p53 pathways are eliminated. However, the partial (as opposed to complete) growth inhibitory effect of p19^{ARF} in a p53 null context indicates that p19^{ARF} may also operate in a p53-independent manner.

2.3.2. Implications for Tumorigenesis. The potential to disrupt two essential growth control pathways through a single genetic hit may provide an explanation for (1) the exceedingly high frequency of *ink4a* gene deletion in many human tumors and their derivative cell lines[3,39–41] (2) the high incidence of spontaneous tumors in mice lacking *ink4a* exon2/3 sequences[29], and (3) the distinctively strong connection between tumorigenesis and the *ink4a* gene in contrast to other genes encoding cyclin-dependent kinase inhibitors such as p21^{CIP1}, p27^{KIP1}, p57^{KIP2} and other *ink4* family proteins[42,43]. Specifically, mice lacking p21^{CIP1} do not show increased rates of spontaneous tumors[44], and although p27^{KIP2}-deficient mice can develop intermediate lobe pituitary hyperplasia or adenoma, these neoplasias rarely progress to malignant pituitary tumors[45–47]. Similarly, in human cancers, the frequent alteration of *ink4a* contrasts sharply with an overall lower rate of *ink4b* mutation/deletion[48,49], infrequent mutations in p21^{CIP1} (apparently found only in prostate and bladder cancers)[50–53], and no reported genetic lesions for p27^{KIP1}or p57^{KIP2}[54,55]. Such biological correlates would not have been anticipated given the highly similar biochemical profiles and cell culture activities of these cyclin-dependent kinase inhibitors. These observations raise the possibility that an anti-tumorigenic role for these other cyclin-dependent kinase inhibitors could be uncovered in a p53-null context—a supposition that is now testable. In particular, the genetic mechanisms leading to these cancers could require the disruption of multiple tumor suppressor pathways by elimination of any one of the functional inhibitory elements positioned along each pathway. Such observations have been made convincingly for the Rb pathway in which the tumor associated inactivating mutations in either Rb or p16^{Ink4a} occur, but not both[56]. Based upon the findings of this study, one would predict that tumors deficient for both $p16^{Ink4a}$ and $p19^{ARF}$ would be less likely to harbor Rb or p53 mutations and that *ink4a* mutations which spare $p19^{ARF}$ would be associated with crippling alterations involving other components of the p53 pathway (e.g., *mdm2* amplification or loss of p53 function). It is important to add that elimination of $p19^{ARF}$ would not preclude p53 mutation since p53 plays multiple roles in suppressing neoplastic growth that are likely to extend beyond the $p19^{ARF}$-p53 pathway. Stated differently, loss of function mutations of $p19^{ARF}$ would be predicted to decrease, not eliminate, tumor-associated p53 mutations. Support for this view has come from an analysis of *ink4a* and *p53* mutations in the same human cancers; this demonstrated a clear reciprocal relationship between these two genes, i.e., *ink4a*-deficient ($p16^{Ink4a}$ + $p19^{ARF}$) cancers rarely exhibit p53 mutant products (Liegeois and DePinho, in preparation).

Although *p53* mutations represent the most common genetic abnormality in human cancers[57,58], human melanomas and those generated in the *tyr-RAS* model are remarkably free of *p53* mutations and deletions. A possible explanation for this phenomenon is that some other component of the *p53* pathway renders melanomas functionally deficient for p53, e.g., *mdm*-2 gene amplification[59,60]. Alternatively, it is possible that some degree of functional overlap in tumor suppressor activity exists between *p53* and *ink4a.* If such senarios are indeed the case, then the very high frequency of *ink4a* deletion could obviate the need for *p53* elimination in such tumors. Along these lines, both *p53* and *ink4a* encode potent growth and tumor suppressive activities and their loss of function correlates with cellular immortalization and transformation by activated RAS[34]. Although a direct mechanistic link between *p53* and *ink4a* pathways has yet to be established, it is intriguing that high levels of $p19^{ARF}$ have been observed in p53-deficient cells[11], leaving open a possible regulatory feedback loop. Moreover, while $p19^{ARF}$ can block transformation by Myc/RAS or E1a/RAS, it has no effect on the capacity of SV40 Large T-antigen to cooperate with RAS to transform primary cells (Liegeois and DePinho, in preparation). This result gains significance in light of the ability of SV40 Large T-antigen to render cells functionally deficient for p53[61]. Although the actions of $p19^{ARF}$ have yet to be positioned along a known tumor suppressor pathway, a functional link between $p19^{ARF}$ and p53 could account for the reciprocal relationship of mutations in these genes in human and mouse melanomas.

3. CONCLUSION

The studies described here demonstrate that *ink4a* is a bona-fide tumor suppressor *in vivo.* Moreover, the Tyr-*RAS* model provides the first *in vivo* experimental evidence to support the observed link between *ink4a* loss and melanoma development. The necessity for the presence of activating *RAS* mutations in order to generate melanoma on the *ink4a* deficient background highlights the importance of the RTK-RAS-MAPK pathway in melanoma pathogenesis. But the long latency in the absence of *ink4a* deficiency, and the lack of a metastatic phenotype in all established tumors, points to a weak oncogenic role for H-RAS^{G12V} in this setting. Finally, evidence is provided in support of anti-oncogenic activity for the *ink4a* alternative product, $p19^{ARF}$. $p19^{ARF}$ shows potent suppression of transformation in vitro which is partially dependent on an intact p53 pathway. This experimental data is supported by the observation that, *in vivo*, Tyr-*RAS* tumors show loss of heterozygosity in which the wild-type *ink4a* allele is deleted in regions encoding both $p16^{INK4A}$ and $p19^{ARF}$ while p53 remains wild-type.

ACKNOWLEDGMENTS

JP is a recipient of a HHMI Medical Student Research Training Fellowship and the Oncogene Obewon Award. RAD is supported by grants from the National Institutes of Health (R01HD28317, R01EY09300, and R01EY11267) and is a recipient of the Irma T. Hirschl Career Scientist Award. Support from the Cancer Core grant P30CA13330 is also acknowledged.

REFERENCES

1. Herlyn M. *Molecular and cellular biology for melanoma* (R.G. Landes, Austin, 1993)
2. Rigel DS, Friedman RJ, Kopf AW. Lifetime risk for development of skin cancer in the U.S. population: current estimate is now 1 in 5. *J.Am.Acad.Derm.* **35,** 1012–1013 (1996)
3. Kamb A, et al. Analysis of the p16 gene (CDKN2) as a candidate for the chromosome 9p melanoma susceptibility locus. *Nature Genet.* **8,** 23–26 (1994)
4. Jen J, et al. Deletion of p16 and p15 genes in brain tumors. *Cancer Res.* **54,** 6353–6358 (1994)
5. Orlow I, et al. Deletion of the p16 and p15 genes in human bladder tumors. *J.Natl Cancer Inst* **87,** 1524–1529 (1995)
6. Flores JF, et al. Loss of the p16INK4a and p15INK4b genes, as well as neighboring 9p21 markers, in sporadic melanoma. *Cancer Res.* **56,** 5023–5032 (1996)
7. Serrano M, Hannon GJ, Beach D. A new regulatory motif in cell cycle control causing specific inhibition of cyclin D/cdk4. *Nature* **366,** 704–707 (1993)
8. Kamb A, et al. A cell cycle regulator potentially involved in genesis of many tumor types. *Science* **264,** 436–440 (1994)
9. Hannon GJ, Beach D. $p15^{INK4b}$ is a potential effector of cell cycle arrest mediated by TGF-β. *Nature* **371,** 257–261 (1994)
10. Quelle DE, et al. Cloning and characterization of murine $p16^{INK4a}$ and $p15^{INK4b}$ genes. *Oncogene* **11,** 635–645 (1995)
11. Quelle DE, Zindy F, Ashmun RA, Sherr CJ. Alternative reading frames of the INK4a tumor suppressor gene encode two unrelated proteins capable of inducing cell cycle arrest. *Cell* **83,** 993–1000 (1995)
12. Cordon-Cardo C. Mutations of cell cycle regulators. Biological and clinical implications for human neoplasia. *Am.J.Pathol.* **147,** 545–560 (1995)
13. Levine AJ. p53, the cellular gatekeeper for growth and division. *Cell* **88,** 323–331 (1997)
14. Hussussian CJ, et al. Germline p16 mutations in familial melanoma. *Nature Genet.* **8,** 15–21 (1994)
15. Gruis NA, et al. Homozygotes for CDKN2 (p16) germline mutation in Dutch familial melanoma kindreds. *Nature Genet.* **10,** 351–353 (1995)
16. FitzGerald MG, et al. Prevalence of germ-line mutations in p16, p19ARF, and CDK4 in familial melanoma: analysis of a clinic-based population. *Proc.Natl.Acad.Sci.USA* **93,** 8541–8545 (1996)
17. Iwamoto T, et al. Aberrant melanogenesis and melanocytic tumour development in transgenic mice that carry a metallothionein/ret fusion gene. *EMBO* **10,** 3167–3175 (1991)
18. Takayama H, et al. Diverse tumorigenesis associated with aberrant development in mice overexpressing hepatocyte growth factor/scatter factor. *Proc.Natl.Acad.Sci.USA* **94,** 701–706 (1997)
19. Wittbrodt J, Lammers R, Malitschek B, Ullrich A, Schartl M. The Xmrk receptor tyrosine kinase is activated in Xiphophorus malignant melanoma. *EMBO* **11,** 4239–4246 (1992)
20. Wilson RE, Dooley TP, Hart IR. Induction of tumorigenicity and lack of in vitro growth requirement for 12-O-tetradecanoylphorbol-13-acetate by transfection of murine melanocytes with v-Ha-ras. *Cancer Res.* **49,** 711–716 (1989)
21. Ramon y, Cajal S, Suster S, Halaban R, Filvaroff E, Dotto GP. Induction of different morphologic features of malignant melanoma and pigmented lesions after transformation of murine melanocytes with bFGF-cDNA and H-ras, myc, neu, and E1a oncogenes. *Am.J.Pathol.* **138,** 349–358 (1991)
22. Albino AP, Sozzi G, Nanus DM, Jhanwar SC, Houghton AN. Malignant transformation of human melanocytes: induction of a complete melanoma phenotype and genotype. *Oncogene* **7,** 2315–2321 (1992)
23. van 't Veer LJ, et al. N-ras mutations in human cutaneous melanoma from sun-exposed body sites. *Mol.Cell.Biol.* **9,** 3114–3116 (1989)

24. Ball NJ, Yohn JJ, Morelli JG, Norris DA, Golitz LE, Hoeffler JP. Ras mutations in human melanoma: a marker of malignant progression. *J.Invest.Dermatol.* **102,** 285–290 (1994)
25. Wagner SN, Ockenfels HM, Wagner C, Hofler H, Goos M. Ras gene mutations: a rare event in nonmetastatic primary malignant melanoma. *J.Invest.Dermatol.* **104,** 868–871 (1995)
26. Jafari M, et al. Analysis of ras mutations in human melanocytic lesions: activation of the ras gene seems to be associated with the nodular type of human malignant melanoma. *J.Cancer Res.&Clin.Oncol.* **121,** 23–30 (1995)
27. Albino AP, et al. Analysis of ras oncogenes in malignant melanoma and precursor lesions: correlation of point mutations with differentiation phenotype. *Oncogene* **4,** 1363–1374 (1989)
28. Powell MB, et al. Hyperpigmentation and melanocytic hyperplasia in transgenic mice expressing the human T24 Ha-ras gene regulated by a mouse tyrosinase promoter. *Mol.Carcin.* **12,** 82–90 (1995)
29. Serrano M, Lee H, Chin L, Cordon-Cardo C, Beach D, DePinho RA. Role of the INK4a locus in tumor suppression and cell mortality. *Cell* **85,** 27–37 (1996)
30. Jacks T, Fazeli A, Schmitt EM, Bronson RT, Goodell MA, Weinberg RA. Effects of an Rb mutation in the mouse. *Nature* **359,** 295–300 (1992)
31. Ganss R, Montoliu L, Monaghan AP, Schutz G. A cell-specific enhancer far upstream of the mouse tyrosinase gene confers high level and copy number-related expression in transgenic mice. *EMBO* **13,** 3083–3093 (1994)
32. Thomson TM, Real FX, Murakami S, Cordon-Cardo C, Old LJ, Houghton AN. Differentiation antigens of melanocytes and melanoma: analysis of melanosome and cell surface markers of human pigmented cells with monoclonal antibodies. *J.Invest.Dermatol.* **90,** 459–466 (1988)
33. Orlow I, Lianes P, Lacombe L, Dalbagni G, Reuter VE, Cordon-Cardo C. Chromosome 9 deletions and microsatellite alterations in human bladder tumors. *Cancer Res.* **54,** 2848–2851 (1994)
34. Serrano M, Lin AW, McCurrach ME, Beach D, Lowe SW. Oncogenic ras provokes premature cell senescence associated with accumulation of p53 and p16INK4a. *Cell* **88,** 593–602 (1997)
35. Newcomb EW, Rao LS, Giknavorian SS, Lee SY. Alterations of multiple tumor suppressor genes (p53 (17p13), p16INK4 (9p21), and DBM (13q14)) in B-cell chronic lymphocytic leukemia. *Mol.Carcin.* **14,** 141–146 (1995)
36. Kinoshita I, et al. Altered p16INK4 and retinoblastoma protein status in non-small cell lung cancer: potential synergistic effect with altered p53 protein on proliferative activity. *Cancer Res.* **56,** 5557–5562 (1996)
37. Heinzel PA, Balaram P, Bernard HU. Mutations and polymorphisms in the p53, p21 and p16 genes in oral carcinomas of Indian betel quid chewers. *Intl.J.Cancer* **68,** 420–423 (1996)
38. Hangaishi A, et al. Inactivation of multiple tumor-suppressor genes involved in negative regulation of the cell cycle, MTS1/p16INK4A/CDKN2, MTS2/p15INK4B, p53, and Rb genes in primary lymphoid malignancies. *Blood* **87,** 4949–4958 (1996)
39. Hunter T, Pines J. Cyclins and cancer II: cyclin D and CDK inhibitors come of age. *Cell* **79,** 573–582 (1994)
40. Nobori T, Miura K, Wu DJ, Lois A, Takabayashi K, Carson DA. Deletions of the cyclin-dependent kinase-4 inhibitor gene in multiple human cancers. *Nature* **368,** 753–756 (1994)
41. Merlo A, et al. 5'CpG island methylation is associated with transcriptional silencing of the tumour suppressor CDKN2/p16 in human cancers. *Nature Med.* **7,** 686–692 (1995)
42. Kamb A. Cell-cycle regulators and cancer. *Trends in Genetics* **11,** 136–140 (1995)
43. Elledge SJ, Winston J, Harper JW. A question of balance: the role of cyclin-kinase inhibitors in development and tumorigenesis. *TICB* **6,** 388–392 (1996)
44. Deng C, Zhang P, Harper JW, Elledge SJ, Leder PJ. Mice lacking *p21*$^{CIP1/WAF1}$ undergo normal development, but are defective in G1 checkpoint control. *Cell* **82,** 675–684 (1995)
45. Fero ML, et al. A syndrome of multiorgan hyperplasia with features of gigantism, tumorigenesis, and female sterility in p27^{Kip1}-deficient mice. *Cell* **85,** 733–744 (1996)
46. Kiyokawa H, et al. Enhanced growth of mice lacking the cyclin-dependent kinase inhibitor function of p27^{KIP1}. *Cell* **85,** 721–732 (1996)
47. Nakayama K, et al. Mice lacking p27^{KIP1} display increased body size, multiple organ hyperplasia, retinal displasia, and pituitary tumors. *Cell* **85,** 707–720 (1996)
48. Stone S, et al. Complex structure and regulation of the p16(MTS1) locus. *Cancer Res.* **55,** 2988–2994 (1995)
49. Hirama T, Keoffler HP. Role of cyclin-dependent kinase inhibitors in the development of cancer. *Blood* **86,** 841–854 (1995)
50. Gao X, et al. Somatic mutations of the WAF1/CIP1 gene in primary prostate cancer. *Oncogene* **11,** 1395–1398 (1995)

51. Bathia K, et al. A mutant p21 cyclin-dependent kinase inhibitor isolated from a Burkitt's lymphoma. *Cancer Res.* **55,** 1431–1435 (1995)
52. Lancombe L, et al. Analysis of p21$^{WAF1\ CIP}$ in primary bladder tumors. *Oncol.Res.* **8,** 409–414 (1997)
53. Vidal M, Loganzo Jr. F, de Oliveira AR, Hayward NK, Albino AP. Mutations and defective expression of the WAF1 p21 tumour-suppressor gene in malignant melanomas. *Melanoma Res.* **5,** 243–250 (1995)
54. Spirin KS, Simpson JF, Takeuchi S, Kawamata N, Willer WM, Koeffler HP. p27/KIP1 mutation found in breast cancer. *Cancer Res.* **56,** 2400–2404 (1996)
55. Orlow I, et al. Cyclin-dependent kinase inhibitor p57KIP2 in soft tissue sarcomas and Wilms'tumors. *Cancer Res.* **56,** 1219–1221 (1996)
56. Otterson GA, Kratzke RA, Coxon A, Kim YW, Kaye FJ. Absence of p16*INK4* protein is restricted to the subset of lung cancer lines that retains wildtype RB. *Oncogene* **9,** 3375–3378 (1994)
57. Hollstein M, Sidransky D, Vogelstein B, Harris CC. p53 mutations in human cancers. *Science* **253,** 49–53 (1991)
58. Harris CC, Hollstein M. Clinical implications of the p53 tumor-suppressor gene. *N.Engl.J.Med.* **329,** 1318–1327 (1993)
59. Gelsleichter L, Gown AM, Zarbo RJ, Wang E, Coltrera MD. p53 and mdm-2 expression in malignant melanoma: an immunocytochemical study of expression of p53, mdm-2, and markers of cell proliferation in primary versus metastatic tumors. *Modern Pathology* **8,** 530–535 (1995)
60. Poremba C, Yandell DW, Metze D, Kamanabrou D, Bocker W, Dockhorn-Dworniczak B. Immunohistochemical detection of p53 in melanomas with rare p53 gene mutations is associated with mdm-2 overexpression. *Oncol.Res.* **7,** 331–339 (1995)
61. Van Dyke TA. Analysis of viral-host protein interactions and tumorigenesis in transgenic mice. *Sem.Cancer Biol.* **5,** 47–60 (1994)

DISCUSSION

Klausner: What is the functional relationship of p19ARF to RB and p53 in tumors?

DePinho: Tumors that sustain a deletion or mutation of the p19ARF open reading frame have a low incidence of p53 mutations. In the few informative tumor cases in which the p16INK4a open reading frame is exclusively affected and p19ARF remains wild-type, p53 mutations have been observed. Another genetic manifestation of this relationship is that 9p21-associated lesions that impact on p15INK4b but spare the p16INKa open reading frame p19ARF is also eliminated. These data point to a potential cooperating tumor suppressor role for these cyclin dependent kinase inhibitors (p15 or p16) and p19ARF (or p53).

Klausner: It seems to me that you would not negatively select for not knocking out this locus in RB minus so that it would be another way to look at it.

DePinho: One approach to experimentally verify this would be to examine the nature of the genetic lesions sustained by p53, RB and INK4a during the spontaneous immortalization of primary fibroblasts. For example, in MEFs derived from p53-deficient mice, does the INK4a gene sustain a deletion? Alternatively, in MEFs derived from INK4a knockout mice, do p53 mutations occur?

Livingston: So what about expression of the p19 gene and/or synthesis of the protein normally, either in developing embryos of the mouse or in human tissues?

DePinho: As organs systems progress through late post-natal development, there is significant up-regulation of both p16 and p19 message. The work of Gordon Peters and others has shown that as normal human cells and rodent cells enter replicative senescence, there is dramatic up-regulation of p16 and p19, suggesting that it might be important for the senescence process. It is intriguing that RB and p53 are among the most important pathways or complementation groups that have been linked to senescence. Since the INK4a gene plays a role in cell mortality, it would be tempting to speculate that the normal function of this gene is to integrate senescence signals to RB and p53.

Hanahan: I have a couple of questions. The first one is, I do not know how representative your data is, but it looked to me like in your RB p53 cells there was very little apoptosis, whereas in the INK4 ones there was less apoptosis. So, is it not possible that p19 is only partially suppressing the p53 pathway which would in part answer Rick's point which is that it is not really completely knocking it out, it is only inhibiting it.

DePinho: The data presented support the view that the activities of p53 are likely to extend beyond that of p19, i.e., p19 may be thought of as a modifier of p53. However, we speculate that there is a relationship somewhat analogous, but not necessarily identical, to that between p16 and RB.

Hanahan: A couple of questions on the melanomas. In the RAS mice that are wild type for INK4A, do you see losses, do those tumors take longer to come up? Do they lose INK4A spontaneously?

DePinho: We see deletion of the locus in INK4a+/+ mice; one sample was available and informative since these are rare tumors in the wild-type setting. The deletion is larger than the ones that we see in the engineered mice.

Hanahan: Can you comment on the issue of metastasis in these mice and what are the prospects for studying them?

DePinho: Metastatic disease kills these patients. There has been some data in the literature suggesting that RAS activation might be very important for invasion and late stage advancement of the disease. What we see in our model is highly aggressive, locally invasive tumors. We see no evidence for micro- or macrometastasis. It would be difficult to completely rule out metastatic disease in these animals since these tumors are S100 negative and de-differentiated but, from a clinical standpoint and histologically, we do not see any gross metastatic disease. This situation now provides us with an experimental system in which to evaluate candidate genes involved in modulating late progression and metastatic spread. There has certainly been a great deal of work in the area in human melanoma and this model provides a context in which to validate experimentally their roles in metastatic behavior.

Helin: I have a couple of questions regarding the selective mutations in p16 INK4A because some of the tumors, as I understand, selectively mutate the p16 without touching the p19. If you analyze those tumors you should see, if your model is correct, that you have p53 mutations, is that something that you looked into?

DePinho: We have seen p53 mutations in a E1α (p16-specific) exon in an established human tumor cell line. We have not looked at primary tumors, but there are very few tumors available with p16-specific mutations. If you look at one hundred tumors, ninety percent or greater will duly affect the sequences encoding both p16 and p19. But on the reverse side, we do not see p53 mutations in p19ARF mutations so there is a correlation but I would not say that the numbers have achieved statistical significance. It is also possible that other levels of p53 pathway could be affected. There is evidence to suggest that MDM2 is disregulated in malignant melanoma and that other aspects of p53 translation are affected in malignant melanoma and so on. Even in the absence of p53 mutations, our observations do not rule out the possibility that there are additional genetic lesions at other levels that are disregulating the p53 pathway.

Helin: My other question relates to the biochemical activity of p19, which I see from the papers which have been published by Sherr's lab is not that informative yet. But one thing that seems like is that if you over express p19 you are basically stopping cell cycle progression in any context. So do you have any, let us say, any mutations or anything that can overcome a p19 block.

DePinho: The studies that you are referring to are those conducted by Dawn Quelle and Chuck Sherr, showing that P19ARF is not an inhibitor of known cyclin dependent kinases. We are looking at p53 deficient primary cells to see if they are refractory to the inhibitory effects of p19 ARF. We have not performed biochemical analyses.

Berns: You mentioned that with the tet-RAS and the INK4 deficient cells, if you transplanted them in SCID mice that you would see the continuous requirement for RAS. How is that if you retract that in tumors which have been initiated in the mice themselves. Do you see regression of the primary tumors in that setting?

DePinho: Those experiments are underway. We do not see immediate regression of the tumor, but we also do not see advancement but I think that at this point the data is far too preliminary.

Visentin: Apparently there is an excess of Non-Hodgkin's Lymphomas produced by loss of the p16 gene/*INK*4A. I wonder whether there is more information about the histology of these NHLs, in terms of possible relationships to overexpression of *bcl*-2 and other antiapoptotic mechanisms.

DePinho: To my knowledge, there has not been a systematic review of the BCL2 family in the context of INK4A.

Visentin: Of course, I was thinking essentially of low-grade follicular NHLs, where the *bcl-2* product does admittedly prevent programmed cell death.

Livingston: Does the basic model recapitulate what happens in the human?

DePinho: In terms of the types of mutations that we see? I guess this issue comes up a lot with animal models of human malignancies. In fact, I find it rather surprising that some genetically engineered mice do show over-lap with the types of tumors that one sees in the human condition. The behavior of mouse cells with respect to their immortalization capacity and so on is quite different than that in human cells. In human cells spontaneous immortalization is rare but in rodent cells escape from replicative senescence occurs at a measurable frequency. So there may be a number of genetic differences with respect to the regulation of very important processes, not the least of which is replicative senescence. Regardless, you would still anticipate that there would be some overlap with respect to the types of tumors that you come up with. Humans are extremely polymorphic. Humans bearing the same genetic lesion can develop different types of tumors as well. There could be tissue-specific modifiers that impact on the spectrum of tumor types seen. Most of the mice that we work with were generated from relatively few mice earlier in this century so, in essence, their genetic background is very similar to one another and so it is possible that those particular genetic modifiers are strongly influencing the phenotype of a genetic manipulation and perhaps if those modifiers were present in a human context there might be better overlap in the tumor type spectrum.

Livingston: Were your mice shaved and maintained, for example, with regular exposure to UB light?

DePinho: We have UVed the mice and they get some benign melanocytic lesions that are not seen in the wild type background. I think we need to understand a little bit more about the role of this gene in melanocytes specifically, before we can understand exactly what is going on with respect to tumor progression and so on. But at this point, all we have done is to show that melanocytes, when taken from the knockout mice do not undergo senescence, but we have not examined other melanocyte specific pathways, UV pathways and the like, to really assess the role of this gene in mediating or modulating a very important and fundamental property of melanocytes which is to replicate and induce pigment *de novo* upon UV exposure. That needs to be done, but we need a purer genetic background before we can conduct those experiments because there are sixty genetic loci that control coat color and melanocytic biology in mice and these genetic variables need to be controlled before we look at that.

Klausner: In the families that do not have the INK4 deletion but rather have the CDK4 mutation, what do you see in terms of p53 and N-RAS?

DePinho: Your questions form the conceptual basis for our current efforts. We have yet to generate meaningful results to allow me to answer your questions.

2

IDENTIFICATION AND CHARACTERIZATION OF COLLABORATING ONCOGENES IN COMPOUND MUTANT MICE

Anton Berns, John Allen, Harald Mikkers, Blanca Scheijen, and Jos Jonkers

Division of Molecular Genetics
The Netherlands Cancer Institute
Plesmanlaan 121, 1066 CX, Amsterdam, The Netherlands

1. INTRODUCTION

Retroviral insertional mutagenesis is a powerful approach to identify genes that can confer a selectable phenotype to cells or an organism. Mechanistically, insertional mutagenesis can cause deregulated expression of genes or disrupt their coding sequence, which could either lead to an altered activity or to inactivation. Since retroviral integration is a relatively random process it has a reach comparable to that of chemical mutagenesis screens. However, it differs in that it leaves a sequence tag at the "site of the crime" and, therefore, permits the swift identification and characterization of the gene involved in conferring the selective advantage to the cell. Since its mutagenesis spectrum is different from chemical mutagenesis it will not necessarily yield the same genes, e.g. insertions will not cause point mutations or large deletions and, consequently, if such a mutation (e.g. a point mutation in the *ras* proto-oncogene) is required, retroviral insertional mutagenesis will not lead to the identification of such a gene or locus. They are capable of initiating, enhancing and/or terminating transcription of host genes depending on the integration site and the transcriptional orientation of the provirus with respect to the cellular gene (proviruses carry transcriptional enhancers and polyadenylation signals in their long terminal repeats (LTR's)). Proviral tagging is a straightforward procedure when used for tumor growth in vivo as the selected phenotype. We will restrict ourselves here to this application (1), even though its utility is much broader. In the vast majority of cases replication competent viruses are used that carry no oncogenes but can activate cellular proto-oncogenes or inactivate tumor suppressor genes. Operationally, tagging is performed by infecting newborn mice with murine leukemia virus (MuLV) or mammary tumor virus (MMTV). This gives rise to a long-term viremia (2) with re-infection and retroviral insertion events continuing throughout the life of the animal. For a recent overview of retroviral insertional mutagene-

The Biology of Tumors, edited by Mihich and Croce
Plenum Press, New York, 1998.

sis we refer to (1). We will confine ourselves to the studies performed with Moloney murine leukemia virus. The Moloney virus is lymphotropic but can infect many cell types (viruses with different tropism, such as MMTV, have been employed in non-lymphoid tumor models). Viral integrations in infected cells occasionally lead to the activation of proto-oncogenes, which, in turn, can provide a selective growth advantage to the cell, eventually resulting in a tumor in which the array of proviruses present in the original cell is preserved. The pattern of integrations is easily observed by Southern analysis. Typically, several to many clonal or near-clonal proviral integrations are seen. When more than one independently induced tumor carries an integration at a particular locus, this is named a "common insertion site" (c.i.s.). Such an insertion is invariably of oncogenic significance since the possibility of this happening by chance alone is extremely small. The majority of integrations do not occur in common insertion sites and are of no significance for tumorigenesis, having occurred before relevant oncogenic events in the parental cell and carried along for the "ride", as it were. Nevertheless, it is not unusual to find MuLV integrations at more than one known c.i.s. in any given tumor. This strongly implies collaboration of the two affected genes in tumorigenesis, since the later occurring of the two integrations would not otherwise have been selected for during tumor outgrowth.

The ability of an integrated provirus to unleash the transforming potential of cellular genes has led to the identification of a large number of cellular genes that, upon aberrant expression or truncation, mediate a selective advantage to cells (1). Well-known proto-oncogenes that are frequently found to be activated by proviral insertion in lymphoid tumors are c-*myc*, N-*myc*, and *Pim1* (1). The reach of insertional mutagenesis has been further extended by utilizing mice bearing oncogenes in the germ line as a starting point. Infection of these mice will preferentially lead to tumors carrying provirally induced alterations that synergize with the action of the oncogene present as a transgene. E.g. the infection of Eμ-*myc* transgenic mice resulted in the outgrowth of tumors carrying proviral insertions near the collaborating oncogenes *Pim1*, *Bmi1*, and *Gfi1* (3,4) but not near *myc*, as no selective advantage is associated with the activation of *myc* in cells that already harbor a highly expressed *myc* transgene. Applying this approach to a range of mutant mice carrying different oncogenes permits assignment of the various oncogenes to distinct complementation groups in transformation.

Further expansions along this theme take advantage of the possibility to prolong the lifespan of primary tumors by transplanting them into syngeneic hosts. This permits the outgrowth of more malignant clones that were arising in the primary tumor as a result of ongoing insertional mutagenesis. Genes frequently targeted later in the tumorigenic process likely contribute to tumor progression and probably belong to the category of genes conferring accelerated growth or metastatic potential to tumor cells. Finally, proviral tagging in compound mutant mice harboring activated proto-oncogenes as well as disrupted proto-oncogenes can be used to specifically search for genes acting in defined signal transduction pathways. This focused search for genes acting in a specific pathway has been called "complementation tagging". The latter two approaches will be discussed in more detail.

2. RESULTS AND DISCUSSION

We have utilized retroviral insertional mutagenesis with the following aims:

1. To identify sets of oncogenes that synergize in the induction of lymphomas and to assign each of these oncogenes to a particular complementation group in transformation.

2. To identify genes involved in later stages of tumorigenesis.
3. To determine the mechanism of action of these genes by the identification of the pathway in which they act in the cell.

2.1. Synergizing Oncogenes

The identification of synergizing oncogenes using proviral tagging methodologies has been most fruitful in mice bearing oncogenes in their germ line (3,5). We have concentrated on transgenic mice overexpressing oncogenes involved in lymphomagenesis. Earlier studies have shown that retrovirus infection in Eμ-*myc* transgenic mice resulted in the frequent activation of the *Pim1*, *Pim2* and *Bmi1* genes in the B cell lymphomas which is the predominant tumor type arising in these mice (3,4,6). In addition, insertions in the *Pal1* locus, frequently found in these tumors, appeared to be associated with the upregulation of the *Gfi1* gene (7). Using different transgenic lines as a starting point, activation of different complementing oncogenes were found. E.g. in Eμ-*Pim1* transgenic mice activation of c-*myc*, N-*myc*, and *Gfi1* was observed, while in *Bmi1* transgenic mice activations of *Pim1*, *Pim2*, c-*myc*, and N-*myc* were predominant. These experiments showed a consistent integration pattern, i.e. mostly the same subset of genes were found, irrespective of whether T or B cell lymphomas arose, the only exception being *Bmi*1, whose activation appears restricted to B cell lymphomas in the mouse. The expression pattern of the transgene determined to a large extent the type of tumor.

Consequently, Eμ-*myc* transgenic mice, which express the *myc* oncogene almost exclusively in the B cell compartment, will primarily yield B lymphomas, whereas H_2K-*myc* transgenic mice, which express the transgene widely in the hematopoietic systems, show a preponderance of T cell lymphomas (see Figure 1). Eμ–*Bmi1* transgenic mice show predisposition to both B and T cell lymphomas. In all these transgenic lines a similar set of oncogene activations appears to join forces to transform the cells. Three groups that can be assigned to different complementation groups in transformation were discerned. This assignment is primarily based upon the observation that proviral insertions near a particular oncogene will not be found in a transgenic background that carries an activated oncogene belonging to the same complementation group. The reasoning is that an insertion near a gene of that same complementation group will usually provide insufficient selective growth advantage to the cell and will therefore not be found. E.g. in Eμ-*myc* transgenic

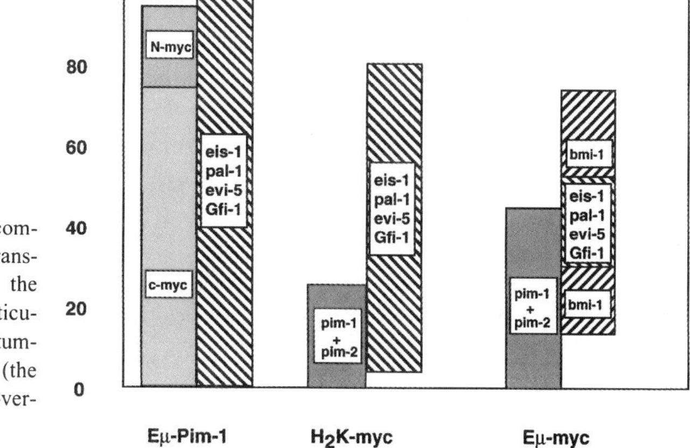

Figure 1. Proviral occupancy of common insertion sites in tumors of transgenic mice. The bars indicate the percent of tumors in which the particular locus carries a provirus. Often tumors carry proviruses in both loci (the regions where the juxtaposed bars overlap along the vertical axis).

mice activation of N-*myc* is not observed, in *Pim1* transgenic mice activation of the *Pim2* gene is not found, and proviral insertional mutagenesis in *Bmi1* transgenic mice failed to yield insertion in the *Pal1* locus (8). This identifies three complementation groups: 1. the *myc* proto-oncogenes, 2. the *Pim1* and *Pim2* proto-oncogenes, and 3. *Bmi1* and *Gfi1* proto-oncogenes. Whereas the assignment of e.g. c- and N-*myc* or *Pim1* and *Pim2* (6) to the same complementation group will not come as a surprise, this situation is different for *Bmi1* and *Gfi1*. These latter two genes encode very different proteins (3,7,9). Interestingly, both appear to play a role in suppressing gene expression: *Bmi1* by virtue of its involvement in modulating chromatin structure (10–12), *Gfi1* by acting as a transcriptional repressor with a distinct DNA binding and transcriptional repression domain (9,13,14). It will be interesting to identify the genes that are downregulated by the overexpression of these proteins.

The application of proviral insertional mutagenesis in transgenic mice has taught us that insertional mutagenesis can be manipulated to a large extent. There is no reason why any other specific selection protocol, *in vivo* in mice or *in vitro* in cell culture, might not be combined with insertional mutagenesis. At this moment the application *in* vivo is limited by the tissue tropism of the slow transforming retroviruses, i.e. MoMuLV and MMTV. It is worth discussing two other applications: the specific identification of genes involved in tumor progression and the search for genes acting in specific signal transduction pathways.

2.2. Tumor Progression

To apply insertional mutagenesis to tumor progression, one should realize that once a mouse has been infected as a newborn (to prevent an immune response and consequently clearance of virus and infected cells) with a replication competent virus such as Moloney murine leukemia virus, a life-long viremia will be established in which infection and re-infection of dividing cells can continue as long as virus entry is not prevented by interference (15). This "interference" appears to be limited by two mechanisms. Firstly, during the replication of the MuLV in mice, recombinant viruses are being formed that carry a different envelope glycoprotein (Mink Cell Focus (MCF) forming virus, for review see (1)) and, therefore, can enter cells through a different receptor. Secondly, due to inaccurate reverse transcription, defective proviruses are frequently produced and, in case the defect resides in the envelope gene, such proviruses will be unable to prevent re-infection by interference. The presence of an average of 5–10 proviruses per tumor cell, both of ecotropic and MCF origin and of which a substantial fraction does not encode a functional envelope glycoprotein (our unpublished results), is in accordance with this notion.

The ongoing insertional mutagenesis therefore creates a situation in which sequential activations of genes that can contribute to the various stages of tumorigenesis can take place. This also explains why more than one oncogene can be efficiently activated by proviral insertional mutagenesis in a single tumor cell clone. One would expect that in the early phases of the tumorigenic process (in)activation of genes directly involved in cell proliferation plays a predominant role, whereas later in the disease genes that might contribute to tumor progression become the primary target. Following this reasoning we argued that transplantation of primary tumors might lead to the proviral marking of genes involved in this latter process.

To address this experimentally, cell suspensions were made of primary lymphomas induced by MuLV infection in Eμ-*Pim1* or H_2K-*myc* transgenic mice, and transplanted subcutaneously to a number of independent syngeneic hosts (Figure 2). Tumors that grew

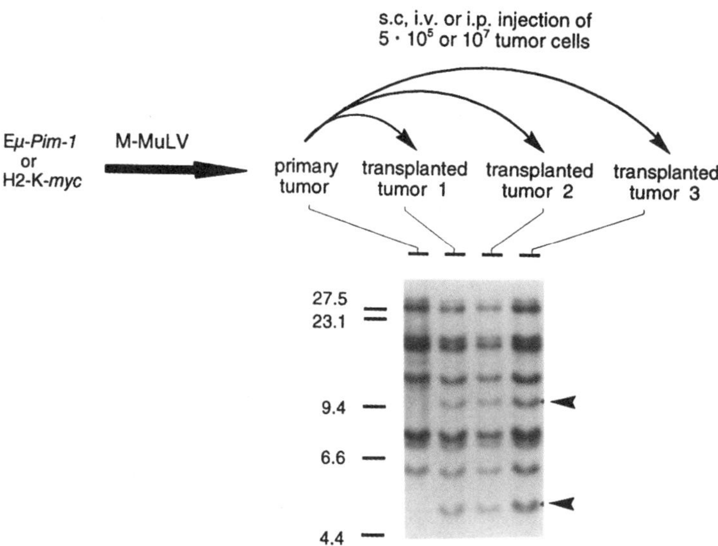

Figure 2. Methodology to identify genes involved in tumor progression. Primary tumors induced by M-MuLV in transgenic mice are made into a cell suspension and equal aliquots transplanted to syngeneic recipients. Tumors that grow out are screened by Southern blot analysis using a MuLV specific probe. Note the additional bands (arrows) found in the transplanted tumor. This indicates that a minor subclone already present in the primary tumor (not yet visible in the Southern of the primary tumor) grew out preferentially.

out were collected and the proviral integration pattern compared with that of the primary tumor by Southern blot analysis. All the transplanted tumors harbored a number of provirus-specific bands that were also present in the primary tumor. However, the transplanted tumors frequently carried additional bands that were common to the independent outgrown transplants of the same primary tumor, but not or hardly detectable in the primary tumor samples (Figure 3). The presence of the same insertion site in independent transplants indicates that this insertion was already present in a small fraction of the primary tumor cells. Apparently, this cell clone which had a selective growth advantage upon transplantation, was generated long after expansion of the primary tumor had started and, therefore, represented a tumor progression event. A series of these "additional" insertions were cloned and we searched whether one of these insertions represented a "common insertion site".

In this way new loci were identified that were specifically involved in tumor progression. In one of these loci we have identified the relevant gene that conferred this selective advantage to transplanted tumor cells. We have named this gene *Frat1* (16). *Frat1* encodes a small protein of 274 amino acids with unknown function. It has no peptide motifs that would suggest a particular activity. However, its role in tumor progression was established by monitoring the growth characteristics of suitable cell lines after retroviral transduction. Tumor cell lines derived from spontaneously arising tumors in Eμ-*Pim1* transgenic mice in which we transduced the *Frat1* gene together with a lacZ marker, showed a significant better growth capacity upon transplantation than cells that were transduced with the lacZ marker only. This has been most clearly demonstrated by transplanting mixtures of transduced and non-transduced cells (16).

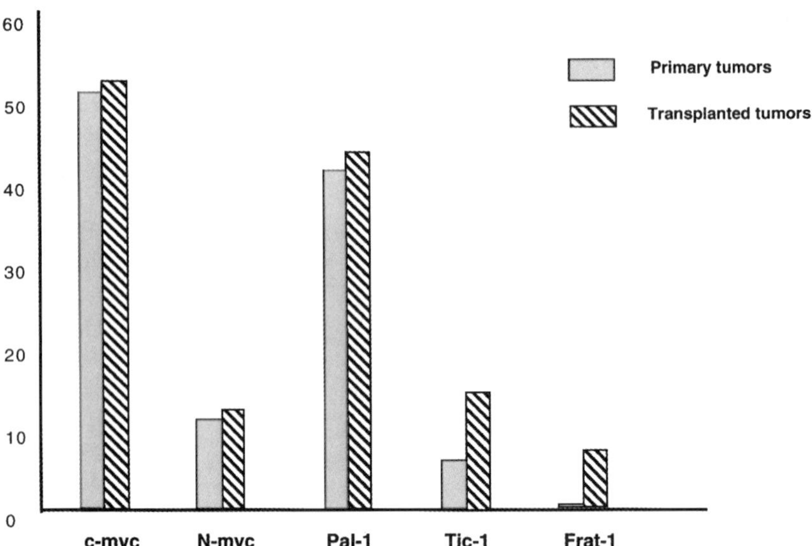

Figure 3. Proviral occupancy in primary and transplanted tumors of Eμ-pim1 transgenic mice. The height of the bars indicates the percentage of tumor cells carrying an insertion in that locus. Some of the common insertion sites are equally represented in primary and transplanted tumors. The loci preferentially found in transplanted tumors denote genes more likely contributing to tumor progression (Frat1, Tic1).

These experiments illustrate that proviral insertional mutagenesis can be employed in different settings permitting identification of genes involved in a defined subset of tumors, e.g. B lymphomas in Eμ-*myc* transgenic mice, T cell lymphomas in H_2K-*myc* transgenic mice, or genes involved in specific stages of tumor development, e.g. in tumor initiation (*myc*, the *Pim*'s, *Bmi1, Gfi1*) or tumor progression (*Frat1*, *Tic1*).

2.3. Mechanism of Action of Oncogenes: Pathway Identification by "Complementation Tagging"

We have subsequently investigated whether proviral insertional mutagenesis could also be employed to identify genes functioning in defined pathways. We aimed at an approach that resembles with suppressor screens performed in Drosophila or C.elegans (17). We reasoned that the strong synergism found between members of the *myc* and *Pim* complementation group (double transgenic mice can succumb from tumors in utero (18)) might be exploited by generating compound Eμ-*myc;Pim*$^{-/-}$ mice and using proviral tagging as an approach to mark genes that compensate for the loss of the wt *Pim* allele. Since we knew that loss of *Pim1* and/or *Pim2* had no detrimental effect on the viability of mice (19) (Allen & Berns, unpublished results), we generated the various compound mutant genotypes and performed an insertional mutagenesis screen. This methodology was named "complementation tagging" as it was expected to yield genes that would complement the loss of function of *Pim1*, *Pim2* or both (see diagram Figure 4).

When this experiment was performed in Eμ-*myc;Pim1*$^{-/-}$ mice, a remarkable increase in the activation of the *Pim2* proto-oncogene was found, as compared to Eμ-*myc* transgenics. Nearly all tumors in Eμ-*myc;Pim1*$^{-/-}$ showed activation of *Pim2*, even though in Eμ-*myc* mice proviral activation of *Pim1* and *Pim2* together was less than 50%. The

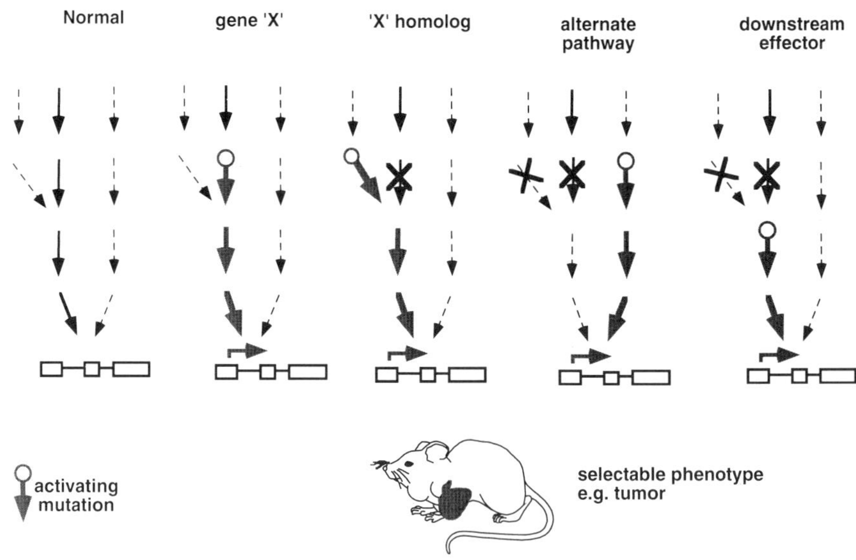

Figure 4. Complementation tagging in proto-oncogene-deficient mice. This diagram depicts the situation in which a particular pathway becomes activated by mutation (fat arrow with open circle). If a particular gene product in the pathway is disabled by targeted disruption of the corresponding gene (crossed out arrow) activation of the pathway has to be achieved by another route. Examples are given in which one or two closely related genes with similar biochemical properties are disrupted (e.g. Pim1 and Pim2).

converse was not found: proviral activation of *Pim1* was not more frequent in the Eμ-*myc*;*Pim2*$^{-/-}$ than in the Eμ-*myc* background. This illustrates an interesting feature of the relative roles of *Pim1* and *Pim2* in lymphomagenesis. Although data from these and other experiments indicate that the two genes are about equally efficient at generating T or B cell lymphomas, it seems that *Pim1* is somehow limiting, while *Pim2* is not. A simple explanation for this might be that *Pim2* is not, or only lowly, expressed in the MuLV target cell population. The lack of difference in tumor latency between the *Pim2* knock-out and wild-type cohorts is likely a reflection of the same phenomenon and so might be the failure of *Pim2* knock-out mice to show any deficiency in growth factor responses like those seen in the *Pim1* knock-outs (20,21).

While the abundant activation of *Pim2* by retroviral insertion in the Eμ-*myc*;*Pim1*$^{-/-}$ background validated this approach, it did not provide us with new candidate genes that might compensate for the function of *Pim* genes. Therefore, we repeated this proviral tagging experiment in an Eμ-*myc*;*Pim1*$^{-/-}$;*Pim2*$^{-/-}$ background. It should be pointed out that the *Pim1/2* double knock-out mice show a similar phenotype as *Pim1* single knock-outs (20,22). The deficiencies in responses to growth factors *in vitro* are somewhat more severe, but this still appears to be of little consequence for the health of the mice or the composition of their hematopoietic cell compartment (Allen, unpublished data). Effects may well be present, but under normal circumstances they must be subtle. An Eμ-*myc* transgene was duly bred onto a *Pim1*/*Pim2* double knockout background using a breeding program that simultaneously generated *myc* transgenics on wild-type and *Pim1* and *Pim2* single knockout backgrounds as controls.

In view of the amount of work involved in identifying, cloning and characterizing common provirus insertion sites, it was worthwhile to spend some time checking for any

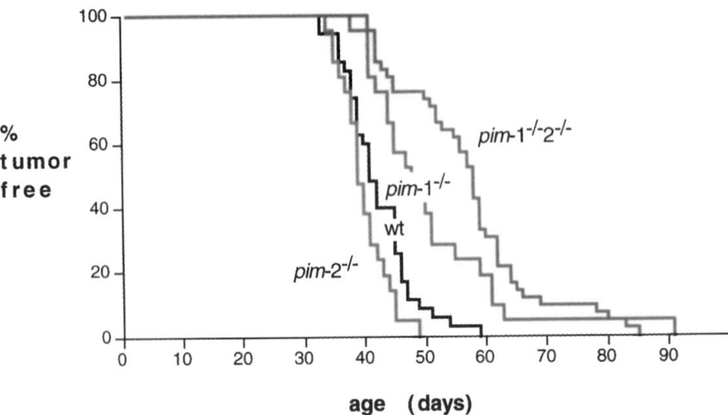

Figure 5. Incidence of MuLV-induced tumorigenesis in Eμ-myc;pim$^{-/-}$ compound mutant mice. The curves indicate the percentage of animals remaining tumor free. The number of animals in the various groups was between 30–40. Note the delay in tumor incidence in the pim1$^{-/-}$;pim2$^{-/-}$ background.

signs that the system might *not* have behaved as hoped. For example, a large increase in tumor latency might indicate utilization of different routes to tumorigenesis. Conversely, if tumorigenesis still involved pathways in which the *Pim*s were normally active, then neither the cell type in which the tumors arise, nor the pattern of collaborating proviral integrations in the tumors should be substantially changed by eliminating the *Pim* genes.

In fact, the median tumor latency in the *Pim1/2* knockout cohort increased by about three weeks (Figure 5). This was large enough to suggest that the system had been "stressed", but not so great an increase that one would fear utilization of an entirely different pathway. In this experiment, a smaller increase in latency was also observed on the *Pim1* knockout background. Analysis of a sample of tumors (10 per genotype) by flow cytometry indicated that the compositions of the tumor panels by cell lineage were comparable. The B-lineage tumors had slightly more mature surface phenotypes in the absence of *Pim1* and 2, fitting a pattern seen in unconnected experiments which point to the possibility that without a functional *Pim1* or *Pim2*, lymphoid cells have a greater tendency to mature.

The greatest confidence that *Pim*-type pathways were still being utilized came from an examination of the patterns of collaborating proviral insertions. The principal complementation group of genes collaborating with both the *Pim* and *myc* families comprises the *Bmi1* gene and the *Gfi1/Pal1/Evi5* locus (3,7,23). The frequency of proviral integrations in these loci did not vary substantially between the wild-type and *Pim1/2* knockout background. This was further supported by the analysis of new common insertion sites (Figure 6). One predominant gene that was targeted appeared to be *Pim3*. This is another homologue of the *Pim* family. However, *Pim3* was never found activated in tumor panels of mice that carried at least one functional *Pim1* and/or *Pim2* allele. Therefore, *Pim3* is a less preferred target for activation. Whether this relates to the fact that insertions near *Pim3* occur at a lower frequency, or whether *Pim3* has a reduced oncogenic activity remains unclear. The other gene now known to be activated in the *Pim1/2* knock-out tumor panel is *Tpl2,* another serine/threonine kinase and a common MuLV insertion site in rats (24). *Tpl2* is the homologue of the human COT oncogene, a carboxy-terminal truncated protein (25). The characteristic mode of activation by MuLV in rats is insertion of the provirus between the seventh and eighth exon, resulting in over-expression of a similar carboxy-terminal truncated protein (26). Also in the *pim1/2* knock-out tumors proviral integrations invari-

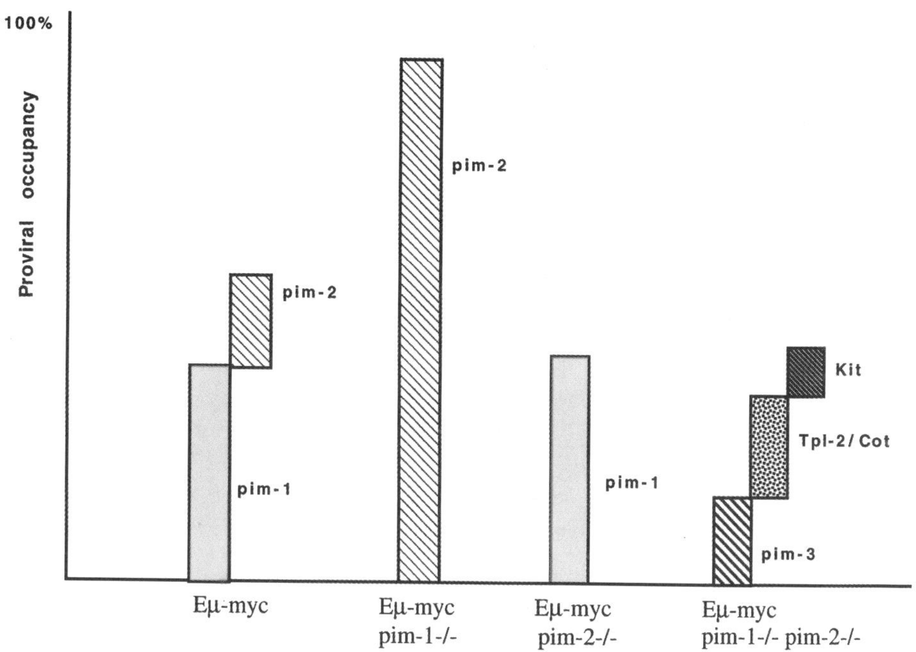

Figure 6. Complementation tagging in Pim deficient mice. The occupancy of proviruses in one of the pim loci or in loci that might substitute for pim are indicated. Integrations near pim3, Tpl2 and kit are only found in the pim1$^{-/-}$; pim2$^{-/-}$ background suggesting that these genes indeed compensate for the lack of pim1 and pim2.

ably occurred in the seventh intron. In rats, MuLV integrations in the *Tpl2* locus were seen only after tumor transplantation or in derived cell lines, therefore the gene was thought of as a transplantation or progression-specific locus. Surprisingly, we detected several *Tpl2* integrations in the *Pim1/2* double knock-out panel, whereas such insertions were not found in primary tumors from the wild-type, *Pim1* knock-out or *Pim2* knock-out panels.

A recent study (27) indicated that *cot* acts in the MAP kinase pathway and that the truncated forms are much more active than the wild-type protein. This is consistent with collaboration between Tpl2 and myc, as potent collaboration with myc has been observed for several other genes acting in the MAPK pathway, notably *ras* and *raf* (28,29). Activation of *Tpl2* in place of *Pim1* and *Pim2* (and superficial biochemical similarities) implied that the *Pim*s might either act in the same pathway or have common downstream targets. COS cell transfection experiments indicate that the former of these possibilities is not the case, at least in the simpler scenarios imaginable—*Pim1* does not activate the *Tpl2*, MEK or ERK kinases.

It is not yet clear from the available data whether the *Pim*s and *tpl2* are true alternatives or to what extent their function in tumorigenesis might overlap—it is still possible that they are sometimes activated together and collaborate. It is at least evident that the absence of *Pim1* and *Pim2* increases the relative selective value of *Tpl2* activations and, given the longer latency of tumor onset in the *Pim* knockout mice, allows more time for cells with such activations to develop into full blown tumors. The requirement for activation of *Tpl2,* namely proviral insertion in the seventh intron, is certainly more stringent than for either *Pim1* or *Pim2*, which have several distinct integration clusters spanning larger regions.

In this screen we also found the proviral activation of another interesting gene, the *kit* receptor (30). Again, no evidence for activation of this gene was found in any of the other tumor panels, suggesting that it might substitute for or act downstream of Pim. Mice carrying mutations in the kit receptor show defects in hematopoietic cells, primordial germ cells and melanocytes (31). Many kit mutant alleles have been identified due to the specific external features of mice carrying these mutations. At present we have no experimental data that explain how activation of the kit receptor might substitute for Pim function. One model that we would like to test is whether Pim proteins play a role in the crosstalk between various classes of hematopoietic receptors. One might envisage that activation of Pims through interleukin receptor signaling permits cells to become responsive to kit ligand mediated growth signals. In that way the specific expansion of distinct hematopoietic compartments might be regulated. This is in good agreement with the observed capacity of *Pim* to reverse the impaired cell growth in IL2-γ Receptor deficient mice (Krimpenfort et al., unpublished results). The future will teach whether *Pim* proteins indeed fulfill such a role in crosstalk between different receptor pathways.

In conclusion, the various tagging strategies in mice make it possible to attack biochemical problems via a genetic route analogous to the powerful techniques employed in invertebrate model organisms. The main difference is that mutagenesis and selection is done in somatic cells. It is practical for oncogenes because their collaboration in tumorigenesis can be exploited to encourage activation of substitutes, and because the complementing mutations are self-selecting. With only a little more ingenuity, the approach is potentially applicable to a much wider range of problems, provided that selection of revertant cellular phenotypes and some expansion of the cells for analysis can be arranged. The more knock-outs become available and the better the tagging systems employed for mutagenesis, the more useful this approach will become for investigating biochemical pathways in the cellular systems of higher organisms. It is to be expected that other, more versatile transposons can be developed that do not require an extracellular phase and also lack some of the other disadvantages of retroviruses. Both retrotransposons (27) and DNA transposons from other organisms (32) do hold promise in this respect.

ACKNOWLEDGMENTS

This work was supported by grants from the Dutch Cancer Foundation (KWF) and Dutch Science Foundation (NWO) to A.B.

REFERENCES

1. Jonkers, J. and Berns, A. Retroviral insertional mutagenesis as a strategy to identify cancer genes. [Review]. Biochim. Biophys. Acta., *1287*: 29–57, 1996.
2. Jaenisch, R., Fan, H., and Croker, B. Infection of preimplantation mouse embryos and of newborn mice leukemia virus: Tissue distribution of viral DNA and RNA and leukemogenesis in the adult animal. Proc. Natl. Acad. Sci. USA., *72*: 4008–4012, 1975.
3. van Lohuizen, M., Verbeek, S., Scheijen, B., Wientjens, E., van der Gulden, H., and Berns, A. Identification of cooperating oncogenes in E mu-myc transgenic mice by provirus tagging. Cell, *65*: 737–752, 1991.
4. Haupt, Y., Alexander, W.S., Barri, G., Klinken, S.P., and Adams, J.M. Novel zinc finger gene implicated as myc collaborator by retrovirally accelerated lymphomagenesis in Eμ-myc transgenic mice. Cell, *65*: 753–763, 1991.

5. van Lohuizen, M., Verbeek, S., Krimpenfort, P., Domen, J., Saris, C., Radaszkiewicz, T., and Berns, A. Predisposition to lymphomagenesis in pim-1 transgenic mice: cooperation with c-myc and N-myc in murine leukemia virus-induced tumors. Cell, *56*: 673–682, 1989.
6. van der Lugt, N.M., Domen, J., Verhoeven, E., Linders, K., van der Gulden, H., Allen, J., and Berns, A. Proviral tagging in E mu-myc transgenic mice lacking the Pim-1 proto- oncogene leads to compensatory activation of Pim-2. EMBO J., *14*: 2536–2544, 1995.
7. Scheijen, B., Jonkers, J., Acton, D., and Berns, A. Characterization of pal-1, a common proviral insertion site in murine leukemia virus-induced lymphomas of c-myc and Pim-1 transgenic mice. J. Virol., *71*: 9–16, 1997.
8. Alkema, M.J., Jacobs, J., van Lohuizen, M., and Berns, A. Perturbation of B and T cell development and predisposition to lymphomagenesis in EμBmi1 transgenic mcie require the Bmi1 RING finger. Oncogene, *15*, : 899–910, 1997.
9. Zweidler-McKay, P.A., Grimes, H.L., Flubacher, M.M., and Tsichlis, P.N. Gfi-1 encodes a nuclear zinc finger protein that binds DNA and functions as a transcriptional repressor. Mol. Cell Biol., *16*: 4024–4034, 1996.
10. Alkema, M.J., Bronk, M., Verhoeven, E., Otte, A., van't Veer, L.J., Berns, A., and van Lohuizen, M. Identification of Bmi1-interacting proteins as constituents of a multimeric mammalian polycomb complex. Genes. Dev., *11*: 226–240, 1997.
11. Alkema, M.J., van der Lugt, N.M., Bobeldijk, R.C., Berns, A., and van Lohuizen, M. Transformation of axial skeleton due to overexpression of bmi-1 in transgenic mice. Nature, *3748*: 724–727, 1995.
12. van der Lugt, N.M.T., Domen, J., Linders, K., van Roon, M., Robanus Maandag, E., te Riele, H., Van der Valk, M., Deschamps, J., Sofroniew, M., van Lohuizen, M., and Berns, A. Posterior transformation, neurological abnormalities, and severe hematopoietic defects in mice with a targeted deletion of the *bmi-1* proto-oncogene. Genes & Dev., *8*: 757–769, 1994.
13. Grimes, H.L., Chan, T.O., Zweidler-McKay, P.A., Tong, B., and Tsichlis, P.N. The Gfi-1 proto-oncoprotein contains a novel transcriptional repressor domain, SNAG, and inhibits G1 arrest induced by interleukin-2 withdrawal. Mol. Cell Biol., *16*: 6263–6272, 1996.
14. Grimes, H.L., Gilks, C.B., Chan, T.O., Porter, S., and Tsichlis, P.N. The Gfi-1 protooncoprotein represses Bax expression and inhibits T-cell death. Proc. Natl. Acad. Sci. U. S. A., *93*: 14569–14573, 1996.
15. Rein, A. Interference grouping of murine leukemia viruses: A distinct receptor for the MCF-recombinant viruses in mouse cells. Virol., *120*: 251–257, 1982.
16. Jonkers, J., Korswagen, H.C., Acton, D., Breuer, M., and Berns, A. Activation of a novel proto-oncogene, Frat1, contributes to progression of mouse T-cell lymphomas. EMBO J., *16*: 441–450, 1997.
17. Gaul, U., Chang, H., Choi, T., Karim, F., and Rubin, G.M. Identification of ras targets using a genetic approach. GTPase. Superfamily., *176*: 85–92, 1993.
18. Verbeek, S., van Lohuizen, M., Van der Valk, M., Domen, J., Kraal, G., and Berns, A. Mice bearing the E mu-myc and E mu-pim-1 transgenes develop pre-B- cell leukemia prenatally. Mol. Cell Biol., *11*: 1176–1179, 1991.
19. Laird, P.W., Vanderlugt, N.M.T., Clarke, A., Domen, J., Linders, K., Mcwhir, J., Berns, A., and Hooper, M. Invivo analysis of pim-1 deficiency. Nucleic Acids Res., *21*: 4750–4755, 1993.
20. Domen, J., Vanderlugt, N.M.T., Acton, D., Laird, P.W., Linders, K., and Berns, A. Pim-1 levels determine the size of early B-Lymphoid compartments in bone marrow. J. Exp. Med., *178*: 1665–1673, 1993.
21. Domen, J., van der Lugt, N.M., Laird, P.W., Saris, C.J., Clarke, A.R., Hooper, M.L., and Berns, A. Impaired interleukin-3 response in Pim-1-deficient bone marrow- derived mast cells. Blood, *82*: 1445–1452, 1993.
22. Domen, J., Vanderlugt, N.M.T., Laird, P.W., Saris, C.J.M., and Berns, A. Analysis of pim-1 function in mutant mice. Leukemia, *7*: S108-S112, 1993.
23. Liao, X., Buchberg, A.M., Jenkins, N.A., and Copeland, N.G. Evi-5, a common site of retroviral integration in AKXD T-cell lymphomas, maps near Gfi-1 on mouse chromosome 5. J. Virol., *69*: 7132–7137, 1995.
24. Patriotis, C., Makris, A., Bear, S.E., and Tsichlis, P.N. Tumor progression locus 2 (Tpl-2) encodes a protein kinase involved in the progression of rodent T-cell lymphomas and in T-cell activation. Proc. Natl. Acad. Sci. U. S. A., *90*: 2251–2255, 1993.
25. Miyoshi, J., Higashi, T., Mukai, H., Ohuchi, T., and Kakunaga, T. Structure and transforming potential of the human cot oncogene encoding a putative protein kinase. Mol. Cell Biol., *11*: 4088–4096, 1991.
26. Makris, A., Patriotis, C., Bear, S.E., and Tsichlis, P.N. Genomic organization and expression of Tpl-2 in normal cells and Moloney murine leukemia virus-induced rat T-cell lymphomas: activation by provirus insertion. J. Virol., *67*: 4283–4289, 1993.

27. Salmeron, A., Ahmad, T.B., Carlile, G.W., Pappin, D., Narsimhan, R.P., and Ley, S.C. Activation of MEK-1 and SEK-1 by Tpl-2 proto-oncoprotein, a novel MAP kinase kinase kinase. EMBO J., *15*: 817–826, 1996.
28. Adams, J.M. and Cory, S. Oncogene co-operation in leukaemogenesis. Cancer Surv., *15*: 119–141, 1992.
29. Alexander, W.S., Adams, J.M., and Cory, S. Oncogene cooperation in lymphocyte transformation: Malignant conversion of E(mu)-myc transgenic pre-B cells in vitro is enhanced by v-H-ras or v-raf but not v-abl. Mol. Cell. Biol., *9*: 67–73, 1989.
30. Majumder, S., Brown, K., Qiu, F.H., and Besmer, P. c-kit protein, a transmembrane kinase: Identification in tissues and characterization. Mol. Cell. Biol., *8*: 4896–4903, 1988.
31. Motro, B., van der Kooy, D., Rossant, J., Reith, A., and Bernstein, A. Contiguous patterns of c-kit and steel expression: analysis of mutations at the W and Sl loci. Development., *113*: 1207–1221, 1991.
32. Vos, J.C., De Baere, I., and Plasterk, R.H. Transposase is the only nematode protein required for in vitro transposition of Tc1. Genes. Dev., *10*: 755–761, 1996.

DISCUSSION

Rauscher: It is always surprising that you get GFI-1 and Bmi1 in the same complementation group. It is hard to reconcile that with their potential biochemical activities as opposed to, like, the myc family. GFI-1 is a zinc finger protein whereas the repressor Bmi1 is certainly a nuclear protein but probably does not bind DNA. So could you expand on that, do they form a complex?

Berns: Bmi1. I did not touch on that aspect. Bmi1 is a member of the Polycomb group of proteins. From what we know of the phenotypes of knockouts and transgenics Bmi1 is directly involved in the repression of gene transcription. But Bmi1 and Gfi1 work probably very differently. Although I expect them to have mostly different target genes, that does not mean that there is not some overlap. So I think that the data are not difficult to reconcile. It is just a matter of finding what the common denominator is. In fact, the data that Gfi1 and Bmi1 belong to the same complementation group is very strong. When either Gfi1 or Bmi1 transgenic mice are infected with murine leukemia viruses this will lead to tumors in which we did not find activation of Bmi1 or Gfi1.

Rauscher: With GFI-1 it is likely that it is highly sequence specific: I mean, it has an extended binding site. There should be very few target genes whereas Bmi1 probably does not bind DNA and has this global repressive effect, probably even as a co-repressor with something like GFI-1.

Berns: Yes, but still, that could mean that some of the same genes are affected. I think that would probably be the answer for this finding.

DePinho: I just want to follow up on the discussion here with respect to Bmi1. Given that yin-yang mechanistic relationship between Myc and its antagonists of the Mad family which are involved in repression of gene expression, at least at the level of chromatin regulation, can you speculate as to the relationship between Myc and Bmi1? It is somewhat counter-intuitive that a gene involved in repression of gene expression at the level of chromatin should co-operate with Myc in leukemogenesis. I know that the critical matter is in the details with respect to what genes it might be regulated, but can you provide us with some kind of genetic argument as to why you see a relationship that is counter intuitive.

Berns: Yes, well I can say something about that although it will remain handwaving. Since Polycomb group proteins are present in large complexes together with other members it is not necessarily obvious that overexpression of one of the members would lead to more repression. It might also lead to the disruption of certain complexes. On the other hand, one can envisage that Bmi1 might act on genes involved in tumor suppression.

DePinho: Have you looked to see if Bmi1 is regulating the negative regulators of Myc, that would be one possible outcome that it is negatively regulating Mad or MXI or something that normally co-operates with Myc?

Berns: No. We have no answer to that.

Anderson: In your retroviral mediated tumor promotion, do you know that only the insertion of the retrovirus itself is the predominant type of genomic event? In other words, do you no longer see extensive deletions, amplifications, etc., or is this a lymphomic-exclusive type of promotion that you are seeing? What I am trying to get at is where you have mismatch repair defects, it looks the genome is surprisingly stable to other types of genomic events, and I was wondering where you have retroviral insertional promotion; is it the same type of thing that is happening?

Berns: Intuitively, I would say yes; that does not mean that one could not find genomic aberrations, for example sometimes you might find duplication of a chromosome carrying an allele which already contains a provirally activated oncogene. So it is not excluded that some additional events occur, as seems to be the case in mismatch repaired deficiency. Overall I would think that if you have a strong insertional mitogen it should behave like mismatch repair deficiency and a stable diploid genome is expected to be retained. All the mutations then occur by insertional mutagenesis. I should add that we have not extensively looked at the karyotype of all the lymphomas.

Anderson: There is no CGH data on any of these then?

Berns: No.

Helin: One thing you mentioned in the overview on the first slide was that pro-viral insertion could be a good method to identify tumor suppressors. As I see your data, I think that is the only thing you do not get. Is there any way to make the method more efficient to identify recessive oncogenes?

Berns: Well, there are a couple of points to make in that respect. The first argument is: why would a retrovirus try to inactivate two alleles of a tumor suppressor gene when it can do the same job by activating one oncogene? That is the first point. However, if you look closer you realize that it has to depend on how much selective pressure is conferred by the event. This argument holds both for the activation of oncogenes as for the inactivation of tumor suppressor genes. So if you have achieved a selective advantage by losing one allele of a tumor suppressor gene you might expand the population sufficiently to increase the chance of hitting the other allele either by proviral insertion or by loss of heterozygocity. So proviral insertion can lead to tumor suppressor gene inactivation as has been shown in the case of NF1 by Neil Copeland and in the case of p53. Well, why then don't we see it more often. One explanation might be that activating strong dominant genes is an easier way to do the job. The second point might be that we missed quite a number of targeted tumor suppressor genes because the target site for inactivation is large. For example, if you want to inactivate Rb you have a couple of hundred Kb to put your retrovirals in and in fact, nobody has looked through the whole 200 Kb of Rb. However, if you look for activation there is usually a very narrow region where the provirus has to integrate. Otherwise it would not achieve the dominant effect. So what could one do about this? One way we are trying to approach this is to use vectors which are less suitable for activation but more designed to inactivate genes by a gene trap strategy. Now the problem with this is that, although it looks good on paper and even might work in tissue culture, it is difficult to achieve *in vivo* in mice. A defective virus *in vivo* seems not very useful *in vivo* when you need a large number of integration events. That almost requires a replication competent entity, at least next to a defective virus.

Hanahan: Ron DePinho actually eluded to an emerging issue which has to do with modifiers that are present in different mouse populations, and even within the in-breed laboratory mouse populations and I am wondering, with regard to what you are seeing here, have you seen genetic background effects? Have you looked in different genetic backgrounds? Do you see the same spectrum of mutations come up?

Berns: Many of the experiments have been performed in a number of backgrounds. Even though the frequency of the specific insertions might slightly vary from one background to the other - to some extent this might relate to the effectiveness of these viruses to replicate - we do see the same spectrum of mutations in the tumors.

Hanahan: Well, actually that is a nice introduction to my next question. You alluded at the beginning to another long-term goal, which is to extend this elegant approach in epithelial cell types. Could you comment, at all, at least on what your vision for the future is and how to get this working in epithelial cells?

Berns: Yes, there are two routes we are attempting at the moment. The first is to have retrotransposons to work for us in mice. I am not necessarily optimistic about it, but nevertheless we are trying them out. One of the worries is that they might be silenced early in embryogenesis and as a consequence do not transpose due to lack of transcription. The second approach is the use of DNA transposons. It is clear from studies done in the laboratory of Ronald Plasterk in the Netherlands Cancer Institute that transposons from C-elegans only require the transposase for transposition. The idea would be that we provide mice with a large number of copies of these DNA transposons, include in the mice an inducible system to activate the transposase at a desired time in a tissue specific fashion. This could be done for example, in oncogene bearing mice. The idea is then to pick up collaborating events that complement the mutations that were put in.

Livingston: First of all, how will you know when you are out of additional genes for this particular lymphoma and what about humans? Are any of these genes naturally activated during the genesis of any known human lymphoma?

Berns: Well, it is the same situation as, for example, in C-elegans with respect to saturation. You would look how often you'd find the same locus. To some extent that is what we are doing here and I do not expect to find all the genes but I do think that we will find quite a number of them. In fact, I am pleasantly surprised by the large number of relevant sites we are finding which are in a 30–40% range of the insertion sites. Looking at how many insertions we have it should not be difficult to collect 20–30 different genes and by the time we have them I can spend the rest of my career to figure out what precisely they are doing. With respect to the involvement in humans, well, TPL2 is known to be the cot oncogene which is found in human tumors. With respect to PIM, we have looked very hard and in a certain subsets of large B lymphomas it is very highly expressed and so, intuitively, I would say that it is fulfilling an important function there. The only problem is that without genetic evidence for a mutation it is unclear what it means. What might help further is to know what the pathway is in which PIM1 and PIM2 are acting. We know that both are regulated by cytokines. Many of the lymphomas in man show an upregulation of both PIM1 and PIM2 and that might argue in favor of mutations in upstream interleukin pathways. However, a very small subset has either PIM1 or PIM2 activated. Those are the ones we will focus on because there the chance is higher that something with the gene itself has happened. Unfortunately, we have not found any changes so far. And the same holds, more or less, for Bmi1. In other words, there is no evidence that Bmi1 is involved in human tumors even though in the chromosomal region where Bmi1 has been mapped aberrations are found.

Livingston: Are any of these genes linked?

Berns: No, actually they are all spread out.

DePinho: We have knocked out MXI-1 and we get a cancer prone condition in which B lymphomas predominate, such a model might provide an opportunity to see if you identify the same set of cooperating genes identified in your Myc screens. It might provide an experimental opportunity to integrate the pathways involved in lymphomagenesis. From a genetic standpoint, we and Bob Eisenman have identified the drosophila homologue and it maps to the diminutive locus. I was wondering whether or not you have gone back to the drosophila genetics to see if there are any enhancers or suppressors of that locus that involve any of the genes that you have identified in your screens.

Berns: No, we have not. Well, for Bmi1 the situation is clear but for the PIM genes we have tried very hard to get Drosophila homologues but without success. In the group of Plasterk they have been able to clone the C-elegans homologues and generated knockouts. Those knockouts have a similar lack of phenotype as mice.

Klausner: Is the limitation of the transposition, either retro-transposition or DNA transposition, truly the frequency of the transposition or is there any effect of the survival of cells into which there are transpositions? For example, is there a difference in the frequency of stable transpositions in p53 negative cells or in Bcl-2 overexpressing cells?

Berns: No, I do not think so. Even though you might see after infection of 3T3 fibroblasts with Moloney murine leukemia virus cytopatic effects and the killing of many cells. But we have no evidence for that *in vivo*. We have been looking in Bcl2 transgenic mice and if you use a retroviral infection of Bcl2 transgenics the transgene actually appears to be rather inert when it comes to insertional mutagenesis because we do not see a significant acceleration in that genetic background. In addition, retroviral insertion will also not lead to the activation of Bcl2. We try to explain this by assuming that there is not sufficient selective advantage to the cell to activate Bcl2. This argues that survival of the cell is not playing a significant role.

Gray: Genes that are over expressed in lymphomas and leukemias show up to be amplified in epithelial cells tumors and I was wondering how much you have surveyed it?

Berns: We have not carefully looked for a role of these genes in human epithelial tumors. In an early stage when we just had found them we have analyzed a limited number of tumors for aberrant expression. At the time we did not find any evidence for their involvement. But it might be time to reexamine that and look at a larger group of tumors. But we have not done it.

Gray: That is in your lymphomas?

Berns: In human lymphomas.

Gray: What I was suggesting is in human epithelial tumors.

Berns: We have not looked there. We had, at an early stage, looked at a number of different human tumors to see if there was any aberrant expression and at that time we did not find much involvement of the gene in the first place. But, it might be time to reconsider that and look to a larger group and more tumors, but we have not done that.

3

THE TRANSCRIPTION FACTOR B-Myb IS PHOSPHORYLATED AND ACTIVATED BY CYCLIN A/Cdk2

Olaf Bartsch,[1,2] Ulrike Ziebold,[2,3] Richard Marais,[4] Karl-Heinz Klempnauer,[3] and Stefano Ferrari[1]

[1]Institute for Experimental Cancer Research
Tumor Biology Center
P.O Box 1120, D-79011 Freiburg, Germany
[2]School of Biology
University of Freiburg
Schänzlestr. 1, D-79104 Freiburg, Germany
[3]Hans-Spemann-Laboratory
Max-Planck-Institute for Immunobiology
Stübeweg 51, D-79108 Freiburg, Germany
[4]Institute for Cancer Research
Chester Beatty Laboratories
Fulham Road, London SW3 6JB, United Kingdom

1. ABSTRACT

The restriction point (R) in the late G_1 phase of the cell cycle is a time at which a decision is made on whether cells enter a round of division or return to quiescence. Transition through R and progression into S-phase has been shown to be promoted by progression factors and driven by the kinase activity associated with cyclin E and A, respectively. To investigate the mechanism by which cyclin-dependent kinases (Cdk) facilitate the execution of these steps we set to identify substrates for Cdk2. We have examined the regulation of B-Myb, a member of the *myb* proto-oncogene family, by the cell cycle machinery. Compelling evidence indicates that B-Myb plays an important role during the late G_1- and early S-phase of the cell cycle. We obtained evidence that B-Myb undergoes cyclin A/Cdk2-mediated phosphorylation at the onset of S-phase. This event triggers

The Biology of Tumors, edited by Mihich and Croce
Plenum Press, New York, 1998.

B-Myb transactivation potential by relieving a constraint conferred by the C-terminal inhibitory domain of the protein. Contrary to the reported inhibition of E2F-DP following phosphorylation by cyclin A/Cdk2, we show here that cyclin A/Cdk2 increases the transactivation potential of B-Myb. Our findings, therefore, provide the first evidence for a positive role of cyclin A/Cdk2 and imply that B-Myb is a cell cycle regulated transcription factor activated at the onset of S-phase.

2. INTRODUCTION

The regulation of cell division maintains homeostatic balance between cell growth, differentiation, survival and death. Emergence from quiescence is controlled by growth stimulators called growth factors. Their effect is mediated by intracellular protein kinase cascades which are engaged within seconds from occupation of growth factor receptors. Such cascades propagate and amplify the signal and ultimately activate transcription factors in order to initiate metabolic processes necessary for growth[1]. Progression through the cell cycle is orchestrated by the timely assembly and activation of cyclins and their catalytic partners, cyclin-dependent kinases. Evidence obtained in recent years indicates that cyclin D-associated kinases support transition through G_1, whereas cyclin E/Cdk2 and cyclin A/Cdk2 complexes appear to be involved with exit from G_1 and progression through S-phase of the cell division cycle, respectively[2–12]. The identification and characterization of some Cdk targets[13], allowed to begin speculating on the role of those kinases. However, to date the vast majority of substrates for Cdks remain to be identified.

In order to expose the mechanism by which cyclin E- and A/Cdk2 drive transition to and progression throughout S-phase, respectively, we set to identify substrates for this kinase. Based on both biological (i.e., cell cycle-dependent expression and function) and biochemical criteria (i.e., presence of sites fulfilling the requirement for phosphorylation by Cdks), we selected a number of potential targets for Cdk2. Among those, we began our analysis with B-*myb,* a member of the *myb* proto-oncogene family[14–17]. B-*myb* encodes a sequence specific DNA binding protein (B-Myb) which displays weak transactivation activity[14–19]. B-Myb is known to play an essential role at the onset of S-phase[15] and contrary to c-Myb, it is expressed in virtually all proliferating cells and tissues[20]. During embryonic development of the mouse expression of B-*myb* is tightly linked to the proliferative activity of cells and tissues[21]. B-*myb* expression is controlled by transcription factor E2F and reaches its maximum at the G_1/S-phase boundary and during the S-phase of the cell cycle[22,23]. Evidence obtained using antisense oligonucleotides indicates that downregulation of B-*myb* in lymphoid cell lines causes a block of their proliferation[24], whereas overexpression of the protein reduces growth factor requirements and induces a transformed phenotype in fibroblasts[25]. Finally, constitutive overexpression of B-*myb* can bypass a p53-induced cell cycle arrest[26]. Interestingly, this effect does not involve restoration of cyclin E/Cdk2 kinase activity[26], raising the intriguing possibility that B-*myb* acts downstream of cyclin E and of the checkpoint controlled by $p21^{Waf1/Cip1}$.

Here we have examined the regulation of B-Myb activity by the cell cycle machinery. We show that the transactivation potential of B-Myb is repressed by a regulatory domain located at the C-terminus of the protein and that the inhibitory effect of this domain is relieved upon phosphorylation by cyclin A/Cdk2. This finding, therefore, identifies B-Myb as a novel target for cyclin A/Cdk2 and indicates that its transactivation activity is regulated by cyclin A-mediated phosphorylation. Furthermore, our work provides the first evidence for a link between the Myb family and the cell cycle machinery.

3. RESULTS

3.1. Transactivation by B-Myb Is Repressed by Its C-Terminus

Repression of transactivation by c-Myb has been shown to be mediated by the C-terminal domain of the protein[27–29]. In order to investigate whether this mechanism may also explain the weak transactivation activity displayed by B-Myb, we employed two C-terminally deleted constructs (Fig. 1A). As compared to full length B-Myb, we found that a construct lacking the entire C-terminus (B-Myb-Δ3) was capable of activating a Myb-responsive reporter gene (Fig. 1B). Further deletion of an acidic region located in the middle of B-Myb (B-Myb-ΔPvu) resulted in an inactive protein.

Considering the cell cycle-dependent expression of B-Myb and the presence of several potential sites of phosphorylation for cyclin-dependent kinases in the C-terminal half of the molecule, we examined the possibility that B-Myb activity could be regulated by phosphorylation. To this aim we ectopically expressed full-length B-Myb and various cyclins and determined the transactivation potential of B-Myb. Interestingly, full-length B-Myb activated a Myb-responsive reporter when co-expressed with cyclin A but not with cyclins B1, B2, D1, or E (Fig. 1C, and data not shown). On the other hand, transactivation by c-Myb was not affected by cyclin A (Fig. 1D), indicating that the stimulation of B-Myb activity by cyclin A is specific. Futhermore, the activity of truncated B-Myb (B-Myb-Δ3) was not increased by cyclin A (Fig. 1E). This last finding indicates that the effect of cyclin A is possibly to counteract the inhibitory effect of the C-terminal domain.

To formally exclude the possibility that cyclin A may instead alter the affinity of B-Myb for DNA and in such way simply cause increased binding of the protein, we decided to test a Gal4/B-Myb fusion protein. In this construct the DNA-binding domain of B-Myb was replaced by the Gal4 DNA-binding domain and transactivation was monitored by employing an adenovirus E4 promoter fused to five copies of the Gal4 binding site. The results presented in Fig. 2 show that Gal4/B-Myb was also strongly stimulated by cyclin A. As a control we performed a similar experiment using a Gal4/Myc fusion protein. In this case the activity of the Gal4/Myc protein was not affected by cyclin A (Fig. 2). This indicates that the effect of cyclin A is specific to B-Myb and is not directed to the DNA-binding domain of the protein.

3.2. B-Myb Is Phosphorylated in a Cyclin A-Dependent Manner

The effect of cyclin A on B-Myb activity raised the possibility that B-Myb might be phosphorylated in a cyclin A-dependent fashion. To address this possibility, B-Myb was co-transfected with various cyclins and analyzed by SDS-PAGE and Western blotting. Immunoprecipitations followed by *in vitro* kinase assays confirmed that all cyclin/Cdk complexes examined displayed activity. However, only cyclin A/Cdk2 was capable of phosphorylating B-Myb *in vitro* (data not shown). Accordingly, only co-expression of B-Myb with cyclin A in Cos-7 cells induced the appearance of a slow-migrating form of B-Myb (Fig. 3A). On the other hand, cyclin A expression did not affect the mobility of C-terminally truncated B-Myb, c-Myb or of C/EBPβ (Fig. 3B-D), confirming the specificity of the effect observed.

During S-phase, the catalytic partner of cyclin A is Cdk2[3,5,7,8]. To assess whether the endogenous Cdk2 present in transfected cells is involved in the effect of cyclin A on B-Myb, we co-expressed B-Myb and cyclin A with increasing amounts of a kinase-negative variant of Cdk2, in which Lys_{33} was mutated to Arg (U.Deuschle and S.F., unpublished).

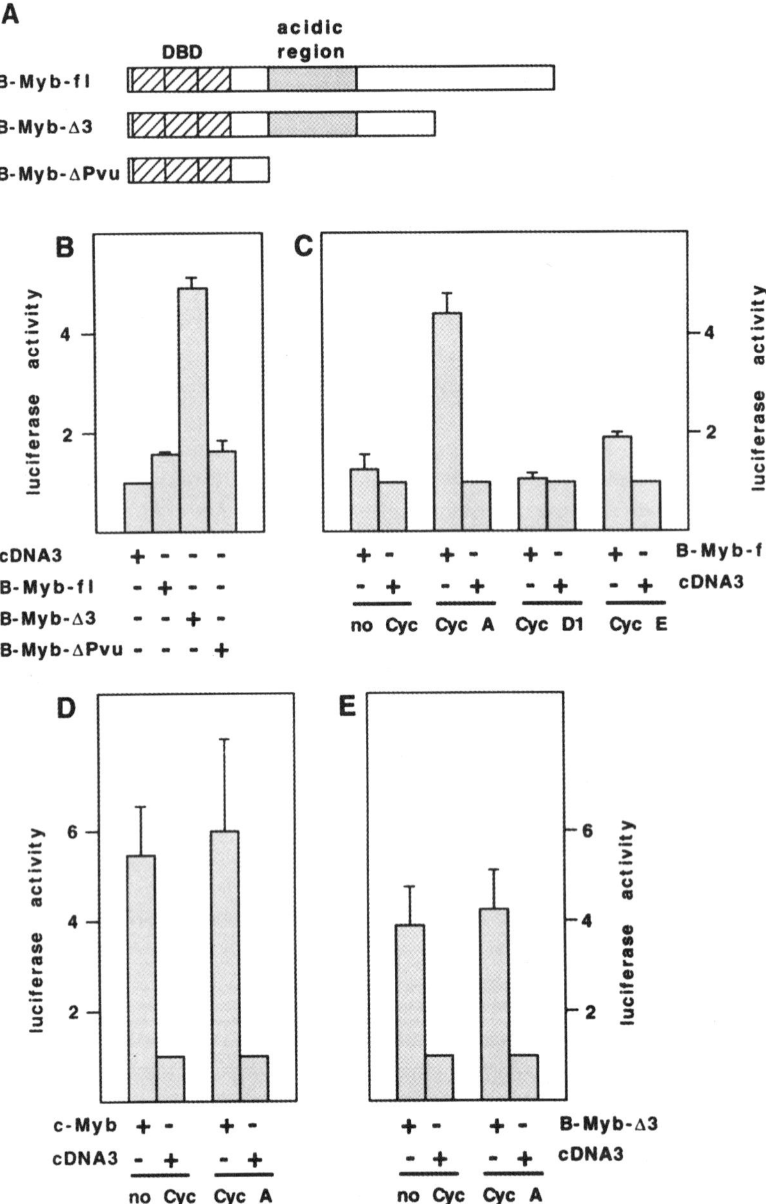

Figure 1. Stimulation of B-Myb transactivation activity by cyclin A. **A**. Schematic structure of full-length and deleted mouse B-Myb proteins. DBD: Myb DNA-binding domain. **B-E**. Cos-7 cells were co-transfected with expression vectors for full-length or deleted mouse B-Myb or c-Myb (5μg), cyclins A, D1 or E (3μg), the Myb-responsive reporter gene p3xATk-Luc (5μg), and the β-galactosidase reference plasmid pCMVβ (to monitor the transfection efficiency). Control transfections contained equivalent amounts of empty expression vector pCDNA3, as indicated. Cells were analyzed for luciferase and β-galactosidase activities 24 hrs after transfection. The luciferase activity in the absence of Myb was designated as 1. Thin lines show standard deviations. (Reprinted in part from ref. 33 with permission).

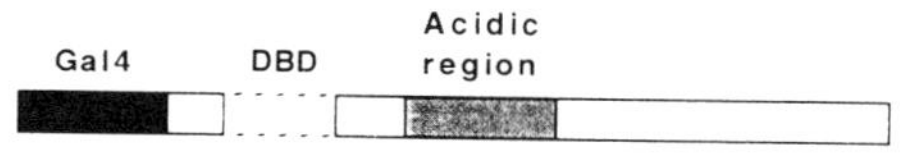

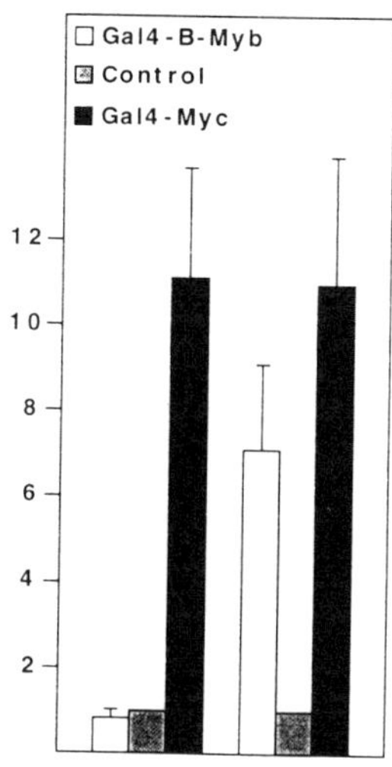

Figure 2. Stimulation of B-Myb transactivation activity by cyclin A does not require the Myb DNA-binding domain. Cos-7 cells were co-transfected with Gal4/B-Myb or Gal4/Myc fusion protein, cyclin A, the Gal4-responsive reporter gene pG5E4–38Luc and the β-galactosidase reference plasmid pCMVβ. Control transfections contained equivalent amounts of empty expression vector pCDNA3. Cells were analyzed as decribed in Fig. 1. (Reprinted from ref. 33 with permission).

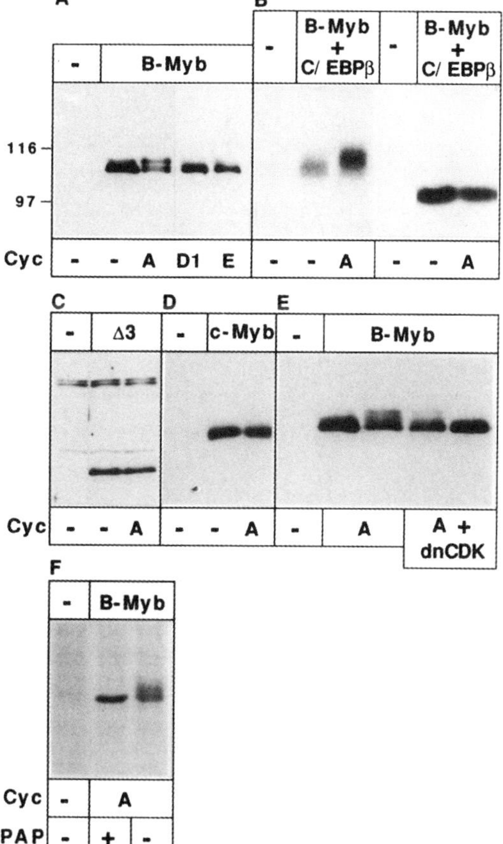

Figure 3. Cyclin A induces a slow-migrating form of B-Myb. **A, C–E**. Cos-7 cells were co-transfected with full-length or truncated mouse B-Myb or c-Myb and cyclins A, D1 or E. **B**. Full length B-Myb was co-transfected with chicken C/EBPβ (1 μg) in the presence or absence of cyclin A. Cells were labeled with ^{32}P-ortophosphate, extracted and immunoprecipitated with antibodies specific to B-Myb (first three lanes) or C/EBPβ (last three lanes). In (**E**), the last two lanes show transfections which contained additionally 1μg (lane 4) and 3μg (lane 5) expression vector for a kinase-negative Cdk2 mutant. **F**. Transfected cells were labeled with ^{35}S-methionine and immunoprecipitated with B-Myb antiserum. Aliquots of the immunoprecipiate were treated with potato acid phosphatase as described[33], and analyzed by SDS-PAGE and autoradiography. Control incubations using phosphatase inactivated by treatment with sodium-orthovanadate and β-glycerol-phosphate or by heat did not lead to a change of the electrophoretic mobility pattern of B-Myb (not shown). (Reprinted in part from ref. 33 with permission).

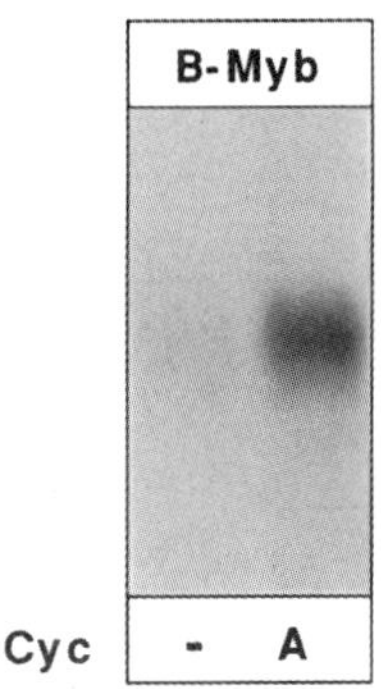

Figure 4. Cyclin A-dependent phosphorylation of B-Myb. Cos-7 cells transfected with B-Myb in the presence or the absence of cyclin A were ^{32}P-labeled and B-Myb was immunoprecipitated and analyzed by SDS-PAGE and autoradiography. (Reprinted in part from ref. 33 with permission).

We observed that kinase negative Cdk2 suppressed the appearance of the slow-migrating form of B-Myb, likely by titrating out the endogenous wild-type kinase (Fig. 3E). To confirm that the altered electrophoretic mobility of B-Myb is the result of cyclin A-mediated phosphorylation, we performed an *in vitro* phosphatase treatment of immunoprecipitated B-Myb. As shown in Fig. 3F, potatoe acid phosphatase (PAP) converted the retarded form of B-Myb into a faster migrating form, confirming that the effect observed on the pattern of electrophoretic mobility was due to phosphorylation. Finally, to directly demonstrate the phosphorylation of B-Myb, B-Myb was expressed in the presence or the absence of cyclin A and immunoprecipitated following ^{32}P-labeling of cells. As shown in Fig. 4, the amount of phosphate incorporated into B-Myb was increased upon co-expression with cyclin A. Taken together, these data suggest that cyclin A induces phosphorylation of B-Myb to high stoichiometry.

As mentioned above, several potential sites fulfilling the requirements for recognition and phosphorylation by cyclin-dependent kinases are located in the C-terminal half of B-Myb. In order to disclose the number of sites affected by the cyclin A-dependent kinase, *in vivo* ^{32}P-labeled B-Myb was immunoprecipitated and subjected to two-dimensional tryptic phosphopeptide analysis. Tryptic digestion generated a complex pattern of phosphopeptides which was clearly evident only when B-Myb was expressed in the presence of cyclin A (Fig 5, panels A and B). A number of phosphopeptides co-migrated with those obtained from an *in vitro* labeled glutathione-S-transferase-B-Myb fusion protein (GST-B-Myb) (Fig. 5C), thus confirming that B-Myb is a direct target for cyclin A/Cdk2.

3.3. B-Myb Is Phosphorylated during S-Phase of Normal Cells

In order to exclude the possibility that the effect reported above may be an artifact caused by massive overexpression of proteins, we next investigated whether phosphorylation of B-Myb also occurs in normal cells. To this aim, Swiss 3T3 fibroblasts were synchronized by serum starvation, labeled with ^{35}S-methionine and endogenous B-Myb was immunoprecipitated. The cells entered S-phase approximately 14 hours after addition of serum and showed a broad peak of B-*myb* mRNA expression between 12 and 24 hours post-stimulation (Fig. 6A). An approximately 110 kD protein was specifically immunoprecipitated with B-Myb antiserum from S-phase cells but not from growth-arrested cells (Fig. 6B). Interestingly, also in this case we observed a slightly retarded protein band which had the same mobility as the slow-migrating form of B-Myb detected after co-transfection with cyclin A. To examine whether also in this case phosphorylation is the cause of retarded mobility, synchronized Swiss 3T3 fibroblasts were labeled in S-phase with ^{32}P-or-

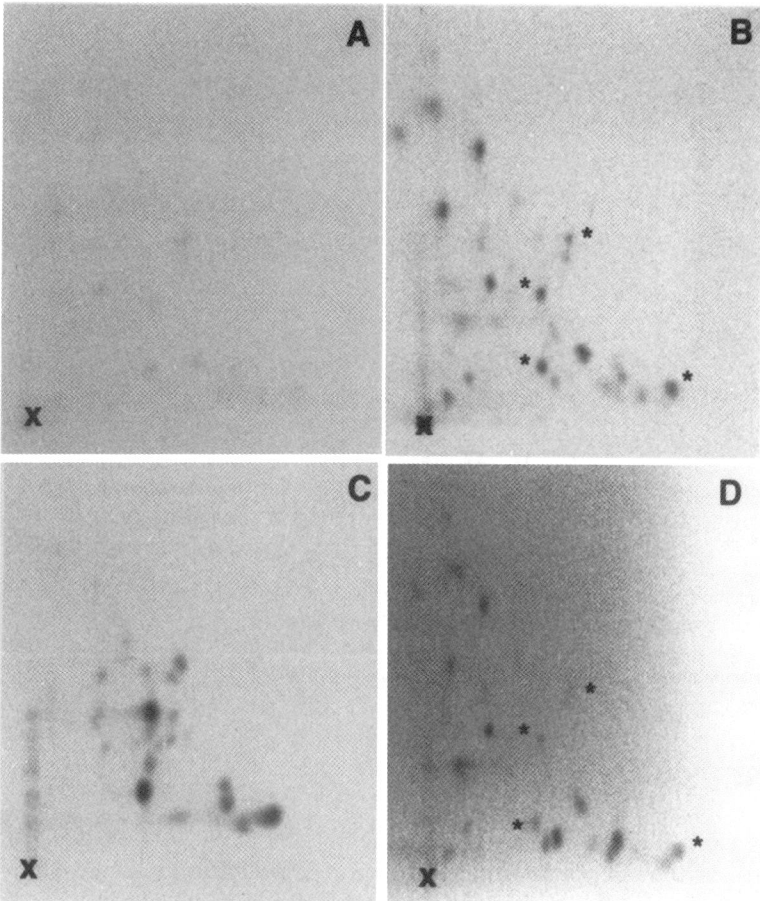

Figure 5. Two-dimensional tryptic phosphopeptide analysis of mouse B-Myb phosphorylated *in vivo* and *in vitro*. Cos-7 cells transfected with B-Myb in the absence (**A**) or presence (**B**) of cyclin A were ^{32}P-labeled for 5 hrs. B-Myb was isolated by immunoprecipitation and analyzed by digestion with trypsin followed by two-dimensional fractionation of the resulting peptides. **C**. Bacterially expressed GST-B-Myb was phosphorylated *in vitro*, using immunopurified cyclin A/Cdk2, and analyzed as described above. **D**. Synchronized Swiss 3T3 fibroblasts were ^{32}P-labeled in S-phase and B-Myb was isolated and analyzed as described above. Radioactive peptides were visualized with a phospho-image analyzer. The peptides marked with asterisks in (**B**) and (**D**) comigrate with the major *in vitro* labeled peptides, as determined by a mixing experiment (data not shown). (Reprinted from ref. 33 with permission).

thophosphate and B-Myb analyzed by immunoprecipitation. A specific protein band appeared in the immunoprecipitate from the phosphate-labeled cells (Fig. 6C), which migrated slightly slower than the major band of ^{35}S-labeled B-Myb. In order to examine whether endogenous B-Myb is phosphorylated at the same sites observed in co-transfection experiments, we performed two-dimensional tryptic phosphopeptide mapping of endogenous B-Myb isolated from ^{32}P-labeled Swiss 3T3 fibroblasts (Fig. 5D). Although the relative intensity of some of the phosphopeptides varied, the pattern obtained corresponded to that of ectopically expressed B-Myb. These data confirm that phosphorylation occurs *in vivo* at the same sites observed with overexpressed B-Myb.

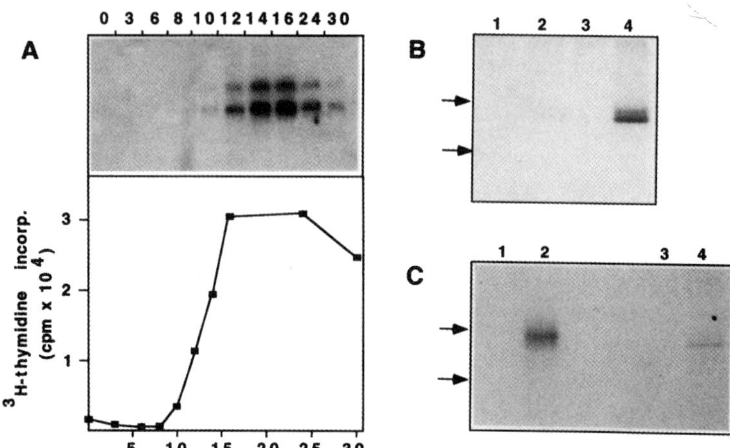

Figure 6. Detection of phosphorylated B-Myb in Swiss 3T3 fibroblasts. **A**. Cells were synchronized by serum starvation followed by addition of serum for the time periods (in hrs) indicated at the top and bottom, respectively. The expression of B-*myb* mRNA was analyzed by Northern blotting (top) and the onset of S-phase determined by ^{3}H-thymidine incorporation (bottom). **B**. Swiss 3T3 cells in G_0 (lanes 1,2) or S-phase (lanes 3,4), were labeled with ^{35}S-methionine for 2 hrs and immunoprecipitated with preimmune serum (lanes 1,3) or B-Myb serum (lanes 2,4). Proteins were then analyzed by SDS-PAGE and autoradiography. **C**. Swiss 3T3 fibroblasts in S-phase were labeled for 5 hrs with ^{32}P-orthophosphate (lanes 1,2) or for 3 hrs with ^{35}S-methionine (lanes 3,4) and analyzed by immunoprecipitation with preimmune serum (lanes 1,3) or B-Myb serum (lanes 2,4).

4. DISCUSSION

The results presented here provide the first evidence that B-Myb, a conserved member of the Myb transcription factor family, is regulated directly by the cell cycle machinery. Our findings indicate that an inhibitory function is associated with the C-terminal domain of B-Myb and this is counteracted by cyclin A/Cdk2-mediated phosphorylation. This suggests that the C-terminus of B-Myb acts as a cell cycle sensor. Although this work did not address in detail the mechanism by which phosphorylation of B-Myb affects its transactivation activity, it clearly establishes that the DNA-binding domain of B-Myb is not involved in the cyclin A-dependent regulation. Therefore, we conclude that the regulatory role of phosphorylation is independent on effects on B-Myb DNA-binding activity.

A second major conclusion of this work is the identification of B-Myb as novel target for cyclin A/Cdk2. We have shown that transactivation by B-Myb is inhibited by the C-terminal domain of the protein and inhibition is specifically relieved by co-expression with cyclin A. This is accompanied by phosphorylation of the protein at multiple sites, as indicated by two-dimensional mapping studies. Some of the sites appear to be phosphorylated by cyclin A/Cdk2 also *in vitro*, implying that B-Myb is a direct target for this kinase. However, the complexity of *in vivo* two-dimensional maps led us to speculate that in addition to direct phosphorylation by cyclin A/Cdk2 other kinases, which are either consitutively active or are activated by cyclin A/Cdk2, may contribute to generate the pattern of phosphopeptides observed. According to this scenario, initial phosphorylation of B-Myb would likely alter the conformation of the protein such that it becomes accessible to kinases other than cyclin A/Cdk2 (Fig. 7). We are at present testing this model by mutating potential sites of phosphorylation and examining the effect of cyclin A on B-Myb transactivation activity as well as on the two-dimensional phosphopeptide pattern.

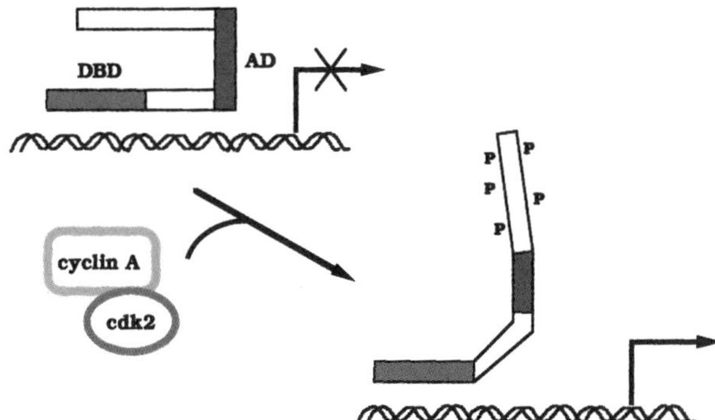

Figure 7. Possible mechanism of B-Myb activation by cyclin A/Cdk2. DBD: DNA-binding domain; AD: acidic domain.

Finally, our results provide the first evidence for a positive role of cyclin A/Cdk2 in S-phase. As of yet, cyclin A/Cdk2 has been shown to phosphorylate certain members of the E2F-DP transcription factor family, thereby causing inhibition of their DNA-binding activity during S-phase[30–32]. Contrary to E2F-DP, our findings indicate that the transactivation potential of B-Myb is increased by cyclin A/Cdk2-mediated phosphorylation. Considering the relevance of B-Myb during proliferation[24,25], the identification of the genes whose expression is regulated by B-Myb might now provide interesting insight into the role of cyclin A during S-phase.

5. MATERIALS AND METHODS

5.1. Expression Vectors and Transient Transfection Assays

Expression vectors for full-length mouse B-Myb and Gal4/B-Myb as well as all other constructs employed in this study have been described[33]. Expression vectors for human cyclins A, D1 and E were obtained from R. Weinberg. A kinase-negative mutant of Cdk2, Cdk2-R_{33}, where K_{33} was mutated to R, was obtained from U. Deuschle and subcloned into pCDNA3. Cos-7 cells were transfected as described[33]. The Myb-responsive reporter gene contained the Herpes simplex virus thymidine kinase promoter fused to three Myb binding sites. To correct for differences in transfection efficiency plates were co-transfected with 0.5 μg of the β-galactosidase reporter gene pCMVβ (Clontech).

5.2. Western Blotting, *in Vivo* Labeling, and Immunoprecipitations

B-Myb was detected by Western blotting as previously described[33]. *In vivo* labelling and immunoprecipitations were carried out as described[33].

5.3. Cell Cycle Analysis

Swiss 3T3 cells were synchronized as described[33]. Synchronization was verified by labeling the cells for 15 min with 5 μCi/ml ^{3}H-thymidine (Amersham, 83 Ci/mmol), followed by precipitation with 10% trichloroacetic acid. Radioactivity incorporated into

DNA was quantified by liquid scintillation counting. Cell cycle-dependent expression of B-*myb* was analyzed by Northern blotting as described[21].

5.4. *In Vitro* Phosphorylation

293 cells overexpressing human cyclin A and Cdk2 were lysed as described[33] and extracts immunoprecipitated using either anti-Cdk2 or anti-cyclin A antibodies (St. Cruz). Immunoprecipitates were used to phosphorylate 5 μg bacterially expressed GST-B-Myb (residues 227 to 704) employing the conditions described in[33].

ACKNOWLEDGMENTS

We thank B. Haenig and B. Mutschler for excellent technical assistance, R. Weinberg, E. Harlow and U. Deuschle for providing plasmids. This study was supported by grants from the German Research Council (DFG) to K.-H. K. and to S.F. and by the Center for Clinical Research (ZKF) of the University of Freiburg.

REFERENCES

1. T. Hunter, Oncoprotein networks. *Cell* 88: 333–346 (1997)
2. F. Girard, U. Strausfeld, A. Fernandez and N.J.C. Lamb, Cyclin A is required for the onset of DNA replication in mammalian fibroblasts. *Cell* 67: 1169–1179 (1991)
3. L.H. Tsai, E. Harlow and M. Meyerson, Isolation of the human *cdk2* gene that encodes the cyclin A- and adenovirus E1A-associated p33 kinase. *Nature* 353: 174–177 (1991)
4. V. Dulic, E. Lees and S.I. Reed, Association of human cyclin E with a periodic G_1-S phase protein kinase. *Science* 257: 1958–1961 (1992)
5. S.J. Elledge, R. Richman, F.L. Hall, R.T. Williams, N. Lodgson and J.W. Harper, *CDK2* encodes a 33-kDa cyclin A-associated protein kinase and is expressed before *CDC2* in the cell cycle. *Proc Natl Acad Sci. USA* 89: 2907–2911 (1992)
6. A. Koff, A. Giordano, D. Desai, K. Yamashita, J.W. Harper, S. Elledge, *et al*, Formation and activation of a cyclin E-cdk2 complex during the G_1 phase of the human cell cycle. *Science* 257: 1689–1694 (1992)
7. M. Pagano, R. Pepperkok, J. Lukas, F. Verde, W. Ansorge and G. Draetta, Cyclin A is required at two points in the human cell cycle. *EMBO J.* 11: 961–971 (1992)
8. J. Rosenblatt, Y. Gu and D.O. Morgan, Human cyclin-dependent kinase-2 is activated during the S and G_2 phases of the cell cycle and associates with cyclin A. *Proc Natl Acad Sci USA* 89: 2824–2828 (1992)
9. M. Ohtsubo and J.M. Roberts, Cyclin-dependent regulation of G_1 in mammalian fibrolasts. *Science* 259: 1908–1912 (1993)
10. C.J. Sherr, Mammalian G_1 cyclins. *Cell* 73: 1059–1065 (1993)
11. G.F. Draetta, Mammalian G_1 cyclins. *Curr Opin Cell Biol* 6: 842–846 (1994)
12. J.A. Knoblich, K. Sauer, L. Jones, H. Richardson, R. Saint and C.F. Lehner, Cyclin E controls S-phase progression and its downregulation during Drosophila embryogenesis is required for the arrest of cell proliferation. *Cell* 77: 107–120 (1994)
13. R.A. Weinberg, The retinoblastoma protein and the cell cycle. *Cell* 81: 323–330 (1995)
14. N. Nomura, M. Takahashi, M. Matsui, S. Ishii, T. Date, S. Sasamoto, *et al*, Isolation of human cDNA clones of *myb*-related genes, A-*myb* and B-*myb*. *Nucleic Acids Res* 16: 11075–11089 (1988)
15. E.W.-F. Lam, C. Robinson and R.J. Watson, Characterization and cell cycle-regulated expression of mouse B-*myb*. *Oncogene* 7: 1885–1890 (1992)
16. G. Foos, S. Grimm and K.-H. Klempnauer, Functional antagonism between members of the *myb* family: B-*myb* inhibits v-*myb*-induced gene activation. *EMBO J* . 11: 4619–4629 (1992)
17. T. Bouwmeester, S. Guehmann, T. El-Baradi, F. Kalkbrenner, I. van Wijk, K. Moelling *et al*.., Molecular cloning, expression and in vitro functional characterization of Myb-related proteins in *Xenopus*. *Mechanisms of Development* 37: 57–68 (1992)

18. G. Mizuguchi, H. Nakagoshi, T. Nagase, N. Nomura, T. Date, Y. Ueno, *et al.*, DNA binding activity and transcriptional activator function of the human B-*myb* protein compared with c-MYB. *J Biol Chem* 265: 9280–9284 (1990)
19. R.J. Watson, C. Robinson, E.W.-F. Lam, Transcription regulation by murine B-myb is distinct from that by c-myb. *Nucleic Acids Res* 21: 267–272 (1993)
20. K. Reiss, S. Travali, B. Calabretta and R. Baserga, Growth regulated expression of B-myb in fibroblasts and hematopoietic cells.*J. Cell. Physiol.* 148: 338–343 (1991)
21. J. Sitzmann, K. Noben-Trauth and K.-H. Klempnauer, Expression of B-*myb* during mouse embryogenesis. *Oncogene* 12: 1889–1894 (1996)
22. E.W-F. Lam and R.J. Watson, An E2F-binding site mediates cell-cycle regulated repression of mouse B-myb transcription. *EMBO J* . 12: 2705–2713 (1993)
23. J. Zwicker, N. Liu, K. Engeland, F.C. Lucibello and R. Müller, Cell cycle regulation of E2F site occupation in vivo. *Science* 271: 1595–1597 (1996)
24. M. Arsura, M. Introna, F. Passerini, A. Mantovani and J. Golay, B-myb antisense oligonucleotides inhibit proliferation of human hematopoietic cell lines. *Blood* 79: 2708–2716 (1992)
25. A. Sala and B. Calabretta, Regulation of BALB/c 3T3 fibroblast proliferation by B-myb is accompanied by selective activation of cdc2 and cyclin D1 expression. *Proc Natl Acad Sci USA* 89: 10415–10419 (1992)
26. D. Lin, M. Fiscella, P.M. O'Connor, J. Kackman, M. Chen, L.L. Luo, *et al.*, Constitutive expression of B-*myb* can bypass p53-induced Waf1/Cip1-mediated G_1 arrest. *Proc Natl Acad. Sci USA* 91: 10079–10083 (1994)
27. H. Sakura, C. Kanei-Ishii, T. Nagase, H. Nakagoshi, T.J. Gonda and S. Ishii, Delineation of three functional domains of the transcriptional activator encoded by the c-myb proto-oncogene. *Proc Natl Acad Sci USA* 86: 5758–5762 (1989)
28. C. Kanei-Ishii, E.M. MacMillan, T. Nomura, A. Sarai, R.G. Ramsay, S. Aimoto, *et al.*, Transactivation and transformation by Myb are negatively regulated by a leucine-zipper structure. *Proc Natl Acad Sci USA* 89: 3088–3092 (1992)
29. J.W. Dubendorff, L.J. Whittacker, T.J. Eltman and J.S. Lipsick, Carboxy-terminal elements of c-Myb negatively regulate transcriptional activation in *cis* and in *trans*. *Genes Dev* 6: 2524–2535 (1992)
30. B.D. Dynlacht, O. Flores, J.A. Lees and E. Harlow, Differential regulation of E2F trans-activation by cyclin/cdk2 complexes. *Genes Dev* 8: 1772–1786 (1994)
31. M. Xu, K.A. Sheppard, C.Y. Peng, A.S. Yee and H. Piwnica-Worms, Cyclin A/CDK2 binds directly to E2F-1 and inhibits the DNA-binding activity of E2F-1/DP-1 by phosphorylation. *Mol Cell Biol* 14: 8420–8431 (1994)
32. W. Krek, M.E. Ewen, S. Shirodkar, Z. Arany, W.G. Kaelin and D.M. Livingston, Negative regulation of the growth-promoting transcription factor E2F-1 by a stably bound cyclin A-dependent protein kinase. *Cell* 78: 161–172 (1994)
33. U. Ziebold, O. Bartsch, R. Marais, S. Ferrari and K.-H. Klempnauer, Phosphorylation and activation of B-Myb by cyclin A/Cdk2. *Current Biology* 7: 253–260 (1997)

4

DISTINCT DYNAMICS AND REGULATORY SIGNAL TRANSDUCTION OF CELL MIGRATION

Lessons from Dendritic Cells, Tumor Cells, and T Lymphocytes

F. Entschladen,* K. Maaser, M. Gunzer, P. Friedl, B. Niggemann, and K. S. Zänker

Institute of Immunology
University Witten Herdecke
58448 Witten, Germany

CURRENT VIEW AND PHENOTYPES OF MIGRATING CELLS

In the adult metazoan organism active migration is an exclusive property of a few specialized cells. In the human body, cells of the immune system are highly active migratory cells. The ability of flexible positioning and migration is a crucial prerequisite for the immune surveillance of tissues and a coordinated cellular immune response. Tumor cell migration on the other hand is a decisive step in the progression of cancer towards metastasis.

The mechanistic principles of migration are supposed to be similar in many cell types: Migration starts with the protrusion of a leading lamella and its subsequent adhesion to the matrix. After that, reorganisation of the cytoskeleton follows and the cell moves forward. In a third step the cell detaches from the matrix from its interaction sites at the trailing edge, which is retracted thereby finishing the moving cycle (Abercombie et al. 1970, Huttenlocher et al. 1995). However, the different signals why cells start to migrate are almost unknown, and the dynamics of migration as well as the morphologies of the migrating cells considerably vary (Stossel 1994). About the molecular events of locomotion less is known, but it is reasonable to argue that cell migration is not as uniform among cells as the three-step model might suggest.

* Corresponding author: Dr. F. Entschladen, Institute of Immunology, University Witten/Herdecke, Stockumer Straße 10, 58448 Witten, Germany. Phone: +49-2302-669-162, Fax: +49-2302-669-158, E-mail: frankent@uwh.de.

The Biology of Tumors, edited by Mihich and Croce
Plenum Press, New York, 1998.

We investigated the migratory behaviour of murine dendritic cells, human melanoma cells (MV3), and human T lymphocytes in a three-dimensional collagen matrix (Fig. 1). All of these cells start spontaneous locomotion immediately after incorporation into the collagen gel. Among these three cell types, T lymphocytes are the fastest moving phenotype (8 μm/min). Thereby they exhibit a round cellbody carrying the nucleus, and a trailing uropod. Dendritic cells are considerably slower in respect to velocity (2 μm/min). Locomoting dendritic cells are morphologically characterized by many dendritic processes emerging from the cell body. Morphological studies of dendritic cells migrating in a three-dimensional collagen matrix suggest an important role of the dendritic processes for paving a way through the fiber network (Gunzer et al. 1997). Melanoma cells are the largest (cell length up to 100 μm versus 7–10 μm of T lymphocytes and 10–12 μm of dendritic cells) but also the slowest migrating cells we have investigated (0.2 μm/min). With respect to velocities (Fig. 1 D-F) and dynamics of morphology (Fig. 1 A-C), the migration of these three cell types is substantially different.

REGULATORY MECHANISMS OF THE LOCOMOTORY MACHINERY

One of the key questions concerning cell locomotion is addressed to molecular mechanisms which start and keep going the locomotory machinery of different cell types. Are there common principles, which are shared by many cell types, or are the strategies of migration fundamentally different from each other? Slow migrating cells like fibroblasts (<1 μm/h) and tumor cells use focal adhesion contacts. These main cellular structures connect the intracellular cytoskeleton to the surrounding extracellular matrix (Burridge et al. 1988, Lauffenburger and Horwitz 1996). Focal adhesions consist of structural proteins like paxillin, vinculin, talin, and α-actinin, which are in contact with the actin cytoskeleton, as well as of regulating enzymes like protein tyrosine kinases and the protein kinase C. This aggregation of proteins allows a dynamic regulation of cellular interactions with the surrounding matrix via modification of binding affinities of the structural proteins by the regulatory enzymes of the locomotory machinery (Stossel 1993, Schaller and Parsons 1995, Mogi et al. 1995). Such multi-protein complexes allow a fine tuned regulation of attachment and detachment but on the other hand results in a slow dynamic of migration because of the complexity of such interactions with the matrix. Therefore it has been hypothesized that only slowly migrating cells do follow migration strategies which include the development of focal adhesions. Fast moving cells, like T lymphocytes, should have higher labile contact sites allowing a higher frequency of contact formation and mantling. These contacts must be provided by interaction sites excluding a tight binding and the development of complex structures (Lee et al. 1993). The molecular composition of such postulated contacts is still unknown.

Looking for molecules, which may contribute to the migrating machinery, we investigated the role of β1 integrins in migrating melanoma cells. These transmembraneous molecules are known to be a component of major importance at focal adhesion contacts connecting the extracellular matrix to the cytoskeleton (Clark and Brugge 1995, Hynes 1992). We used the well characterized monoclonal anti-β1 integrin antibodies 4B4 (Coulter, Miami, Florida, USA) and 8A2 (kind gift from N.L. Kovach, M.D., Washington University, Seattle, USA) to either inhibit or activate β1 integrin-mediated adhesion of these tumor cells to the collagen matrix. Interestingly, treatment with both antibodies led to a reduced migratory activity of the cells compared to untreated control cells (K. Maaser,

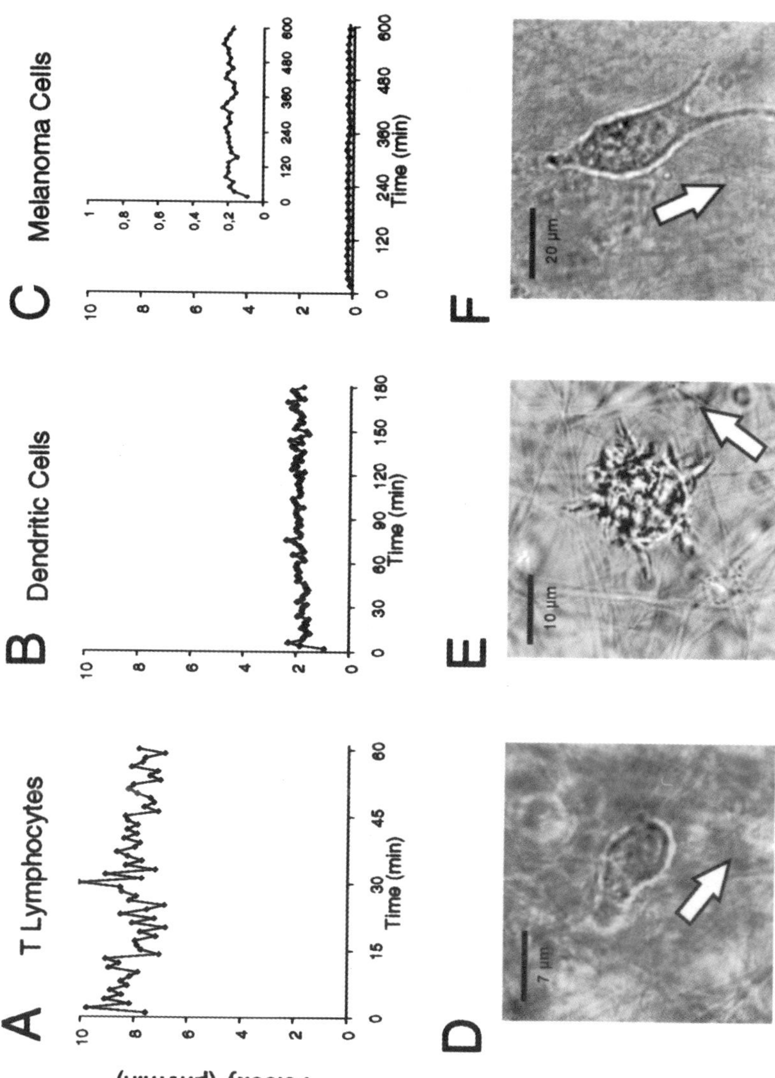

Figure 1. Mean migration velocity (A-C) and morphology of T lymphocytes (D), dendritic cells (E), and melanoma cells (MV3; F) migrating in a three-dimensional collagen matrix. The arrows indicate the direction of migration.

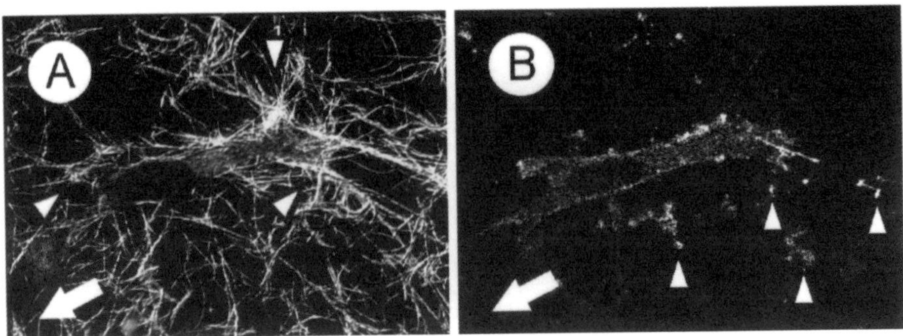

Figure 2. Matrix interaction of a migrating melanoma cell (MV3). Interaction leads to irreversible rearrangement of the matrix (reflection light; A) and shedding of β1 integrins (B; β1 integrin-bound fluorescence). Direction of migration is indicated by the arrows.

manuscript in preparation). This ability of migration inhibition of β1 integrin mAbs is in striking contrast to results obtained for dendritic cells and T lymphocytes investigated under the same experimental conditions. Treatment of these cells with various anti β1 integrin antibodies (4B4, TS2/16, K20, DE9, LIA1/2) alone or in combination did not affect locomotory activity (Friedl et al. submitted). Likewise, the migration of murine and human dendritic cells could not be blocked by different anti-β1 integrin mAbs. These experiments suggest that migration of dendritic cells and T lymphocytes occurs independent of β1 integrin-mediated adhesion. The view of differential involvement of β1 integrins in different cells is supported by the finding that migrating MV3 cells released β1 integrins from the trailing edge (Fig. 2 B, arrowheads), which is considered to support the disruption of cell-matrix interactions. Interestingly, the shedding of cellular material was not found for the fast migration of dendritic cells and T lymphocytes. Furthermore, interaction of the melanoma cell with the surrounding environment led to a rearrangement of the collagen fibers (Friedl et al. 1997), whereas the investigated cells of the immune system did not irreversibly change the structure of the collagen matrix.

ADHESION MOLECULES, FOCAL ADHESIONS, AND CELL MIGRATION IN DIFFERENT CELL TYPES

Histochemical analysis of the cellular distribution of β1 integrins in MV3 cells revealed, that integrins are clustered on the cell surface at contact sites to individual collagen fibers (Fig. 2 A). This finding in tumor cells supports the migration model which postulates the development of focal adhesions. To address the question, whether focal adhesions play a functional role in T lymphocyte locomotion, we examined the intracellular distribution of protein tyrosine kinases, especially the focal adhesion kinase, and the protein kinase C. These signal transduction elements are known to colocalize generally in focal adhesions and are involved in the regulation of assembly and disassembly of the protein-complexes by changing binding affinities via phosphorylation/dephosphorylation of the substrates (Jaken et al. 1989, Luna and Hitt 1992). In migrating T cells, the focal adhesion kinase was found to be predominantly located at the leading edge, which is not permanently in contact with matrix fibers. Surprisingly, the protein kinase C was detected

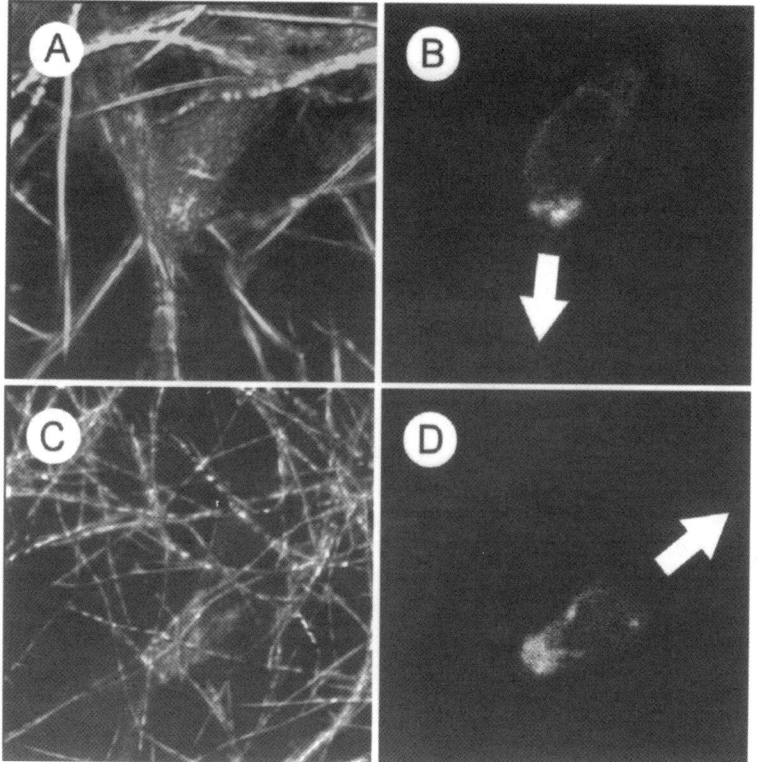

Figure 3. Reflection imaging of a moving T lymphocyte in a three-dimensional collagen matrix (A and C). In 3 B, the focal adhesion kinase is located at the leading edge of the cell, whereas, in D the protein kinase C is distributed at the trailing edge of the cell. The white arrows indicate the direction of movement.

at the trailing edge (Fig. 3 D), and in close spatial proximity to a number of functionally important adhesion molecules located in the uropod, including ICAM-1 and -3 as well as CD43, CD44 and CD29 (del Pozo et al. 1995, del Pozo et al. 1996, Friedl et al. submitted). Thus, we conclude that migration of T cells does not involve the development of focal adhesions in a classical meaning. These results are supported by functional studies on the role of protein tyrosine kinases and the protein kinase C (Entschladen et al. in press). After incorporation into a three-dimensional collagen matrix, about 25 percent of the T cell population develops a spontaneously migrating phenotype. Adding broad range protein tyrosine kinase inhibitors like genistein and tyrphostin A23 significantly reduced the spontaneous locomotion. Subsequent activation of the protein kinase C by the phorbol ester PMA induced locomotion in both inhibitor-treated and untreated T cell populations. Spontaneous locomotion does not depend on the activity of the protein kinase C, because addition of the highly specific protein kinase C inhibitor calphostin C did not affect spontaneous locomotion. We can distinguish a spontaneous protein tyrosine kinase-dependent locomotion, which is independent of protein kinase C activity, from an induced migratory activity, which depends on the activity of the protein kinase C. The distinct function of the different kinases is reflected in their distinct cellular distribution (leading edge focal adhesion kinase versus trailing uropod protein kinase C). It is reasonable to speculate that T cells use different signalling cascades, depending on their functional mode of action, on the one hand reflecting spontaneous migration (immune surveillance), patrolling through tissues and organs and on the other hand carrying out induced locomotion triggered by cytokines/chemokines (Dorner et al. 1997) and engagement of the T cell receptor. It will be of great interest wether in dendritic cells different or similar signalling hierarchies govern the onset of migration.

CONCLUSIONS

Melanoma cells are slowly migrating cells. This type of locomotion is characterized by β1 integrin dependence and clustering of these molecules at extracellular matrix contact sites, which supports the model of an involvement of focal adhesions. During migration the cells tightly interact with the surrounding matrix. The interaction of the cells with the matrix leads to irreversible reorganisation of the collagen fibers and shedding of β1 integrin containing material. In contrast, a cocktail of mAbs directed against β1 integrins did not block locomotion of the two investigated cell types of the immune system. We also did not observe irreversible rearrangment of the matrix structure or shedding of cellular material during T cell and dendritic cell migration. The distinct distribution of the focal adhesion kinase and the protein kinase C in migrating T lymphocytes does not support the focal adhesion model for these fast moving immune competent cells. Thus, migration of distinct cell types differs not only in the velocity of migration and the morphology of the migrating cells but is also very likely to be characterized by a different mode of action of the migrating machinery.

ACKNOWLEDGMENTS

Supported by the Fritz-Bender-Foundation, Munich, Germany.

REFERENCES

Abercombie, M., Heaysman, J.E.M., and Pegrum S.M. (1970). The locomotion of fibroblasts in culture. II. Ruffling. Exp. Cell Res. 68, 437–444.

Burridge, K., Fath, K., Kelly, T., Nuckolls, G., and Turner, C. (1988). Focal adhesions: transmembrane junctions between the extracellular matrix and the cytoskeleton. Ann. Rev. Cell Biol. 4, 487–525.

Clark, E.A., and Brugge, J.S. (1995). Integrins and signal transduction pathways: the road taken. Science 268, 233–239.

del Pozo, M.A., Sánchez-Mateos, P., and Sánchez-Madrid, F. (1996). Cellular polarization induced by chemokines: a mechanism for leukocyte recruitment? Immunol. Today 17, 127–131.

del Pozo, M.A., Sánchez-Mateos, P., Nieto, M., and Sánchez-Madrid, F. (1995). Chemokines regulate cellular polarization and adhesion receptor redistribution during lymphocyte interaction with endothelium and extracellular matrix. Involvement of cAMP signaling pathways. J. Cell Biol. 131, 495–508.

Dorner, B., Müller, S., Entschladen, F., Schröder, J.M., Franke, P., Kraft, R., Friedl, P., Clark-Lewis, I., and Kroczek, R. (1997). Purification, structural analysis, and function of natural ATAC a cytokine secreted by $CD8^+$ T cells. J. Biol. Chem. 272, 8817–8823.

Entschladen, F., Niggemann, B., Zänker, K.S., and Friedl, P. Differential requirement of protein tyrosine kinases and protein kinase C in the regulation of T cell locomotion in three-dimensional collagen matrices. J. Immunol., in press.

Friedl, P., Maaser, K., Klein, C.E., Niggemann, B., Krohne, G., and Zänker, K.S. (1997). Migration of highly agressive MV3 melanoma cells in 3-dimensional collagen lattices results in local matrix reorganization and shedding of α2 and β1 integrins and CD44. Cancer Res. 57, 2061–2070.

Friedl, P., Entschladen, F., Conrad, C., Niggemann, B., and Zänker, K.S. CD4+ T lymphocytes migrating in three-dimensional collagen lattices lack focal adhesions and utilize β1 integrin-independent strategies for polarization, interaction with collagen fibers, and locomotion. J. Immunol., submitted.

Gunzer, M., Kämpgen, E., Bröcker, E.B., Zänker, K.S., and Friedl, P. (1997). Migration of dendritic cells in 3D-collagen lattices: visualisation of dynamic interactions with the substratum and the distribution of surface structures via a novel confocal reflection imaging technique. In Dendritic cells in fundamental and clinical immunology, Vol. 3, P. Ricciardi-Castagnoli, ed., Plenum Publishing Corporation, NY, in press.

Huttenlocher, A., Sandborg R.R., and Horwitz, A.F. (1995). Adhesion in cell migration. Curr. Opin. Cell Biol. 7, 697–706.

Hynes, R.O. (1992). Integrins: versatility, modulation, and signaling in cell adhesion. Cell 69, 11–25.

Jaken, S., Leach, K., and Klauck, T. (1989). Association of type 3 protein kinase C with focal contacts in rat embryo fibroblasts. J. Cell. Biol. 109, 697–704.

Lauffenburger, D.A., and Horwitz, A.F. (1996). Cell migration: a physically integrated molecular process. Cell 84, 359–369.

Lee, J., Ishihara, A., and Jacobson K. (1993). How do cells move along surfaces? Trends Cell Biol. 3, 366–370.

Luna, E.J., and Hitt, A.L. (1992). Cytoskeleton-plasma membrane interactions. Science 258, 955–969.

Mogi, A., Hatai, M., Soga, H., Takenoshita, S., Nagamachi, Y., Fujimoto, J., Yamamoto, T., Yukota, J., and Yaoi, Y. (1995) Possible role of protein kinase C in the regulation of intracellular stability of focal adhhesion kinase in mouse 3T3 cells. FEBS letters 373, 135–140.

Schaller, M.D., and Parsons, J.T. (1995). $pp125^{FAK}$-dependent tyrosine phosphorylation of paxillin creates a high-affinity binding site for crk. Mol. Cell. Biol. 15, 2635–2645.

Stossel, T.P. (1994) The machinery of blood cell movement. Blood 84, 367–379.

Stossel, T.P. (1993) On the crawling of animal cells. Science 260, 1086–1094.

5

GENOMIC INSTABILITY IN SPORADIC COLORECTAL CANCER

A Destabilized Genome Producing Accelerated Cellular Evolution as the Fundamental Nature of Cancer

Garth R. Anderson,[1,2] Daniel L. Stoler,[1] Morton S. Kahlenberg,[2] and Nicholas J. Petrelli[2]

[1]Department of Molecular and Cellular Biology
[2]Department of Surgical Oncology
Roswell Park Cancer Institute
Buffalo, New York 14263

Two major forms of genomic alterations are seen abundantly in solid tumors (1). The predominant form is intrachromosomal genomic instability, which manifests itself as deletions, insertions, amplifications and translocations (2). The molecular basis of this form of instability has not been established, and recent evidence indicates p53 is unlikely to be involved here (3). A common feature of this major type of instability is an early role for DNA breakage, suggesting nuclease involvement. Aneuploidy produces gains or losses of entire chromosomes, in a process involving p53 and inappropriate meiotic-like segregation (5). Microsatellite instability represents a less common third type of genomic instability which produces oligonucleotide insertions or deletions within repetitive sequences; this is most often seen in hereditary cancer syndromes (e.g. HNPCC), where it somehow tends to preclude the appearance of the other types of instability (6).

Two very basic questions exist. First, exactly how unstable is the cancer cell genome; how many total genomic events usually occur during tumor progression, and what fraction actually contribute to malignancy? Second, when is the onset of genomic instability; is it an early event which facilitates tumor progression, or is it a late event which is a secondary consequence of malignancy itself? And if the onset of genomic instability should be an early event facilitating tumor progression, can therapeutic agents be identified which suppress it, in turn suppressing the occurrence of cancer itself?

To quantitate the extent of genomic rearrangements in tumor biopsy specimens, we have developed the genome-wide sampling procedure named inter-(simple sequence repeat) PCR (7). This technique exploits the abundant scattering of simple repeat sequences

The Biology of Tumors, edited by Mihich and Croce
Plenum Press, New York, 1998.

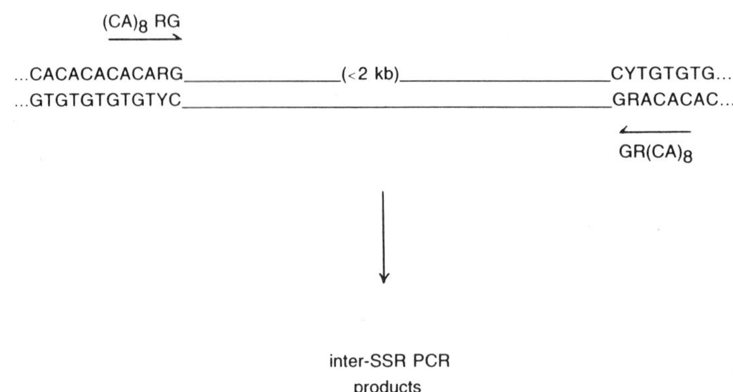

Figure 1. Inter-(simple sequence repeat) PCR provides a convenient means of sampling numerous dispersed areas of the genome. $(CA)_8$ RG or $(CA)_8$ RY are both effective for this purpose, by amplifying regions between (CA) repeats inverted in fairly close proximity. By comparing tumor DNA to normal DNA from the same patient, the overall extent of genomic rearrangements can be quantified. In turn, this provides a measure of the major known forms of genomic instability, associated with insertions, deletions, translocations and amplifications, along with aneuploidy.

throughout the human genome; for example (CA) repeats number approximately 100,000, spacing them on average about 30 kb apart. Using a single PCR primer corresponding to these repeats, but anchored at the 3′ end to preclude detection of alterations within the repeats themselves (e.g. $(CA)_8$ RG), we can amplify those sequences located between the corresponding repeat sequences in these cases where the primer binding sites are facing one another and within about 2 kb (Figure 1). Alterations in the intervening sequences, such as arising from insertions or translocations, produce tumor PCR DNA products of different sizes than those generated from corresponding adjoining normal tissue DNA. Deletions produce either smaller products, or no product at all if primer binding sites themselves are lost; amplifications generate increased band intensity. A representative example is shown in Figure 2, where DNAs from colon carcinomas were compared to that of adjoining normal mucosa from the same patients.

This methodology allows us to produce an estimate of how many genomic events have occurred in the tumor cells. By adding up the total size of the PCR products (64 kb), which represents that fraction of the genome which we are sampling, and determining that on average three bands are altered in each tumor (range = 0 to 10), we can compute that for the entire 3×109 bp human genome, about 1.4×105 alterations would have occurred. But a single large deletion such as loss of an entire chromosomal arm may eliminate multiple bands; thus the apparent number of alterations could easily exceed the actual number of individual genomic events, and 1.4×105 would be a maximal number. To estimate the minimal number of individual genomic events which must have occurred, we can limit ourselves to the set of new tumor-specific bands for which there is no normal tissue counterpart. This set would have to reflect genomic events occurring between the two primer binding sites; a single genomic event would not produce several altered bands. Of the 171 bands we found to be altered in a set of 60 colorectal tumors, between one tenth and one fifth appeared to meet the criterion of having no visible normal tissue counterpart. Thus at minimum one tenth of the total number of altered bands would have to reflect individual genomic events. Accordingly, our estimated minimal number of genomic events per cancer cell becomes 1.4×104. Thus the actual number of genomic events per tumor cell would

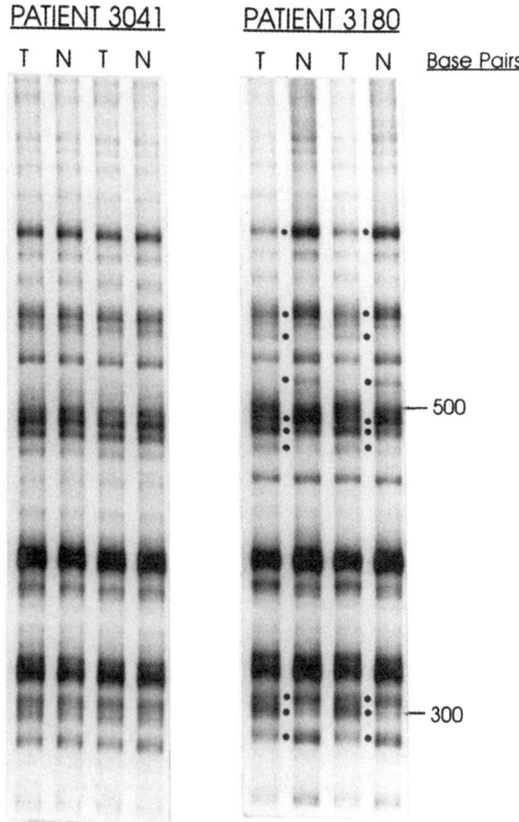

Figure 2. Examples of colorectal cancers with low (patient 3041) and high (patient 3180) degrees of genomic instability, as determined by inter-(simple sequence repeat) PCR. A ^{32}P-end labeled $(CA)_8$ RY primer was used as described (7), and bands altered either in intensity or position are denoted by dots.

fall somewhere in the range between 14,000 and 140,000. This surprisingly large number points to the cancer cell genome as being considerably more fluid than previously thought. Earlier genome-wide sampling techniques such as comparative genomic hybridization on mitotic spreads (8) have suffered from a practical resolution limit of about ten megabases; this may help explain why previous estimates have been lower.

Our calculated numbers reflect the assumption that genomic events occur in the proximity of simple sequence repeat sequences at rates equal to that throughout the entire genome, and that inverted repeat sequences within two kilobases also have no effect. This may not be the case (9); the alternative method of high resolution comparative genomic hybridization on microarrays may be able to examine these issues. Another concern is that late genomic events which are not present in most tumor cells would not be scored in our assay, since they would be counterbalanced by those cells not containing the alteration.

When does genomic instability begin in tumor progression? Again using colorectal cancer as our system of study, we have examined seven adenomatous polyps. Their mean genomic instability index was 5.7, meaning that 5.7 percent of the polyp PCR products migrated anomalously compared to their corresponding normal tissue PCR products. This is slightly higher than that seen in our much larger set of colon carcinomas, where the mean index value was 3.9. In one case of a polyp resected along with a synchronous carcinoma, of seven bands altered in the carcinoma, three were also altered in the polyp. Together these results lead us to conclude that genomic instability has already commenced by the very early stages of tumor progression, with additional genomic events accumulating during the course of progression.

We have previously reported how expression of an anoxia-inducible endonuclease correlates with genomic instability in cultured fibroblasts experiencing low oxygen tensions, and have shown how this same endonuclease is active during the inflammatory phase of wound healing, when low oxygen tensions prevail (10,11). This nuclease is also found in sporadic colorectal cancers, where its expression correlates with the extent of genomic instability (10; Kahlenberg et al., submitted). We have evaluated the long-shot possibility that anti-inflammatory drugs may inhibit the nuclease, and in turn suppress genomic instability. Initial experiments with the SW620 colorectal tumor line and several nonsteroidal anti-inflammatory drugs (NSAIDS), indicate the nuclease is suppressed to about one-fourth of its level in untreated cells, and that genomic instability as measured by PALA selection of CAD gene amplification can be suppressed by about three-fold (Kahlenberg et al., in preparation).

Our findings have clear implications regarding the basic, fundamental nature of cancer. In contrast to hematologic malignancies, where genomic alterations apparently are few, sporadic colorectal cancers (and presumably other solid tumors) are associated with rampant genomic instability, which is an early event in tumor progression. In other words, this suggests cancer arises from cells whose genome has become highly unstable. While an extreme genomic instability obviously would be lethal, as most genetic information would no longer be usable, a somewhat lesser degree of instability would be expected to almost inexorably lead to malignancy. For while the human genome contains roughly 100,000 genes, simple eukaryotic organisms such as yeast need only about 6,000 genes to metabolize, grow and divide (12). Most of the human genome thus must be involved in coordination and regulation of multicellular organismic growth, differentiation and development. As the genome becomes scrambled, such regulatory genes become lost or altered, although the cell itself remains viable. Accelerated evolution at the cellular level ensues, with natural selection giving rise to the ordered steps of tumor progression. Along with the small number of relevant genomic events which directly contribute to malignancy, there appear to be much greater numbers of marginally- or non-contributing genomic events also arising during tumor progression. Many of these presumably generate altered gene products, but natural selection would be expected to select for other mutations which effect evasion of host immunological defenses. The net result of this genomic instability is that slow, gradual evolution at the organismic level has been supplanted by rapid evolution at the cellular level. Natural selection for genomic events in the unstable cells' interests produces a large population of self-interested cells with specific "advantageous" mutations, but at the ultimate expense of the host.

REFERENCES

1. Kastan, M.B., ed. Genetic Instability and Tumorigenesis. Current Topics in Microbiology and Immunology. Volume 221, 1997.
2. Barrett, J.C., Tsutsui, T., Tlsty, T. and Oshimura, M. Genetic mechanisms in carcinogenesis and tumor progression. C.C. Harris and L.A. Liotta, eds. Wiley-Liss, NY. pp. 97–114, 1990.
3. Kahlenberg, M.S., Stoler, D.L., Basik, M., Petrelli, N.J., Rodriguez-Bigas, M. and Anderson, G.R. p53 tumor suppressor gene status and the degree of genomic instability in sporadic colorectal cancers. J. Natl. Cancer Inst. 88: 1665–1670, 1996.
4. Fishel, R. Genomic instability, mutators and the development of cancer: is there a role for p53? J. Natl. Cancer Inst. 88: 1608–1609, 1996.

5. Fukasawa, K., Choi, T., Kuriyama, R., Rulong, S. and Vande Woude, G.F. Abnormal centrosome amplification in the absence of p53. Science 271: 1744–1747, 1996.
6. Liu, B., Nicolaides, N.C., Markowitz, S., Willson, J.K.V., Parsons, R.E., Jen, J., Papadopolous, N., Peltomaki, P., de la Chapelle, A., Hamilton, S.R., Kinzler, K.W. and Vogelstein, B. Mismatch repair gene defects in sporadic colorectal cancers with microsatellite instability. Nature Genetics 9: 48–55, 1995.
7. Basik, M., Stoler, D.L., Kontzoglou, K.C., Rodriguez-Bigas, M.A., Petrelli, N.J. and Anderson, G.R. Genomic instability in sporadic colorectal cancer quantitated by inter-simple sequence repeat PCR analysis. Genes, Chromosomes and Cancer 18: 19–29, 1997.
8. Kallioniemi, O.P., Kallioniemi, A., Sudar, D., Rutovitz, D., Graw, J.W., Waldman, F. and Pinkel, D. Comparative genomic hybridization: a rapid new method for detecting and mapping DNA amplification in tumors. Sem. Cancer Biol. 4: 41–46, 1993.
9. Gordenin, D.A., Lobachev, K.S., Degtyareva, N.P., Malkova, A.L., Perkins, E., Resnick, M.A. Inverted DNA repeats: a source of eukaryotic genomic instability. Mol. Cell. Biol. 13: 5315–5322, 1993.
10. Russo, C.A., Weber, T.K., Volpe, C.M., Stoler, D.L., Petrelli, N.J., Rodriguez- Bigas, M., Burhans, W.C. and Anderson, G.R. The anoxia-inducible endonuclease and enhanced DNA breakage as contributors to genomic instability in cancer. Cancer Res. 55: 1122–1128, 1995.
11. Anderson, G.R., Volpe, C.M., Russo, C.A., Stoler, D.L. and Miloro, S.M. The anoxic fibroblast response: a cellular program utilized during early stage wound healing. J. Surg. Res. 59: 666–674, 1995.
12. Mewes, H.W., Albermann, K., Bahr, M., Frishman, D., Gleissner, A., Hani, J., Heumann, K., Kleine, K., Maierl, A., Oliver, S.G., Pfeiffer, F. and Zollner, A. Overview of the yeast genome. Nature 387: 7–9, 1997.

DISCUSSION

De Pinho: What is the relationship between your assay and cytogenetic abnormalities, level of ploidy, karyotypic chaos that you can detect?

Anderson: Well, karyotypic cytogenetics is a relatively crude procedure. We have not yet attempted to correlate what we are seeing with cytogenetic anomalies. We have shown that there is absolutely no correlation with mismatch repair defects. We are looking at various correlates.

Dalla Favera: Are these tumors that have no defect in DNA mismatch repair genes?

Anderson: Out of the first set of sixty tumors that we have looked at, I think five had mismatch repair defects. Those with mismatch repair defects showed negligible instability of the inter-SSR type. In other words, when you have a mismatch repair defect, you do not see the break related type of instability. We did not pre-screen for tumors without mismatch repair defects.

Dalla Favera: So what kind of instability do you think you are measuring?

Anderson: We have actually cloned several altered products, and what we are finding is that you are getting deletions, you are getting insertions, you are getting translocations.

Dalla Favera: The second part of the question is that you are probably seeing a lot of subclonal alterations, right? One common assumption is that as much as 20% of the genome of these tumors is undergoing rearrangement of any sort. Do you think that what you are seeing is clonally represented in every cell in the malignant tumor?

Anderson: If you microdissect-out small chunks of the tumor, it appears you might see more events. In other words, if you have a relatively large mass, tumor heterogeneity will average some out that occurred late in tumor progression.

Dalla Favera: Just to understand, when you try clone-out to the alteration, you think that has biological significance?

Anderson: In most cases, no. In fact, we feel that the cancer cell genome is tremendously dirty, it has all kinds of alterations in it, most of which do not mean anything. You have a lot of rearrangements of which a few, occasionally, will screw up various important genes such as p53, but meanwhile you are getting all of these irrelevant mutations. Basically you have this highly evolved genome, and suddenly genomic entropy kicks in and it degrades down to a more primitive level. It is sort of like when a kid cuts up an audio tape in a tape cassette then randomly glues it back together, the noise is still there but instead of being a symphony it is like country-western.

Livingston: In your assay in one of your early slides, you demonstrated that after PCR, x number of tumors, 2 or more, reveal new bands. Can you tell us exactly how that assay is done? I was wondering how two different tumors could have the same new PCR band?

Anderson: The assay methodology was published in *Genes, Chromosomes and Cancer*, in January, 1997. It looks like there are some advantageous, or recurrent events that we have seen that have attracted our interest, that may be at sites particularly vulnerable to genomic instability. In other words, there are few bands which show up recurrently and we have mapped these to certain portions which, when studied cytogenetically, are altered frequently.

Livingston: Can you show that these bands, for example, exist by those sequences that are allegedly new; can they be demonstrated by some technique other than PCR?

Anderson: The sequence match--one of them matched the creative transporter in Gen-bank. Is that the type?

Livingston: My point is, you see two different tumors and there is the same new PCR band, can you demonstrate that those bands are not a product of PCR for example? Can you show some independent assay for the presence of those abnormal sequences?

Anderson: At this point, I do not know the answer to your question. We have not yet shown it.

6

APC AND THE EARLY EVENTS OF COLON CANCER

Raymond L. White

Huntsman Cancer Institute
University of Utah
15 North 2030 East, Suite 7410
Salt Lake City, Utah 84112-5330

Colon cancer continues to be one of the major killers of adults in the developed world. Almost 100,000 new cases and 40,000 deaths are expected in the U.S. this year. Neither the age specific incidence nor mortality of this disease has improved significantly over the last thirty years. The same picture holds for most of the adult cancers and has led to the charge that our heavy investment in cancer research, promised to yield new therapies for these cancers, has not succeeded and has therefore been wasted. This is a very premature judgement. There have been dramatic improvements in the treatment of some cancers; the vast majority of childhood leukemia is now a curable disease and testicular cancer now has an excellent prognosis.

Furthermore, there is now strong evidence that the research investment of the last 30 years has not been wasted. Indeed, this investment has led to the development of a massive and detailed database of specific knowledge of the molecular basis of cell regulation. Based on the new knowledge this research investment has created, we now have a detailed molecular description of the etiology of many of the major forms of cancer, the cell-regulatory genes whose activities are altered in cancer initiation and progression. This has created a strong and robust foundation for the development of a wide variety of new and innovative approaches to cancer prevention and cure. As was the case with microbes and infectious disease, detailed knowledge of the primary genetic etiology of cancer will provide a pathway for major progress in diagnosis and treatment.

Colon cancer and its initiating APC gene have become a paradigm for application of this knowledge to the development of new approaches to reducing morbidity and mortality and will be the primary focus of this discussion. We understand now, for example, that a cell is constantly responding, in a manner based on its own history, to a barrage of chemical information presented at its surface. The outcome of the cell's deliberation is often a decision either to divide and create daughter cells, as is the case with a stem cell, or to differentiate into a specialized cell that becomes non-replicative. The epithelial cells that line

The Biology of Tumors, edited by Mihich and Croce
Plenum Press, New York, 1998.

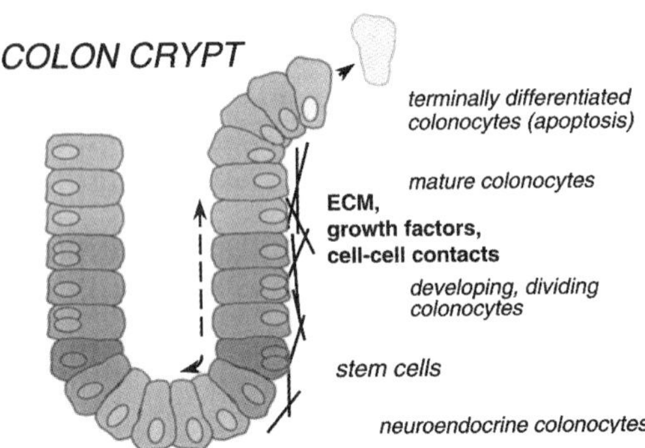

Figure 1

the crypts of the colon provide an excellent example of this process. As illustrated in Figure 1, the stem cells of the colonic crypt lie at the base of the crypt. Daughter cells produced by the stem cells migrate up the walls of the crypt, dividing only once or twice in their two to three day lifetime before reaching the top of the crypt, undergoing a programmed cell death and shedding into the lumen. Thus the replicative potential of the stem cell must be very great, as each must oversee the production of as many as 10,000 differentiating colonocytes in order to maintain the population of the normal crypt over an average lifetime. Assuming each progeny cell undergoes an average of 1.5 doublings, the stem cell itself will have undergone more than 3,000 doublings in an average lifetime—it is effectively an immortal cell.

A great deal is known about the molecules that interact with and govern the decisions of the developing colonocyte. Polypeptide growth factors, steroid hormones, cell-cell contacts and contact with the extracellular matrix each play key roles in informing the colonocyte as it migrates up the wall of the crypt. The daughter cell finally determines that it will divide no more, produces its mucin droplet and undergoes apoptotic death, giving literal meaning to the term "terminal differentiation". These cells are a beautiful developmental system, physically displaying, as do the oocytes of the silk moth, the full temporal history of the developing colonocyte.

Genetics has played a key role both in ascertaining new genes important to the developmental regulation of the colonocyte and in identifying those already known that play and etiologically significant role in colon carcinogenesis. However, application of the genetic gestalt to analysis of the aberrant colonocyte development that leads to cancer has depended on the development of an abundant set of genetic markers based on the direct detection of DNA sequence variation.

The history of this component has moved relatively quickly, driven by opportunities created by new technologies. The first generation genetic markers were hypothesized to be detectable by polymorphism in the length of locus-specific restriction fragments (RFLPs)[1]. Such variants were soon found and shown to be sometimes due to base pair changes that created or eliminated a restriction site, yielding a two-allele polymorphic marker system (2, 3). Two-allele RFLP marker systems, however, had only limited utility as genetic markers for the human as relatively few parents were found to heterozygous, and therefore, informative for these limited systems. Credibility for the idea[1] that DNA-based genetic markers could be used to create human genetic maps and localize human disease genes depended upon the discovery of the first multi-allelic locus, PAW 101[4]. Ultimately, we came to understand that the majority of these multi-allelic marker systems were due to variations in the number of tandem, short sequence repeats internal to the restriction fragment. The multi-allelic markers proved to be especially effective as they were much more genetically informative—almost all progeny would give information for linkage studies. Large sets of these markers were soon identified[(5)]. The most recent generation of VNTR markers has moved to PCR amplification of loci containing tandem repeats of a very short sequence—a di, tri or tetra nucleotide. These markers are both multi-allelic and suitable for automated genotyping systems. Literally thousands of human genetic markers based on DNA sequence variation in the human population have now been characterized.

The next stage in developing a generalized tool for the mapping of human genes was to arrange the new genetic markers into ordered, chromosomal subsets[(6)]. This was a critical step in developing the marker sets spaced every 10 cM to 20 cM required for efficient mapping of new disease genes. Furthermore, mapped high-density marker sets make it possible to efficiently home in on a precise location for the disease gene. Once an initial mapping has been obtained, other markers mapped to the immediate area can be typed in the families. In order to serve as a basis for the mapping of many genetic markers identified by many laboratories around the world, a single set of highly efficient families was impaneled, primarily from our Utah family resources, and the DNA distributed to collaborators. The resultant international collection of mapped marker sets, available for each chromosome, now constitutes a very fine-mesh genetic net, capable of capturing the precise location of virtually any major gene that might underlie a genetic disease.

A large number of linkage studies in families have led to the successful mapping of literally hundreds of disease genes based on these mapped marker sets. In brief, the pattern of inheritance of the disease allele, determined by clinical phenotype, is matched against the patterns of inheritance of the alleles of the marker loci. If a marker locus is physically very close to the disease gene, its inheritance pattern will match very closely the inheritance pattern of the disease gene, revealing the chromosomal location of the disease gene. Precise mapping is very important as the final step of gene identification as each gene in the mapped region, by virtue of its location, becomes a candidate for the disease gene. The real gene is ultimately determined by sequencing of the candidates at the locus, as only one should have mutations that associate with the presence of disease.

An early target of this approach in our laboratory was the APC (adenomatous polyposis coli) gene, the etiologic agent in familial polyposis. Familial polyposis is an extremely interesting gene from the perspective of cancer initiation, as its major phenotype is the development of hundreds or thousands of adenomatous polyps of the colon at a very early age. This is a dangerous condition as adenomatous polyps are the precursors to colon cancer. Prophylactic removal of the colon at a relatively early age prevents the colon cancer which is otherwise highly likely to emerge by the early forties or sooner among these patients. Linkage mapping placed the APC gene on the long arm of chromosome 5 (7, 8) and was followed by a detailed physical analysis of the region (9) which defined a number of candidate genes and provided the genetic markers which revealed the presence of small deletions in two families (10). These turned out to be the Rosetta stone for this disease, ultimately leading to the discovery of the gene (10–13).

The first important finding following cloning of the gene was that it was also frequently found to be mutant in sporadic polyps and colon cancer. Analysis of the spectrum of mutations found in both families with polyposis and in sporadic disease showed that virtually all were base pair changes or deletions that resulted in prematurely terminated peptides.

However, the major challenge following positional cloning is to determine the gene's function. Analysis of the predicted amino acid sequence and a search for similari-

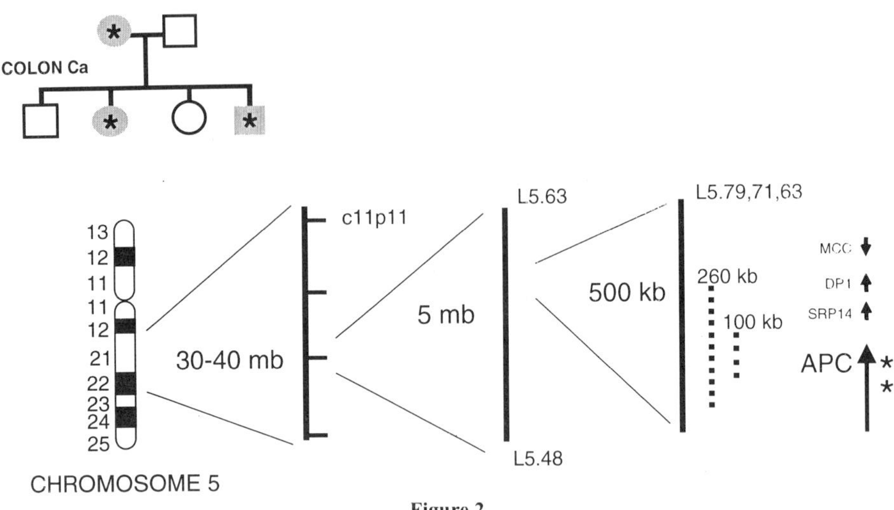

Figure 2

ties to APC domains in the databases revealed little beyond several repeat motifs, which, however, included multiple clusters of heptad repeats. These led to the hypothesis that at least some forms of the APC protein might dimerize, which was verified experimentally[14,15].

The most revealing observation, however, was that immunoprecipitation of the APC protein also brought down β-catenin protein[16,17]. This was an important clue, because β-catenin was known to bind to E-cadherin and mediate its attachment to the actin cytoskeleton. Furthermore, new experiments revealed that β-catenin could also translocate to the nucleus in the company of Tcf/Lef protein to become part of a transcription complex. Much of this had been foreshadowed by genetic experiments in Drosophila, which had demonstrated that armadillo, the Drosophila version of β-catenin, was a key developmental regulator, induced by the Wingless ligand. APC's role was shortly shown to be that of modulating the levels of free β-catenin protein by stimulating its proteasome mediated proteolytic degradation[18]. Wingless signaling prevents this degradation by inhibiting the activity of GSK3-β kinase suggested as responsible for the efficient binding of APC to β-catenin[19] as outlined in Figure 2.

Another approach to understanding the role of APC protein is to examine its intracellular location. We were especially interested in whether APC could be found in the nucleus. Several C-terminal antibodies gave the same result—APC staining could be seen as a particulate distribution in the cytoplasm, with strong foci of labeling at the leading edge of migrating cells[20,21], and in the nucleus[21]. The nuclear staining showed a pattern reminiscent of nucleoli, which was confirmed by joint staining with two nucleolar-specific probes, antisense rRNA and Topoisomerase II. Furthermore, cell fractionation experiments confirmed the presence of APC protein in the nuclear compartment and also indicated a strong association with the nuclear scaffold. These observations have led to additional immunocytochemical observations following extraction of the microtubule and actin cytoskeletal elements, as well as soluble proteins, that support the idea that APC protein is associated with the intermediate filament network.

The studies of the intracellular localization of APC protein are consistent with its function of regulating β-catenin protein levels. We may speculate, for example, that APC is present at the leading edge of migrating cells in order to enable the reorganization of the actin cytoskeleton necessary for cell movement through regulating β-catenin protein necessary to link the cytoskeleton to the cell adhesion molecule E-cadherin. APC protein in cytoplasmic granules may well play an important role in modulating the levels of the free, cytoplasmic β-catenin available for translocation to the nucleus. APC protein in the nucleus could function to modulate the levels of nuclear β-catenin protein, for example, to turn off a signal received from the cytoplasm.

Where does all this lead in our understanding of the role of APC in colon carcinogenesis? A major focus of interest is now in the transcripts that are induced by the β-catenin/Tcf transcription factor. A significant candidate is the inducible prostaglandin synthesis enzyme COX-2, known to be present at high levels in most adenomas (RR). This is a very interesting enzyme as it is the target of the nonsteroidal anti-inflammatory drugs, Sulindac and aspirin, that have been implicated as either causing polyps to resorb (Sulindac), or reducing mortality due to colon cancer.

Taken together, these observations have led to a model for colon cancer initiation through loss of APC function leading to increased β-catenin signaling and induction of COX-2, which in turn would lead to prostaglandin signaling. Finally, the recent discovery of two drugs that target COX-2 and have important anti-tumor activity, Sulindac and aspirin, leads to the important possibility that colon cancer may be preventable by drugs that interfere in the β-catenin signaling pathway and inhibit or cause regression of the colon cancer precursor, the adenomatous polyp.

REFERENCES

1. Botstein, D., White, R. L., Skolnick, M., and Davis, R. W. Construction of a genetic linkage map in man using restriction fragment length polymorphisms, American Journal of Human Genetics. *32:* 314–331, 1980.
2. Kan, Y. and Dozy, A. Polymorphism of DNA sequence adjacent to the human β-globin structural gene: relationship to sickle mutation., Proc Natl Acad Sci USA. *75:* 5631–5635, 1978.
3. Jeffreys, A. DNA sequence variants in the ^{G}gamma, ^{C}gamma, delta, and β-globin genes of man., Cell. *18:* 1–10, 1979.
4. Wyman, A. and White, R. L. A highly polymorphic locus in human DNA, Proceedings of the National Academy of Sciences. *77:* 6754–6758, 1980.
5. Nakamura, Y., Leppert, M., O'Connell, P., Wolff, R., Culver, M., Martin, C., Fujimoto, E., Hoff, M., Kumlin, E., and White, R. Variable number of tandem repeat (VNTR) markers for human gene mapping, Science. *235:* 1616–1622, 1987.
6. Drayna, D. and White, R. L. The genetic structure of the human X chromosome, Science. *230:* 753–758, 1985.
7. Bodmer, W. F., Bailey, C., Bodmer, J., Bussey, H., Ellis, A., Gorman, P., Lucibello, F., and al., e. Localization of the gene for familial polyposis on chromosome 5, Nature. *328:* 614–616, 1987.
8. Leppert, M., Dobbs, M., Scambler, P., O'Connell, P., Nakamura, Y., Stauffer, D., Woodward, S., Burt, R., Hughes, J., Gardner, E., Lathrop, M., Wasmuth, J., Lalouel, J.-M., and White, R. The gene for familial polyposis coli maps to the long arm of chromosome 5, Science. *238:* 1411–1413, 1987.
9. Nakamura, Y., Lathrop, M., Leppert, M., Dobbs, M., Wasmuth, J., Wolff, E., Carlson, M., Fujimoto, E., Krapcho, K., Sears, T., Woodward, S., Hughes, J., Burt, R., Gardner, E., Lalouel, J.-M., and White, R. Localization of the genetic defect in familial adenomatous polyposis within a small region of chromosome 5, American Journal of Human Genetics. *43:* 638–644, 1988.
10. Joslyn, G., Carlson, M., Thliveris, A., Albertsen, H., Gelbert, L., Samowitz, W., Groden, J., Stevens, J., Spirio, L., Robertson, M., Sargeant, L., Krapcho, K., Wolff, E., Burt, R., Hughes, J. P., Warrington, J., McPherson, J., Wasmuth, J., Le Paslier, D., Abderrahim, H., Cohen, D., Leppert, M., and White, R. Identification of deletion mutations and three new genes at the familial polyposis locus, Cell. *66:* 601–613, 1991.
11. Groden, J., Thliveris, A., Samowitz, W., Carlson, M., Gelbert, L., Albertsen, H., Joslyn, G., Stevens, J., Spirio, L., Robertson, M., Sargeant, L., Krapcho, K., Wolff, E., Burt, R., Hughes, J. P., Warrington, J.,

McPherson, J., Wasmuth, J., Le Paslier, D., Abderrahim, H., Cohen, D., Leppert, M., and White, R. Identification and characterization of the familial adenomatous polyposis coli gene, Cell. *66:* 589–600, 1991.

12. Kinzler, K., Nilbert, M., Vogelstein, B., Bryan, T., Levy, D., Smith, K., Presinger, A., and al., e. Identification of a gene located at chromosome 5q21 that is mutated in colorectal cancers, Science. *251:* 1366–1370, 1991.
13. Nishisho, I., Nakamura, Y., Miyoshi, Y., Miki, Y., Ando, H., Horii, A., Koyama, K., Utsunomiya, J., Baba, S., Hedge, P., Markham, A., Krush, A. J., Petersen, G., Hamilton, S. R., Nilbert, M. C., Levy, D., Bryan, T. M., Preisinger, A. C., Smith, K. J., Su, L.-K., Kinzler, K., and Vogelstein, B. Mutations of chromosome 5q21 genes in FAP and colorectal cancer patients, Science. *253:* 665–668, 1991.
14. Joslyn, G., Richardson, D., White, R., and Alber, T. Dimer formation by an N-terminal coiled coil in the APC protein, Proceedings of the National Academy of Sciences. *90:* 11109–11113, 1993.
15. Smith, K. J., Johnson, K. A., Bryan, T. M., Hill, D. E., Markowitz, S., Willson, J. K., Paraskeva, C., Petersen, G. M., Hamilton, S. R., Vogelstein, B., and Kinzler, K. The APC gene product in normal and tumor cells, Proceedings of the National Academy of Sciences. *90:* 2846–2850, 1993.
16. Su, L. K., Vogelstein, B., and Kinzler, K. W. Association of the APC tumor suppressor protein with catenins, Science. *262:* 1734–1737, 1993.
17. Rubinfeld, B., Souza, B., Albert, I., Muller, O., Chamberlain, S. H., Masiarz, F. R., Munemitsu, S., and Polakis, P. Association of the APC gene product with beta-catenin, Science. *262:* 1731–1734, 1993.
18. Aberle, H., Bauer, A., Stappert, J., Kispert, A., and Kemler, R. β-catenin is a target for the ubiquitin-proteasome pathway., EMBO Journal. *16:* 3797–3804, 1997.
19. Rubinfeld, B., Albert, I., Porfiri, E., Fiol, C., Munemitsu, S., and Polakis, P. Binding of GSK3b to the APC-b-catenin complex and regulation of complex assembly, Science. *272:* 1023–1026, 1996.
20. Nathke, I. S., Adams, C. L., Polakis, P., Sellin, J. H., and Nelson, J. The Adenomatous Polyposis Coli Tumor Suppressor Protein Localizes to Plasma Membrane Sites Involved in Active Cell Migration, J. Cell Biol. *134:* 165–179, 1996.
21. Neufeld, K. L. and White, R. L. Nuclear and cytoplasmic localizations of the Adenomatous Polyposis Coli protein., Proc Natl Acad Sci USA. *94:* 3034–3039, 1997.

DISCUSSION

Anderson: I am wondering what is the evidence that APC directly regulates COX-2 expression, and also in the adenomatous polyps? Is it the microenvironment that may be contributing? In other words, if you established cell lines based on the adenomatous polyps?

White: Experiments are ongoing as we speak to find out how direct the effect is. I tried to be very careful in that representation. PSH2 is one of the earliest messengers transcripts that is induced. The mechanism of induction is not known at this point, it is under investigation but not clear.

Anderson: Are there adenoma lines that still express it?

White: Adenomas have been, colon cells in general from humans are extremely difficult to grow. There have been a few adenoma lines developed. I actually do not know what their PSH2 status is, we were never able to do that. Even putting in fairly potent transgenes, it is still hard to get these cells to grow.

DePinho: Does β-catenin also co-localize to the nuclei and, specifically, is there any impact on Pal-1 transcriptional regulation that APC might be involved in? What is it doing in the nucleus? Are these normal expression levels, or forced expression studies?

White: No, all the APC's that I have been representing to you, the staining has been endogenous genes, it is not transgenes. If you put it in transgene, that simply coats the microtubules, that is the most abundant label. We have not yet got good enough β-catenin staining in the nucleus to see it. We are working on that, but have not made that yet, so we cannot comment. I do not know anything about PAL1.

Rauscher: Along the same lines, are there nuclear export signals in APC? Is that regulated at any level?

White: There are candidate nuclear export.

Pierotti: If you look at the desmoid tumors for APC distribution, relationship to β-catenin and so on, is it the same picture that emerges in the colon cancer cells or lesion?

White: In the desmoids? You know I have not seen the reports on the molecular biology of β-catenin yet in the desmoid tumors. So these are fibroblast tumors that grow in the scar tissue after the colons are removed, associated with polyposis. I would expect it to be, because others have shown that there is loss of heterozygosity for APC. So these cells would be deficient homozygos mutant for APC and by our theory β-catenin levels should go up. But I do not know that for a fact.

Livingston: Does GFP-APC appear to be nuclear?

White: Clinically relevant mutants, that is one slide that I showed you where there is no labeled DLD1 stained with the same antibodies, show no labeling anywhere. The self-fractionation experiments showed that in the nuclei, when fractionated, there is very little, if any, APC. That is NDLD1.

Livingston: Is the total level of the protein in the cell depressed?

White: No, again this is tricky. This is a very hard protein to get out of cells. It always seems that there is more protein when you look at the truncated peptide of a carcinoma cell line. We are not absolutely sure how much of that is extraction, but there is certainly no diminution in the amount of APC. So, going one step back, the green fluorescent protein experiments are still in process.

Klausner: The question was, if you introduce the protein, wild type versus mutant GFP labeled, can you demonstrate that there is a sequence present in the GFP labeled proteins required to get at the nucleus, if indeed, it signals?

White: Those, again, are part of the same set of experiments that are underway.

Pierotti: Do you think that it could be a shattering protein for β-catenin? Have you tried, for example, to study the localization of the protein of the β-catenin in the presence of the mutants? They should modify the localization.

White: That is easier to do. We find less β-catenin in the nucleus than one might expect. Well, I do not want to stress that because that is what we see when we do the fractionation.

Bankert: Since you said these cells were hard to grow in culture, I just want to make the comment that they can be maintained for up to 18 months in SCID mice, placing them either intraperitoneally or subcutaneously. It might be a good model to follow up on.

Klausner: Ray, what is limiting, free β-catenin or LEF1, and can you, in cells, transfect β-catenin and LEF1 if both of those are limiting to actually try to get a sense for the target genes?

White: I think that is going to depend on the cell type that you are talking about.

Klausner: There are colon cell lines that can be transfected whether they are good or not.

White: There are some cells where you can get a fairly significant increase in a reporter by putting in LEF1. I do not know yet whether β-catenin will have the same effect in those cells. It is not obvious what to expect for any specific cell line. In most cell lines almost all the β-catenin is associated with the nucleus. So, depending on whether the cells are induced, not induced (APC +, APC -), you would make different predictions on which one should be, whether β-catenin or LEF1 should be limiting.

Klausner: There is apparently a very good LEF-1 site in the COX-2 promoter.

White: That is what we think too.

Klausner: In the COX-2 knockout mice, what is it, sixty to seventy percent reduction in formation adenomas, or maybe even eighty percent? Do you know if those mice have been treated with NSA?

White: I have asked that same question. It is interesting because, if I understand the point that you are raising, all cells have a constitutive level of COX-1. You might expect that those residual twenty percent would still be, to some extent, sensitive to a drug inhibitor. The more I think about it, the more I think that that experiment has not been done.

Klausner: Ray, just to go back to this theme that there might be some sort of interaction in the nucleus, it is a prejudice of mine, obviously, to be in favor of such a thing happening.

Livingston: But if that were true, then a mammalian 2 hybrid experiment might work. Has that been tried?

White: Not to my knowledge.

Draetta: Do you know whether the mutant β-catenin proteins found in tumors affect any of the protein-protein interactions you describe?

White: The mutations that I am remembering at this moment, are mutations that affect the aminoterminus that need to be phosphorylated in order to degrade the protein. So, when those are mutated, no longer can you phosphorylate, no longer can you degrade the protein. That is what gives you the resistance to APC in elevated levels.

Draetta: What is the effect of β-catenin overexpression on apoptosis like for the APC?

White: In some cells, these are experiments of Steve Prescott's and I do not remember the cell type, but they put LEF-1 into those cells and with a reporter for β-catenin LEF1 saw an induction.

7

GENOME SCANNING AND GENE DISCOVERY IN BREAST AND OVARIAN CANCER

Joe W. Gray,[1] Colin Collins,[3] Daniel Pinkel,[1] Laleh Shayesteh,[1] Yiling Lu,[3] and Gordon Mills[3]

[1]UCSF Cancer Center
University of California
San Francisco, California
[2]MD Anderson Cancer Center
University of Texas
Houston, Texas
[3]Life Sciences Division
Lawrence Berkeley National Laboratory
Berkeley, California

INTRODUCTION

Genome scanning techniques are becoming increasingly important as tools for the discovery of aberrations that are important in the genesis and progression of tumors. Our entry into genome scanning began in 1992 with the development of Comparative Genomic Hybridization (CGH)[1]. This technique compares DNA sequence copy number between two genomes; typically, a tumor genome and a normal genome. In CGH, DNA isolated from the tumor is labeled so that it fluoresces green and DNA from the normal reference genome is labeled so that it fluoresces red. Typically, label is incorporated by nick translation. These DNA samples, plus unlabeled Cot-1 (to block hybridization of interspersed repeat sequences) are hybridized to normal metaphase chromosomes under conditions in which their rates of deposition at each point in the genome are proportional to their relative copy numbers. After this process, the ratio of green to red fluorescence along each of the target metaphase chromosomes is measured as indication of the relative DNA sequence copy number at that point in the genome. Regions that appear relatively green are regions in the tumor that are present at increased copy number while regions that appear relatively red are regions in the tumor that are present at decreased copy number. The power of this technique is that it maps these changes on to a normal representation of the genome so that they can easily be interpreted in the context of genomic information that is increasingly available through world wide genome databases. Furthermore the technique

The Biology of Tumors, edited by Mihich and Croce
Plenum Press, New York, 1998.

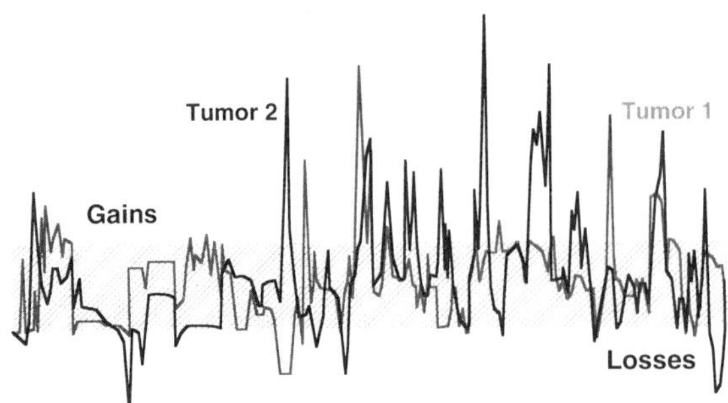

Figure 1. CGH "genome scans" of two clinically similar breast cancers. The Y axis is the green:red ratio measured after CGH with tumor DNA labeled so that it fluoresces green and normal DNA labeled so it fluoresces red. The data from the various chromosomes are arranged from 1pter to 22qter. The hatched band indicates the region of peak heights typically observed when the test and reference samples used for CGH are normal. Peaks that extend above or below this region are considered significant.

gives an overall genomic assessment of relative copy number abnormalities. Figure 1 compares genome scans of two clinically similar human breast tumors made using CGH. The data are represented left to right from the short arm chromosome 1 through the long arm of chromosome 22. The figure is remarkable in two respects: (a) a large number of regions of increased and decreased relative copy number relative to normal are apparent and (b) the differences between the two tumors are substantial. These tumors are not atypical in their degree of abnormality. In fact, it is common to find as much as 30% of solid tumor genomes to be present at abnormal copy number in CGH analysis. Of course this is a function of the genetic abnormalities that underlie tumor progression so that not all tumors are equally rearranged. For example, tumors carrying mutations that lead to microsatellite repeat instability seem to progress through mechanisms that do not lead to chromosome evolution and typically have few copy number abnormalities when analyzed using CGH[2]. On the other hand, tumors with mutations in p53 or other cell cycle check point genes seem to evolve via chromosomal reorganization and typically accumulate multiple copy number abnormalities (Chen et al., manuscript in preparation). The differences between the two tumors also are remarkable. With this degree of genomic change, it is not surprising to find that tumors progress at different rates and respond differently to therapy. We believe that such genetic signatures eventually will be understood sufficiently that they can be used to assess probability of tumor progression and/or response to therapy.

IDENTIFICATION OF RECURRENT ABNORMALITIES

Although there are many abnormalities in solid tumors, some recur frequently in tumors of the same type as well as in tumors of different types. One goal of our research program is to identify these recurrent abnormalities and to locate genes in these regions that contribute to cancer progression. This information can contribute to our understanding of the basic biology of tumor progression and also to provide targets for therapeutic attack.

Our first step was to apply CGH to scan many breast and ovarian cancers. Results from the breast cancer studies are shown in Figure 2[3]. Those from the ovarian cancer studies are shown in Figure 3[4]. Numerous regions of recurrent abnormality are clear in these data. We have focused on two abnormalities; one a region of increased copy number on chromosome 20q especially apparent in breast cancers, and the other a region of increased copy number at chromosome 3q26 especially apparent in ovarian cancers.

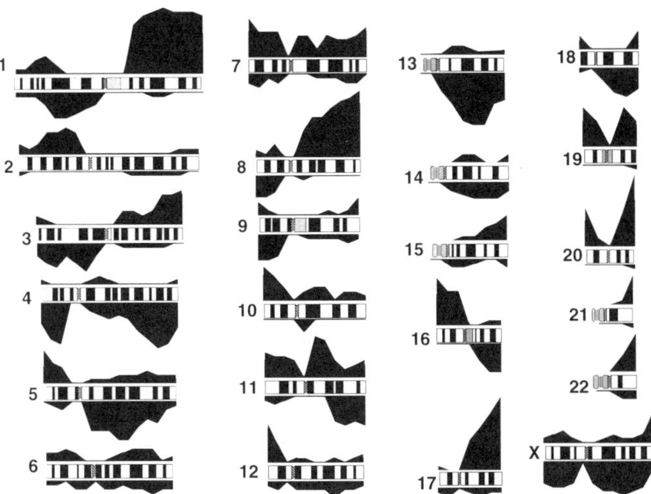

Figure 2. Distributions showing the frequencies with which copy number abnormalities were observed in multiple breast cancer samples. The height of the black regions above the chromosomes are measures of the frequencies with which those parts of the tumor genome were determined to be abnormally increased in copy number. The heights of the black regions extending below the ideograms indicate the frequencies with which those parts of the tumor genomes were found to be reduced in copy number.

Increased copy number at 20q13. Our interest in 20q13 stems from its frequent involvement in breast cancer[3] and a variety of other solid tumors, especially those of the ovary[4], brain[5], head and neck[6], pancreas[7] and colon[2]. Increased copy number in this region also has been associated with poor clinical outcome in node-negative breast cancers[8] and with immortalization and genomic instability in bladder epithelial[9,10] and keratinocytes[11] after transfection of HPV16. We applied fluorescence in situ hybridization (FISH) with

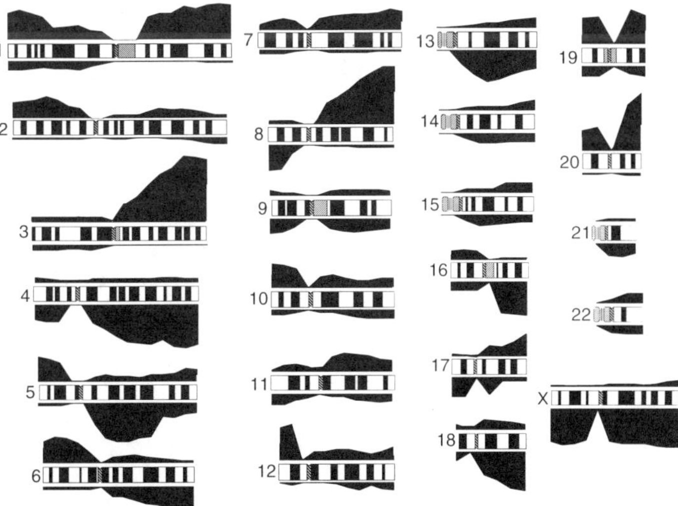

Figure 3. Distributions showing the frequencies with which copy number abnormalities were observed in multiple ovarian cancer samples. The data are represented as described in Figure 2.

probes mapped along chromosome 20q to define the region of increased copy number. These studies defined a 1 to 2 Mb wide region of recurrent amplification in breast cancer cell lines and in ~10% of primary breast tumors[12]. High level amplification (>3×) was associated with high tumor grade, high S-phase fraction, and poor clinical outcome[13] . Analysis of genome databases revealed no obvious candidate genes that might be selected for during amplification of this region so we embarked on a gene discovery program to identify such genes in this region. Our approach was to first develop a sequence-ready physical map across the region comprised of P1 and BAC clones, and to use these clones as substrates for gene discovery. Specifically, we employed exon trapping[14], direct cDNA selection[15] and genome scale sequencing (in collaboration with the Berkeley National Laboratory Human Genome Program,[16] . This process resulted in the discovery of two significant genes designated *ZABC1* and AIBC1[17]. Both genes were amplified and overexpressed in a subset of human breast cancers. *ZABC1* was particularly interesting since it was located in a region of most frequent amplification and was expressed in all tumors and cell lines in which it was amplified and in some in which it was not amplified. Sequence analysis of *ZABC1* suggested it to be a classic Krupple-like zinc finger transcription factor. *AIBC1* was located just outside the region of most frequent amplification, and was overexpressed in most but not all of the tumors in which it was amplified. We were unable to identify any sequence homology between *AIBC1* and known genes. We are now engaged in the elucidation of the function of *ZABC1* and assessment of its importance in the progression of breast cancer and in other solid tumors.

Increased copy number at 3q26. Our interest in a region of increased copy number at chromosome 3q26 was stimulated by the finding of this abnormality in a broad range of human solid tumors; especially those originating in the ovary,[4] lung[18], head and neck[19] and cervix[20]. In addition, this aberration appeared to be an early event in ovarian[4] and cervical[20] cancer. Our approach to analysis of this region was similar to that employed for chromosome 20. That is, we assessed copy number across the chromosome 3q26 region using FISH. In this case, a candidate gene search identified an interesting gene which guided our further studies. Specifically, we noted that PIK3CA, the 110 KD catalytic subunit of PI3-kinase was located at 3q26.3[21]. FISH with probes containing this gene showed increased copy number in almost all ovarian cancer cell lines and tumors tested. The protein encoded by PIK3CA, designated p110α binds through an 85 KD protein to activated tyrosine kinase receptors to initiate PI3-kinase signaling. This pathway has been implicated in a variety of cellular functions including proliferation, apoptosis, adhesion and vesicle transport. Our finding of increased PIK3CA copy number using FISH prompted an assessment of the PIK3CA expression and PI3-kinase activity. Western analysis showed that p110α was overexpressed in all ovarian cancer cell lines and tumors in which it was present at increased copy number relative to expression in one ovarian cancer line and normal ovarian epithelial cells in which PIK3CA was not present at increased copy number. Subsequent studies showed that increased p110α expression was associated with increased PI3-kinase activity, and also with increased apoptotic response to the PI3-kinase inhibitor LY294002[21]. These results suggest that events leading to increased copy number 3q26 play a role in the activation of the PI3-kinase pathway and to overexpression of p110α . The involvement of the PI3-kinase pathway in ovarian cancer is not unexpected given that a possible down stream effector of PI3-kinase, AKT2, also has been found to be amplified in human ovarian cancers[22]. In addition, recent studies have shown that PI3-kinase or it's homologues can be tumorigenic when over expressed in both birds and mammals[23,24]. It now remains to be determined whether a simple copy number increase of PIK3CA is sufficient to activate the pathway or whether additional structural and genomic changes are required.

FUTURE DIRECTIONS

Genome scanning techniques have revealed a remarkable number and diversity of DNA sequence copy number abnormalities in human solid tumors including breast and ovarian cancer. Most of the genes in these regions that contribute to cancer when present at altered copy number remain to be discovered. However, elucidation of the gene organization and DNA sequence of the human genome is now proceeding rapidly and this information will substantially speed the discovery of the genes in these regions of abnormality as they are defined. Rapid and precise definition of these regions using point-by-point analysis techniques such as analysis of LOH or FISH now is rate limiting. We anticipate that this limitation will be removed as high throughput techniques for analysis of LOH[25] or array based CGH[26] are developed. Proof of principle studies have already been completed in these areas and full scale implementation is underway. When completed, application of these techniques followed by focused genomic analysis should provide an increasingly accurate picture of the genetic events that contribute to the initiation and progression of specific human tumors.

REFERENCES

1. Kallioniemi, A., *et al.* Comparative genomic hybridization for molecular cytogenetic analysis of solid tumors. *Science.* **258**, 818–821 (1992).
2. Schlegel, J., *et al.* Comparative genomic in situ hybridization of colon carcinomas with replication error. *Cancer Res.* **55**, 6002–6005 (1995).
3. Kallioniemi, A., *et al.* Detection and mapping of amplified DNA sequences in breast cancer by comparative genomic hybridization. *Porch Natl. Acad Sci U S A.* **91**, 2156–2160 (1994).
4. Iwabuchi, H., *et al.* Genetic analysis of benign, low-grade, and high-grade ovarian tumors. *Cancer Research.* **55**, 6172–6180 (1995).
5. Mohapatra, G., Kim, D.H. & Feuerstein, B.G. Detection of multiple gains and losses of genetic material in ten glioma cell lines by comparative genomic hybridization. *Genes Chromosomes Cancer.* **13**, 86–93 (1995).
6. Bockmuhl, U., Petersen, I., Schwendel, A. & Dietel, M. Genetic screening of head-neck carcinomas using comparative genomic hybridization (CGH). *Laryngorhinootologie.* **75**, 408–414 (1996).
7. Solinas-Toldo, S., *et al.* Mapping of chromosomal imbalances in pancreatic carcinoma by comparative genomic hybridization. *Cancer Res.* **56**, 3803–3807 (1996).
8. Isola, J.J., *et al.* Genetic aberrations detected by comparative genomic hybridization predict outcome in node-negative breast cancer. *Am J Pathol.* **147**, 905–911 (1995).
9. Savelieva, E., *et al.* 20q gain associates with immortalization: 20q13.2 amplification correlates with genome instability in human papillomavirus 16 E7 transformed human uroepitelial cells. *Oncogene.* **14**, 551–560 (1997).
10. Reznikoff, C.A., *et al.* Long-term genome stability and minimal genotypic and phenotypic alterations in HPV16 E7-, but not E6-, immortalized human uroepithelial cells. *Genes Dev.* **8**, 2227–2240 (1994).
11. Solinas-Toldo, S., Durst, M. & Lichter, P. Specific chromosomal imbalances in human papillomavirus-transfected cells during progression toward immortality. *Proc Natl Acad Sci U S A.* **94**, 3854–3859 (1997).
12. Tanner, M.M., *et al.* Increased copy number at 20q13 in breast cancer: defining the critical region and exclusion of candidate genes. *Cancer Res.* **54**, 4257–4260 (1994).
13. Tanner, M.M., *et al.* Amplification of chromosomal region 20q13 in breast cancer: prognostic implications. *Clinical Cancer Research.* **1**, 1455–1461 (1995).
14. Church, D.M., *et al.* Isolation of genes from complex sources of mammalian genomic DNA using exon amplification. *Nat Genet.* **6**, 98–105 (1994).
15. Lovett, M., Kere, J. & Hinton, L.M. Direct selection: a method for isolation of cDNAs encoded by large genomic regions. *Proc Natl Acad Sci U S A.* **88**, 9628–9632 (1991).
16. Kimmerly, W.J., Kyle, A.L., Lustre, V.M., Martin, C.H. & Palazzolo, M.J. Direct sequencing of terminal regions of genomic P1 clones. A general strategy for the design of sequence-tagged site markers. *Genet Anal Tech Appl.* **11**, 117–128 (1994).
17. Collins, C., *et al.* Postitonal Cloning of ZABCI and AIBCI: Genes Amplified at 20q13.2 and Overexpressed in Breast Carcinoma. *Proceedings of the National Academy of Science (USA).* (submitted), (1998).

18. Levin, N., Brzoska, P., Warnock, M., Gray, J. & Christman, M. Identification of novel regions of altered DNA copy number in small cell lung tumors. *Genes, Chromosomes, and Cancer*. **13**, 175–185 (1995).
19. Speicher, M., *et al.* Comparative genomic hybridization detects novel deletions and amplifications in head and neck squamous cell carcinomas. *Cancer Research*. **55**, 1010–1013 (1995).
20. Heselmyer, K., *et al.* Gain of chromosome 3q defines the transition from severe dysplasia to invasive carcinoma of the uterine cervix. *Proc Natl Acad Sci USA*. **93**, 479–484 (1996).
21. Shayesteh, L., *et al.* PIK3CA is implicated as an oncogene in ovarian cancer. *Nature Genetics*. (submitted), (1998).
22. Cheng, J., *et al.* AKT2, a putative oncogene encoding a member of a subfamily of protein-serine/threonine kinases, is amplified in human ovarian carcinomas. *Proc Natl Acad Sci USA*. **89**, 9267–9271 (1992).
23. Marte, B. & Downward, J. PKB/Akt: connecting phosphoinositide 3-kinase to cell survival and beyond. *Trends Biochem Sci*. **22**, 355–358 (1997).
24. Chang, H., *et al.* Transformation of chicken cells by the gene encoding the catalytic subunit of PI 3-Kinase. *Science*. **276**, 1848–1850 (1997).
25. Dietrich, W.F., *et al.* Genome-wide search for loss of heterozygosity in transgenic mouse tumors reveals candidate tumor suppressor genes on chromosomes 9 and 16. *Proceedings of the National Academy of Science (USA)*. **91**, 9451–9455 (1994).
26. Pinkel, D., *et al.* Quantitative high resolution analysis of DNA copy number variation in breast cancer using comparative genomic hybridization to DNA microarrays. *Nature Genetics*. (submitted), (1998).

DISCUSSION

Livingston: In the 20q 13.2 region, where you detect multiple peaks, is it possible to do fine structure mapping, both in DCIS and in the final malignant tumor, to ask whether there are some amplifications that are present early and in others late?

Gray: Yes, that is the right experiment to do. We would not have attempted that with fish but I think with the ELISA we will do it. So that is clearly to be done. We have got material in the refrigerator, we just need to get the technology a little bit more robust before we do it.

Livingston: If you do *in-situ* analysis of ZABC1 expression, is it selectively expressed?

Gray: So far what we have apparently does appear to be expressed in breast epithelial cells, and it does appear to be expressed somewhat higher in the tumor than in the epithelium.

Helin: Is there any way in your system to level out differences between the tumors to identify singular regions which are, let us say, more amplified or more mutated? One puzzling thing about your system is that, you say if you take two different tumors with several types of mutations you see very many differences in the way you analyze them. Can you take fifty tumors and analysis them in one go against normal patients and identity regions which are altered consistently in all the tumors?

Gray: The way we do that is really to ignore the noise. In other words, a lot of those peaks do not recur in another tumor type and those, in a sense, we ignore when we try to analyze these tumors. What we focus on are those abnormalities that are present in twenty, thirty, forty percent of the tumors, not those that occur in one or two of them and so we sort of computationally average it out if you will.

Helin: So that number, twenty, thirty, forty percent, how did you end up getting that number?

Gray: What we find is that you can obviously pick whatever threshold you would like. If you lower the threshold too much then you end up with hundreds of regions of the G numbers to study and it probably becomes increasingly uninteresting just because of the small fraction of tumors in which it is involved. So we picked a high threshold just because we cannot work on all the regions that we have out there in any case, so it was just the first step in making a cut.

Rauscher: The AP2 story might be really interesting. It is clearly a key regulator of ERBb2; it is also expressed highly in breast epithelials, you will look at that I am sure.

Gray: Yes, we are going to look at that, absolutely.

Anderson: Based on the size of 20q you are looking at and the number of events you are seeing with the finer structure of the chip based analysis, can you estimate a ballpark number of how many total amplification and deletion events are occurring in a typical tumor cell? In other words you see a recurrent region of 20q 13.2, but then you also have seen secondary events that appear to be random. I was just wondering if these random events are occurring at similar rates throughout the genome. Based on the size of the region you are looking at on 20q, can you extrapolate to get some ballpark figure on how many total genomic events are likely to be occurring in the average tumor cell?

Gray: Not easily. I think we will be able to do that much better when we have the higher resolution technique working over the whole genome. Right now, we can bias ourselves towards regions of increased copy number that involve fairly large stretches of the genome just because CGH is more sensitive to those. And what we find when we go into those, is that they start resolving into multiple amplicons. At this point, I do not have a good sense for how many different peaks are out there.

Azizkhan: Do you have any idea as to what might be the transcriptional targets of ZABC1?

Gray: Not yet.

DePinho: I was wondering what the basis was for eliminating GLU2 from consideration. Why is it not an important amplicon? Glucose transporters are very important for the high metabolic requirements of tumor cells and could be involved in selecting for an enhanced nutritional state of those cells?

Gray: Well, more than that. It is not as attractive. We have not eliminated it. It is still there, we just have not pursued it seriously and we will probably work on PI3K before we get to that. But there is evidence for the other glucose transporters that are in fact upregulated by P13 kinase and so these things could be acting concordantly together, although there is no evidence that GLU2 is under the control of PI3 kinase.

Klausner: You mentioned that Tom Reed was looking at a similar region on 3, looking at the progression of cervical cancer which would be a nice model. Do you know whether Tom found P13 kinase as the amplicon?

Gray: I do not know. We have talked about exchanging probes or material and it has not happened yet, so I suspect that he does not know.

Klausner: If you overexpress just the 110 subunit of P13 kinase in the absence of P85, do you stimulate the pathway?

Gray: We have not done that. My understanding is that a constitutive activation of the 110 subunit does, in fact, activate the pathway by over expression.

Pelicci: To my knowledge p110 does not regulate the levels of p85. So why should they both be over expressed in tumors with amplification of that chromosome region? You showed blots with overexpression of p85 and p110. Which mechanism do you think is responsible for the overexpression?

Gray: Well, I do not have a good mechanism for that. I am a little bit mystified by that.

Pelicci: I have another question. If you go back to the cytogenetics, the cases showing amplifications by the CGH have normal karyotypes in that region. Is 3q normal? Should we think that tumors with normal karyotypes might contain a lot of these genetic abnormalities?

Gray: Most of these tumors are far from karyotypically normal. It is generally the case that the regions of amplification or increased copy number do not reside on the normal chromosomes, they tend to be off on marker chromosomes. In fact, they are generally associated with other regions of amplification in the genomes. So it is quite common, for example, to find two or three regions of chromosome 20 and Myc and they are all co-associated together in the same HSR.

Pelicci: In those cases where CGH is evident for amplification, there was no HSR in the chromosome?

Gray: There was no HSR. In fact, you probably would not have picked this up cytogenetically. By banding analysis you would not pick it up because these regions of increased copy number are often sprinkled in other chromosomes and you just do not see them.

Meltzer: Regarding the problem that you so nicely described of multiple amplicons in a given chromosomal region which is a problem I think which is just going to get worse as we look at more and more of the regions that have yet been characterized in detail. There are basically two broad conceptual models that you might make; one would be, let us say, that the entire 20q arm undergoes a copy number increase and that due to selection, various subregions are retained as the tumor evolves. Alternatively, multiple subregions might be amplified as independent events. One way that one might resolve those two alternative explanations would be by looking at polymorphic markers on the chromosome and trying to decide if these multiple regions are all ultimately derived from the same parental chromosome. I wonder if you have thought about doing this in any of these complex systems?

Gray: We have talked about doing it. We have not done it. A further comment about that though is that in studies of HPV E16 E7 immortalization, the first step there seemed to be a 10, 20q translocation. So that you actually had most of the entire chromosome 20 intact down to 10 and then from then on it evolved into finer pieces. Now that still does not say whether or not they were all derived from that one originally translocated chromosome, but it does suggest that, I think.

8

FAILURE OF TUMOR IMMUNITY RESULTING FROM INACCESSIBILITY OF ACTIVATED LYMPHOCYTES TO SOLID TUMORS

The Possible Role of the Endothelium

Ruth Ganss and Douglas Hanahan

Department of Biochemistry and Biophysics
Hormone Research Institute
University of California, San Francisco
San Francisco, California 94143-0534

1. LYMPHOCYTE-TUMOR INTERACTION IN A MODEL FOR MULTI-STEP TUMORIGENESIS

Tumor-associated antigens can be demonstrated on different tumors in mice and humans[1]. However, immune responses to tumor antigens have typically shown low efficacy and do not prevent tumor growth in immunocompetent hosts. Many factors contribute to the escape of tumors from immunological control, including defects in antigen-presentation and expression of immunosuppressive molecules. In addition, tumors might have developed alternative intrinsic mechanisms against increased exposure to leukocytes.

To study interactions between the host immune system and cancer progression in a spontaneously arising tumor, we are exploiting transgenic mice expressing the oncoprotein SV40 large T antigen (Tag) under the control of the rat insulin gene promoter (RIP). Tag expression in β cells of the pancreas elicits focal hyperproliferation, tumor angiogenesis with the growth of new capillaries, and eventually, formation of solid tumors[2]. Despite this common pathway of multistep tumorigenesis, different RIP-Tag lines show alternative immunological phenotypes depending on the onset of Tag expression during ontogenesis[3]. In the RIP1-Tag2 transgenic line, transgene expression during embryonic development in pancreatic β cells and thymus results in systemic tolerance towards Tag[4]. In contrast, RIP1-Tag5 transgenic mice express the oncogene at 10–12 weeks of age and fail to induce self-tolerance (Table 1). Within 4–6 weeks, preneoplastic lesions become infiltrated by helper ($CD4^+$) and cytotoxic ($CD8^+$) T cells, B cells and $RelB^+$ cells; among the T cells are ones recognizing the Tag oncoprotein. Despite the autoimmune response against Tag and β

The Biology of Tumors, edited by Mihich and Croce
Plenum Press, New York, 1998.

Table 1. Expression of wildtype T antigen results in different immunological phenotypes

	Onset of Tag oncogene expression	Appearance of tumors	Immune status
RIP1-Tag2	embryonic day 10–12	8–10 weeks	tolerant
RIP1-Tag5	10–12 weeks	18–20 weeks	autoreactive

cells, islet cell carcinomas develop in RIP1-Tag5 mice and lead to premature death at the age of 30 weeks. In striking contrast to hyperplastic islets, Tag-expressing solid tumors in RIP1-Tag5 mice escape the immune surveillance and are devoid of lymphocytic infiltration (Figure 1). Thus, the host immune response in a nontolerant transgenic line with a well defined tumor antigen (Tag) is not efficient enough to prevent tumor outgrowth.

It has been shown in many studies that tumor cells can fail to function as antigen-presenting cells due to an inability to process and present the antigen, or to the absence of costimulatory molecules, or by secretion of inhibitory cytokines[5–7]. Recently, it has also been demonstrated that tumors can induce apoptosis of activated T cells by expression of Fas ligand (FasL)[8]. For a tumor antigen to be recognized by cytotoxic T cells, it has to be intracellularly processed into peptides and presented via major histocompatibility complex (MHC) classI molecules at the cell surface. In the RIP-Tag model, we sought to determine MHC classI expression on β cell lines derived from different stages during tumor progression. Cell lines from preneoplastic stages show relatively high levels of MHC classI expression and are efficient targets for killing by cytotoxic T cells primed against Tag[9]. Different cell lines from insulinomas display a more variable level of MHC complexes, which may indicate an immune escape mechanism[9,10]. However, immunohistochemical analysis in RIP1-Tag5 mice demonstrated that tumors and hyperplastic islets express similar levels of MHC classI molecules (Philippe Hartl, unpublished). Thus, downregulation

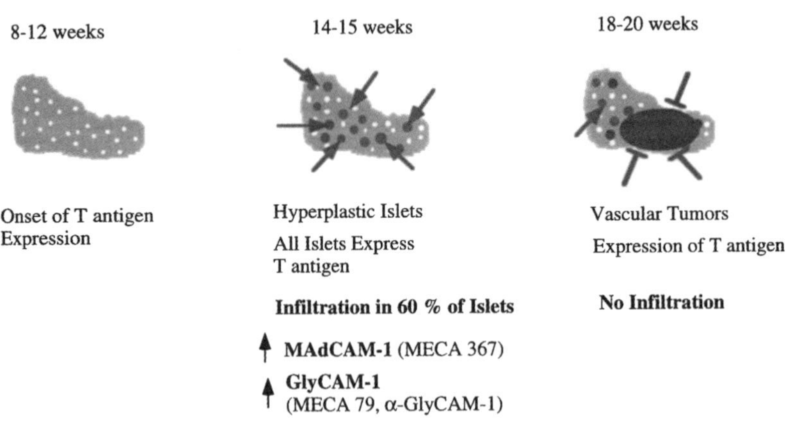

Figure 1. Multistep tumorigenesis in RIP1-Tag5 mice. Tag expression in RIP1-Tag5 mice starts at 10–12 weeks. Hyperplasia is observed in 50% of the islets 4–6 weeks later, and 60% of the islets become infiltrated with lymphocytes. The inflammatory response to Tag is correlated with induction of MAdCAM-1 (mucosal addressin cell adhesion molecule-1) and PNAd (peripheral lymph node addressin: GlyCAM-1, CD34, Sgp200), which are recognized by the monoclonal antibodies MECA367 and MECA79, respectively. Highly vascularized tumors eventually develop in 1–2 % of islets by 18–20 weeks. No MAdCAM-1 or PNAd expression could be detected within β cell tumors.

of MHC classI may not represent an obligatory step in the process of β cell transformation *in vivo*.

Further, solid tumors in the RIPTag5 model do not express FasL, excluding the selective elimination of tumor-surrounding lymphocytes as an escape mechanism (R.G., unpublished).

Besides antigen recognition, the development of an effective immune response depends on the capacity of antigen-specific lymphocytes to migrate into tissue. Similarly, adoptive therapy of solid tumors with antitumor effector cells will depend on their successful delivery to the malignant tissue. The critical step in the process of lymphocyte extravasation involves lymphocyte binding to the vascular endothelium[11]. To assess lymphocyte-vascular endothelium interaction in RIP1-Tag5 transgenic mice, we examined expression of various endothelial adhesion molecules in infiltrated hyperplastic islets versus solid tumors. ICAM-1, PECAM-1 and VCAM-1 were found to be expressed on blood vessel endothelium both in infiltrated islets and non-infiltrated tumors[12]. However, a unique set of surface molecules, defined by the monoclonal antibody MECA79[13], is exclusively induced in areas of infiltration in RIP1-Tag5 islets[12] (Figure 2). MECA79 recognizes specialized postcapillary vascular sites, called high endothelial venules (HEV), which mediate lymphocyte adherence and transendothelial migration in secondary lym-

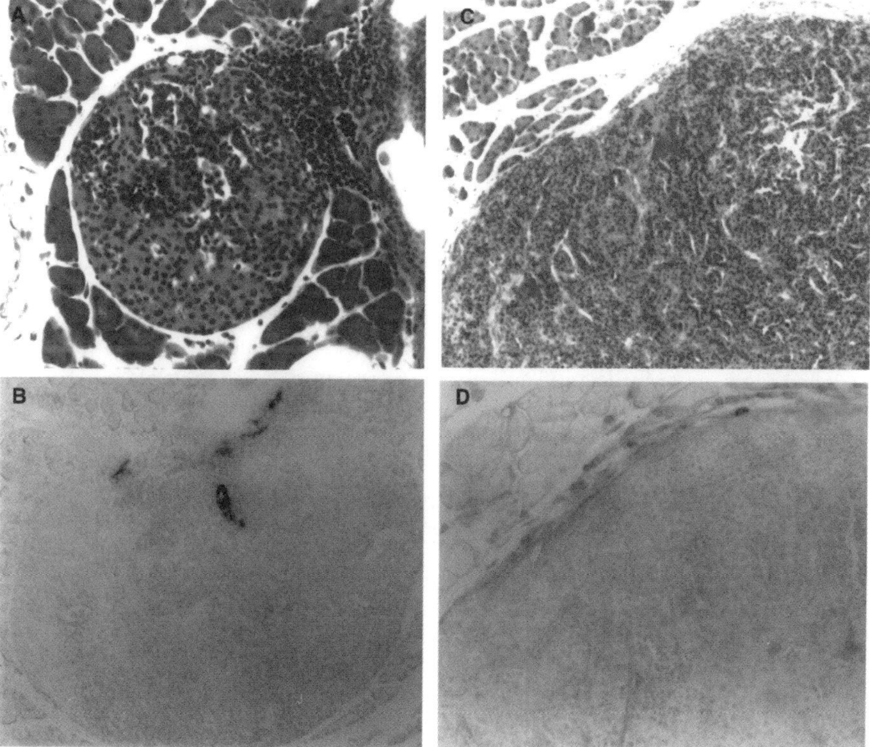

Figure 2. Histology and immunohistochemistry of RIP1-Tag5 transgenic mice. A, B: infiltrated, hyperplastic islets of a 16 week old RIP1-Tag5 mouse. Paraffin sections were stained with H&E, 20x (A) or MECA79, 20x (B). C, D: paraffin sections of a tumor from a 26 week old RIP1-Tag5 mouse, stained with H&E, 10x (C) and MECA79, 20x (D). All immunohistochemistry was counterstained with methyl green. In contrast to hyperplastic islets, the tumor is devoid of lymphocytes and HEVs.

phatic organs. HEVs also play an important role in the pathogenesis of inflammatory diseases by mediating abnormal lymphocyte recruitment[14]. In contrast to infiltrated islets, the endothelium in noninfiltrated, highly vascularized tumors and islets of the tolerant RIP1-Tag2 line does not express HEV-specific markers.

Thus, tumor progression is associated with a downmodulation of addressin expression, suggesting that suppression of endothelial permeability for leukocytes is part of a deliberate tumor escape mechanism.

2. FAILURE OF TUMOR IMMUNITY

2.1. Inaccessibility of Activated Lymphocytes to Solid Tumors

The failure of tumor immunity in RIP1-Tag5 mice is especially striking because of the presence of autoreactive T cells and the high levels of the Tag-oncoprotein in growing tumors. In analogy, many patients with cancer also have detectable immunity to oncogenic proteins; the immune response, however, is in general not sufficient to eradicate the tumor.

One explanation for tumor escape is the infrequency of potentially reactive lymphocytes extravasating during normal immune surveillance. Therefore, adoptive transfer of tumor-specific, autologous T cells that have been expanded *in vitro* is one of the strategies considered for tumor therapy in humans.

To test the efficacy of an increase in the abundance of Tag-reactive lymphocytes, we generated double transgenic mice coexpressing the Tag oncogene and a rearranged Tag-specific T cell receptor (RIP1-Tag2 × TagTCR2; RIP1-Tag5 × TagTCR2, Table 2). TagTCR2 transgenic mice were generated from a $CD4^+$ T cell clone and express Tag-reactive peripheral T cells with a frequency of 80%[15].

Remarkably, the high abundance of Tag-reactive T cells in the periphery does not alter the preferential infiltration of preneoplastic lesions. Due to the overwhelming numbers of Tag-specific T cells in RIP1-Tag2 × TagTCR2 double transgenic mice, tolerance induction fails[15] and a dramatic infiltration of islets is observed. Tumors, however, remain free of lymphocytes. In analogy, islet infiltration is enhanced in RIP1-Tag5 × TagTCR2 double transgenic mice, but tumors are devoid of lymphocytes.

Thus, the combination of a potent tumor antigen (Tag) and activated lymphocytes is not sufficient to promote infiltration and immune-mediated destruction of solid tumors.

2.2. Manipulation of Tumor Cells to Enhance Antitumor T Cell Response

The activation of T lymphocytes is dependent on two signals from antigen-presenting cells: one signal is the antigenic peptide-MHC complex that occupies the T cell recep-

Table 2. Inaccessibility of activated lymphocytes to solid tumors

	Degree of T cell infiltration		
	Hyperplastic islets	Solid tumors	Life span (weeks)
RIP1-Tag2	–	–	18
RIP1-Tag5	+	–	30
RIP1-Tag2 × TagTCR2	++	–	16
RIP1-Tag5 × TagTCR2	++	–	32
RIP1-Tag5 × RIP1-B7-1	+++	–	60

tor; the other is a costimulatory signal, e.g. provided by B7, which binds to CD28 or CTLA-4 on T cells[6]. Costimulatory signals determine whether TCR engagement results in a productive immune response or unresponsiveness of $CD4^+$ T cells[16]. In this context, one explanation why immunogenic tumors can escape host defense is that tumor-reactive T cells receive inadequate costimulation. The importance of B7-1 as a costimulator involved in generating an antitumor immune response has been demonstrated in a number of implantation experiments[17]. In each of these systems, tumor cells were transduced to express B7-1 and subsequently injected s.c. in immunocompetent hosts. Expression of B7-1 causes tumor regression in most of the model systems, which is attributed to a direct activation of tumor specific, MHC classI-restricted $CD8^+$ T cells. The prevailing evidence indicates, however, that the induction of systemic immunity by B7-1 is largely limited to highly immunogenic tumors[18,19].

To assess the effect of the costimulator B7-1 on a spontaneously arising tumor, we crossbred RIP1-Tag mice with a transgenic line, expressing human B7-1 under the control of the RIP promoter (RIP1-B7-1, kindly provided by R. Flavell). It has been shown previously that the ectopic expression of B7-1 on islet β cells does not predispose to diabetes[20]. Our recent findings demonstrate that coexpression of B7-1 in a tolerant Tag-transgenic line does not lead to autoimmunity (R. G. unpublished). In double transgenic RIP1-Tag5 × RIP1-B7-1 mice, however, the already existing autoreactivity is enhanced and results in severe insulitis. The inflammatory response is concomitant with an increased induction of HEV-like vessels. Although tumor formation in these double transgenic mice is considerably delayed, tumors are not attacked by the immune system and remain devoid of lymphocytes (Table 2).

3. TUMOR ANGIOGENESIS AND TUMOR IMMUNE EVASION

It is well established that rejection of cancer cells requires tumors to be antigenic and to be recognized by $CD8^+$ T cells. The concept of modern tumor immunology anticipates that tumor immunity can be induced and will be useful as a strategy for cancer therapy. The identification of new tumor antigens/peptides and the generation of T cell clones are considered as promising strategies to develop cancer vaccines.

Here, we have described a mouse model for a spontaneously arising tumor, where the tumor antigen recognized by T cells as well as the antigen-presenting MHC classI molecule are retained during tumor progression. Although autoreactivity is observed in preneoplastic lesions, highly vascularized tumors escape the host immune defense. In double transgenic mice, we have both increased the abundance of reactive T cells and expressed a potent costimulator on oncogene-expressing β cells. Neither of these strategies, however, elicits immunity towards solid tumors, suggesting that tumor antigenicity and activated T cells are not sufficient for the rejection of primary tumors.

It has been demonstrated that neovascularization is a prerequisite for the outgrowth of solid tumors[21]. Tumor angiogenesis is per se a multistep process involving degradation of the extracellular matrix, proliferation and migration of endothelial cells, and capillary sprout and tube formation. The remodeling and expansion of blood vessels is largely regulated by a balance of angiogenic inducers, e.g. basic fibroblast growth factor (bFGF) and vascular endothelial growth factor (VEGF), and angiogenic inhibitors, e.g. thrombospondin-1.

Cytotoxic T cells represent a fundamental threat to tumors. Therefore, it is attractive to postulate that in the process of angiogenesis endothelium is induced to become an effective barrier that restrains cytotoxic T cells from entering the tumor parenchyma and

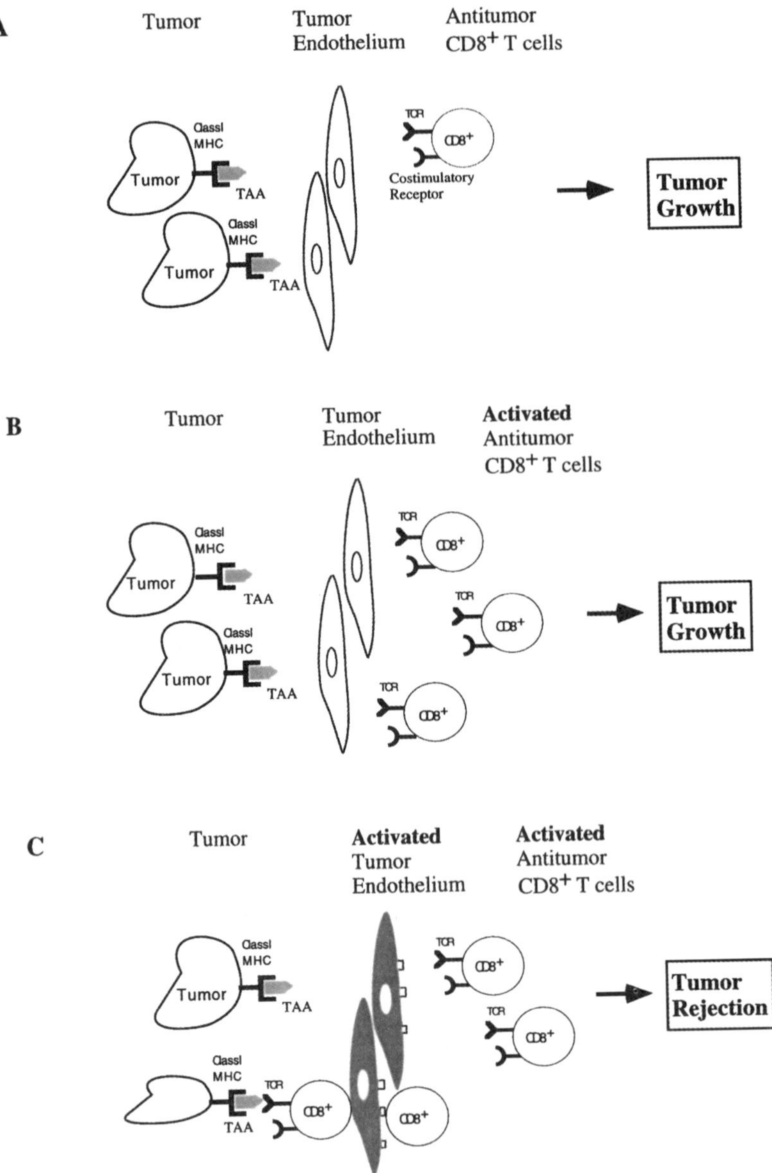

Figure 3. The potential role of endothelial cells in tumor progression. A) Immunogenic tumor cells escape immune surveillance by a subset of antitumor $CD8^+$ T cells, due to a 'non-permissive' tumor endothelium. B) Activation of $CD8^+$ T cells (e.g. priming with tumor antigen/peptide, e.g. costiumlation), or an increase in the abundance of antitumor T cells (shown) fails to elicit longterm tumor immunity, further implicating the endothelium as a barrier. C) We postulate that successful tumor rejection may therefore require immunogenic tumor cells, tumor-responsive lymphocytes, and 'activated' tumor endothelium. According to this hypothesis, induction of addressin expression and other changes in tumor endothelium facilitate extravasation of antitumor lymphocytes and effect consequent tumor cell destruction. TAA: tumor-associated antigen; TCR: T cell receptor.

killing tumor cells. Recent reports confirm a role of angiogenic factors in the modulation of adhesion molecules expressed on tumor endothelium. For example, bFGF can decrease the sensitivity of endothelial cells to inflammatory cytokines, resulting in impaired leukocyte adhesion *in vitro*[22]. Further, in an *in vivo* rat tumor microcirculation model the density of adhering leukocytes was significantly reduced in tumor microvessels compared to normal microvessels[23].

Consistent with these data, the induction of HEVs in the RIP1-Tag5 model is restricted to infiltrated hyperplastic islets, suggesting the tumor may escape host defense by interfering with the mechanism of effector cell extravasation.

Based on these observations, we propose that the tumor endothelium is refractory to lymphocyte extravasation, either intrinsically, or as a result of paracrine influences within the solid tumor microenvironment. Thus, effective tumor immune strategies may well require both activated anti-tumor lymphocytes and conditions that render the tumor endothelium permissive for infiltration by those immune cells (Figure 3). We are currently testing the effects of various angiogenesis inhibitors on lymphocyte-tumor interaction in the autoimmune RIP-Tag5 model. In the future, antiangiogenic therapy may eventually be used synergistically with T cell therapy in the successful treatment of cancer.

REFERENCES

1. E. M. Jaffee and D. M. Pardoll. Murine tumor antigens: is it worth the search?. Cur. Opin. Immunol. 8: 622–627 (1996)
2. D. Hanahan. Heritable formation of pancreatic beta cell tumours in transgenic mice expressing recombinant insulin/simian virus 40 oncogenes. Nature. 315: 115–122 (1985)
3. T. E. Adams, S. Alpert and D. Hanahan. Non-tolerance and autoantibodies to a transgenic selfantigen expressed in pancreatic β cells. Nature. 325: 223–228 (1987)
4. C. Jolicoeur, D. Hanahan and K. M. Smith. T-cell tolerance toward a transgenic β-cell antigen and transcription of endogenous pancreatic genes in thymus. Proc. Natl. Acad. Sci. USA. 91: 6707–6711 (1994)
5. B. E. Elliot, D. A. Carlow, A.-M. Redricks and A. Wade. Perspectives on the role of MHC antigens in normal and malignant cell development. Adv. Cancer Res. 53: 181–244 (1989)
6. C. Lieping, P. S. Linsley and K. E. Hellström. Costimulation of T cells for tumor immunity. Immunol. Today. 14: 483–486 (1993)
7. B. Mukherji and N. G. Chakraborty. Immunobiology and immunotherapy of melanoma. Curr. Opin. Oncol. 7:175 (1995)
8. M. Hahne, D. Rimoldi, M. Schröter, P. Romero, M. Schreier, L. E. French, P. Schneider, T. Bornand, A. Fontana, D. Lienard, J.-C. Cerottini, J. Tschopp. Melanoma cell expression of Fas (Apo-1/Cd95) ligand: implications for tumor immune escape. Science. 274: 1363–1366 (1996)
9. F. Radvanyl, S. Christgau, S. Baekkeskov, C. Jolicoeur and D. Hanahan. Pancreatic β cells cultured from individual preneoplastic foci in a multistage tumorigenesis pathway: a potentially general technique for isolating physiologically representative cell lines. Mol. Cell. Biol. 13: 4223–4232 (1993)
10. K. Hamaguchi and E. H. Leiter. Comparison of cytokine effects on mouse pancreatic α cell and β cell lines. Viability, secretory function, and MHC antigen expression. Diabetes. 39: 415–425 (1990)
11. T. A. Springer. Adhesion receptors of the immune system. Nature. 346: 425–434 (1990)
12. S. V. Onrust, P. Hartl, S. D. Rosen and D. Hanahan. Modulation of L-selectin ligand expression during an immune response accompanying tumorigenesis in transgenic mice. J. Clin. Invest. 97: 54–64 (1996)
13. P. R. Streeter, B. T. N. Rouse and E. C. Butcher. Immunohistologic and functional characterization of a vascular addressin involved in lymphocyte homing into peripheral lymphnodes. J. Cell. Biol. 107:1853–1862 (1988)
14. J.-P. Girard and T. A. Springer. High endothelial venules (HEVs): specialized endothelium for lymphocyte migration. Immunol. Today. 16: 449–457 (1995)
15. I. Förster, R. Hirose, J. M. Arbeit, B. E. Clausen and D. Hanahan. Limited capacity for tolerization of $CD4^+$ T cells specific for a pancreatic β cell neo-antigen. Immunity. 2: 573–585 (1995)

16. D. L. Müller, M. K. Jenkins, R. H. Schwartz. Clonal expansion vs. functional clonal inactivation : a costimulatory pathway determines the outcome of T cell receptor occupancy. Annu. Rev. Immunol. 7: 445–480 (1989)
17. J. P. Allison, A. A. Hurwitz and D. R. Leach. Manipulation of costimulatory signals to enhance antitumor T-cell response. Curr. Opin. Immunol. 7: 682–686 (1995)
18. T.C. Wu, A. Y. C. Huang, E. M. Jaffee, H. I. Levitzky and D. M. Pardoll. A reassessment of the role of B7-1 expression in tumor rejection. J. Exp. Med. 182: 1415–1421 (1995)
19. L. Chen, P. McGowan, S. Ashe, J. Johnston, Y. Li, I. Hellström and K. E. Hellström. Tumor immunogenicity determins the effect of B7 costimulation on T cell-mediated tumor immunity. J. Exp. Med. 179: 523–532 (1994)
20. S. Guerder, D. E. Picarella, P. S. Linsley and R. A. Flavell. Costimulator B7-1 confers antigen-presenting-cell function to parenchymal tissue and in conjunction with tumor necrosis factor α leads to autoimmunity in transgenic mice. Proc. Natl. Acad. Sci. USA. 91: 5138–5142 (1994)
21. J. Folkman, K. Watson, D. Ingber and D. Hanahan. Induction of angiogenesis during the transition from hyperplasia to neoplasia. Nature. 339: 58–61 (1989)
22. A. W. Griffioen, C.A. Damen, G. H. Blijham and Groenewegen. Tumor angigogenesis is accompanied by a decreased inflammatory response of tumor-associated endothelium. Blood. 88: 667–0673 (1996)
23. N. Z. Wu, B. Lkitzman, R. Dodge and M. Dewhirst. Diminished leukocyte-endothelium interaction in tumor microvessels. Cancer Res. 52: 4265–4268 (1992)

DISCUSSION

Livingston: Is there any evidence for or against the notion that the FVB characteristic is a single gene characteristic?

Hanahan: The evidence is suggestive that there is at least a major locus there and that is when we first started seeing invasive cancers in F1 crosses. However, the frequency did seem to increase with further back crossing so there may be several modifiers, but there appears to be one predominant susceptibility locus.

Boon: I am fascinated by your RIP-Tag transgenic model and the notion that the immune system does not work on the tumor cells that express the T antigen. I would like to ask for these kinds of mice, about how many hyperplastic nodules do you have?

Hanahan: Well they start off, essentially there are four hundred nodules that are expressing the oncogene to begin with. Then, over time about half of those become hyperplastic angiogenic. About one to two percent progress to tumors.

Boon: What I am aiming at is, if your immune system was eliminating ninety percent of hyperplastic nodules, maybe the mice would still die about at the same day.

Hanahan: Well, no, indeed the immune system is eliminating a number of those nodules but it is still pretty dramatic when you look, you will see some huge tumor. I did not show you the slides, but you can see some nice little islet here in this tumor. The immune system is basically eating up the little nodule and is totally ignoring the tumor. But we actually see dramatic attack of the islet, particularly in the B7 crosses. Because the cells make insulin these mice become transiently diabetic because there is so much destruction going on. Yet, still within that is selected a tumor which grows out which is protected from the immune system.

Boon: I fully agree that this is a remarkable finding. But it is still possible that you have about ninety-nine percent successes and one failure, and there you get a tumor. Your model is fairly atypical in that you have a tumor that is totally multi-focal to begin with. In humans, maybe we start with one hyperplastic nodule and there, if we have a ninety-nine percent rate of success it might be good enough.

Hanahan: Right. The key question there is though, and I think you are right in terms of targeting preneoplastic lesions. But the real question is if the tumor vascolature is refractory to infiltration. The question is what about when you present with a tumor and if that tumor endothelium is refractory to the endothelium extravasation then you are going to have a protection unless we can understand how to turn that around. That is the thought, but I think it is a very good point.

Mihich: Following this up, if you take your large resistant tumor, call it that way, and you create a disruptive situation whereby you take a piece of the tumor and auto-transplant would, in that situation, the lymphocytes work?

Hanahan: We have done transplants and those ,again, seem to be resistant. We have also taken activated lymphocytes and injected them intratumorally to try to see if we can activate locally the endothelium. So far we have, I do not know if we have done it definitively, tried those experiments; as well as immunizing the mice with T-antigen protein, for example, to see whether any mechanism that we use to sort of activate the immune system has not succeeded in getting into the tumors.

White: I am just a little bit worried that maybe you are being a bit too rigorous in your use of genetic tests to discard the significance of some of these high levels of expression that you find and I appreciate the elegance of it. Nevertheless, it may be true that these amplifications or inductions that you are seeing are important for those specific tumors even though, in general, in the tumor genesis process for that class of tumor, there are other ways around so that they are not very limiting. If that is the case then these systems that are commonly induced might either be good diagnostic targets or even therapeutic targets if, for example, by blocking the FGF receptor you could induce apoptosis.

Hanahan: I think that is a very good point. I mean another one that Anton actually alluded to this morning, is that this could be a gene, that actually if the mice did not die of their tumors, that maybe this is a gene that is being turned on as part of a subsequent progression; it never actually happens here because of the time course of the disease. That is a fair statement, but it does raise the issue that some genes may be markers. I agree with you: I am not ready to completely conclude that this is meaningless, but it is at least food for thought that we need to think about overexpressed genes as being potentially markers in addition to functional contributors.

Pierotti: I am referring to the chromosome 16LOH. Is there some parental preferential loss?

Hanahan: Good question—there was no preference. We were with either of the loses.

Pierotti: Finally, I would like to comment on the involvement of the new system in the HPV related tumors. It is quite well appreciated that in immuno-compromised women there is a dramatic increase in cancer, like in HIV patients.

Hanahan: That is a very good point. I left out that slide, but it is very true that HIV women have a substantially increased incidence. To her credit, which is what motivated Lisa to do this experiment, her reading had revealed that fact and she said let's knock out the immune system and see what happens.

Melief: I have a few questions: one is on the RIP Tag model where the question is, if the TCR from the transgenic was derived from a T cell that was capable of successful adoptive transfer, if you did not have the exact proper T cell receptor for aggressive invasion into tissue, it might be not the right T cell to effectively home into the tissue.

Hanahan: Yes, we have done adopted transfer experiments and, again, you get infiltration of the islets but not the tumors. I did not mentioned that, but we have also done adopted transfers.

Melief: The other question is: in that situation, the tumors may produce factors, like TGF-β, which we have found to completely eliminate the effectiveness of CTL's. Have you looked at that?

Hanahan: We have not seen TGF-β itself, although I think that line of approach is an important one and I do not think that we have laid it to rest. But TGF-β1 is not dramatically expressed.

Melief: The final question is: in the HPV16 model, we have identified powerful CTL epitopes in immunocompetent black 6 mice. I wonder what happens if you do the same experiment in black 6, or if you put the HPV16 genes on the nude or SCID background in these mice?

Hanahan: That is a good question but we have not done that.

Klausner: When you do the IGF knockout and you only get small adenomas—when you look at those, do you no longer get the increase in the FGFR4?

Hanahan: Have we looked at that? I am not sure. I believe that it is still turned on, but I am not absolutely certain that we have done that experiment.

Klausner: One of the questions with the model is that, as opposed to the cervical model, where the inducing pathway genetically is pretty similar, or it smells similar to ninety percent of human cervical cancers, in the pancreatic model the T-antigen is not. We do not know what pathway that represents and so distinguishing if it is a marker of the particular genetic inducer that you are choosing, as opposed to necessary progression of carcinogenesis in that particular tissue, may be difficult. You may actually just see that even without any sort of progression you get a level of a pattern of gene expression that is a marker, as Ray said, but it may not be a marker of anything other than the system that you have chosen to perturb.

Hanahan: Yes, correct. That is certainly an interesting thought that this may be activated as part of a progression in this cell type, but the presence of T antigen obviates it.

Klausner: Can you clarify about the HIV infected women and increased incidence of cervical cancer?

Hanahan: I did not bring all that data and I am not that close to it. Essentially, not only in HIV women but also in immuno-suppressed patients that have received transplants, there is also a statistically significant increase in the frequency of cervical cancer. So it is both HIV infection and transplants. In fact, one of the things we are doing now is that we have put the mice on cyclosporin which is exactly what they put these transplant patients on, to see whether that also increases malignant progression.

Klausner: With HIV women, when they get cervical cancer it is much more aggressive. I think it remains very controversial whether the incidence of cervical cancer is up and vice versa in studies that are ongoing now, where you look at the frequency in the population of women who have cervical cancer who are HIV positive, you are not seeing a higher percentage than in the general population. The jury may not be totally in on that but I think it is more the aggressiveness once you get it.

Hanahan: My understanding was, and I could be wrong, that there was increased incidence. The other thing that one might have thought was that by overexpressing it throughout the pathway, that you would accelerate. In fact, I did not show that data but we also have another gene that is up-regulated, that is BCLX long which is a protector of apoptosis. We overexpressed that in exactly the same way that we overexpressed R4, and in that case, we get dramatic acceleration. We get more tumors and the tumors grow faster, again, consistent with the notion that it is contributing to progression. So, I probably should have put that on there and I will. I guess we are really two for three, rather than one for one, in terms of this approach. How are you going to test the importance of P13 kinase in an ovarian cancer? Is that overexpression instructive? I am just raising that this is going to be a challenge as we get into dealing with hundreds of these genes and how are we going to try to figure out who is a target and whether it is a marker? That is going to be important diagnostically but in terms of therapeutic intervention, obviously, we want to know which is important.

Zanker: Concerning your HPV16 transgenic mice model, can you block the onset of the tumor development using tamoxifen?

Hanahan: We have not tried that. We are just starting to think about playing around with different kinds of pharmacological interventions. But, that is an interesting thought.

Zanker: May I come back to your immune surveillance. Did you look at the expression of the integrins on the lymphocytes where the lymphocytes in your mouse model are not able to migrate into the tumors?

Hanahan: Well, not specifically. All I can say is, I did not show these pictures, but we have examples where right next to a tumor is a lymphocyte. it is an islet that is just full of T cells, B cells, macrophages; I mean everything is right there and literally right next door is a tumor where there is nothing. So, clearly, they have got what it takes to infiltrate those islets. We have not done an exhaustive study on the lymphocyte integrins.

Evans: Following up on that question, as I recall, your published data suggested that there was some specificity in terms of the endothelium cell integrins that were expressed in that tumor versus the inflamed tissues. I think specifically, VCAM, ICAM and PECAM were expressed in both, but MEDCAM is expressed in the inflamed tissue but not in the tumor, is that correct?

Hanahan: Yes, that is correct.

Evans: Have you looked at all to see if any pro-inflammatory cytokines, either administered systemically or actually expressed in the oncogen expressing cells, can modify that endothelial cell expression of ligand?

Hanahan: No, I think that is a very good question. It is known that some cytokines can do that in certain situations. Ruth Gants, who has been performing these studies, is trying to deliver GMCSF into the tumor vasculature. But, we have not done it. It is a good experiment. I think this could be an important component of tumor immunal therapies, finding a way to unlock the tumor endothelium.

Visentin: If I have correctly understood, you are suggesting that the islet tumor model could reflect local resistance to immunocompetent cell infiltration rather than failure of systemic immunosurveillance. Do you plan to make some local and/or systemic cytokine assay?

Hanahan: What we are trying to do is deliver, by sort of a gene therapy approach, cytokines into the tumors or into the tumor endothelium.

Visentin: This is clearly linked to the suggestion of antigen expression.

Hanahan: Yes, that is a good experiment.

Melief: We made an interesting observation in adoptive transfer of HPV16 tumors in mice in that we could not cure them with intravenous infusion of CTL's. The only way we could get rid of these tumors was to inject, simultaneously, CTL's, both intratumorally, which by itself did not cure the tumors, and intravenously in combination with subcutaneous interleukin-2. Maybe the intratumoral T cells created an environment by which the infiltration to remain in residual tumor of the intravenously injected CTL was more effective.

Hanahan: Interesting. We have tried one or the other, but not both.

Melief: It was the only thing that was successful. Whereas in other tumor models, there was complete eradication of the tumors by intravenous injection of CTL alone.

9

DECIPHERING MOLECULAR CIRCUITRY USING HIGH-DENSITY DNA ARRAYS

David H. Mack,[1] Edward Y. Tom,[1] Mamatha Mahadev,[1] Helin Dong,[2] Michael Mittmann,[2] Suzanne Dee,[1] Arnold J. Levine,[3] Thomas R. Gingeras,[2,*] and David J. Lockhart[2]

[1]Program in Cancer Biology
[2]Department of Genomics Research
Affymetrix, 3380 Central Expressway
Santa Clara, California 95051
[3]Department of Molecular Biology
Lewis Thomas Laboratory
Princeton University
Princeton, New Jersey 08544

The immense amount of sequence data available from expressed sequence tag (EST) databases (1,2), together with the development of technologies for the highly parallel analysis of gene expression (3–5) have created the opportunity to interrogate biochemical pathways and gene function on an unprecedented scale. We describe here a set of high-density DNA arrays containing oligonucleotides complementary to more than 6,500 human EST's. These arrays were used to generate normal and breast cancer specific gene expression profiles. More than 1,500 expressed genes were detected in both cell types examined with 85% of all gene expression observed in the range of 1–50 copies per cell. Over 300 genes demonstrated significantly different levels of expression between normal and transformed breast cells. Increased mRNA levels were observed for the Her2/neu oncogene and genes involved in its signal transduction pathway, including Grb-7, Ras, Raf, Mek and ERK. In addition, a simple categorization of the expression changes revealed patterns characteristic of loss of wild-type p53 function. Genotyping of the p53 locus using a DNA re-sequencing array revealed inactivating mutations in the p53 DNA binding domain and loss of heterozygosity. These data demonstrate a general array-hybridization-based approach to deciphering biochemical pathways and generating testable hypotheses concerning the mechanisms of cell growth and differentiation.

---

* Correspondence and requests for materials should be addressed to T.R.G. (e-mail: tom_gingeras@affymetrix.com)

The Biology of Tumors, edited by Mihich and Croce
Plenum Press, New York, 1998.

We previously reported a method that combines photolithography and solid phase chemistry to directly synthesize specified oligonucleotides on derivitized glass at high density, and demonstrated that the arrays can be used to monitor the expression levels of a large number of genes in parallel (5,6). In this study, we have used arrays with oligonucleotides complementary to more than 6,500 EST gene clusters derived from the dbEST public database (1). Among these gene sequences are approximately 3,200 full-length human cDNAs (GenBank) and 3,400 ESTs that have some similarity to other eukaryotic genes.

Each of the mRNAs being monitored is represented on the array by a set of approximately 20 probe pairs. The synthetic oligonucleotides on the arrays are referred to as 'probes' because they serve to probe or interrogate the labeled sample. Each probe pair is composed of a 25-mer oligonucleotide that is designed to be completely complementary to a region of sequence from the specific mRNA, and a partner probe that is identical except for a single base substitution in a central position. The intentionally mismatched probes serve as internal controls for hybridization specificity, and the use of pairs allows for sensitive quantitation of even weak signals in the presence of cross-hybridization and background signals (5). The use of 20 probe pairs per mRNA allows for redundancy in the detection, greatly improving the quantitation of message levels and reducing the effects of potential problems due to EST sequencing errors, polymorphisms and occasional cross-hybridization. Probes for each mRNA were selected on the basis of sequence uniqueness and hybridization properties (see Methods). The aim was to choose probes that hybridize with high specificity and sensitivity.

Labeled RNAs for initial array hybridization experiments were derived from the malignant breast cell line BT-474 and a normal primary breast culture. BT-474 was isolated from a solid, invasive ductal carcinoma of the breast and is tumorigenic in athymic nude mice (7). The primary breast cell culture was obtained from normal breast tissue peripheral to an infiltrating ductal carcinoma (8). Messenger RNA was isolated from normal and malignant cells, converted into double stranded cDNA (ds cDNA), labeled by an *in vitro* transcription (IVT) reaction and hybridized to the oligonucleotide array. Following a washing step, the fluorescence hybridization pattern generated by the complete expressed message population was imaged using an argon ion laser confocal scanner. The fluorescence intensity images were processed and quantitated by image and data analysis software.

Figure 1A (top panel) shows representative fluorescence patterns following hybridization of total mRNA from normal and malignant breast cells to arrays containing probes for approximately 1,650 gene sequences (1 of a set of 4 arrays encompassing more than 6,500 human genes). Clear examples of altered and unchanged gene expression can be observed directly by visual comparison of the patterns for the two samples (Fig. 1) where each detected gene appears as a horizontal "strip" on the array. A systematic comparison for all genes being monitored can be performed by quantitating the fluorescence hybridization data and directly comparing the signal intensities. The quantitative analysis of hybridization patterns is based on the observation that for a specific mRNA the perfect-match (PM) probes hybridize more strongly on average than their mis-matched (MM) partners (Fig. 1B and ref. 5). Consistent patterns of specific hybridization (PM intensity > MM intensity) indicate the presence of a given mRNA, and patterns failing to meet the decision criteria result in mRNAs scored as undetectable. Previous work has demonstrated that the average difference in intensity between sets of the PM and MM hybridization is quantitatively related to mRNA concentration (5) and reflects the relative copy number of a detected message within a factor of 2 (Wodicka et al., manuscript submitted).

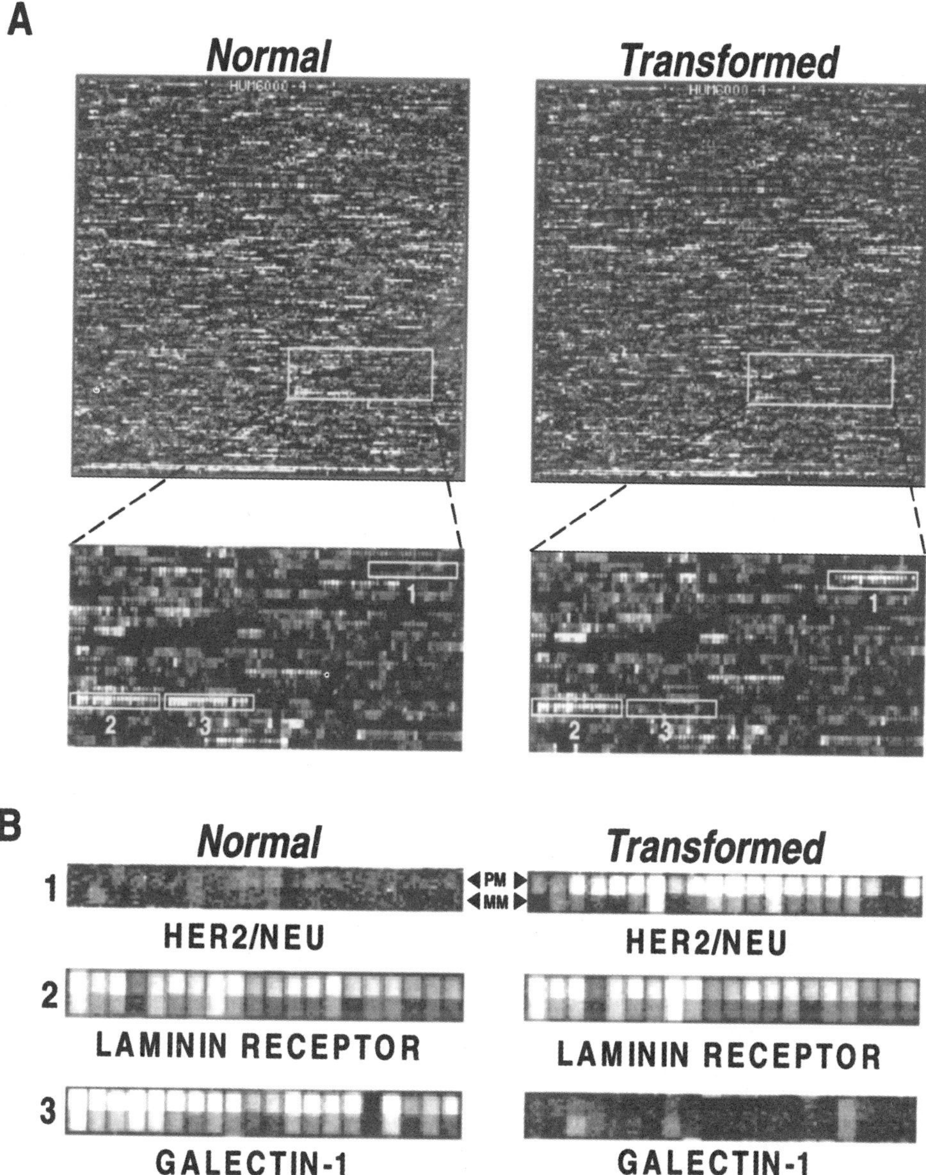

Figure 1. Fluorescence images of oligonucleotide arrays monitoring approximately 1,650 genes in parallel (1 of a set of 4 arrays covering more than 6,500 genes). **A** (top panel), representative hybridization patterns of fluorescently labeled cRNA from normal and transformed breast cells. The images were obtained after hybridization of arrays with fragmented, biotin-labeled cRNA and subsequent staining with a streptavidin-phycoerythrin conjugate (see Methods). Bright rows indicate messages present at high levels. Low level messages (1–10 copies/cell) are unambiguously detected based on quantitative analysis of hybridization intensity patterns (see text). The lower panel contains a magnified view of a portion of the array highlighting examples of altered gene expression between normal and transformed breast cells. Box 1, induced (>10-fold change in hybridization intensity); Box 2, unchanged (< 2-fold change in hybridization intensity); Box 3, repressed (> 10-fold change in hybridization intensity). **B**, Expanded view of the probes for genes 1, 2, and 3 in (A) as 20 probe pairs of perfect-matched (PM) and single base mis-matched (MM) oligonucleotide probe cells. Fluoresence intensity values observed for normal versus transformed breast cells (normalized to β-Actin and GAPDH signals): gene 1 (Her2/neu oncogene) 110 versus 5,130; gene 2 (laminin receptor) 3,500 versus 6,100; and gene 3 (galectin-1) 7,950 versus undetected.

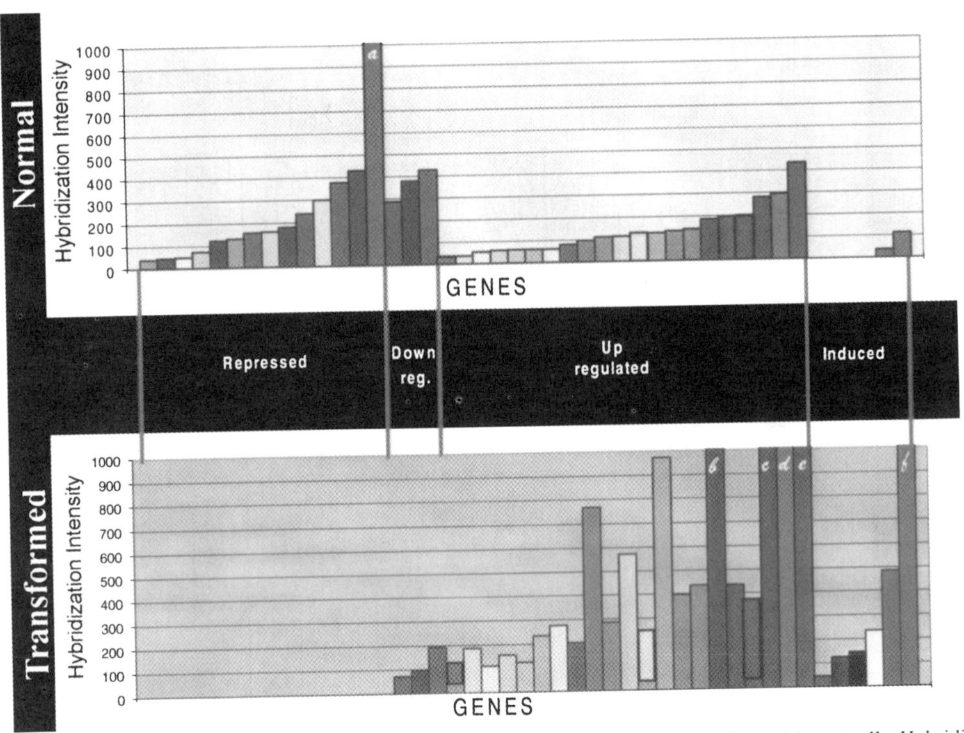

Figure 2. Expression profiles for a subset of monitored genes in normal versus transformed breast cells. Hybridization intensities (normalized to β-Actin and GAPDH signals) are plotted for the genes highlighted in Figure 1A (boxed area) that demonstrated greater than a 2-fold difference in hybridization signals between normal and transformed breast cells. Genes in the repressed and induced categories demonstrated greater than 10-fold differences in expression. Off scale hybridization intensity values are as follows: ***a***, 7,950; ***b***, 5,130; ***c***,1,390; ***d***,1,290; ***e***, 1,370; ***f***, 2,070.

Biotinylated control cRNAs (*E. coli* biotin synthetase genes *bioB, bioC, bioD* and bacteriophage P1 *Cre* recombinase), were added at known concentrations into mRNA samples, to serve as internal quantitation standards. Endogenous mRNAs (e.g., β-Actin and glyceraldehyde 3-phosphate dehydrogenase (GAPDH)) were used to normalize overall hybridization intensities between experiments on different arrays. Previous experiments have shown that mRNAs present at relative frequencies of 1:300,000, corresponding to a few copies per cell, are readily detected (5). A comparison of the hybridization intensities revealed expression changes ranging from relatively subtle (2-fold) increases and decreases to very large differences (more than 50-fold) between normal and malignant cells (see Fig. 2). To investigate the reproducibility of the system three independent preparations of labeled sample were generated from normal and transformed cell mRNA and examined on the DNA arrays. In these experiments, of the over 1,500 genes detected only 2% demonstrated between a 2- to 3-fold difference in hybridization intensity, with none of the genes detected varying more than 3-fold on the arrays from one preparation to the other (normalized to endogenous control genes β-Actin and GAPDH).

Genes that are induced or repressed provide good starting points for the investigation of molecular pathways involved in tumorigenesis. We choose to concentrate on those genes that demonstrated a 5-fold or greater difference in mRNA levels between normal and transformed breast cells, as well as those that were detected uniquely in one of these two cell types. In this study, we identified 305 genes whose mRNA levels had been either induced/increased (137 genes) or repressed/decreased (168 genes) (Table 1). Of these differentially expressed mRNAs, 183 corresponded to human full-length cDNAs listed in GenBank and 122 to EST entries only. Consistent with these numbers was the observation that of the over 1,500 genes detected in both cell types examined, slightly more than half (59%) corresponded to full length GenBank entries.

One of the genes showing the largest changes in expression level was the Her2/neu oncogene (also known as c-erbB-2) which increased by approximately 50-fold in BT-474 versus normal breast cells (hybridization intensities of 5,130 versus 110, see Fig. 1B). Her2/neu belongs to the epidermal growth factor receptor family of receptor tyrosine kinases (RTKs) (9), and has been previously shown to be amplified in BT-474 tumor cells (10). The oncogenic activation of RTKs is commonly achieved by overexpression, resulting in the ability to dimerize in the absence of ligand (11). Overexpression of Her2/neu is observed in 20–30% of all human breast cancers (12,13) and ovarian cancers (14). Elevated expression of related RTK family member Her3 (c-erbB-3) has also been implicated in the development and progression of human malignancies (15) including breast cancer (16). Consistent with these observations, significant levels of Her3 message were also detected in BT-474 (undetectable in normal breast cells, see Table 1). Taken together, these data implicate an involvement of RTKs and RTK heterodimerization (11,17) in the development of this breast carcinoma. Underscoring the involvement of this signaling pathway is the observation that GRB-7 is also upregulated (Table 1). GRB-7 is an SH2 domain protein and component of RTK signal transduction pathways that is linked to the Her2/neu locus and is found tightly bound to the Her2/neu protein (18).

Of the approximately 1,500 genes with detectable expression levels, prominent down-regulation was observed for approximately 11% of these genes (168 out of 1,500), including numerous cell adhesion and extracellular matrix molecules as well as a dramatic reduction of message levels for caveolin-1 and caveolin-2 genes (Table 1). Caveolins are integral membrane proteins and principal components of caveolae (non-clathrin-coated invaginations of the plasma membrane (19)). Recent evidence suggests that caveolae are involved in G protein-coupled receptor (GPCR) signaling events, and specifically, that caveolins interact directly with multiple G protein α subunits to negatively regulate the activation state of heterotrimeric G proteins (19,20). This proposed function is consistent with observations of reduced levels of caveolins in oncogenically transformed cells (21) as well as the results presented here.

Ras proteins have been established as critical intermediates between upstream RTKs (22,23) and GPCRs (24,25), and downstream signaling components involved in cellular transformation (including mitogen activated protein kinases (MAPK)). Although Ras mutations are seen in less than 5% of breast cancers, a large body of evidence implicates deregulation of the Ras pathway in breast carcinomas (26 and refs. therein). The concurrent up-regulation of RTKs and down-regulation of caveolins in BT-474 strongly implicate a convergence of multiple upstream mitogenic signaling events on the Ras pathway. Interestingly, our analysis also revealed up-regulation of Ras, Raf, Mek and ERK (Table 1) which together highlight a deregulated Ras/MAPK pathway (27,28). Collectively, these results begin to elucidate a network of gene expression changes responsible for uncontrolled cell-cycle proliferation.

Further insights into the genetic defects of the BT-474 carcinoma cells came from the expression patterns of genes transcriptionally regulated by p53. p53 is the most commonly mutated gene associated with neoplasia, and mutations are found in over 50% of all human cancers. The p53 gene product is a nuclear phosphoprotein that functions in cell-

Table 1

Accession		Functional category
	Genes transcriptionally induced in BT-474 vs HT-125	
Nominal copy number: 500–1,000 copies/cell		
X00474	PS2 protein precursor	other
M11730	Her2/neu tyrosine kinase receptor	signal transduction
R10066	cDNA clone 128808 similar to Prohibitin	other
Nominal copy number: 100–500 copies/cell		
J05068	Transcobalamin I	other
M22382	Mitochondrial matrix protein P1 precursor	mitochondrial
T54303	cDNA clone 69022 similar to Keratin, type II cytoskeletal 8	structural/cytoskeletal
L33930	CD24	signal transduction
T51499	cDNA clone 71494 similar to Succinate dehydrogenase	metabolic enz.
T70062	cDNA clone 80945 similar to NF45	transcription factor
X70940	Elongation factor 1 alpha-2	translational mach.
M88279	Human immunophilin	other
D45370	Adipose specific protein	other
H52207	cDNA clone 209710 similar to Matrix GLA-protein precursor	other
L08044	Intestinal trefoil factor	secr. gr. fact.
M31627	X Box binding protein-1	transcription factor
U37519	Aldehyde dehydrogenase	metabolic enz.
U39840	Hepatocyte nuclear factor-3 alpha	transcription factor
M26481	KS1/4 antigen	other
H09351	cDNA clone 46019 similar to MCM3 homolog	DNA replication
T51961	Proliferating cell nuclear antigen	DNA replication
T59162	cDNA clone 74635 similar to Selenium-binding protein	other
Nominal copy number: 50–100 copies/cell		
T67986	cDNA clone 82030 similar to Clusterin precursor	cell ad./cell-cell comm.
U21931	Fructose-1,6-biphosphatase	metabolic enz.
Z13009	E-cadherin	cell ad./cell-cell comm.
M18216	Nonspecific crossreacting antigen	other
U31875	Hep27 protein	other
L02547	Cleavage stim.factor	translational mach.
R89566	cDNA clone 195338 similar to Lamin B receptor	other
T98835	cDNA clone 122341 similar to 80.7 KD alpha trans-inducing protein	other
R00451	cDNA clone 123283 similar to Clathrin coat assembly protein AP47	other
M34309	Human epidermal growth factor receptor, HER3	signal transduction
X73424	Propionyl-COA carboxylase beta chain	metabolic enz.
X89985	BCL7B	other
H00645	cDNA clone 149859 similar to Ribonuclease inhibitor	other
H41017	cDNA clone 192957 similar to Creatine kinase, ubiquitous mitochondiral precursor	mitochondrial
R85558	cDNA clone 180221 similar to Inorganic pyrophosphatase	metabolic enz.
Nominal copy number: 10–50 copies/cell		
H41006	cDNA clone 177323 similar to Quinol oxidase polypeptide I	other
X51801	OP-1, osteogenic protein	secr. gr. fact.
M69238	Aryl hydrocarbon receptor nuclear translocator	transcription factor
X69392	Ribosomal protein L26.	translational mach.
R88418	cDNA clone 166353 similar to Cleavage stimulation factor, 50 KD subunit	translational mach.
Y00815	Leukocyte antigen related protein	cell ad./cell-cell comm.
H41406	cDNA clone 192943 similar to mouse Gamma-adaptin	protein process.
X92814	HREV107-like protein.	other
X68194	h-Sp1	other
R42070	cDNA clone 30646 similar to Verprolin	structural/cytoskeletal

Table 1. (*Continued*)

Accession		Functional category
H19467	cDNA clone 172249 similar to Ankyrin R	other
R10620	cDNA clone 128901 similar to Tyrosine-protein kinase CSK	signal transduction
L29254	L-iditol-2 dehydrogenase gene	metabolic enz.
M34458	Lamin B1	other
R53942	cDNA clone 40026 similar to ADP,ATP carrier protein, heart/skeletal muscle isoform	other
T96666	cDNA clone 121357 similar to Cyclin-dependent kinase interactor 1	cell-cycle
R44863	cDNA clone 33477 similar to Phenylalkylamine binding protein	other
D43772	GRB-7	signal transduction
X67334	BM28	DNA replication
T78606	cDNA clone 113492 similar to Translational activator GCN1	translational mach.
D42047	KIAA0089 Human mRNA gene for Mouse glycerophosphate dehydrogenase-related	metabolic enz.
X58072	hGATA3	transcription factor
M34344	Platelet glycoprotein IIb	cell ad./cell-cell comm.
R46731	cDNA clone 36517 similar to Transcriptional regulatory protein RPD3	transcriptional mach.
U33819	Zinc-finger DNA binding protein	transcription factor
R23246	cDNA clone 131094 similar to Goliath protein	other
H64489	cDNA clone 238846 similar to Leukocyte antigen CD3	cell ad./cell-cell comm.
T47377	cDNA clone 571035 similar to S-100P protein	other
H05814	cDNA clone 44151 similar to Putative ATP-dependent RNA helicase	translational mach.
X83618	3-hydroxy-3-methylglutaryl coenzyme A synthase	metabolic enz.
U14550	Sialyltransferase SThM	metabolic enz.
R55185	cDNA clone 154654 similar to Galactoside 3(4)-L-fucosyltransferase	metabolic enz.
V00568	C-myc	transcription factor
L47738	Inducible protein	other
M20560	Lipocortin-III	other
T47964	cDNA clone 71549 similar to Purine nucleoside phosphorylase	metabolic enz.
T90192	cDNA clone 110564 similar to DNA-binding protein MEL-18	transcription factor
U20285	GPS1	signal transduction
M81637	Grancalcin	other
H19201	cDNA clone 508873 similar to Guanine nucleotide diss. stim.	signal transduction
X03635	Oestrogen receptor	transcription factor
X52426	Cytokeratin 13	structural/cytoskeletal
R37799	cDNA clone 270833 similar to Protein phosphatase PP2A	signal transduction
X03484	Raf oncogene	signal transduction
L11284	ERK activator kinase, MEK1	signal transduction
X70683	SOX-4	transcription factor
R72644	cDNA clone 156489 3' similar to Choline kinase	other
M97370	Adenosine receptor (A2)	signal transduction
X76180	Lung amiloride sensitive Na+ channel protein	cell ad./cell-cell comm.
R45583	cDNA clone 35271 similar to NG-CAM related cell adhesion molecule	cell ad./cell-cell comm.
M13665	C-myb	transcription factor
D26018	Human mRNA (KIAA0039) for ORF	other
H54752	cDNA clone 203275 similar to Activator 1 37 KD subunit	other
H04239	cDNA clone 151776 similar to Developmental protein seven in absentia	signal transduction
D21852	Human mRNA (KIAA0029) for ORF	other
M57730	Immediate early response protein B61 precursor	secr. gr. fact.
M16768	Human T-cell receptor gamma chain VJCI-CII-CIII region	other
X13956	12S RNA induced by poly(rI), poly(rC) and Newcastle disease virus	other
U07349	Human B lymphocyte serine/threonine protein kinase	signal transduction
D00726	Ferrochelatase	metabolic enz.
M54968	K-ras oncogene	signal transduction
D31661	ERK tyrosine kinase	signal transduction

(*continued*)

Table 1. *(Continued)*

Accession		Functional category
U21108	Human dual specific protein phosphatase	signal transduction
X52228	Secreted epithelial tumour mucin antigen	other
X13425	Pancreatic carcinoma marker GA733–1	other
R52511	cDNA clone 39763 similar to Gamma-interferon-inducible protein IP-30 precursor	other
X83425	Lutheran blood group glycoprotein.	cell ad./cell-cell comm.
U13045	Human nuclear respiratory factor-2 subunit beta 1	transcription factor
M20132	Androgen receptor	transcription factor
L07615	Neuropeptide Y receptor Y1	signal transduction
M86699	Human kinase, TTK	signal transduction
R43936	cDNA clone 32974 similar to Serine/threonine-protein kinase STE20	signal transduction
X60787	Transcription factor ILF	transcription factor
X74570	Gal-beta(1–3/1–4)GlcNAc alpha-2.3-sialyltransferase	metabolic enz.
U30255	Pphosphogluconate dehydrogenase	metabolic enz.
D11466	For Pig N-acetylglucosaminyl-phosphatidylinositol	other
M64347	Novel growth factor receptor	signal transduction
R43064	cDNA clone 32719 similar to E2 protein	transcription/replication
M60724	p70 ribosomal S6 kinase alpha-1	translational mach.
U34074	A kinase anchor protein S-AKAP84	mitochondrial
H87205	cDNA clone 220505 similar to cell division protein kinase 5	cell-cycle
X57348	cDNA clone 9112 similar to H.sapiens mRNA	other
X07696	Keratin, type I cytoskeletal 15	structural/cytoskeletal
X13482	U2 small nuclear ribonucleoprotein A'	translational mach.
Nominal copy number: 1–10 copies/cell		
D63879	Human mRNA for ORF (KIAA0196)	other
M90656	Gamma-glutamylcysteine synthetase (GCS)	metabolic enz.
H13822	cDNA clone 148291 similar to Zinc finger autosomal protein	other
X83928	Transcription factor TFIID subunit TAFII28	transcriptional mach.
H52655	cDNA clone 235891 similar to DNA polymerase beta	DNA replication
R42095	Tyrosine kinase SEK receptor precursor	signal transduction
D13633	Human mRNA for ORF (KIAA008)	other
U13616	Human ankyrin G	other
M74558	SIL mRNA	other
X63468	Transcription factor TFIIE alpha	transcriptional mach.
M30704	Amphiregulin	other
U11700	Copper transporting ATPase	metabolic enz.
H78807	cDNA clone 230127 similar to D-beta-hydroxybutyrate dehydrogenase precursor	metabolic enz.
M58411	p80-coilin	translational mach.
U25975	Serine kinase, hPAK65	signal transduction
R42898	cDNA clone 32410 similar to JNK activating kinase 1	signal transduction
J03258	Vitamin D3 receptor	signal transduction
	Genes transcriptionally repressed in BT-474 vs HT-125	
Nominal copy number: 500–1,000 coopies/cell		
H41129	cDNA clone 175539 similar to Galectin-1	cell ad./cell-cell comm.
T51261	cDNA clone 70133 similar to Glia derived nexin precursor	protein process.
Nominal copy number: 100–500 copies/cell		
R91912	Placental calcium-binding protein	other
M77349	Transforming growth factor-beta induced gene product	other
M35878	Clone HL 1006d similar to IGFBP3	other
T92451	cDNA clone 118219 similar to Tropomyosin	structural/cytoskeletal
M95787	Smooth muscle protein 22-alpha	structural/cytoskeletal
J03210	Collagenase type IV mRNA, 3' end.	ECM

Table 1. (*Continued*)

Accession		Functional category
L16895	Human lysyl oxidase	cell ad./cell-cell comm.
X57351	Interferon-inducible protein 1–8D	other
T99774	cDNA clone 122884 similar to Follistatin precursor	cell ad./cell-cell comm.
J03464	Clone pHcol2A1 similar to Procollagen alpha 2(I)chain precursor	ECM
T61333	cDNA clone 78034 similar to Metalloproteinase inhibitor 3 precursor	protein process.
X78947	Connective tissue growth factor	secr. gr. fact.
R49855	cDNA clone 152637 similar to Coagulation factor V precursor	other
T54767	cDNA clone 73802 similar to Sparc precursor	cell ad./cell-cell comm.
L01131	Human decorin	cell ad./cell-cell comm.
L16510	Cathepsin B	protein process.
H87343	cDNA clone 252394 similar to Protein-lysine 6-oxidase precursor	protein process.
M69066	Moesin	structural/cytoskeletal
X91911	RTVP-1	other
T95046	cDNA clone 120085 similar to Protease do precursor	protein process.
X15882	Collagen VI alpha-2 C-terminal globular domain.	structural/cytoskeletal
M69054	IGFBP6	other
X55740	Placental cDNA coding for 5' nucleotidase	metabolic enz.
H80342	cDNA clone 241122 similar to Tubulin beta chain	structural/cytoskeletal
U21090	DNA polymerase delta	DNA replication
Z18951	Caveolin-1	signal transduction
Nominal copy number: 50–100 copies/cell		
U39050	Human mitogen-responsive phosphoprotein	other
H62466	cDNA clone 209654 similar to Collagen alpha 3(VI) chain	structural/cytoskeletal
X14787	Thrombospondin	cell ad./cell-cell comm.
H45474	cDNA clone 183508 similar to Peripheral myelin protein 22	signal transduction
H29761	cDNA clone 186308 similar to Annexin I	cell ad./cell-cell comm.
M11718	Alpha-2 type V collagen	structural/cytoskeletal
X12369	Tropomyosin alpha chain, smooth muscle	structural/cytoskeletal
T54317	cDNA clone 69040 similar to RAS-related protein	signal transduction
U20280	Cathepsin X	protein process.
H23235	cDNA clone 52096 similar to APDGFR precursor	signal transduction
R67072	cDNA clone 140763 similar to GAP junction alpha-1 protein	cell ad./cell-cell comm.
R72104	cDNA clone 155771 similar to Bone morphogenetic protein 1 precursor	protein process.
U21128	Lumican	ECM
M23419	Initiation factor 5A	translational mach.
L34056	Cadherin-11	signal transduction
M64110	Caldesmon	structural/cytoskeletal
X72012	Endoglin	signal transduction
R70383	cDNA clone 155241 similar to Vitellogenin II precursor	other
T69425	cDNA clone 82963 similar to Alpha-2 macroglobulin precursor	other
H02540	cDNA clone 151270 similar to Cathepsin L precursor	protein process.
X05232	Stromelysin	protein process.
T54278	cDNA clone 69199 similar to Tissue inhibitor of metalloproteinases II precursor	protein process.
M24486	Prolyl 4-hydroxylase alpha subunit	ECM
H80975	cDNA clone 240954 similar to Plasma protease C1 inhibitor precursor	protein process.
H06706	cDNA clone 44616 similar to RAS-related protein ORAB-1	signal transduction
M33210	Colony stimulating factor 1 receptor	signal transduction
T96364	cDNA clone 121066 similar to M-phase inducer phosphatase 3	metabolic enz.
Nominal copy number: 10–50 copies/cell		
J02814	Chondroitin sulfate proteoglycan core protein	cell ad./cell-cell comm.
D14874	Adrenomedullin	other
L35594	Autotaxin	cell ad./cell-cell comm.

(*continued*)

Table 1. *(Continued)*

Accession		Functional category
R23249	cDNA clone 131100 similar to Preprotein translocase secy subunit	other
L13923	Fibrillin	ECM
T91563	cDNA clone 118372 similar to CD44 antigen, epithelial form precursor	signal transduction
J05017	Aldose reductase	other
M55210	(LAMB 2) gene Laminin B2 chain	ECM
M13755	Interferon-induced 17 KD protein	other
U32114	Caveolin-2	signal transduction
X17097	PSG9	other
X52679	Sphingomyelin phosphodiesterase	protein process.
Z48481	Membrane-type matrix metalloproteinase 1	protein process.
U24389	Lysyl oxidase-like protein	cell ad./cell-cell comm.
M85289	Heparan sulfate proteoglycan	cell ad./cell-cell comm.
D29992	Placental protein 5	other
Y00711	L-lactate dehydrogenase H chain	other
R50499	Fibrinogen beta chain precursor	cell ad./cell-cell comm.
U09278	Fibroblast activation protein	signal transduction
H15811	cDNA clone 159533 similar to Platelet membrane glycoprotein IIB precursor	cell ad./cell-cell comm.
X82494	Fibulin-2	ECM
H37925	cDNA clone 190993 similar to Interferon-activatable protein 204	other
J02685	Plasminogen activator inhibitor-2	protein process.
U03106	Wild-type p53 activated fragment-1, p21$^{WAF1/CIP1}$	cell-cycle
D13665	Osteoblast specific factor 2	cell ad./cell-cell comm.
T55741	cDNA clone 73645 similar to Myosin light chain kinase	cell ad./cell-cell comm.
U18467	Pregnancy-specific beta 1-glycoprotein 7	other
X06700	Pro-alpha1(III) collagen	ECM
H56627	cDNA clone 231424 similar to Glutathione S-transferase P	other
T41199	cDNA clone 62328 similar to Laminin gamma-1chain precursor	ECM
T59895	cDNA clone 76290 similar to Ubiquitin carboxyl-terjminal hydrolase isozyme L1	other
T68426	CD81 antigen	cell ad./cell-cell comm.
X13810	OTF-2	transcription factor
T50086	cDNA clone 70213 similar to Alpha-1-antiproteinase F precursor	protein process.
D28137	Human mRNA for BST-2	cell ad./cell-cell comm.
Z37976	Latent TGF-beta binding protein	other
X02492	Interferon-induced protein 6–16 precursor	other
T66960	cDNA clone 66508 3' similar to Nuclear factor 1 clone PNF1/X	other
H88938	Keratinocyte growth factor precursor	secr. gr. fact.
M29150	Interleukin 6	secr. gr. fact.
X05231	Collagenase	protein process.
X68148	p66shc	signal transduction
D31762	KIAA0057 for ORF	cell-cycle
T61090	cDNA clone 83635 similar to Endoglin precursor	signal transduction
T53889	cDNA clone 78017 similar to Complement C1R component precursor	other
X58288	hR-PTPu gene for protein tyrosine phosphatase	signal transduction
T51913	cDNA clone 72466 similar to Alpha crystallin B chain	other
X17042	Hematopoetic proteoglycan core protein	cell ad./cell-cell comm.
M55621	N-acetylglucosaminyltransferase I	protein process.
H85111	cDNA clone 220183 similar to EBNA-2 nuclear protein	transcription factor
X63613	Pentaxin	other
M24594	Interferon-induced 56 KD protein	other
R36467	cDNA clone 136821 similar to Transforming growth factor beta 1 precursor	secr. gr. fact.
M87503	Transcriptional regulator ISGF3 gamma subunit	transcription factor
X13097	Tissue type plasminogen activator	cell ad./cell-cell comm.

Table 1. (*Continued*)

Accession		Functional category
X60201	Brain-derived neurotrophic factor	other
M63509	Human glutathione transferase M2	other
M92642	Alpha-1 type XVI collagen	ECM
M80469	MHC class I HLA-J	other
T60855	cDNA clone 76721 similar to Calcineurin B subunit isoform 1	signal transduction
X16707	Fra-1	transcription factor
H54425	cDNA 203126 similar to Metallothionein-II	other
L04733	Kinesin light chain	structural/cytoskeletal
R46493	cDNA clone 36200 similar to Tissue factor precursor	ECM
M29696	Interleukin-7 receptor alpha chain precursor	signal transduction
T97948	cDNA clone 121916 similar to Calponin H2, smooth muscle	structural/cytoskeletal
R43953	cDNA clone 33191 similar to Opioid binding protein	cell ad./cell-cell comm.
R16659	cDNA clone 129622 similar to Tissue factor pathway inhibitor	cell ad./cell-cell comm.
U37019	Smooth muscle cell calponin	structural/cytoskeletal
M96322	Gravin	cell ad./cell-cell comm.
T78624	cDNA clone 113518 similar to Carbonyl reductase	other
M60974	GADD45	cell-cycle
X60708	pcHDP7	protein process.
M62424	Thrombin receptor precursor	signal transduction
R94513	cDNA clone 197667 similar to STAT2	signal transduction
V00523	HLA-DR alpha	other
H65605	cDNA clone 209780 similar to Prothrombin precursor	signal transduction
X76488	Lysosomal acid lipase	protein process.
R12676	cDNA clone 129303 similar to Complement factor H-like protein 1 precursor	other
X06985	Heme oxygenase	metabolic enz.
R60908	cDNA clone 42188 similar to protein-tyrosine kinase	signal transduction
U25779	Human chromosome 17q21 mRNA clone 694:2	other
X12496	Erythrocyte membrane sialoglycoprotein beta	other
T71207	cDNA clone 110162 similar to RAS-related C3 Botulinum toxin substrate 2	signal transduction
R52961	cDNA clone 40303 similar to Shc transforming protein 46.8 KD and 51.7 KD precursor	signal transduction
M97676	Homeobox protein	transcription factor
R95874	cDNA clone 199264 similar to LTR5 Retrovirus-related ENV polyprotein	other
T86469	cDNA 114641 similar to Collagen alpha 3(VI) chain	ECM
R48243	cDNA clone 153769 similar to RAS-related protein RHA1	signal transduction
U19969	Two-handed zinc finger protein ZEB	transcription factor
Nominal copy number: 1–10 copies/cell		
R99935	cDNA clone 201757 similar to DNA-directed RNA pol I largest subunit	transcription mach.
T64142	cDNA clone 80145 similar to Proteolipid protein PPA1	protein process.
H74007	cDNA clone 232946 similar to Serine/threonine-protein kinase SGK	signal transduction
H47650	PTS system, sucrose-spec. IIABC component	other
J04080	Complement component C1r	other
D90226	OSF-1	secr. gr. fact.
X03348	Human mRNA for beta-glucocorticoid receptor	signal transduction
R32804	cDNA clone 135146 similar to Glucose transporter type 3	other
M60618	Nuclear autoantigen SP-100	other
L13463	Human helix-loop-helix basic phosphoprotein	other
M26683	Human interferon gamma treatment inducible mRNA.	other
M63838	Interferon-gamma induced protein	other
X73608	Testican	cell ad./cell-cell comm.
M88338	Serum protein MSE55	other

(*continued*)

Table 1. (*Continued*)

Accession		Functional category
T52698	cDNA clone 67233 similar to Protein-tyrosine phosphatase 1B	protein process.
U16307	Glioma pathogenesis-related protein	other
R35976	cDNA clone 137079similar to CD30 ligand	signal transduction
X16665	HOX2H	transcription factor
R42244	cDNA clone 30543 similar to Antigen peptide transporter 1	protein process.
U19718	Human microfibril-associated glycoprotein	ECM
T85165	cDNA clone 111345 similar to SRC substrate P80/85 proteins	signal transduction
R45687	Cyclin G	cell-cycle
L22473	Bax	cell-cycle

Message quantities are expressed as nominal copy number per cell relative to hybridization intensities of known copy number spiked controls.
Abbreviations: secr. gr. fact., secreted growth factor; ad., adhesion; comm., communication; mach., machinery; enz., enzyme; process., processing; ECM, extracellular matrix.

cycle regulation and the preservation of genetic integrity (reviewed in ref. 29). The p53 protein possesses numerous biochemical properties necessary to carry out these functions, including sequence-specific DNA binding activity, transcriptional activation and transcriptional repression. The observation that p53 mutations found in human cancers overwhelmingly result in p53 gene products that have lost the ability to bind DNA and transcriptionally regulate target genes, strongly suggests that these properties are crucial for the ability of p53 to regulate cell proliferation and apoptosis.

Wild-type p53 protein transcriptionally activates a number of genes that are linked with its tumor-suppressor activity and these genes are responsible in part for p53-dependent functions in the cell (29). Messenger RNA for many of these target genes was detectable in normal breast cells but undetectable or dramatically down regulated in the BT-474 carcinoma cells (Table 2). In addition, significant up-regulation of mRNAs that are normally transcriptionally repressed by wild-type p53 activity (30) was observed in BT-474

Table 2. Altered expression of genes transcriptionally regulated by p53

		Normal breast	Transformed breast (BT-474)
Transcriptionally activated by p53			
Bax	Inducer of apoptosis (bcl-2 associated protein)	40 ± 5	undetected
Cyclin G	Cell-cycle component	50 ± 5	undetected
GADD45	Growth arrest and DNA damage inducible gene	300 ± 15	25 ± 5
IGF-BP3	Insulin growth factor pathway inhibitor	4,000 ± 400	undetected
$p21^{WAF1/CIP1}$	Cyclin-CDK and DNA replication inhibitor	350 ± 30	undetected
Thrombospondin	Inhibitor of angiogenesis	800 ± 45	undetected
Transcriptionally repressed by p53			
c-myc	Cellular oncoprotein	30 ± 5	350 ± 25
PCNA	DNA polymerase processivity factor	90 ± 10	1,000 ± 70

Gene expression levels are given as the average fluorescence intensity (arbitrary units, rounded to nearest significant figure) after normalization of hybridization intensities to β-Actin and GAPDH signals.
30-134 intensity value = 1-10 copies/cell; 135-499 intensity value = 10-50 copies/cell; 500-949 intensity value = 50-100 copies/cell; 950-4,499 intensity value = 100-500 copies/cell; 4,500-9,000 intensity value = 500-1,000 copies/cell or more.
Intensity values and uncertainty estimates are based on data from a minimum of 3 independent experiments.
Undetected indicates that the hybridization pattern was not sufficiently consistent and strong to assign the mRNA as clearly present.

cells (see Table 2). Taken together, these expression results suggested a loss of wild-type p53 function.

To investigate the genetic basis of the aberrant p53 transcriptional regulation, we used a high-density DNA array to scan the p53 gene from the BT-474 cells for mutations. The strategy for rapid, simultaneous analysis of large amounts of genetic information using high-density oligonucleotide arrays has been described (31). The DNA array used for genotyping p53 allowed for simultaneous analysis of both sense and anti-sense sequence of p53 coding exons 2–11, including 10 base pairs of intronic flanking sequence (to identify splice donor-acceptor mutations). In addition, the array contained allele-specific probes for over 300 characterized p53 mutations and every possible single base deletion (Dee et al., manuscript in prep.) The re-sequencing portion of the DNA array consisted of sets of four 18-mer oligonucleotides, identical except at a central position which is either A, C, G or T. In each set of four probes, the perfect complement to the sample sequence is expected to hybridize more strongly than the associated probes that contain single base mismatches (see Fig. 3A). The hybridization patterns are analyzed using specially designed software that produces an assignment of nucleotide identity at each position.

Substitution mutations in the p53 gene are identified by the sequence analysis software based on two major effects that a single base change has on the array hybridization: 1) the probe containing the substitution base displays the strongest signal of the four in the probe set; and 2) the neighboring probes that overlap the position display a characteristic loss of signal or "footprint", as these probes have a single base mismatch to the mutated target sequence distinct from the query base (see Fig 3A, and ref. 31).

Sequence analysis of p53 in BT-474 cells revealed a single base substitution of G to A in exon 8 (DNA binding domain), resulting in an amino acid change at position 285 from glutamic acid to lysine (Fig. 3B). The analysis identified a homozygous mutant p53 genotype in BT-474 cells (see Fig. 3B, bottom panel) indicating that this carcinoma had undergone a loss of heterozygosity at the p53 locus (confirmed by dideoxy sequence analysis). These data strongly suggest that the loss of wild-type p53 regulatory function is due to the absence of wild-type p53 protein. We have applied this analysis to other breast carcinoma cell lines and have observed a similar correlation between p53 genotype and patterns of expression for genes transcriptionally regulated by p53 (Table 3).

In this study, oligonucleotide arrays designed and synthesized based on EST and GenBank sequence information were used to explore the changes in gene expression between normal and transformed breast cells. The expression analysis identified more than 300 differentially expressed genes and highlighted altered molecular pathways involved in cell proliferation and apoptosis. Moreover, the pattern of expression changes observed suggested loss of wild-type p53 activity which was confirmed by identification of mutant genomic p53 sequence. The combination of gene expression monitoring of p53 transcriptional targets and array-based p53 re-sequencing may provide a method to characterize the cellular consequences of different p53 mutations as well as mutations in other genes that are involved in cancer.

Recently, Zhang et al. (32) reported relatively small gene expression differences (1.5% changes) between normal and neoplastic colon tissue using the SAGE technique (4). Applying the same criteria for defining expression differences to our data set revealed 3.0% of total genes detected displaying altered expression patterns. This analysis criteria does not allow for scoring 5- to 10-fold changes in expression for genes present at low copy number per cell where we observed the vast majority of expression differences occurring between normal and transformed tissues. We are currently in the process of applying expression and re-sequencing arrays to generate molecular profiles of tumor

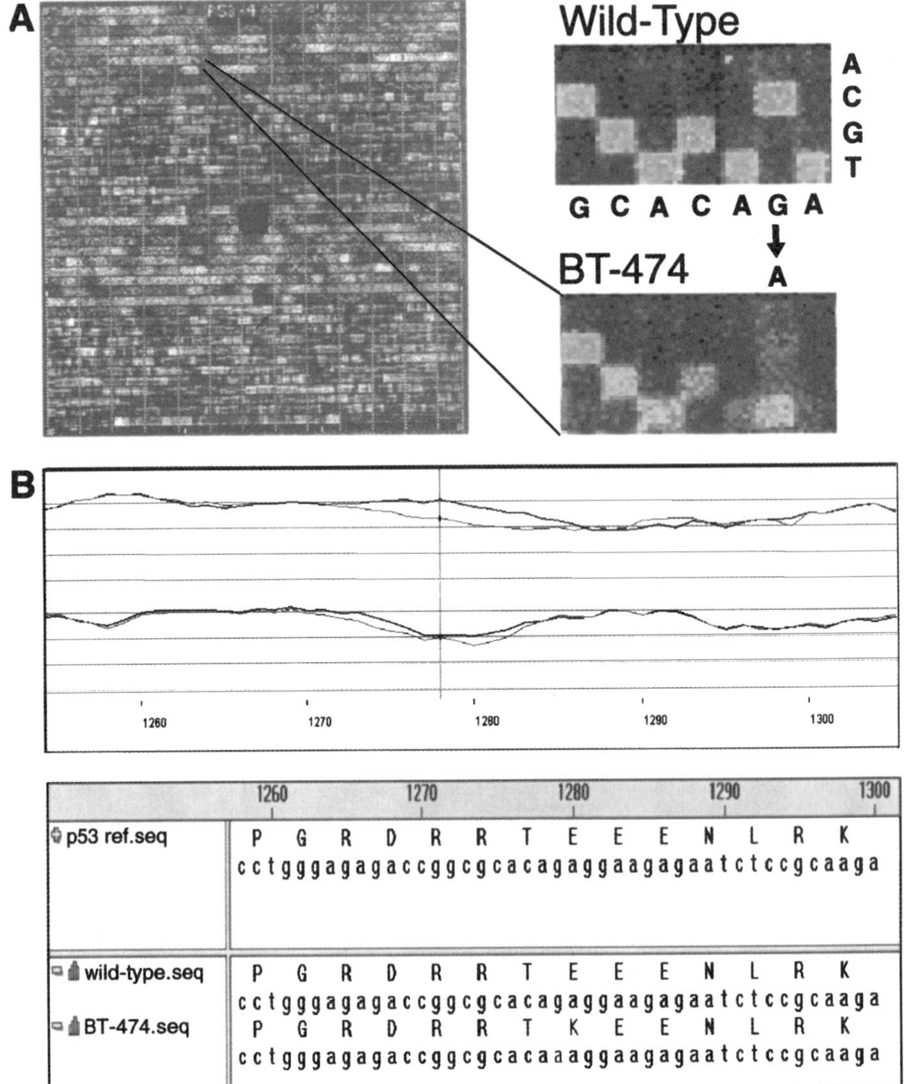

Figure 3. p53 sequencing and mutation detection by hybridization. **A**, an image of the p53 genotyping array hybridized to 1,490 bp of the BT-474 breast carcinoma p53 gene (left). Expanded view of hybridization patterns for the p53 wild-type reference and BT-474 DNA in a region of a G→A single-base mutation in BT-474 (right). In each column are 4 identical probes except for an A, C, G or T substituted at a central position. The hybridized target sequence can be read from left to right as the complement of the substitution base with the brightest signal. The G→A transition seen in BT-474 is accompanied by a loss of signal at flanking positions as these neighboring probes have a single-base mismatch to the target distinct from the query position. B (top), comparison of the wild-type reference (black) and BT-474 (red) p53 gene hybridization intensity patterns from sense (above) and antisense strands (below) in the region containing a mutation. The area shown demonstrates the "footprint" and detection of a single-base difference between the samples (vertical green line). GeneChip data analysis output is shown (bottom) that unambiguously identifies a G→A base change at nucleotide 1,279 of p53 in BT-474 resulting in a glutamic acid to lysine amino acid change in exon 8 (DNA binding domain). Upper portion of the analysis output displays the p53 wild-type reference sequence, and below are aligned outputs for wild-type p53 control and BT-474 samples.

Table 3. Altered p53-mediated gene expression and genotypic status

		Normal Breast	Transformed Breast	
			MDA468[a]	MDA231[b]
Transcriptionally activated by p53				
Bax	Inducer of apoptosis (bcl-2 associated protein)	40 ± 5	undetected	undetected
Cyclin G	Cell-cycle component	50 ±5	undetected	undetected
GADD45	Growth arrest and DNA damage inducible gene	300 ± 15	30 ± 5	undetected
IGF-BP3	Insulin growth factor pathway inhibitor	4,000 ± 400	undetected	undetected
p21$^{WAF1/CIP1}$	Cyclin-CDK and DNA replication inhibitor	350 ± 30	undetected	undetected
Thrombospondin	Inhibitor of angiogenesis	800 ± 45	undetected	undetected
Transcriptionally repressed by p53				
c-myc	Cellular oncoprotein	30 ± 5	350 ± 20	500 ± 20
PCNA	DNA polymerase processivity factor	90 ± 10	1,000 ± 50	1,000 ± 75
p53 gene status (GeneChip)		Wild-type	R→H aa 273 (g→a)	R→K aa 280 (g→a)

[a]MDA468 and [b]MDA231 are human cell lines derived from breast carcinomas.
See legend to Table 2.

development using primary tissues to gain a better understanding of the expression changes underlying cancer initiation and development. Arrays of even higher density will soon be used to monitor the more than 50,000 human genes for which there are publicly available ESTs (more than 40,000 of which have no known biological function). Highly parallel methods such as those described here, combined with the rapid increase in available sequence information, provide an opportunity to investigate and understand the genetic, cellular and molecular bases of disease.

THESE EXPERIMENTS WERE CARRIED IN THE FOLLOWING MANNER

Gene Expression Array Oligonucleotide Probe Selection and Array Design

The probes for the human 6,600 gene arrays were selected from the 600 bases at the 3′-end of sequences chosen from the dbEST database (1) using a probe selection program as described (5). Probes for inclusion on the arrays were identified based on criteria of uniqueness and sequence composition. Uniqueness was assessed by comparing potential probes with the sequence of all genes that were considered for inclusion on the arrays. If any potential probe matched 22 out of 25 nucleotides of another sequence, it was discarded. Selection of probes for sequence composition was done using heuristic rules and a neural net derived rules developed from previous expression experiments (5).

Data shown in Tables 2 and 3 include expression results from an array designed to identify alternatively spliced forms of a smaller collection of human genes. This array surveys 250 genes from functional categories including oncogenes, tumor suppressors, DNA mismatched repair genes and genes involved in apoptosis. Probe pairs for this design were chosen such that each exon for a given gene was represented on the array. In this way, specific loss of signal from a sub-set of probes corresponding to a particular exon would indicate a splice variant form. None were detected in this study.

Cell Culture

BT-474, MDA468 and MDA231 cells were maintained in RPMI-1640 (Gibco/BRL) containing 10% Fetal Bovine Serum, 10 μg/ml bovine insulin, 2 mM glutamine, 100 units/ml Penicillin and 100 μg/ml Streptomycin. Primary normal breast cells (HTB-125, passage 6) were cultured in Modified Dulbecco's Medium (Gibco/BRL) with 10% Fetal Bovine Serum, 30 ng/ml epidermal growth factor, 10 μM non-essential amino acids, 100 units/ml Penicillin and 100 μg/ml Streptomycin. Cell lines were kept at 37°C, 5% CO_2 and split 1 to 3 at approximately 60–70% confluency.

mRNA Preparation and Labeling for Gene Expression Monitoring

Poly A^+ RNA was isolated from cells using an Oligotex Direct mRNA Kit (Qiagen) following both standard and batch protocols according to the manufacturer's instructions. 0.5–1 μg of mRNA was then converted into ds cDNA using a Superscript Choice System cDNA Synthesis Kit (Gibco/BRL) and an oligo dT primer incorporating a T7 RNA polymerase promoter sequence on its 5′-end (5). The resultant ds cDNA was purified by one step of phenol/chloroform extraction using Phase Lock Gel (5 Prime to 3 Prime) followed by EtOH precipitation. The ds cDNA product then served as target in an *in-vitro* transcription labeling reaction using T7 RNA polymerase (Ambion T7 Megascript Transcription Kit), 1.875 mM biotin-CTP and 1.875 mM biotin-UTP for a final concentration of 7.5 mM each NTP. After a 6 hour incubation at 37°C, the total labeled cRNA transcripts were purified by Chromaspin-100 columns (Clontech), followed by ProCipitate treatment (Affinity Binding) and EtOH precipitation to remove unincorporated nucleotides and protein contaminants.

Gene Expression Array Hybridization and Scanning

10 μg of biotinylated cRNA target was fragmented to an average size of 50 nucleotides in 10 μl of magnesium fragmentation buffer (40mM Tris-acetate (pH 8.1), 100mM KOAc, 30mM MgOAc) at 95°C for 35 min. The fragmented samples were brought up to a final volume of 200 μl with hybridization buffer (0.9 M NaCl, 60 mM NaH_2PO_4, 6 mM EDTA and 0.005% Triton X-100, pH 7.6 (6xSSPE-T)) containing 0.1 mg/ml Herring Sperm DNA, 50 pM biotin-labeled control oligo (5′-GTCAAGATGCTACCGTTCAG-3′) and biotinylated cRNA quantitation standards bioB (1.5 pM), bioC (5.0 pM), bioD (25 pM) and Cre (100 pM). Samples were denatured at 95°C for 10 min, chilled on ice for 5 min and equilibrated to room temperature (5 min) before being applied to the array flow cell. Arrays were hybridized at 40°C for 14–16 hr with rotation at 60 rpm, followed by 10 wash cycles (2 drain-fills/cycle) at room temperature with 6xSSPE-T in the GeneChip Fluidics Station (RELA), and a stringent wash at 40°C in 0.5X SSPE-T for 15 min with rotation (60 rpm). For staining of hybridized target cRNA, arrays were incubated with 2 μg/ml of phycoerytherin-streptavidin conjugate (Molecular Probes) in 6xSSPE-T containing 1 mg/ml of acetylated-bovine serum albumin at 40°C for 10 min. Prior to scanning, the arrays were washed at room temperature with 6xSSPE-T for 5 cycles (2 drain-fill/cycle) in the fluidics station. The hybridized stained arrays were scanned using an argon-ion laser GeneChip scanner 50 (Molecular Dynamics) with a resolution of 7.5 μm/pixel (~45 pixels/probe cell) and wavelength detection setting of 560 nm. Fluorescence images and

quantitative analysis of hybridization patterns and intensities were performed using analysis software and gene expression data analysis programs as previously described (5).

p53 PCR and Labeling for Re-Sequencing by Array Hybridization

The p53 gene was genotyped by amplifying coding exons 2–11 in a 100 μl multiplex PCR reaction using 100 ng of genomic DNA extracted from cells using a QIAmp Blood Kit (Qiagen). PCR Buffer II (Perkin-Elmer) was used at 1X along with 2.5mM MgC12, 200 μM of each dNTP and 10 units of Taq Polymerase Gold (Perkin-Elmer). The multiplex PCR was performed using 10 exon-specific primers (Table 4) with the following cycling conditions: 1 cycle at 94°C (5 min), 50 cycles of 94°C (30 sec), 60°C (30 sec) and 72°C (30 sec), followed by 1 cycle at 72°C (7 min). 45 μl of the PCR reaction was then fragmented and dephosphorylated by incubation at 25°C for 15 min with 0.25 units of Amp Grade DNAse I (Gibco/BRL) and 2.5 units of Calf Alkaline Phosphatase (Gibco/BRL), followed by heat-inactivation at 99°C for 10 min. The fragmented PCR products were then labeled in a 100 μl reaction using 10 μM flourecein-N6-ddATP (Dupont-NEN) and 25 units of terminal transferase (Boehringer Mannheim) in 200 μM K-Cacodylate, 25 mM Tris-HCl (pH 6.6), 0.25 mg/ml BSA and 2.5 mM CoC12. The labeling reaction was incubated at 37°C for 45 min and heat-inactivated at 99°C for 5 min. Primers used in multiplex PCR reaction of genomic p53 exons 2–11:

Ex 2 (+), 5′-TCATGCTGGATCCCCACTTTTCCTCTTG-3′
Ex 2 (–), 5′-TGGCCTGCCCTTCCAATGGATCCACTCA-3′
Ex 3 (+), 5′AATTCATGGGACTGACTTTCTGCTCTTGTC-3′
Ex 3 (–), 5′-TCCAGGTCCCAGCCCAACCCTTGTCC-3′
Ex 4 (+), 5′-GTCCTCTGACTGCTCTTTTCACCCATCTAC-3′
Ex 4 (–), 5′-GGGATACGGCCAGGCATTGAAGTCTC-3′
Ex 5 (+), 5′-CTTGTGCCCTGACTTTCAACTCTGTCTC-3′
Ex 5 (–), 5′-TGGGCAACCAGCCCTGTCGTCTCTCCA-3′
Ex 6 (+), 5′-CCAGGCCTCTGATTCCTCACTGATTGCTC-3′
Ex 6 (–), 5′-GCCACTGACAACCACCCTTAACCCCTC-3′
Ex 7 (+), 5′-GCCTCATCTTGGGCCTGTGTTATCTCC-3′
Ex 7 (–), 5′-GGCCAGTGTGCAGGGTGGCAAGTGGCTC-3′
Ex 8 (+), 5′-GTAGGACCTGATTTCCTTACTGCCTCTTGC-3′
Ex 8 (–), 5′-ATAACTGCACCCTTGGTCTCCTCCACCGC-3′
Ex 9 (+), 5′-CACTTTTATCACCTTTCCTTGCCTCTTTCC-3′
Ex 9 (–), 5′-AACTTTCCACTTGATAAGAGGTCCCAAGAC-3′
Ex 10 (+), 5′-ACTTACTTCTCCCCCTCCTCTGTTGCTGC-3′
Ex 10 (–), 5′-ATGGAATCCTATGGCTTTCCAACCTAGGAAG-3′
Ex 11 (+), 5′-CATCTCTCCTCCCTGCTTCTGTCTCCTAC-3′
Ex 11 (–), 5′-CTGACGCACACCTATTGCAAGCAAGGGTTC-3′

All p53 genotyping results were confirmed using a DNA autosequencer (ABI model 373A).

p53 Re-Sequencing Array Hybridization and Scanning

The fragmented, labeled PCR reaction was hybridized to the p53 re-sequencing array in 6xSSPE-T containing 2mg/ml BSA and 1.67 nM fluorescein-labeled control oligo (5′-CTGAACGGTAGCATCTTGAC-3′) at 45°C for 30 min. The array was then washed with 3X SSPE-T at 35°C for 4 cycles (10 drains-fills/cycle) in the GeneChip Fluidics Station (RELA). The hybridized p53 array was scanned using an argon-ion laser scanner (Hewlett-Packard) with a resolution setting of 6.0 μm/pixel (~70 pixels/probe cell) and wavelength detection setting of 530 nm. A fluorescence image was created, intensity information analyzed and nucleotide determination made by GeneChip Analysis Software (Affymetrix). Footprint analysis was done using Ulysses Analysis Software (Affymetrix) essentially as described (31).

ACKNOWLEDGMENTS

We thank A. Aggarwal for data analysis software support, L. Stein for EST cluster analysis, H. Matsuzaki for technical support and advice on p53 re-sequencing, and M. Chee, J. Oliner, S. Fodor and D. Shoemaker for helpful discussions and comments on the manuscript. We extend special thanks to L. Stryer for his invaluable advice, insights and encouragement.

REFERENCES

1. Boguski, M.S., Lowe, T.M. & Tolstoshev, C.M. dbEST-database for 'expressed sequence tags'. *Nature Genetics* **4**, 332–333 (1993).
2. Adams, M.D., Kerlavage, A.R., Fleischmann, R.D., Fuldner, R.A., Bult, C.J., Lee, N.H., Kirkness, E.F., Weinstock, K.G., Gocayne, J.D., White, O., et al. Initial assessment of human gene diversity and expression patterns based upon 83 million nucleotides of cDNA sequence. *Nature* **377**, 3–174 (1995).
3. Schena, M., Shalon, D., Davis, R.W. & Brown, P.O. Quantitative monitoring of gene expression patterns with a complementary DNA microarray. *Science* 270, 467–470 (1995).
4. Velculescu, V.E., Zhang, L., Vogelstein, B. & Kinzler, K.W. Serial analysis of gene expression. *Science* 270, 484–487 (1995).
5. Lockhart, D.J., Dong, H., Byrne, M.C., Follettie, M.T., Gallo, M.V., Chee, M.S., Mittmann, M., Wang, C., Kobayashi, M., Horton, H. & Brown, E.L. Expression monitoring by hybridization to high-density oligonucleotide arrays. *Nature Biotechnology* **13**,1675–1680 (1996).
6. Fodor, S.P.A., Read, J.L., Pirrung, M.C., Stryer, L., Lu, A.T. & Solas, D. Light-directed, spatially addressable parallel chemical synthesis. *Science* **251**, 767–773 (1991).
7. Lasfargues, E.Y., Coutinho, W.G. & Redfield, E.S. Isolation of two human tumor epithelial cell lines from solid breast carcinomas. *Journal of the National Cancer Institute* **4**, 967–978 (1978).
8. Hackett, A.J., Smith, H.S., Springer, E.L., Owens, R.B., Nelson-Rees, W.A., Riggs, J.L. & Gardner, M.B. Two syngeneic cell lines from human breast tissue: the aneuploid mammary epithelial (Hs578T) and the diploid myoepithelial (Hs578Bst) cell lines. *Journal of the National Cancer Institute* **6**, 1795–1806 (1977).
9. Coussens, L., Yang-Feng, T.L., Liao, Y.C., Chen, E., Gray, A., McGrath, J., Seeburg, P.H., Libermann, T.A., Schlessinger, J., Francke, U., et al. Tyrosine kinase receptor with extensive homology to EGF receptor shares chromosomal location with neu oncogene. *Science* **4730**, 1132–1139 (1985).
10. Styles, J.M., Harrison, S., Gusterson, B.A. & Dean, C.J. Rat monoclonal antibodies to the external domain of the product of the c-erbB-2 proto-oncogene. *International Journal of Cancer*, **2**, 320–324 (1990).
11. Earp, H.S., Dawson, T.L., Li, X. & Yu, H. Heterodimerization and functional interaction between EGF receptor family members: a new signaling paradigm with implications for breast cancer research. *Breast Cancer Research and Treatment* **1**,115–132 (1995).
12. King, C.R., Kraus, M.H. & Aaronson, S.A. Amplification of a novel v-erbB-related gene in a human mammary carcinoma. *Science* **4717**, 974–976 (1985).
13. Slamon,D.J., Clark, G.M,, Wong, S.G., Levin, W.J., Ullrich, A. & McGuire, W.L. Human breast cancer: correlation of relapse and survival with amplification of the HER-2/neu oncogene. *Science* **4785**, 177–182 (1987).
14. Slamon, D.J., Godolphin, W., Jones, L.A., Holt, J.A,, Wong, S,G., Keith, D.E., Levin, W.J, Stuart, S.G., Udove, J., Ullrich, A., et al. Studies of the HER-2/neu proto-oncogene in human breast and ovarian cancer. *Science* **4905**, 707–712 (1989).

15. Kraus, M.H., Issing, W., Miki, T., Popescu, N.C. & Aaronson, S.A. Isolation and characterization of ERBB3, a third member of the ERBB/epidermal growth factor receptor family: evidence for overexpression in a subset of human mammary tumors. *Proceedings of the National Academy of Sciences of the United States of America* **23**, 9193–9197 (1989).
16. Lemoine, N.R., Barnes, D.M., Hollywood, D.P., Hughes, C.M,, Smith, P., Dublin, E., Prigent, S.A., Gullick, W.J. & Hurst, H.C. Expression of the ERBB3 gene product in breast cancer. *British Journal of Cancer* **6**, 1116–1121 (1992).
17. Wallasch, C., Weiss, F.U., Niederfellner, G., Jallal, B., Issing, W. & Ullrich, A. Heregulin-dependent regulation of HER2/neu oncogenic signaling by heterodimerization with HER3. *EMBO J.* 14, 4267–4275 (1995).
18. Stein, D., Wu, J., Fuqua, S.A., Roonprapunt, C., Yanjnik, V., D'Eustachio, P., Moskow, J.J., Buchberg, A.M., Osborne, C.K. & Margolis, B. The SH2 domain protein GRB-7 is co-amplified, overexpressed and in tight complex with HER2 in breast cancer. *EMBO J.* 13, 1331–1340 (1994).
19. Lisanti, M.P., Tang, Z., Scherer, P.E., Kubler, E., Koleske, A.J. & Sargiacomo, M. Caveolae, transmembrane signalling and cellular transformation. *Molecular Membrane Biology*, **1,** 121–124 (1995).
20. Li, S., Okamoto, T., Chun, M., Sargiacomo, M., Casanova, J.E., Hansen, S.H., Nishimoto, I. & Lisanti, M.P. Evidence for a regulated interaction between heterotrimeric G proteins and caveolin. *Journal of Biological Chemistry* **26**, 15693–15701 (1995).
21. Koleske, A.J., Baltimore, D. & Lisanti, M.P. Reduction of caveolin and caveolae in oncogenically transformed cells. *Proceedings of the National Academy of Sciences of the United States of America* **5**, 1381–1385 (1995).
22. Van Biesen, T., Hawes, B.E., Luttrell, D.K., Krueger, K.M., Touhara, K., Porfiri, E., Sakaue, M., Luttrell, L.M. & Lefkowitz, R.J. Receptor-tyrosine-kinase- and G beta gamma-mediated MAP kinase activation by a common signalling pathway. *Nature* **6543**, 781–784 (1995).
23. Li, N., Batzer, A., Daly, R., Yajnik, V., Skolnik, E., Chardin, P., Bar-Sagi, D., Margolis, B. & Schlessinger, J. Guanine-nucleotide-releasing factor hSos1 binds to Grb2 and links receptor tyrosine kinases to Ras signalling. *Nature* **6424**, 85–88 (1993).
24. Alblas, J., Van Corven, E.J., Hordijk, P.L., Milligan, G. & Moolenaar, W.H. Gi-mediated activation of the p21ras-mitogen-activated protein kinase pathway by alpha 2-adrenergic receptors expressed in fibroblasts. *Journal of Biological Chemistry* **30**, 22235–22238 (1993).
25. Winitz, S., Russell, M., Qian, N.X., Gardner, A., Dwyer, L. & Johnson, G.L. Involvement of Ras and Raf in the Gi-coupled acetylcholine muscarinic m2 receptor activation of mitogen-activated protein (MAP) kinase kinase and MAP kinase. *Journal of Biological Chemistry* **26**, 19196–19199 (1993).
26. Clark, G.J. & Der, C.J. Aberrant function of the Ras signal transduction pathway in human breast cancer. *Breast Cancer Research and Treatment* **1**, 133–144 (1995).
27. Sivaraman, V.S., Wnag, H., Nuovo, G.J. & Malbon, C.C. Hyperexpression of mitogen-activated protein kinase in human breast cancer. *J. Clin Invest.* **99(7)**, 1478–1483 (1997).
28. Marshall, M.S. Ras target proteins in eukaryotic cells. *Faseb Journal* **13**, 1311–1318 (1995).
29. Levine, A.J. p53, the cellular gatekeeper for growth and division. *Cell* **3**, 323–331 (1997).
30. Velculescu, V.E. & El-Deiry, W.S. Biological and clinical importance of the p53 tumor suppressor gene. *Clinical Chemistry* **6 Pt 1**, 858–68 (1996).
31. Chee, M., Yang, R., Hubbell, E., Berno, A., Huang, X.C., Stern, D., Winkler, J., Lockhart, D.J., Morris, M.S. & Fodor, S.P. Accessing genetic information with high-density DNA arrays. *Science* **5287**, 610–614 (1996).
32. Zhang, L., Zhou, W., Velculescu, V.E., Kern, S.E., Hruban, R.H., Hamilton, S.R., Vogelstein, B. & Kinzler, K.W. Gene expression profiles in normal and cancer cells. *Science* **276**, 1268–1272 (1997).

DISCUSSION

Anderson: Can I start by asking, just from a commercial point of view, is it feasible to strip these chips after they have been used to wash them and re-use them? If so, how many times can they be re-used?

Gingeras: You can appreciate that if you are going to use these arrays, particularly the expression arrays, and expect to see two-fold levels of difference, then any stripping process that you might apply to those chips will undoubtedly compromise some of the probes on those chips, and we know, from lots of repeated experiments, that achieving the levels of sensitivity down to one to ten copies per cell and reproducibility of two-fold differences will be compromised in those situations. For the genotyping situations, I think it is certainly possible to strip those chips, you did not hear me say that, but it is certainly possible. I would not use those in a clinical setting to do that, only because I would not know what a negative result meant if I had been using the chip, but we certainly in times where the supply of chips have been limited and we have wanted to understand something very rapidly, we would strip those. My guess is that you could strip those a couple of times for genotyping and get a general sense of what is happening.

Gray: A couple of questions. One of them is, right now you are working down at, let us say, twenty micron feature sizes for that 6500K expression chip. How much is your dynamic range and sensitivity limited by the feature size at this point?

Gingeras: The dynamic range will go down appreciably at that 6500, we are still at about four logs of dynamic range and at one to ten. All of those experiments that I showed you were done with the 6500 feature chip. We do not know how much of a hit will take when we go to the 250 probe arrays, my guess is that we will take some and I really do not know quantitatively what that will be.

Gray: What is the status of the availability of that 6500 chip?

Gingeras: Those arrays are available to a whole variety of commercial/pharmaceutical like companies and they have been very interested and anxious to get those chips. The problem for us is that it is a very lucrative business. They are willing to pay a lot of money for those chips. We understand and we would like to get these arrays out to the scientific community as broadly and as rapidly as possible. We have to figure out some way to do that without compromising that business strategy. We have been talking to a variety of groups that use NIH grants of ways in which we can, perhaps, get this technology out. One approach would be to have some sort of shared value in the use of those chips and some of the intellectual property that might come from that, as a way to go back to the commercial partners that we have and say: Although these institutions did not pay as much money as you did, the differential is being made up in another form, namely, as shared in the downstream intellectual property. That is one strategy that is being investigated. The other strategy is that less complex arrays, of course, can be made. These 6600 gene chips really represent about 85% of the fully sequenced genes that are in genbank and about 35000 of EST's who have a very strong homology to non-human genes. You can make a collection of two to three hundred oncogenes tumor suppressor genes, DNA repair genes as a sub-section of those and those might be more readily available on a commercial level for very specific kinds of things. So I think that kind of strategy is really how that is going to be done specifically for the human expression sets.

Draetta: At present, you are analyzing differences in mRNA expression between different tumor cell lines. What you detect might not have any relevance to what could be seen comparing mRNA expression in primary tumors? If you were to provide access to your technology to research institutes throughout the world, both the academic and industrial research communities could benefit.

Gingeras: Right, and I do not disagree with you. It is not that the concept of the value is not understood; at the window of time that we are in right now there is a high value placed on the technology. Because we are a commercial concern, we have to capture as much of that value as possible. So that window will change with time, other technologies will come into place, the technology will get less expensive to use, and so forth. But, it is not the fact that people do not understand the value of disseminating the technology, rather it is the point of trying to get that return while the window is open.

Pierotti: One of the most challenging diagnosis problem is the detection of tumor cells in the context of normal. The biggest limitation is background noise introduced by PCR, etc. I was wondering whether the comparison of the pattern between normal versus pathological samples is somehow ameliorating this 20% limitation of the normal PCR?

Gingeras: Well, the experiments I showed are cell lines so they are basically homogeneous populations. In a tumor situation, I think the answer is obvious: In order to see what the real patterns are, there is going to have to be significant effort in order to segment out those transformed cells, either through a micro-dissection process or through some selection process to capture that profile.

Pierotti: Yes, that is one part of the problem. The other part is when you are basically analyzing a normal sample, let us say, bone marrow, and the question you are putting is how many cancer cells are there? That is my point. If this methodology can somehow bypass the limitation of the 20% of background noise that is known when the PCR is introduced.

Gingeras: Only if you know what patterns you are looking for, I think in that case it would be so. If we do not know what those patterns are, then we will not know how to differentiate what is normal and what is abnormal in that population. I think we will wait until that data set is collected.

Livingston: Do you have a strategy for serving the myriad of needs of the non-commercial research market without pricing out small scientific labs?

Gingeras: Yes, I think the strategy is easier to understand when you have a generic set of chips. Where you do not have to do custom design and synthesis and then make X number of chips for each of those mass and each of those synthesis. Because then you really are segmenting your efforts many times over. So, it is easy to understand that for the human genome expression chips you can either build a very high density array, as many chips as possible, or some sub-segment of that. So the answer for the expression I think will be that, with time, the strategy will be to move these arrays into centers, specific designated centers, in which we can support them, because we will know what the volume is and there will be a fixed number distributed throughout the United States and Europe. The price for those will reflect the decreasing costs of making those chips and probably what the market will actually bear in paying for those. The genotyping chips are a different issue, because the genotyping chips require, as you have just seen from the very small example of the p53 chip, a whole different strategy where each of those will have to be individually designed and so, I think, there is going to be, in terms of technology development, another leap that needs to go forward from our point of view in order to make that custom chip business related. So that any chip, anytime, anywhere, can be done.

Livingston: Like oligonucleotides. For a long time, oligonucleotides were synthesized by a few companies. Now you can make them on your benchtop. So, the question is whether there is a segment of Affymetrix that is devoted to benchtop chip development/preparation?

Gingeras: Most certainly. It goes to the whole issue of making generic chips. These chips are now one of the reasons that they need this kind of attention and focus because you are actually designing very specific oligos to interrogate every position based on the known sequence. I have not given you any data here, but my real life is to actually think about this problem in terms of pattern recognition. How patterns can be informative from both the genotyping and expression levels. In doing pattern recognition you do not need to have specific complements to the targets. So there is a whole group of people who are actually focused on this as a way to try to get this out.

Melief: Has this technology been used to type highly polymorphic regions, such as HLA, where you would use an array of different oligonucleotides for closely related gene products?

Gingeras: We have not made an HLA chip, but a variety of other people have been very interested in making that chip and it is very suitable to do that. So making such a chip would be very straightforward, although we have not made that array.

White: You showed normal breast epithelium versus a breast tumor epithelium sample. What were the conditions from which those cells came? Were those tissue samples from individuals?

Gingeras: These are just cell lines that we obtained from ATCC.

White: So ATCC grown on the same medium--is that the idea?

Gingeras: Yes. This is not meant to be illustrative of what we would see in breast epithelium kinds of conditions, but rather just the kind of data that would be generated from that.

Zanker: Just to understand the technology -- what is the difference between your technology and that of BioCor? Because both use a design chip, both use molecular information binding to the chip, both use substrates binding its ligands to this biological information and both have the same readouts to look at some arrays coming out?

Gingeras: I am not so familiar with BioCor, but, my understanding of that process is that it uses essentially a different detection modality. These rays are actually, I think, the highest density rays that have been available. These oligos are not being spotted on to the substrates; these are actually being synthesized *in situ* on these substrates. The way the assay is being configured here for commercial use, you are not doing real time dynamics as you would see in a BioCor reaction. So, you are not looking at the kinetics of on and off rates on these oligos; you are actually looking at the end stage process.

Gray: Back to the expression rays for a minute. I was wondering if you could say a few words about the kinds of software that you are thinking of developing that will enable the interpretation of these things. I mean, a pattern recognition is easy to say, but it seems to me that you are going to have to extend well beyond that in order to make these things maximally useful.

Gingeras: Yes, again, I do not have time, but I would be glad to show you some of that work. We have spent a lot of time looking at this problem. We have a little more history in looking at this kind of high density data set analysis, looking at the genotyping of HIV and correlating the genotype of HIV to resistance, and trying to understand what the pre-existing polymorphisms that are present in the quasi species, how that may determine which way the virus moves in terms of selection as it is put under drug selection. To do that we have literally sequenced tens of thousands of samples now and we have obtained IC50 values, resistance levels, determinations, viral load data, clinical data, on all of those samples. In order to understand which of those polymorphisms is or may be interesting or informative as it relates to resistance. So, a lot of that work is involved using principle component analysis, a variety of cluster analysis oligorhythyms in trying to segment subcompartments of the data set with each defined genotypes that we have seen illustrated in the time course of experiments with patients that have been under drug therapy. Those kinds of analytical tools are being used to do those kinds of analysis and there is a whole selection of such tools that are ongoing.

Hanahan: Looking ahead at the alternative technologies, in terms of being able to use your technology, which obviously you are quite well immersed in, being able to design custom chips, do you see the evolution of this technology, being that that is going to be realistic, or would you envision that this would actually be complementary to, say, the ink jet printing options. If you decide that one lab for custom purpose needs a good kind of chip, is it going to be realistic, down the road, for you to do that, or do you really think that these alternative technologies will step up to the play here?

Gingeras: No, I think the technology will be, again, competitive in that regard. Remember the ink jet technologies themselves will have, I think, their own challenges in that respect, namely, in disseminating the technology in a way, for example, to have bench top oligonucleotide synthesis, if that is the approach you would want to do. I think in that kind of circumstance there may be room for that approach in the short term because that technology, if it can be gotten and put together in a fairly robust manner, will probably allow you to build fairly small numbers of arrays and low density arrays. But, I do think what will happen in the long term is that these arrays, the ability to make custom chips, will become available either through the fact that the process can be diversified on a single wafer so that you can make many different kinds of chips on a single wafer. We can do that now, but it is harder to quality control that kind of thing. Or, we can make more generic arrays where the probes, as I indicated, are not specific for the targets that you want, but use an oligorhythym approach to get the information that you want. So I think either of those two methods is possible. I do not know how that development is occurring, in terms of the ink jet printing, and how rapidly that is moving along, but it may be that it will not fit that niche currently.

Hanahan: The second question which follows a little bit on Ray's question: Take, for example, your p53 chip, how do you imagine that is going to be used if you have a tumor and you want to look for p53 mutations given that there are stromal cells and epithelial cells and perhaps heterogeneity in the tumor; are you doing any work on that issue of delectability in a complex population?

Gingeras: Yes, I think I indicated earlier that the cleanest data will come from that which have the samples prepared in such a manner that you enrich as much as possible for the cells of interest. Barring the ability to do that, this array, I think will come with the capacity to be able to see mixtures down to about twenty-five percent in a population. So it will be in a pattern which has both wild type and mutant. The ability to see mutant at about a twenty-five percent level will be detectable in this particular array. But, that is not going to solve the problem for a whole variety of other kinds of systems.

Livingston: Are you in discussions with at least one or more institutes of the NIH with the idea that public money would be used for the purpose of supporting research to be done in laboratories with your technology in which there would be a pass-through of intellectual property rights from the laboratory to the company?

Gingeras: It is slightly different. The thought would be that, given the relatively high cost of the technology, the relatively limited capacity to meet the demand out there, that there would be technology centers set up. They would not only include this technology, but also clone spotting, and a variety of other approaches. Those centers would seek money, through a competitive strategy, through the NIH, or through Hughes, or whatever.

Livingston: Would federal money go to run these centers and your technology be used there? If so, would an investigator bring an experiment to such a center, have it performed, and the results of that experiment, if applicable, then be shared with the company? Is that the idea?

Gingeras: It is not clear that that latter part is the case, in the case of the intellectual property. That is what the big hang up is. We are stuck at this position. We have to figure some *quid pro quo* that we can justify in order to be able to see that our current business is not undercut by this strategy. I do not know the solution. I am not suggesting that is the solution.

Comoglio: Have you calculated an oligorhythym for a possible genetic sequencing chip? When you have so many mutations, so many deletions for some regions, what you are going to do is possibly a complete sequencing. Of course, not at the actual limits of resolution. Have you ever thought of a generic sequencing chip?

Gingeras: Yes, this is something that we are very actively pursuing and trying to understand how one can build a set of oligos that can look through a sequence whose identity we have no idea because of the number of polymorphisms that might be present. Based on the fact that we can use that pattern to go back to a set of sequences whose identity we know and fits most closely with that pattern and build a library set of sequences compared with patterns in order to provide at least some insight into what those collection of polymorphisms might be. So that work is on-going.

Comoglio: Have you ever thought of combining PCR with this high density hybridization, when you do not know the sequence, if you just have random oligo_nucleotides say, for example, a differential hybridization from two samples?

Gingeras: Well, one thing I neglected to indicate here is that these syntheses occur such that the syntheses goes from three prime to five prime on the surface of the chip. So, a PCR reaction would have to be done as a sandwich kind of reaction, where you would use the chip to capture the target and then run the PCR off the captured target in a separate way. You cannot do it from the chip itself, since the orientation of the oligos are in the wrong direction.

10

USE OF cDNA MICROARRAYS TO ASSESS DNA GENE EXPRESSION PATTERNS IN CANCER

Paul S. Meltzer, Michael Bittner, Mervi Heiskanen, Tiffany Hoffman, Yidong Chen, and Jeffrey M. Trent

Laboratory of Cancer Genetics
National Human Genome Research Institute
Bethesda, Maryland 20892

1. INTRODUCTION

It is no longer controversial to state that genomic alterations are of fundamental importance in carcinogenesis. However, it remains difficult to determine the spectrum of genomic changes in any given tumor and even more difficult to determine the impact of these changes on gene expression. As a result of progress in the Human Genome Project, significant advances have been made which offer the potential to solve this problem. These techniques utilize human genome maps, cloned resources, and sequence data in conjunction with fluorescence based technologies to enable the genome wide analysis of tumor cells. One such technique, cDNA microarray hybridization[1,2], has the potential to provide large scale analysis of gene expression. This novel system for massively parallel cDNA hybridization is based upon robotic printing of DNAs on glass slides and simultaneous two-color fluorescence hybridization. For studies of gene expression, it is possible to utilize the vast resource of arrayed cDNA clones which have been developed as part of the Expressed Sequence Tag (EST) project to produce microarrays containing tens of thousands of genes[3–6]. When fluorescent probes generated by reverse transcription of tumor cell mRNA are hybridized to the array, the level of expression of every array element is simultaneously determined with respect to a reference probe labeled in a second color. Although the processing of such large quantities of data is challenging in itself, microarray technology will provide a far more detailed picture of gene expression than has previously been possible. The availability of this information has ramifications which will affect virtually all areas of cancer biology. This chapter will provide an overview of the current status of cDNA microarray technology and discuss its potential applications to problems in cancer biology.

The Biology of Tumors, edited by Mihich and Croce
Plenum Press, New York, 1998.

2. ESSENTIAL ASPECTS OF MICROARRAY TECHNOLOGY

In order to carry out microarray hybridizations, two instruments are required: a microarray printer and a scanner to read out the hybridization data. Appropriate software must also be available to process the hybridization results. The overall process is outlined in Figure 1. Since, at present, no commercial source can provide these instruments, they must be constructed by laboratories who wish to develop this technology. In addition to these requirements, appropriate cloned cDNAs must be available to serve as hybridization targets, and strategies must be developed to generate fluorescent probes from biological materials. In considering the following discussion, it is important to bear in mind that hybridization experiments are always designed to detect relative expression of two probes labeled with distinct fluorophors.

2.1. Robotic Printer

Microarray printing requires a simple X,Y,Z robot which can transfer droplets of DNA onto ordinary glass microscope slides which have been coated with poly-L-lysine. By providing a platform with a tray of slides, numerous identical slides can be printed simultaneously. The cDNAs to be printed are assembled in 96 well microtiter plates which are delivered to the robot's print head. A simple print tip resembling a fine tipped fountain pen picks up a 1–2 μl of DNA solution and deposits a small droplet on each slide. By designing the print head to carry multiple printing pins, it is possible to print the contents of a 96 well plate on 48 slides in several minutes. Spot size of 200μ is readily achievable and

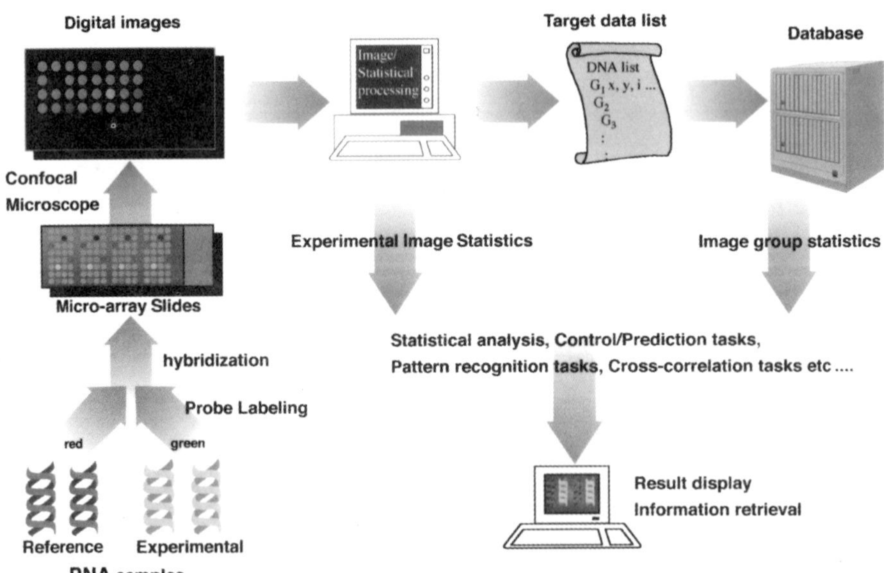

Figure 1. Overview of the microarray hybridization process. Probes from reference and experimental RNA sources are labeled with distinct fluors and hybridized to the robotically printed arrays. After stringency washes, the array is scanned generating an image file corresponding to the hybridization signal generated by each probe. The images are processed to produce quantitative ratio information for each array element and global array statistics. This information is then linked to the array element identifiers and external databases for data interpretation.

allows the deposition of approximately 10,000 array elements on a single slide. Each slide is intended for a single hybridization.

2.2. Array Scanner

Although various principles might be employed to obtain fluorescence images of the hybridized microarrays, the most straightforward design can be described as a laser scanning confocal epifluorescence microscope. At least two lasers are required to excite spectrally distinct fluors. The hybridized microarray is rastered through the laser's optical path and the emitted light is captured by a photomultiplier. The device is equipped with dichroic filter cubes similar to those used in standard fluorescence microscopy. The output of the photomultiplier is transferred to a computer which stores the data as an image file for each wavelength scanned. Although pixel size is arbitrary 20μ pixels provide adequate resolution for the arrays described above. In 20 to 40 minutes, the raw data can be acquired which provides flourescence intensities for each array element in two wavelengths.

2.3. Image Analysis Software

Although it is possible to identify the most extreme changes in hybridization ratio by visual inspection of the image files on a computer monitor, this is cumbersome and would ignore most of the quantitative data inherent in the hybridization images. Because the individual array elements are not perfect circular spots, and because background may vary over the surface of the slide, image analysis software is necessary to process the data. Ideally, in processing the images such software identifies the pixels which represent significant hybridization signals, adjusts for local background, calculates the mean signal intensity for each array element, the ratio of the two scanned wavelengths, and reports this data in spreadsheet format. Of course, this output should also provide the sequence identity of each hybridization target spot in order to facilitate data interpretation. Statistical analysis of the entire array generates a confidence interval for the ratio data so that those array elements are readily identified which yield significantly different signal intensities between the two channels (corresponding to experimental and reference probes).

2.4. Selecting Array Elements

Over 700,000 human cDNAs have been partially sequenced in the various EST (expressed sequence tag) projects which have been reported in the public sequence databases. There is considerable redundancy in this data, and selecting a non-redundant subset of EST clones for the construction of microarrays poses a significant informatics problem. Several clustering systems have been devised to reduce the EST sequence database to unique clusters. One such clustering algorithm, the Unigene system, is publicly accessible, and consists of over 60,000 clusters[3]. An additional advantage of the Unigene system is its utilization in the mapping effort which has assigned 16,000 of these to loci on the genome delimited by radiation reduction hybrid mapping panels[6]. Further, these clones have been archived through the efforts of the IMAGE consortium and are readily available[5].

Because the microarray printer requires purified DNA for the printing process, the cDNA inserts must be prepared from each of the clones selected for use. This is most readily accomplished by PCR amplification of individual bacterial cultures using vector specific primers. By judiciously selecting cDNA clones from libraries constructed in similar vectors, the same primer pair can be used to amplify the insert DNA from numerous clones. When utilizing clones retrieved from stored libraries, there will be inevitable errors in clone identity which might arise at multiple stages during the processing of the libraries. It is therefore of importance to obtain a measure of clone authenticity by generating sample sequence data from a subset of the clones to be arrayed. Additional problems arise because of some clones which fail to grow in culture or which contain admixtures of a second clone. Because of the significant efforts involved in clone handling and authentication, the most ambitious cDNA microarray efforts to date plan to print no more than 15,000 clones per array.

2.5. Informatics Issues

Although the ability to analyze the expression of numerous genes simultaneously is novel and exciting, the need to examine such massive quantities of data poses a problem in itself. To facilitate data interpretation, a system is necessary which allows rapid retrieval of information concerning each gene represented on a given microarray. Ideally, the processed fluorescence image data will be presented in a format which allows the investigator to easily query existing internet accessible public databases such as the Entrez system which provides links to literature, sequence, and protein structure databases.

2.6. Fluorescent Probe Generation

In order to use microarrays for gene expression analysis, it is necessary to prepare a probe from cellular mRNA which is quantitatively representative of the mRNA population in a given biological sample. Micorarray hybridization experiments reported to date have primarily utilized cDNA probes generated from purified polyadenylated mRNA. A fluorophor conjugated nucleotide is incorporated during a single round of reverse transcription. In order to prepare the several hundred nanograms of probe for hybridization, microgram quantities of purified mRNA are required. This poses a significant limitation on the types of experiments which can be performed because of the quantity and quality of RNA required for each hybridization. It is likely that alternative labeling strategies such as the use of total RNA might simplify probe production. The introduction of an amplification step, whether via an RNA polymerase promoter introduced with the reverse transcriptase primer or PCR, is possible in principle, but carries the risk of altering the representation of mRNA species. Although most studies reported to date have utilized directly labeled fluorophor conjugated nucleotides, it is also conceivable that hapten conjugated nucleotides might be used which would allow various signal amplification strategies. Modifications of probe preparation protocols will be important in order to facilitate the application of microarray technology to small biological samples. It is also important to note that cDNA hybridization probes contain repetitive sequences of various types which necessitates the use of excess unlabelled repetitive DNA in the hybridization mixtures to suppress non-specific hybridization. Sensitivity and accuracy comparable to standard Northern blot hybridizations to total RNA are readily achievable.

3. APPLICATION OF MICROARRAY TECHNOLOGY TO PROBLEMS IN CANCER BIOLOGY

The feasibility of using this novel system for parallel expression analysis of multiple genes has been demonstrated in human melanoma cells using RNAs from the highly tumorigenic melanoma cell line (UACC 903) and its chromosome-6 suppressed subline [UACC 903 (+6)][7,8]. The experiment examined the expression patterns of 870 arrayed human cDNAs. These cDNAs included oncogenes and tumor suppressor genes as well as housekeeping and tissue specific genes (for positive and negative controls). Fluorescent probes were generated by reverse transcription of mRNA from UACC 903 and 903 (+6). Results indicated that excellent specificity of hybridization was be obtained, enabling the identification of cDNAs whose expression is correlated with tumor suppression. A cutoff of three standard deviations from the mean ratio of 90 housekeeping genes was utilized to identify those genes whose expression was significantly altered in UACC903 cells by introduction of chromosome 6. According to this criterion, 1.7% of the arrayed genes decreased in expression and 7.3% increased. The array hybridization data was corroborated with Northern analysis in the 16 genes tested. The functional significance of many of the genes with altered expression in this experiment was readily interpretable in terms of pathways known to be associated with tumor suppression or cell differentiation. One of the significant observations to emerge from this study is the recognition that only a subset of the genes assayed actually were altered in this experiment. This is important because it implies that the analysis of large array hybridization patterns may not be as overwhelming as might be expected based on the sheer number of genes included. This also suggests the possibility of creating smaller arrays directed at particular biological questions using genes identified on larger scale experiments. Since smaller arrays are relatively easy to print, multiple samples could readily be analyzed.

The study described above provides a tantalizing look at the potential for microarray hybridization to provide new insights in cancer biology. Similar sized arrays have been utilized to study heat shock and phorbol ester-regulated gene expression in human T cells[9] and in inflammatory disease[10]. These studies further validate the enormous possibilities for applying microarray technology gene expression analysis, and directly demonstrate the powerful potential of this technology. Larger arrays are now being generated containing up to 15,000 cDNAs on a single microscope slide. Software has been developed which can rapidly extract quantitative hybridization information from the computer files generated by imaging these arrays. Computational tools are under development for the problems which will emerge as multiple hybridization experiments are performed to facilitate the comparison of several hybridization results and the recognition of interactions between array elements. Such large scale analysis of gene expression has the potential to identify crucial cellular pathways in oncogenesis, to assist in the classification of tumors, and to examine the impact of therapeutic interventions.

It is perhaps worthwhile to compare the cDNA microarray technology which has been described above with the closely related alternative technology of oligonucleotide array hybridization. A unique method of directly synthesizing high density oligonucleotide arrays has been developed which also has the potential for use in expression analysis[11]. This technology has certain advantages compared to cDNA printing techniques. Because clone handling is not involved, the arrays are sequence perfect. The manufacturing process also produces geometrically perfect arrays which pose an relatively straightforward image analysis problem and which allow the incorporation of hybridization specificity controls containing an arbitrary number of mismatches. However, there are intrinsic problems in oligonucleotide hybridization because rule have not yet been developed which enable the selection of sequences which will provide excellent hybridization targets. Thus it is not possible to predict that any given oligonucleotide will function as an effective hybridization target, and some proportion of array elements will not be useful. Oligonucleotide arrays also currently have a high production cost, and have an absolute requirement for sequence data for array design. In contrast, cDNA microarrays have a high level of flexibility at low cost. It is easy to envision using large arrays for the initial investigation of a given question and then construct smaller arrays composed of a few hundred of the most relevant genes identified by large arrays. Similarly, arrays containing all members of a given functional class or biochemical pathway might be constructed. Another important difference is the possibility to print arrays from cDNA libraries which might not have been sequenced from any biological source desired. Because of the longer target size, cDNA microarrays also are amenable to high stringency hybridization and washing conditions, significantly obviating problems in cross-hybridization. It should be noted that both oligonucleotide and cDNA arrays have a common limitation in that both are confined to analysis of previously cloned cDNAs and are therefore inherently incomplete until all human genes are discovered. It is likely that both oligonucleotide and cDNA array systems will co-exist for sometime as the best applications for each of these systems are identified by experience.

Many key genes and pathways relevant to tumor initiation and progression have been recognized in the last several years. Alterations in these key regulatory genes may cause a profound change in cellular phenotype which must involve alterations in the expression of multiple genes. However, the process of defining the downstream consequences of alterations in regulatory genes is still difficult. cDNA microarray hybridization offers a new view of gene expression which is tremendously broad in scope. This novel window into cellular function has the potential to identify new tumor markers and therapeutic targets, but perhaps of greatest importance it provides a tool to determine the specific consequences at the level of gene expression of any genetic alteration, extracellular signal or change in metabolic state.

REFERENCES

1. Schena M, Shalon D., Davis RW and Brown PO. Quantitative monitoring of gene expression patterns with a complementary DNA microarray. Science 270: 467–470 (1995).
2. Shalon D, Smith SJ, Brown PO, A DNA microarray system for analyzing complex DNA samples using two-colorfluorescent probe hybridization, Genome Res 6: 639–645 (1996).
3. Boguski MS and Schuler GD. ESTablishing a human transcript map. Nat Genet 10: 369–371 (1995).
4. Soares MB, Bonaldo MF, Jelene P, Su L, Lawton L Efstratiadis A. Construction and characterization of a normalized cDNA library. Proc Natl. Acad. Sci. U S A 91: 9228–9232 (1994).
5. Lennon G, Auffray C, Polymeropoulos M, Soares MB, The I.M.A.G.E. Consortium: an integrated molecular analysis of genomes and their expression. Genomics 33: 151–152 (1996).
6. Schuler GD, Boguski MS, Stewart EA, Stein LD, Gyapay G, Rice K, White RE, et al., A gene map of the human genome, Science 274: 540–546 (1996).
7. Trent JM, Stanbridge EJ, McBride HL, Meese EU, Casey G, Araujo DE, Witkowski CM, Nagle RB. Tumorigenicity in human melanoma cell lines controlled by introduction of human chromosome 6. Science 247: 568–71 (1990).
8. DeRisi, J, Penland, L, Brown, PO, Bittner, B, Meltzer PS, Ray, M, Chen, Y, Su, YA, Trent, JM, Use of a cDNA microarray to analyze gene expression patterns in human cancer. Nature Genet. 14:457–460 (1996).
9. Schena M, Shalon D, Heller R, Chai A, Brown PO, Davis RW. Parallelhuman genome analysis: microarray-based expression monitoring of 1000 genes. Proc. Natl. Acad. Sci. U.S.A. 93, 10614–10619 (1996).
10. Heller, R.A., Schena, M., Chai, A., Shaon, D., Bedilion, T., Gilmore, J., Woolley, D.E., and Davis, R.W., Discovery and analysis of inflammatory disease-related genes using cDNA microarrays, Proc. Natl., Acad, Sci, USA, 94:2150–2155 (1997).
11. Chee M, Yang R, Hubbell E, Berno A, Huang XC, Stern D, Winkler J, Lockhart DJ, Morris MS, Fodor SP, Accessing genetic information with high-density DNA arrays Science 274: 610–614 (1996).

DISCUSSION

Gingeras: Some of the more interesting phenomenon we have been looking at involves changes which occur at levels, roughly from two to five fold and then, of course, the ones that are really obvious are the ones that are greater than that. In the clone arrays, what is your sense about the ability to work in that area where you are having changes in that dynamic range of about two to five fold?

Meltzer: I think that the data is going to be pretty good, certainly at four or five fold, and whether it will go down to 1.5 to 2, we do not really know yet. I think we are at the point where we have spent the last couple of years sort of building the instruments, getting them working, getting the pipeline open and, now, the types of things that we are doing are answering exactly the kind of questions you are asking, which is taking the same pair of samples and doing a whole bunch of hybridizations and repeats, and really determining accurate statistics. So, I am not really in a position to make a lot of claims, quantitatively, as to what we can do with this system. We want to get really good accurate information on that before we can say. You have to ask what your standard of reference is going to be anyway, whether it is really even valid to compare two hybridization based assays. I think down to two to three fold is probably going to be pretty reasonable.

Gray: I have got a couple of questions. One is, you mentioned authentication--what do you do about that?

Meltzer: Well, it is a really a tough problem. We do not want to repeat the entire sequencing of all of these EST's. Our approach has been to sample sequence out of the library and determine a percent accuracy. It gets complicated here, because there are a lot of different EST libraries; the ones we are working with were primarily from Livermore and Gregg Lennon, and then they have been copied and duplicated and, in that process, errors have occurred and what is more, in the early days of this whole project there were some serious problems in lane tracking and things like that, that led to absolute actual mismatches between the database sequence and the clone id that is linked to that. So, there were actual errors in there to start with. What we have been doing is looking library by library, picking the ones that are giving us the cleanest data and then going back and polishing the arrays as we go. We are about ninety percent accurate at this point. I think what is going to end up happening, I hate to say this but, you will do experiments and if you come up with a certain number of genes that look really interesting, you are probably going to go back and re-sequence those to make sure that they are what you thought they were. We will build up gradually a higher degree of confidence in the sequence accuracy. One point that I want to make is that, in a lot of experiments that you might do, even though you might be looking at thousands of genes, the number of genes that are actually going to show alterations, that are going to be interesting to you is going to be perhaps relatively limited, and in the case of anonymous EST's, where you may have primarily sequenced from the three prime UTR, that is not much good to you to figure out gene function anyway independently, and someone is going to have to go now and get the full-length sequence information anyway and figure out what that gene is, so . . .

Gray: But it seems to me that, to the extent that this is going to become a public resource, you may as well just go ahead and have them sequenced. I mean this is not a formidable project by IMAGE consortium standards.

Meltzer: No, probably not. I think we may end up at that point but, before getting there, we probably want to go through a phase of optimization and make sure we are using the libraries so we can have a lowest error rate. Because that creates another problem which is sort of retro-fitting the array to correct the errors.

Gray: O.K. One other question. In your hands, right now, what fraction of the arrays that you print are useable?

Meltzer: Actually the printing process is quite robust, but it is where the greatest black art to the whole thing resides and, when you get into it, it turns out that there is very little hard science in the surface chemistry of DNA attachment to glass coated with poly-L-lysine. We have gone through quite a lot of pain in terms of optimizing the conditions and variables that you might not think would be important, but turn out to be important. Once you have a handle on all of those things, in a batch which prints and is good, every slide in that batch is fine. Can you get a bad batch? Yes, that is possible, but as we have gained experience that seems to be dropping pretty sharply, so, we do not think that is going to be a big problem, but we have paid a high price in terms of fiddling with variables that we did not think were going to be important at first:, like the humidity in the printer and the ionic composition of the solution that DNA is kept in. I do not have it with me, but you can take AFM pictures of spots that have been made and actually you end up with a very nice image, more or less as predicted theoretically from the amount of DNA of completely packed, stretched out DNA molecules, filling up the spot.

Hanahan: To follow with a couple of technical questions. Do these pro cDNA's all have different sizes, and, presumably, different GC contents and I am wondering how you deal with issues about hybridization across the board of different GC contents and then in particular, cross-hybridization because I think this is an elegant idea of following pathways, but if these are multi-gene families, how are you going to discriminate?

Meltzer: There are a couple of factors in there that work on our behalf, and maybe do not completely solve all the problems that you are talking about, which I think are extremely important to bear in mind. One is that these are approximately 1KB fragments from the early EST libraries which are three prime oriented as are our probes which tends to help with the signal, but there is a lot of three prime untranslated sequence here, which is not, as a rule, very GC rich. So there is a similar base composition, they are fairly small and many of them do not contain coding sequence from genes that would cross hybridize on that basis. They do contain repeats, and I did not mention that, actually in all the hybridizations we have Cot 1 DNA and polyA and so on, and the suppression can be actually excellent, so that is not a big problem. But, it is, I think, expected that not all spots will behave equally. This is a new technology and we are just going to learn through experience.

11

EH, A NOVEL PROTEIN

Protein Interaction Domain

Margherita Doria, Anna Elisabetta Salcini, Stefano Confalonieri, Elisa Santolini, Gioacchin Iannolo, Pier Giuseppe Pelicci, and Pier Paolo Di Fiore

Department of Experimental Oncology
European Institute of Oncology
20141 Milan, Italy

1. PROTEIN-PROTEIN INTERACTION DOMAINS IN SIGNALING

Intracellular signaling pathways activated by growth factors, hormones and cytokines, are commonly subverted in cancer cells. Thus, knowledge of signaling molecules and of their function not only allows for understanding of cellular regulatory mechanisms, but also offers the opportunity for improvements in diagnosis and in therapeutic intervention. Critical to the physiology of signal transduction is the existence of a complex intracellular network of protein:protein interactions, mediated by the presence, in signal transducers, of binding domains[1]. These binding domains mediate dimerization and multiprotein complex formation, bring substrates to enzymes, modulate the activity of enzymes, and direct complexes to particular cellular locations. The two best studied binding domains are the SH2 (for Src homology module) domains[1], which recognize phosphotyrosine containing motifs, and the SH3 domains[1,2], which bind proline-rich regions. Ligand specificity of both SH2 and SH3 domains is determined by the variable amino acids that surround invariant residues (phosphotyrosines and prolines, respectively). Thus, every SH2 and SH3 domain binds to a distinct ligand. These specificities, however, are not absolute and there may be more than one binding domain within the cell with high affinity for a particular ligand. Hence, *in vivo*, binding may depend on other critical factors such as the local concentration of proteins and the modulating effect of other domains expressed on interacting proteins.

More recently, the PH (Pleckstrin homology) domain[1], the WW (for two conserved tryptophan, W, residues) domain[3] and the PTB (phosphotyrosine binding) domain[4] have been shown to be specific protein:protein binding domains. Frequently, binding domains are in variable number and in different location within a single protein. More importantly,

The Biology of Tumors, edited by Mihich and Croce
Plenum Press, New York, 1998.

a given binding domain is not restricted to a particular type of signal transducers, but may be present in proteins with different enzymatic and effector activities, as well as in several structural proteins, transcription factors and adaptors.

The complexity of signaling networks is only beginning to be appreciated. In particular high affinity protein:protein interactions appear to mediate many different biological functions and are involved in all the transduction mechanisms known so far. It is thus likely that new types of binding domains will continuously be identified, improving our understanding of the intracellular signaling.

2. EH IS A NOVEL PROTEIN-PROTEIN INTERACTION DOMAIN

The EH domain is a novel protein-protein interaction domain originally identified as a motif present in three copies at the N-terminus of the tyrosine kinase substrate eps15[5], hence the nomenclature EH, for Eps15 Homology[6]. The same motif was found in several heterogeneous proteins from yeast (including End3p and Pan1p) and nematodes, thus establishing its evolutionary conservation[6]. The EH domains of eps15 were shown to be *bona fide* protein:protein interaction domains by means of *in vitro* binding assays[7]. Subsequently, a protein highly related to eps15, called eps15R (for eps15-Related)[6], and other mammalian proteins of unknown function, were shown to contain at least one EH binding domain[7,8]. The domain is about 100 amino acid long and approximately 50% of the positions are conserved when a plurality of more than 50% is used to calculate overall homology (Fig. 1). EH domains are frequently, but not obligatorily, present in multiple copies (Fig. 2), and may contain calcium-binding motifs of the EF-hand type; however, the modification of calcium ion concentration does not seem to affect the binding properties of these domains *in vitro*.

EH-containing proteins appear to have heterogeneous characteristics and functions. However, several observations suggest that most of them are involved in the mechanisms of intracellular trafficking of molecules, such as endocytosis and transport of proteins to specific subcellular locations. Thus, a comprehensive understanding of the broad function of the EH domains relies on the identification of specific binding partners and on the characterization of the biological role of each EH-mediated interaction.

3. EH-CONTAINING PROTEIN IN MAMMALS: EPS15 AND EPS15R

The eps15 protein was discovered as a phosphorylation substrate of the epidermal growth factor receptor (EGFR) and other receptor tyrosine kinases[5]. Eps15R was originally identified as a protein homologous to eps15[6] and subsequently shown to be also phosphorylated upon EGF stimulation[9]. Eps15 and eps15R share a modular structure composed of a N-terminal portion, which contains three EH domains, a central putative coiled-coil region composed of several contiguous heptads, and a C-terminal domain displaying multiple copies of an Aspartate-Proline-Phenylalanine (DPF) motif and a proline-rich region capable of interacting with the SH3 domain of the Crk proto-oncogene[10].

The functions of eps15 and eps15R are still unknown. A role in cell proliferation is suggested by the fact that overexpression of eps15 can transform NIH-3T3 cells, although with low efficiency[5]. Furthermore, the *eps15* gene is rearranged with the *HRX/ALL1* locus in the t(1;11)(p32-q23) translocation of acute leukemias[11,12]. Additional clues to the role of eps15 and eps15R derive from the observation that a fraction of both proteins is associated

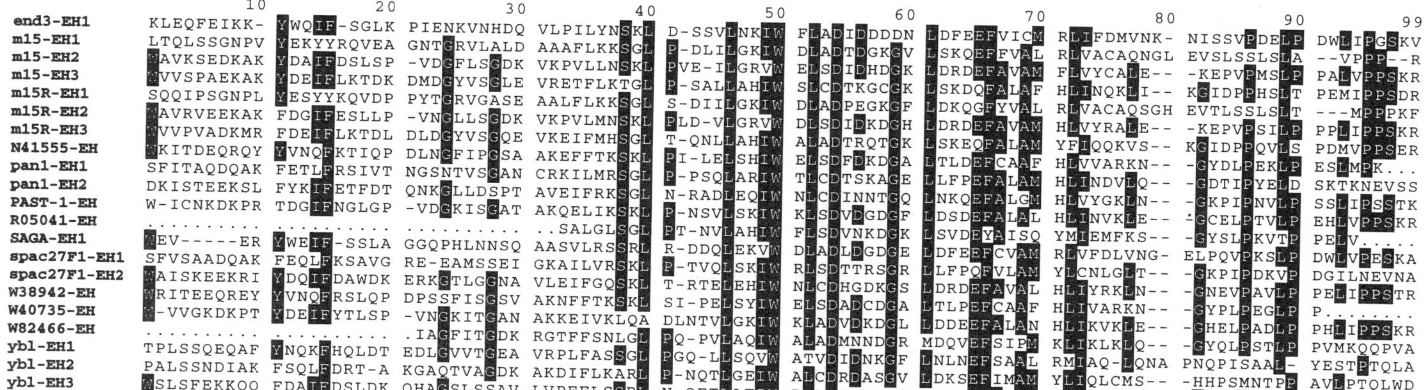

Figure 1. Alignment of various EH domains. EH domains from the indicated proteins were initially aligned by a Higgins-Sharp algorithm followed by optimization by eye. Amino acid positions, in the isolated EH domains, are indicated on the top. In the alignment, amino acids are displayed with reverse print if identical at a plurality of $\geq 50\%$ (i.e. 11 of 21 reported sequences). Proteins in the alignment include: *S. cerevisiae* End3p (end3), mouse eps15 (m15), mouse eps15R (m15R), *S. cerevisiae* Pan1p (pan1), *D. melanogaster* ORF PAST-1 (GenBank U70135), *E. nidulans* ORF SAGA (GenBank Z50037), *S. Pombe* ORF SPAC27F1.01c (spac27F1, GenBank Z69368), *S. cerevisiae* ORF YBL047c (ybl, GenBank Z35808), the predicted products of several ORFs of EST sequences, indicated by their GenBank accession number and including *H. sapiens* N41555 and W38492, *M. musculus* W40735 and W82466, and *C. elegans* R05041. A few more divergent EH domains were not included in the alignment including the 2nd EH domain of End3p (see also Fig. 2), EH domains from *S. cerevisiae* ORFs YKR019c, YNL271c, YJL083w (GenBank Z28244, Z71547, Z49358, respectively), the EH domain from the C. elegans R10E11.6 protein (predicted ORF of the YNJ6 locus, GenBank P34550) and the 2nd EH domain of *E. nidulans* ORF SAGA. The EH domains of human eps15 and of *C. elegans* ORF Zk1248 (most probably the homolog of eps15) were also not computed in order not to bias the alignment.

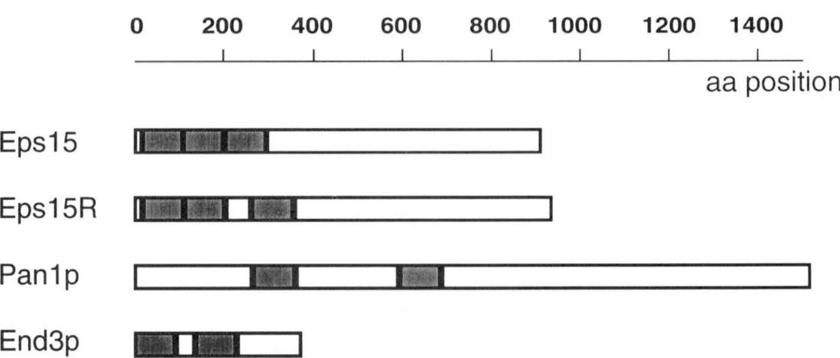

Figure 2. Multiple EH domains in various proteins. The EH domains are indicated with filled boxes. Amino acid positions are indicated on the top.

in vivo with the clathrin adapter protein complex AP-2[9,13–16]. AP-2 is a ubiquitous heterotetrameric complex which is thought to be involved in receptor-mediated endocytosis by virtue of its ability to bind to clathrin, thereby promoting clathrin lattice assembly and association to the plasma membrane, and to tyrosine-containing sequences of certain receptors[17]. The C-terminal of the α-adaptin subunit of AP-2, the so called α-ear, was shown to bind to a 70 amino acid region contained in the C-terminus of eps15 and eps15R[9,14,15]. Moreover, by immunofluorescence analysis, both eps15 and eps15R have been shown to be in part associated with plasma-membrane clathrin coated pits and vesicles, where they colocalize with AP-2 and clathrin[9,18]. These observations support a possible role of eps15 and eps15R in coated pits-mediated endocytosis. Since eps15 is distributed mostly at the rim of coated pits and at the neck of budding vesicles[18], it is possible that the protein regulates the initial steps of coated pits formation, for instance by facilitating the formation of the clathrin lattice or activating the membrane budding process. In particular eps15 and eps15R may activate the coated-pit mediated endocytosis of the EGFR, which is an important step in downregulation that follows receptor activation. The complex between AP-2 and the eps15 or eps15R proteins is constitutive, but presumably it could be recruited to the activated EGF receptor. Data indicate that upon cell stimulation with EGF, eps15 is not only phosphorylated but it also binds to the EGF receptor[16], albeit this evidence is controversial at present. The binding could be either direct or mediated by other proteins like AP-2 or Crk.

The identification of the EH protein:protein interaction domain disclosed a series of novel molecular interactions in which eps15 and eps15R are involved. By direct screening of expression libraries with a portion of eps15 encompassing the three EH domains, several putative interactors were isolated[7]. Among these interactors were the human homologue of NUMB, a developmentally regulated protein of Drosophila; NUMB-R, a protein sharing 57% identity with NUMB; RAB, a cellular cofactor of the HIV-1 Rev protein; RAB-R, a protein related to RAB; and other three novel proteins of unknown function. In a parallel study, the EH domains of eps15R were shown to be able to bind RAB, RAB-R, NUMB-R and other cellular proteins distinct from those binding to eps15 and that still await characterization.

It was subsequently demonstrated that eps15 can interact with NUMB and RAB *in vivo*[7]. NUMB is a membrane-associated protein that, during Drosophila neurogenesis, determines cell fate by segregating at cell division into only one of the two daughter cells[19,20]. Several lines of evidences, suggest a similar role for NUMB in mammals[21]. The biological function of NUMB could be mediated by its ability to inhibit the nuclear translocation, hence the activity, of the Notch transmembrane receptor which plays a key role in cell-cell communication during neurogenesis[22,23].

RAB, on the other hand, is a cellular cofactor of Rev, a regulatory protein of HIV-1 which promotes the nucleocytoplasmic transport of viral mRNAs containing intronic sequences and coding for late structural proteins required for viral replication[24]. The Rev protein is constantly shuttling between the nucleus and the cytoplasm, due to the presence of two regulatory domains: a nuclear/nucleolar localization sequence (NOS) and a nuclear export sequence (NES)[25]. RAB binds to the NES of Rev and activates Rev function through an unknown mechanism[26,27]. Since RAB displays some features that are typical of a class of nucleoporines, it was suggested that RAB can facilitate the nuclear translocation of Rev by interacting with other proteins at the nuclear pore[28]. The biological significance of the eps15:RAB and eps15:NUMB is presently under investigation.

It appears therefore as if eps15 and eps15R contract numerous interactions with other cellular proteins, through specialized regions in their N-termini (EH domains) and C-termini (Crk and AP2 binding sites). It is not known whether all binders can simultaneously associate with eps15 and eps15R, or whether a hierarchy of interactions exists *in vivo*. Besides, a more complex network of interactions could be predicted, since both eps15 and eps15R proteins can form dimers via the central domain that contains heptads repeats characteristic of the coiled-coil structure. Eps15/eps15 homodimers, as well as eps15/eps15R heterodimers, are in fact readily detected *in vivo*[9,29]. It is possible that dimerization is important for the functional interactions of the EH and the C-terminal domains of eps15 and eps15R.

4. EH-CONTAINING PROTEINS IN YEAST: End3p AND Pan1p

The End3p and Pan1p proteins of *Saccharomyces cerevisiae* are the best characterized yeast EH-containing proteins. The gene encoding for End3p was cloned by complementation of a temperature-sensitive mutant, *end3*, defective in the internalization step of the α-factor endocytosis[30,31]. End3p was subsequently shown to be required for internalization of several other yeast plasma membrane proteins, such as Ste6p[32], and the uracil and inositol permeases[33,34]. In addition, End3p was found to be essential for normal organization of the cortical actin cytoskeleton and for proper budding pattern[31]. Analysis of the structure-function relationship of End3p demonstrated that the first of the two EH domains (Fig. 2) and two repeated regions at the C-terminus are required for the protein function[31].

Like End3p, Pan1p is also involved in the organization of the actin cytoskeleton as well as in endocytosis. Loss of activity or overproduction of Pan1p results in an abnormal distribution of the actin cytoskeleton on the cell cortex[35]. Moreover, a mutant of the PAN1 gene, *pan1–20*, exhibits defects both in actin localization and in internalization of the α-factor and a lipophilic dye[36]. In this latter analysis, it was also observed that, compared to the wild type strain, *pan1–20* cells accumulate vesicles and tubulo-vesicular structures as well as plasma membrane invaginations. At the structural level, the Pan1p protein share extensive similarity with eps15 (Fig. 2): it contains two EH domains at the N-terminus, a central region rich in heptads repeats which has high probability of forming coiled-coils, and a C-terminal proline-rich domain. The second EH domain has been shown to be essential for Pan1p function[37]. More recently, this domain was found to be able to bind to the C-terminal repeats of End3p[38]. In fact, Pan1p and End3p form a complex *in vivo*, although the physiological role of this interaction must be elucidated. Since the N-terminal EH domain of End3p is also required for the protein function, it is possible that this region is contributing to the interaction with Pan1p *in vivo* or, alternatively, that it binds to other cellular proteins required for endocytosis or actin organization.

It is not clear whether the Pan1p and End3p proteins are required for endocytosis simply because they are essential for the organization of the actin cytoskeleton, or whether they are able to play a dual direct role in both processes. In support of the first hypothesis, the actin cytoskeleton has been demonstrated to play an essential role in endocytosis in yeast and in certain mammalian cells[39,40]. On the other hand, Pan1p localizes not only at actin patches, but also at other punctuate membrane structures[35]. These structures could correspond to the plasma membrane invaginations that accumulate in *pan1–20* mutant and that are distinct from actin cortical patches[36]. Since actin filaments and actin binding proteins use invaginated membrane structures for their organization, proteins that function at those membrane locations are expected to be important for the organization of the cortical actin cytoskeleton as well. Pan1p and End3p display the appropriate features to be such proteins.

5. BINDING PROPERTIES OF EH DOMAINS

The molecular basis of EH-mediated interactions was studied in detail in the case of eps15 and eps15R[7]. Data obtained by screening of combinatorial peptide libraries show that the three EH domains of eps15 and eps15R recognize a short peptide sequence, NPF (Asparagine-Phenilalanine-Proline). This result was confirmed by the ability of NPF-containing peptides to bind native eps15 and eps15R from cellular lysates. Besides, all of the proteins that were shown to bind to the EH domains of eps15 and eps15R *in vivo,* invariably contained at least one NPF motif. More importantly, it was demonstrated that the *in vivo* binding of eps15 to NUMB requires the presence in the NUMB protein of an intact NPF motif.

A mutational analysis demonstrated that an intact NPF motif is necessary but not sufficient for binding, and that optimal binding is conditioned by the presence of certain amino acids in the surrounding positions. Whether this is due to impact on conformation or directly on binding remains to be established. By analogy with other binding domains, it is conceivable that ligand specificity of each EH domain will be determined by variable residues that surround the NPF motif. In addition, since a single EH domain is sufficient for binding to an NPF-containing protein, a protein containing multiple EH domains might have diversified binding abilities. Identification of peptides binding to different EH domains and determination of the crystal structures of these domains, both with and without the cognate peptides bound, should aid in the resolution of these issues.

6. A POSSIBLE GENERAL ROLE FOR EH-CONTAINING PROTEINS

Many of the observations discussed earlier indicate a role in endocytosis for EH-containing proteins. Eps15 and eps15R, through the binding to the clathrin adapter protein complex AP-2, might function in coated pits-mediated endocytosis[9,13–16]. Besides, the yeast proteins End3p and Pan1p are necessary for endocytosis of a lipophylic dye as well as for the receptor mediated internalization of the α-pheromone[31,36]. Moreover, in a recent report an NPFxD sequence, which satisfies the criteria of binding to the EH domain, has been shown to function as a clathrin-dependent endocytosis signal in yeast[41]. Thus, EH-containing proteins appear to regulate some general step(s) of endocytosis, although their precise role is not known. It is possible that EH-containing proteins function as adapter molecules, recruiting other proteins that regulate endocytosis like for example those involved in coat formation, in membrane budding or in organization of the actin cytoskeleton. The modular structure and the ability to oligomerize of EH-containing proteins support this hypothesis.

Some characteristics and functions of EH-containing and EH-binding proteins indicate that these proteins might have other biological roles distinct from or additional to the one in endocytosis. First, already mentioned evidence suggests that eps15 has a role in cell proliferation[5,11,12]. Moreover, both eps15 and eps15R are not only localized at the clathrin coated pits, but they are also distributed in a perinuclear/Golgi area and in the nucleus respectively[9,18]. In addition, eps15 interacts *in vivo* with cellular proteins such as NUMB and RAB for which no immediate role in endocytosis is evident[7]. Interestingly both NUMB and RAB appear to serve a role in processes connected with sorting of molecules within the cell, thus raising the intriguing possibility that the whole EH-based network is involved in the regulation of these events. This hypothesis is indirectly supported by the characteristics of other putative EH interactors. A protein data-bank search was conducted to identify such proteins, by looking at sequences displaying multiple NPFs. In fact, *bona fide* EH-binders frequently have more than one NPF, a feature that mirrors the presence of multiple EH domain in their binding partners. This search yielded candidates like SCAMP37[42], which is part of a family of membrane molecules that functions in cell surface recycling, and synaptojanin[43], which is involved in synaptic vesicle recycling. Indeed, direct interaction *in vivo* between eps15 and synaptojanin has been recently demonstrated[44]. Thus EH-containing proteins might serve as centers of organization of many cellular proteins that regulate various aspects of protein and/or organelle sorting and transport.

REFERENCES

1. Cohen, G. B., R. Ren, and D. Baltimore. 1995. Modular binding domains in signal transduction proteins. *Cell* **80**: 237–248.
2. Musacchio, A., M. Wilmanns, and M. Saraste. 1994. Structure and function of the SH3 domain. *Prog. Biophys. Mol. Biol.* **61**: 283–297.
3. Bork, P. and M. Sudol. 1994. The WW domain: a signaling site in dystrophin?. *Trends Biochem Sci* **19**: 531–533.
4. van der Geer, P. and T. Pawson. 1995. The PTB domain: a new protein module implicated in signal transduction. *Trends Biochem. Sci.* **20**: 277–280.
5. Fazioli, F., L. Minichiello, B. Matoskova, W. T. Wong, and P. P. Di Fiore. 1993. Eps15, a novel tyrosine kinase substrate, exhibits transforming activity. *Mol. Cell. Biol.* **13**: 5814–5828.
6. Wong, W. T., C. Schumacher, A. E. Salcini, A. Romano, P. Castagnino, P. G. Pelicci, and P. P. Di Fiore. 1995. A protein-binding domain, EH, identified in the receptor tyrosine kinase substrate Eps15 and conserved in evolution. *Proc. Natl. Acad. Sci. U S A* **92**: 9530–9534.
7. Salcini, A. E., S. Confalonieri, M. Doria, E. Santolini, E. Tassi, O. Minenkova, G. Cesareni, P. G. Pelicci, and P. P. Di Fiore. 1997. Binding specificity and *in vivo* targets of the EH domain, a novel protein:protein interaction module. *Genes & Development* **11**: 2239–2249.
8. Di Fiore, P.P., P.G. Pelicci and A. Sorkin. EH: a novel protein:protein interaction domain potentially involved in intracellular sorting. *TIBS*, in press.
9. Coda, L., A. E. Salcini, S. Confalonieri, G. Pelicci, T. Sorkina, A. Sorkin, P. G. Pelicci, and P. P. Di Fiore. 1977. Eps15R is a tyrosine kinase substrate with characteristics of a docking protein possibly involved in coated pits-mediated internalization. Submitted.

10. Schumacher, C., B. S. Knudsen, T. Ohuchi, P. P. Di Fiore, R. H. Glassman, and H. Hanafusa. 1995. The SH3 domain of Crk binds specifically to a conserved proline-rich motif in Eps15 and Eps15R. *J. Biol. Chem.* **270**: 15341–15347.
11. Bernard, O. A., M. Mauchauffe, C. Mecucci, H. Van Den Berghe, and R. Berger. 1994. A novel gene, *AF-1p*, fused to HRX in t(1;11)(p32;q23), is not related to AF-4, AF-9, nor ENL. *Oncogene* **9**: 1039–1045.
12. Rogaia, D., F. Grignani, R. Carbone, D. Riganelli, F. LoCoco, T. Nakamura, C. M. Croce, P. P. Di Fiore and P. G. Pelicci. 1997. The localization of the HRX/ALL1 protein to specific nuclear subdomains is altered by fusion with its eps15 translocation partner. *Cancer Research* **57**: 799–802.
13. Benmerah, A., J. Gagnon, B. Begue, B. Megabarne, A. Dautry-Varsat and N. Cerf-Bensussan. 1995. The tyrosine kinase substrate eps15 is constitutively associated with the plasma membrane adapter AP-2. *J.Cell Biol.* **131**: 1831–1838.
14. Benmerah, A., B. Bègue, A. Dautry-Varsat and N. Cerf-Bensussan. 1996. The ear of α-adaptin interacts with the COOH-terminal domain of the eps15 protein. *J. Biol. Chem.* **271**: 12111–12116.
15. Iannolo, G., A. E. Salcini, I. Gaidarov, O. B. Goodman Jr., J. Baulida, G. Carpenter, P. G. Pelicci, P. P. Di Fiore, and J. H. Keen. 1997. Mapping of the molecular determinants involved in the interaction between eps15 and AP-2. *Cancer Research* **57:** 240–245.
16. van Delft, S., R. Govers, G. J. Strous, A. J. Verkleij and P. M. P. van Bergen en Henegouwen. 1997. Association and colocalization of eps15 with adaptor protein-2 and clathrin. *J. Biol. Chem.* **272**: 14013–14016.
17. Sorkin, A., and G. Carpenter. 1993. Interaction of activated EGF receptors with coated pit adaptins. *Science* **261**: 612–615.
18. Tebar, F., T. Sorkina, A. Sorkin, M. Ericsson, and T. Kirchhausen. 1996. Eps15 is a component of clathrin-coated pits and vesicles and is located at the rim of coated pits. *J. Biol. Chem.* **271**: 28727–28730.
19. Rhyu, M. S., L. Y. Jan and Y. N. Jan. 1994. Asymmetric distribution of numb protein during division of the sensory organ precursor cell confers distinct fates to daughter cells. *Cell* **76**: 477–491.
20. Knoblich, J.A., L. Y. Jan, and Y.N. Jan. 1995. Asymmetric segregation of Numb and Prospero during cell division. *Nature* **377**: 624–627.
21. Zong, W., J. N. Feder, M. Jiang, L. Y. Jan and Y. N. Jan. 1996. Asymmetric localization of a mammalian Numb homolog during mouse cortical neurogenesis. *Neuron* **17**: 43–53.
22. Frise, E., J. A. Knoblich, S. Younger-Shepherd, L.Y. Jan, and Y.N. Jan. 1996. The *Drosophila* Numb protein inhibits signaling of the Notch receptor during cell-cell interaction in sensory organ lineage. *Proc. Natl. Acad. Sci. U S A* **93**: 11925–11932.
23. Guo, M., L. Y. Jan, and Y. N. Jan. 1996. Control of daughter cell fates during asymmetric division: interaction of Numb and Notch. *Neuron* **17**: 27–41.
24. Cullen, B. R.. 1992. Mechanism of action of regulatory proteins encoded by complex retroviruses. *Microbiol. Rev.* **56**: 375–394.
25. Meyer, B. E., and M. H. Malim. 1994. The HIV-1 *trans*-activator shuttles between the nucleus and the cytoplasm. *Genes & Development* **8:** 1538–1547.
26. Bogerd, H. P., R. A. Fridell, S. Madore and B. R. Cullen. 1995. Identification of a novel cellular cofactor for the Rev/Rex class of retroviral regulatory proteins. *Cell* **82**: 485–494.
27. Fritz, C. C., M. L. Zapp, and M. R. Green. 1995. A human nucleoporin-like protein that specifically interacts with HIV Rev. *Nature* **376**: 530–533.
28. Stutz, F., E. Izaurralde, I. W. Mattaj and M. Rosbash. 1996. A role of nucleoporin FG repeat domains in export of human immunodeficiency virus type 1 Rev protein and RNA from the nucleus. *Mol. Cell. Biol.* **16**: 7144–7150.
29. Tebar, F., S. Confalonieri, R. E. Carter, P. P. Di Fiore, and A. Sorkin. 1997. Eps15 is costitutively oligomerized due to homophilic interaction of its coiled-coil region. *J. Biol. Chem.* **272**: 15413–15418.
30. Raths, S., J Rohrer, F. Crausaz, and H. Riezman. End3 and end4: two mutants defective in receptor-mediated and fluid-phase endocytosis in *Saccaromices cerevisiae*. 1993. *J. Cell Biol.* **120**: 55–65.
31. Benedetti, H., S. Raths, F. Crausaz, and H. Riezman. 1994. The END3 gene encodes a protein that is required for the internalization step of endocytosis and for actin cytoskeleton organization in yeast. *Mol. Biol. Cell.* **5**: 1023–1037.
32. Kolling, R., and C. P. Hollenberg. 1994. The ABC-transporter Ste6 accumulates in the plasma membrane in an ubiquitinated form in endocytosis mutants. *EMBO J.* **13**: 3261–3271.
33. Lai, K., C. P. Bolognese, S. Swift, and P. McGraw. 1995. Regulation of inositol transport in *Saccaromices cerevisiae* involves inositol-induced changes in permease stability and endocytic degradation in the vacuole. *J. Biol. Chem.* **270**: 2525–2534.
34. Volland, C., D. Urban-Grimal, G. Geraud, and R. Haguenauer-Tsapis. 1994. Endocytosis and degradation of the yeast uracil permease under adverse conditions. *J. Biol. Chem.* **269**: 9833–9841.

35. Tang, H-Y., and M. Cai. 1996. The EH-containing protein Pan1 is required for normal organization of the actin cytoskeleton in saccaromyces cerevisiae. *Mol. Cell. Biol.* **16**: 4897–4914.
36. Wendland B., J. M. McCaffery, Q. Xiao, and S. D. Emr. 1996. A novel fluorescence-activated cell sorter-based screen for yeast endocytosis mutants identifies a yeast homologue of mammalian eps15. *J. Cell. Biol.* **135**: 1485–1500.
37. Sachs, A. B., and J. A. Deardorff. 1992. Translation initiation requires the PAB-dependent poly(A) ribonuclease in yeast. *Cell* **70**: 961–973.
38. Tang, H. Y., A. Munn, and M. Cai. 1997. EH domain proteins Pan1p and End3p are components of a complex that plays a dual role in organization of the cortical actin cytoskeleton and endocytosis in *Saccaromyces cerevisiae*. *Mol. Cell. Biol.* **17**: 4294–4304.
39. Riezman, H. 1993. Yeast endocytosis. *Trends Cell Biol.* **3**: 273–277.
40. Gottlieb, T. A., I. E. Ivanov, M. Adesnik, and D. D. Sabatini. 1993. Actin microfilaments play a critical role in endocytosis at the apical but not the basolateral surface of polarized epithelial cells. *J. Cell Biol.* **120**: 695–710.
41. Tan, P. K., Howrad, J. P., and Payne, G. S.. 1996. The sequence NPFXD defines a new class of endocytosis signal in saccharomyces cerevisae. *J. Cell. Biol.* **135**: 1789–1800.
42. Brand, S.H. and J.D. Castle. 1993. SCAMP 37, a new marker within the general cell surface recycling system. *EMBO J.* **12**: 3753–3761.
43. McPherson, P.S., E.P. Garcia, V.I. Slepnev, C. David, X. Zhang, D. Grabs, W.S. Sossin, R. Bauerfeind, Y. Nemoto, and P. De Camilli. 1996. A presynaptic inositol-5-phosphatase. *Nature* **379**: 353–357.
44. Haffner, C., K. Takei, H. Chen, N. Ringstad, A. Hudson, A.E. Salcini, P.P. Di Fiore and P. De Camilli. 1997. Eps15, an EGF receptor substrate associated with the clathrin adaptor AP2, interacts via its EH domain containing region with the 170 kDa isoform of synaptojanin. Submitted.

12

DEREGULATION OF CYCLIN D1 IN CANCER

Rob Michalides*

Division of Tumor Biology
The Netherlands Cancer Institute
Plesmanlaan 121, 1066CX Amsterdam, The Netherlands

1. INTRODUCTION

Cell cycle progression in eukaryotic cells is governed by a series of cyclins and cyclin dependent kinases. Individual cyclins act at different phases of the cell cycle by binding and stimulating the activities of cdk's. Because these cyclins and cdk's are pivotal to cell cycle control and thereby cell proliferation, mutational changes and alterations in expression of the corresponding genes play a critical role in transformation and tumor progression (for reviews, see Sherr, 1996; Hunter, 1997). In epithelial cells most of these alterations affect the requirement of cells to respond to external growth factors and to adhere onto extracellular matrix components for proliferation, and involve the regulatory circuits controlling the transitions of the G_1 and S phase of the cell cycle.

2. G_1-S CYCLINS

The commitment of mammalian cells in late G_1 to replicate in response to mitogenic factors depends ultimately on phosphorylation of the retinoblastoma protein, pRb, a process controlled by cyclin D1, cyclin D1-associated cyclin-dependent kinases, cdk's, and their inhibitors, cdi's (Figure 1). Mitogenic signal transduction pathways from three different classes of receptors, the membrane tyrosine kinase receptors activated by serum mitogens or EGF, estrogen receptors triggered by estradiol, and the cyclic AMP-dependent signalling from G-protein coupled receptors, all converge and strictly require cyclin D-cdk activity to induce S-phase in epithelial cells such as MCF-7 and primary dog thymocytes (Lukas et al., 1996). A transient accumulation of cyclin D1 protein in response to mitogenic stimulation results in binding to, and activation of its cdk partner, predominantly cdk-4, and subsequently in phosphorylation of its major target, pRb. Phosphorylation of

* tel. +31 20 5122022; fax. +31 20 5122029

The Biology of Tumors, edited by Mihich and Croce
Plenum Press, New York, 1998.

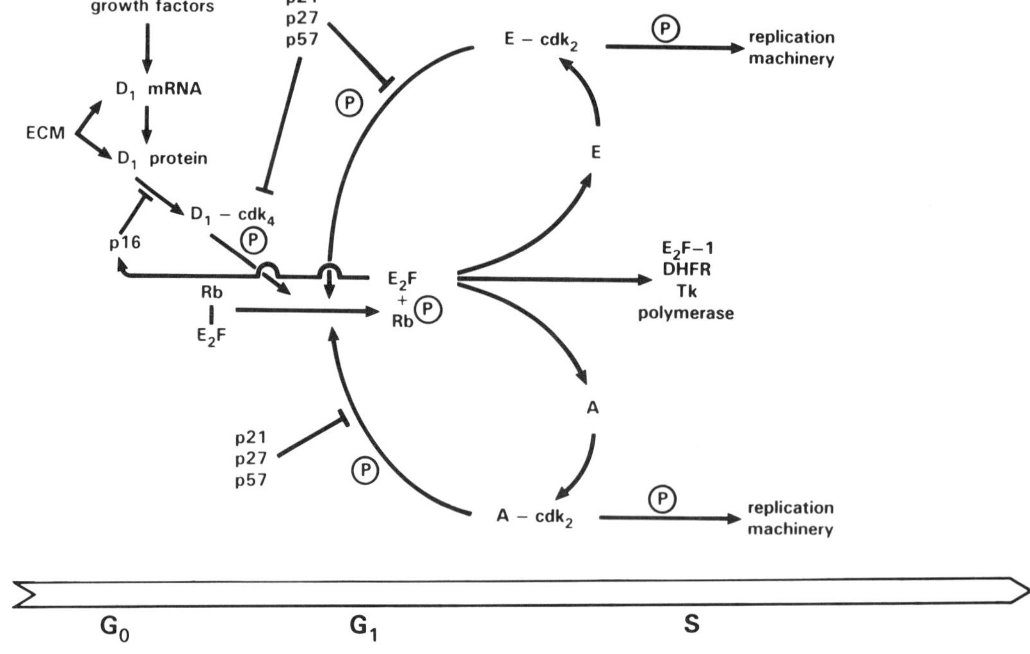

Figure 1. Regulation of transition through G_1 and S phase of the cell cycle by cyclin-cdk's and cki's.

pRb results in the release of E2F transcription factor(s) that mediate(s) transcription of genes essential for further progression through the cell cycle (Weinberg, 1995). pRb appears unique among the pocket proteins which also include p107 and p130, since elimination of the function of pRb alone is sufficient to liberate abundant E2F activity and to provide significant growth alterations to the cells (Herrera et al., 1996; Cobrink et al., 1996).

Cyclin D1 does complex with cdk4 or cdk6 to regulate the early to mid G_1 transition of the cell cycle (Ewen et al., 1993), whereas cyclin E-cdk2 and cyclin A-cdk2 complexes control the G_1-S and S phase transition, respectively (Ohtsubo et al., 1995; Sherr 1996). Cyclin E and A are both E2F responsive genes (Schulze et al., 1995; Ohtani et al., 1995), implying that once a cyclin D1-cdk4 activity has set the G_1 regulatory circuit into motion, cyclin E- and A-cdk2 activity is induced and acts on progression through the cell cycle.

All the three G_1-S cyclins, D, E and A, phosphorylate pRb *in vitro*, with cyclin D1 being the first one in action *in vivo* and being rate-limiting for G_1-phase progression. The effects as well as the sites of pRb phosphorylation by cyclin D1-cdk4 are different from those by cyclin E- and cyclin A-cdk2 (Resnitzsky et al., 1995; Ohtsubo et al., 1995; Kitagawa et al., 1996), indicating a division of labour among the G_1 cyclins. Indeed, cyclin E-cdk2 and, more prominently, cyclin A-cdk2 activity is in late G_1 phase involved in activation of the DNA replication machinery.

Three closely related human D-type cyclins have been identified, all of which interact with, and activate cdk4 and cdk6, although cyclin D2 also activates cdk2 (Sweeney et al., 1996). Cyclins D2 and D3 may have specialized functions in distinct cell types. These cyclins behave similar to cyclin D1, but appear, however, not to be involved in tumor development (Bates and Peters, 1995).

3. REGULATION OF ACTIVITY OF THE G_1-S CYCLINS

The serine-threonine kinase activity of cyclin-cdk's becomes manifest when the cyclin-cdk's have overcome inhibitory thresholds set by inhibitors of cyclin dependent kinase, cki's (Sherr and Roberts, 1995). Some of the cki's, including p16, p18, and p19 (taken together as the INK4 family), specifically inhibit cyclin D1-cdk4/6 activity by binding to either cdk4 or cdk6, thereby preventing association between cyclin D1 and its catalytic partner. The free form of cyclin D1 is degraded much faster than de cdk-bound form, resulting in lower levels of cyclin D1 in cells with an overexpression of p16 (Bates et al., 1994). This situation is met in tumor cells with a functional inactivation of pRb resulting in overexpression of E2F responsive p16 and in low levels of cyclin D1 protein (for review, see Bartek et al., 1996).

The other class of cki's, including p21, p27 and p57, inhibit all of the cyclin-cdk's. (Sherr and Roberts, 1995). Since cyclin D-cdk complexes are formed early during G_1 and bind to cki's p21 and p27 prior than the cyclins E-cdk2 and A-cdk2 do, cyclin D-cdk complexes are perceived to titrate out the inhibiting effects of cki's p21 and p27. Once cyclin E- and cyclin A-cdk2 complexes are formed upon cyclin D-cdk activity, p21 and p27 cki's now act subsequently upon these newly formed complexes, thus regulating the order of cyclin-cdk activities during G_1-S.

Cyclin D1 mRNA and subsequent protein synthesis is induced when arrested cells are released from G_0 to G_1 by growth factors (Baldin et al., 1993). Protein levels of cyclin D slightly fluctuate during the cell cycle, they contain a PEST destruction motif which accounts for their short half-lives (Bates and Peters, 1995). Accumulation of cyclin E, however, is highly periodic, with a sharp decline during S phase, which is due to autophosphorylation of cyclin E on threonine 380 by cyclin E-cdk2, rendering cyclin E now a target for ubiquitin-dependent degradation (Won and Reed, 1996).

Adhesion onto extracellular matrix (ECM) components, such as laminin, fibronectin or collagen, is mandatory for G_1 progression of normal epithelial cell. Adhesion affects induction of cyclin D1 mRNA and protein and results in a yet undissolved manner into degradation of p27 when this becomes complexed to cyclin E-cdk2 (Fang et al., 1996; Schulze et al., 1996; Assoian, 1997, Michalides and Muris, non-published data). Degradation of p27 during that step results in cyclin E.cdk2 activity yielding progression of the cell cycle.

Although the level of cyclin D1 protein, and subsequently of cyclin D1-cdk4 activity, is influenced by adhesion of cells onto extracellular matrix components, a major control over adhesion restricted proliferation of cells is exerted by cyclin A and E-cdk2 activity. Overexpression of cyclin A or E, or reduced levels of p27 enable normal adherent cells to proliferate in suspension (Guadagno et al., 1993). Overexpression of cyclin E and reduced expression of p27 have recently been observed in breast cancer and were found to be associated with poor prognosis (Porter et al., 1997; Catzavelos et al., 1997).

The overwhelming information over the last years on the function of cyclins has led to the qualification of cyclin D1-cdk4 as a "sensor" of cells for growth factor conditions, whereas cyclin D1-cdk4 together with cyclin E-cdk2 act as "sensors" of cellular attachment onto ECM. Deregulation of these cyclins releases the restrictions imposed by the external regulators on cell proliferation and affects the determinative phosphorylation of pRb which permits progression of the cell cycle. In the scenario of increasing cellular autonomy, functional inactivation of pRb represents the most drastic genetic alteration leading to autonomy, whereas deregulation of cyclins and cki's do in part overcome the restrictions on proliferation imposed by external factors.

4. CANCER

Cancer arises as a consequence of multiple genetic alterations in the cell resulting in a continuing progression of tumor cells which evolve from previous stages of tumor development. These changes during tumor progression lead to a diminished control over cellular proliferation and reduce the ability of cells to differentiate and to enter apoptosis under less favorable conditions. The major achievement in tumor biology over the last years has been the identification of molecular events that coincide with the various stages of tumor progression. The challenge ahead lies in how to use these genetic alterations themselves as a direct target for anti-tumor therapy or as markers for prognosis and to identify tumor stages for a better treatment.

A coordinated control by growth factors and the extracellular matrix is required for progression of the G_1 phase of the cell cycle in normal epithelial cells. This control is disturbed in tumor cells. In most of the different tumor types genetic alterations have been found that affect the cyclin D1/cdk4/p16/pRb regulatory circuit of G_1 progression (for review, see Bartek et al., 1996). Tumor cells with alterations in this pathway become less dependent on external growth factors. These genetic alterations in tumors involve either amplification of cyclin D1, mutation of p16 or mutation of pRb (the latter coincides with an increase of p16 since the expression of p16 depends on free E2F) and, although less frequently observed, amplification of cdk4, see Figure 2. Increased levels of p16 result in a decrease of cyclin D1, since p16 competes with cyclin D1 in binding to cdk4 and free cyclin D1 is more prone to degradation (Bates et al., 1994; Parry et al., 1995). Because there have been many recent excellent reviews on genetic alterations in cancer, including on cyclin D1 (Hall and Peters, 1996), treatment of the subject in this chapter will restrict itself to prognostic value of deregulation of cyclin D1, the selective advantage that may be conferred by it, and to the application of overexpression of cyclin D1 to identify tumor stages for treatment.

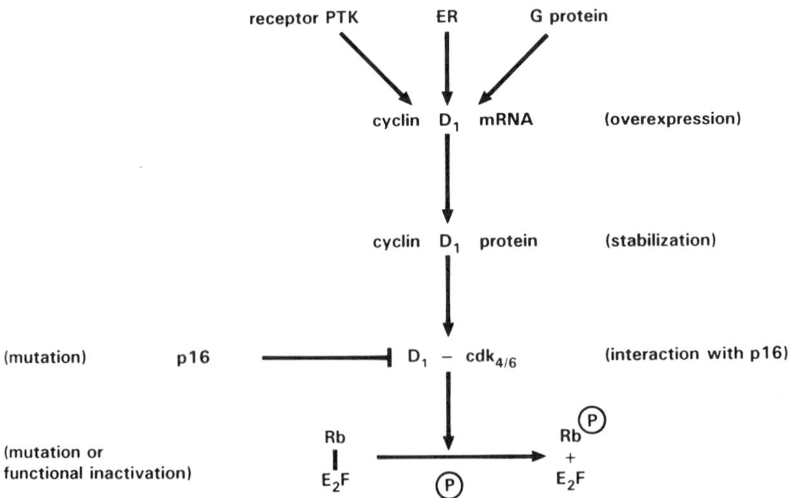

Figure 2. Activation of cyclin D1-cdk 4/6 activity. Mode(s) of deregulation of cyclin D1-cdk4/6 kinase activity is in parentheses.

5. DEREGULATION OF CYCLIN D1 IN CANCER CELLS

The first case of overexpression of cyclin D1 was found in parathyroid adenomas, where an inversion of part of chromosome 11, inv(11)(p15:13), placed cyclin D1 adjacent to the promotor of the parathyroid hormone gene (Rosenberg et al., 1991). This rather infrequent alteration led to the identification of cyclin D1 gene, that was also found to be overexpressed as a result of t(11; 14) translocations in a relative low number of chronic B cell lymphomas, but in practically all mantle cell lymphomas (de Boer et al., 1993; Swerdlow et al., 1995).

In a few B cell lymphomas, overexpression of cyclin D1 is ascribed to interstitial deletion or rearrangements that affect the 3' untranslated region of exon 5, resulting in removal of the AUUUA destabilizing motifs (Rimokh et al., 1994). By far, most cases of overexpression of cyclin D1 are associated with amplification of the 11q13 amplicon that encompasses cyclin D1 (for review, see Hall and Peters, 1996). These tumors include carcinoma of the breast, liver and bladder, squamous cell carcinoma of the head and neck, esophagus, and lung, sarcoma and intestinal adenoma of patients with familial adenomatous polyposis. In the latter, concurrent overexpression of cyclin D1 and cdk4 was found in proliferative tumor areas (Zhang et al., 1997).

Cyclin D1 overexpression is observed in early stages of development of breast tumor (Weinstat-Saslow et al., 1995; Peters and Michalides, unpublished data), of squamous cell carcinoma of the head an neck (Michalides et al., 1997) and of intestinal adenoma (Zhang et al., 1997). Abnormal upregulation of this important G_1 regulator is therefore a relatively early event in progression of these tumors and may define a transition from a benign stage to commitment to carcinoma. Cyclin D1 overexpression is found in 30–50% of infiltrating breast carcinomas and in 40–50% of squamous cell carcinomas of the head and neck, SCC-HN. In approximately half of the breast carcinomas with overexpression of cyclin D1, this is due to amplification of cyclin D1 gene, whereas almost all SCC-HN cases with overexpression of cyclin D1 are associated with amplification.

The clinical relevance of overexpression of cyclin D1 is different for the various tumor types, and may depend largely on variance in the size of tumor panels analyzed, the composition of the tumor panels, the methodology used, and the negation of other genetic alterations in conjunction with cyclin D1 overexpression. In case of mantle cell lymphomas, where practically all of the pathologically diagnosed mantle cell lymphomas carry a (11; 14) translocation resulting in overexpression of cyclin D1, this genetic alteration is apparently diagnostic for, and instrumental in progression of this particular form of lymphomas. Such an intimate association between cyclin D1 overexpression and a particular tumor subtype has not yet been found in other tumors. Either sufficient parameters to distinguish these subtypes are lacking, or overexpression of cyclin D1 is just one of the multi-factorial genetic alterations contributing to cancer development.

6. GENERAL TRENDS IN TUMORS WITH OVEREXPRESSION OF CYCLIN D1

For *breast cancer*, various studies indicated that amplification and corresponding overexpression of cyclin D1 is associated with a more advanced stage of disease, but no such association was found when overexpression of cyclin D1 was judged by immunohistochemistry alone (Gillett et al., 1994, 1996; Michalides et al., 1996; McIntosh et al.,

1995; Hall and Peters, 1996). Moreover, most of the larger studies found a significant association between cyclin D1 overexpression and positivity for estrogen receptor (Gillett et al., 1996; Michalides et al., 1996; Barbareschi et al., 1997). Since approximately half of the breast tumor cases with cyclin D1 overexpression as judged by immunohistochemistry, result from cyclin D1 amplification, these findings might well suggest that cyclin D1 overexpression as a result of DNA amplification may contribute to cancer development in a different manner than overexpression of cyclin D1 associated with estrogen receptor activation without a cyclin D1 gene amplification. The limit of detection of cyclin D1 protein by immunohistochemical methods is comparable to a level found in tumors with a three fold amplification of the cyclin D1 gene (Gillet et al., 1994; Michalides et al., 1996). Immunohistochemistry makes no distinction between overexpression of cyclin D1 as a result from amplification or from other mechanism(s). This seeming contradiction in the contribution of overexpression of cyclin D1 to breast cancer prognosis may well have to do with cyclin D1 expression being influenced by activation of the estrogen receptor, ER. Elevated cyclin D1 expression reached by ER activation is likely of no relevance to breast cancer development, since the expression levels are similar to those present during normal mammary gland development (Sutherland et al., 1993). But, since estrogen receptor positivity is viewed as a marker for better prognosis in breast cancer, a coinciding increase in cyclin D1 expression would then also be associated with better prognosis, whereas overexpression of cyclin D1 as a result of gene amplification may well indicate the opposite. The two mechanisms of overexpression of cyclin D1, amplification versus estrogen induced expression, might, when taken together, therefore neutralize any effect of cyclin D1 overexpression on tumor progression.

Amplification of cyclin D1 occurs preferentially in breast tumor that are positive for the estrogen receptor, with, in general, a better outlook on prognosis for its carriers. In the few cases where overexpression of cyclin D1 was observed in ER-negative breast tumors, this seems to contribute to a better prognosis (Gillet et al., 1996). This finding vitiates the hypothesis that overexpression of cyclin D1 would be involved in progression of hormone-dependent breast tumor cells to become hormone-independent. Evaluation of these studies is, however, complicated by the way of measuring functionality of the ER and by the negation of other genetic alterations which may interfere with cyclin D1 activity. ER status is usually determined by either estradiol-binding or by immunohistological methods, which does not always reflect a functionally active ER. Determination of functionality of ER by measuring, for instance, levels of progesterone receptors that are induced by activated ER would be more indicative than measuring presence of ER alone. The lack of information on status of pRb in the tumors is another shortcoming in these studies. Functional inactivation of pRb would overrule any deregulation of cyclin D1. Similarly, the contribution of mutation in cki p16 to breast tumor development is still unclear.

In *squamous cell carcinomas of the head and neck*, SCC-HN, many groups have reported a correlation between overexpression of cyclin D1 and more advanced stage of disease, lymph node involvement and reduced overall survival time or time to recurrence (Meredith et al., 1995; Michalides et al., 1995,1997; Masuda et al., 1996; Bellacosa et al., 1996; Fracchiolla et al., 1997). However, in most of these studies, the association between cyclin D1 overexpression and advanced stage of disease barely reached statistical significance, indicating that other factors together with overexpression of cyclin D1 co-determine the fate of the tumor. It is not yet clear how and which of the other genetic alterations that are frequently observed in SCC-HN, including p53 and p16 mutation and amplification of EGF-R, do in combination with overexpression of cyclin D1 contribute to poor prognosis in SCC-HN.

It is evident that many breast tumors and SCC-HN have acquired genetic alterations which do affect the cyclin D1/cdk4/p16/pRb regulatory pathway. With a better methodology and a combination of data in a large tumor series, one may even wonder whether deregulation of this circuit is not a prerequisite for development of these and other tumor types. Since most of the deregulations in this pathway take already place at an early stage of tumor development, this deregulation may well provide the means for a less restricted proliferation of cells from which invasive descendants arise.

7. SELECTIVE ADVANTAGE(S) BESTOWED ON TUMOR CELLS BY OVEREXPRESSION OF CYCLIN D1

Cyclin D1 cooperates with activated oncogenes such as *ras* and *myc* in transformation of embryo fibroblast cells (Hinds et al., 1994). Enforced overexpression of cyclin D1 alone in these and other cell types does not lead to transformation, but results in a more rapid transition through G_1 and in a reduced dependency on growth factors and on adhesion for cellular proliferation. These events may either alone or in combination provide a selective advantage to tumor cells with overexpression of cyclin D1.

7.1. Overexpression of Cyclin D1 and Reduced Dependency on Growth Factors

In order to study any effect of overexpression of cyclin D1 on cell growth, we, and others, have generated stably transfected cells with a regulatable overexpression of cyclin D1. We used in these experiments MCF7 breast tumor cells and a tetracycline regulatable promotor-cyclin D1 construct, and generated cell clones in which the expression of cyclin D1 was maximally eight-fold increased. Inducible overexpression of cyclin D1 in these MCF-7 breast cancer cells led to mitogen independent proliferation (Zwijsen et al., 1996). This mitogen independent proliferation resulted from a reduced exit from the G_1 to G_0 phase of the cell cycle in cells with an overexpression of cyclin D1. This modulation of G_1 to G_0 exit by overexpression of cyclin D1 adds a novel activity to the illustrious list of properties of cyclin D1, see Table 1. From previous studies, it was concluded that overexpression of cyclin D1 shortened the G_1 phase and enhanced exit from quiescent phase, G_0, to enter G_1 (Resnitzky et al., 1994; Sherr, 1995). Moreover, in estrogen responsive T47D breast tumor cells overexpression of cyclin D1 was shown to reduce the requirements for growth factors, including estrogens, in proliferation of cells (Mushgrove et al., 1993,1994).

The combined effects of overexpression of cyclin D1, i.e. a reduced exit from G_1 to G_0, an increased exit from G_0 to G_1 and reduced requirements for growth factors, yield a greater growth fraction of tumor cells when they are exposed to limiting amounts of growth factors, providing under those circumstances a selective advantage to tumor cells with overexpression of cyclin D1. We assume that this selective advantage does not only apply to the experimental MCF7 cells, but to *in vivo* tumors as well.

7.2. Cyclin D1, an Activator of Estradiol Mediated Proliferation

The immediate downstream event in activation of the estrogen receptor, ER, is an increased expression of ER responsive genes, such as progesterone receptor, Cathepsin D or

Table 1. Effects of overexpression of cyclin D1*

In vitro:
• transformation of primary cells in conjunction with another activated oncogene, such as *ras* or *myc*
• reduces growth factor dependency
• abrogates adhesion-restricted proliferation in NRK cells
• shortens G_1 phase of the cell cycle
• induces DNA amplification
• induces apoptosis under GF-depleted conditions
• induces cyclin kinase inhibitors, p16, p27 and p21
• stimulates quiescent cells to enter G_1
• prevents entry into G_0
• activates ER mediated transcription.
In vivo:
• induces hyperproliferation in cyclin-D1 transgenic mice.
• is associated with advanced stage of disease in some tumor types.

*References are given in the text.

pS2 (Klijn et al., 1993; Foekens et al., 1994), but their role in cell proliferation control is mostly limited. Estradiol-mediated proliferation is likely to be mediated via cyclin D1, since estradiol mediated proliferation of breast epithelium during pregnancy is abrogated in cyclin D1 knock-out mice (Sicinski et al., 1995; Fantl et al., 1995). Furthermore, cyclin D1 overexpression can also bypass a cell cycle arrest that is induced by deprivation of estrogens (Zwijsen et al., 1996; Musgrove et al., 1994).

Despite its crucial role in mediating estradiol-induced proliferation, it is still not yet clear how activated ER induces cyclin D1 expression. The structure of the cyclin D1 promoter contains various binding sites for transcription factors, among which a perfect AP1 *fos/jun* binding site, but it lacks a *bona fide* ERE sequence to wich an activated ER would bind (Herber et al., 1994). Estradiol- mediated activation of cyclin D1 could well occur via *fos-jun* activation, one of the first nuclear events following growth factor stimulation. Estrogen stimulation results in the induction of *fos* and *c-myc*, reaching maximal levels of *fos* within 30 minutes (Weisz and Bresciani, 1988; Mushgrove et al., 1993; Bonapace et al., 1996). The response of *fos* to estrogen results from a direct interaction between ER and an estrogen responsive element (ERE) upstream of the transcription start site in the human fos gene (Weisz and Rosales, 1990). Induction of cyclin D1 by activated ER may, however, also occur via the release of a negative control factor from the cyclin D1 promotor under the action of an activated ER (Altucci et al., 1996). By either mechanism, cyclin D1 appears to be an immediate target for activated ER.

A link between estrogen mediated proliferation and ER mediated transcription is strongly indicated by the inhibition of both by anti-estrogens. Anti-estrogens bind to the ER in a manner that is competitive with estrogen but fail to effectively activate gene transcription. Anti-estrogen binding to ER does not inhibit binding of ER to ERE-DNA sequences, however (Katzenellenbogen et al., 1985; Thompson et al., 1989; Jordan and Murphy, 1990; Reese and Katzenellenbogen, 1991). The anti-estrogen 4-hydroxytamoxifen has mixed agonistic and antagonistic activities and inhibits the AF-2 transactivation domain of ER, whereas ICI 164.384 and ICI 182.780 act as a pure anti-estrogen and block activities of both transactivation domains AF-1 and AF-2 (Berry et al., 1990; DeFriend et al., 1994; Metzger et al., 1995). Short term treatment of MCF-7 cells with ICI 182.780 caused a significant reduction in the expression of proliferation marker Ki67, and of estrogen-regulated genes such as the progesterone receptor and pS2 (DeFriend et al., 1994). Moreover, treatment of T47D breast cancer cells with pure anti-estrogen ICI 164.382 reduced expression of cyclin D1 and is associated with inhibition of cyclin dependent kinase activity and decreased pRb phosphorylation (Musgrove et al., 1993; Watts and al, 1995).

These findings strongly indicate that ER mediated proliferation of cells is exerted via induction of cyclin D1.

7.3 Activation of ER by Cyclin D1

The similarity in actions generated by ER mediated transcription and cyclin D1 (both stimulate proliferation of breast epithelium) led us to investigate a possible relationship between these two activities. In these studies, ER mediated transcription was activated in ER-containing T47D breast tumor cells by treatment with 17β-estradiol and was read from the activity of a chloramphenicol acetyl transferase (CAT) reporter gene containing an upstream ERE site, that was transfected into these cells. Using this system, we found that introduction of cyclin D1 via transfection resulted in a strong enhancement of ERE-responsive gene transcription. In the presence of estradiol, transcription of the ERE-containing reporter gene increased linearly with increasing amounts of cyclin D1 expression vector. Cyclin D1 did not induce ER activity when the reporter gene was lacking an ERE sequence, or when the same experiments were performed in U2OS cells, lacking an ER. These results demonstrated that cyclin D1 induction of ER required a functional ER and ERE sequences in the transcription of the reporter gene (Zwijsen et al., 1997).

Striking was that cyclin D1 did not require its "normal" cdk4 kinase partner to activate ER mediated transcription. This was also apparent by assaying a mutant of cyclin D1 (cyclin D1-KE), which carries a mutation in the cyclin box and fails to bind to cdk's. Just like wild type cyclin D1, the cyclin D1-KE mutant potentiated transcriptional activation in a dose-dependent manner. Significantly, co-transfection of cyclin D1 and cdk4 with p16, a cdk inhibitor which competes with cyclin D1 for binding to cdk4, enhanced transcription to a level comparable with those obtained with cyclin D1 alone. Taken together, these data indicate that cyclin D1 activates ER independently of its cdk4 kinase partner.

Cyclin D1 appeared to be able to activate ER in the absence of estradiol, but to synergize with estradiol in stimulating ER mediated response in the presence of estradiol. Therefore, cyclin D1 can even substitute for estrogen in activating ER.

In vitro binding studies and mammalian two hybrid binding studies *in vivo*, in which transcription of a reporter gene construct depends on interaction between two proteins, provided evidence for a direct interaction between cyclin D1 and ER proteins which involved the ligand binding domain of ER (Zwijsen et al., 1997). The interaction between cyclin D1 and ER was not depending on 17β-estradiol, suggesting that cyclin D1 interacts with liganded as well as with unliganded ER. These results indicate that cyclin D1 interacts with the ER *in vivo* and can regulate ER-mediated transcription through protein-protein interactions. These data are summarized in Figure 3. They highlight a novel role of cyclin D1 in growth regulation of estrogen-responsive tissues and provide also a possible mechanism for the abrogation of proliferation of mammary epithelium in cyclin D1 knock-out mice, as has recently been reported (Sicinski et al., 1995; Fantl et al., 1995). In these experiments, targeted disruption of the cyclin D1 gene prevented proliferation of the mammary gland during pregnancy, suggesting that cyclin D1 is either the mediator of ER induced growth regulation, or that the growth promoting effects that result from activation of ER are only manifest when they become magnified by cyclin D1.

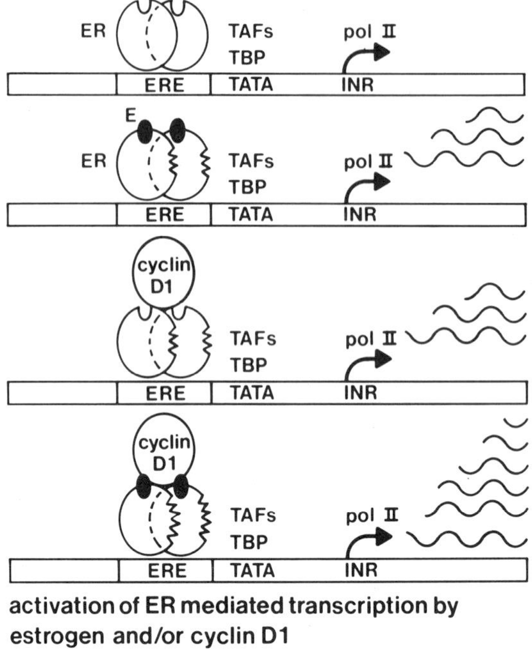

Figure 3. Activation of ER mediated transcription by estrogen and/or cyclin D1.

It is tempting to implement from these findings that hormone dependent breast tumor cells may progress into hormone-independent cells and thereby acquire a more advanced tumor phenotype by amplification and/or overexpression of cyclin D1. In seeking support for this hypothesis, the following points of concern arise:

- retrospective clinical studies, thus far, showed no effect of overexpression of cyclin D1 on progression in neither ER positive- or ER negative breast tumors (Gillet et al., 1996; Michalides et al., 1996). The conclusions from these studies may however, be weakened from shortcomings in measuring ER functionality and negation of pRb and p16 status as mentioned above.
- the effect of ER activation by cyclin D1 is observed in transient transfections in which extremely high expression levels of cyclin D1 protein are obtained in the transfected cells. Under those conditions the transfected cells easily acquire a hundred or more fold overexpression of cyclin D1, which is hardly ever obtained in stably transfected cells, and most likely also not in breast tumor cells *in vivo*. In these transient transfection studies, a massive overexpression of cyclin D1 finally results in apoptosis, reason why these were short-term experiments in which the data were related to expression of a non-estrogen responsive reporter construct as an internal reference. Furthermore, one must take into account that the interaction between cyclin D1 protein and its catalytic cdk4 partner is preferred three fold over binding to ER (Zwijsen et al., 1997).

These considerations render it difficult to just measure expression of cyclin D1 in breast tumor specimen by immunohistochemistry alone and to relate these semi-quantitative data with clinical effect(s).

These problems may, however, be overcome by studying tamoxifen resistance in breast cancer in relation to cyclin D1 overexpression. In our studies, we found surprisingly that anti-estrogen 4-hydroxytamoxifen did not impede activation of ER by cyclin D1, whereas it does so in the presence of estradiol. Activation of ER dependent transcription may also occur in the absence of its cognate estradiol ligand by EGF or TGF-α, by the neurotransmitter dopamine, and by activators of protein kinase A (Ignar-Trowbridge et al., 1993; Kato et al., 1995; Bunone et al., 1996; Smith et al., 1993; Aronica et al., 1994). Estrogen-independent activation of ER by EGF or TGF-α involves the MAPK kinase pathway and a direct phosphorylation of serine residue at position 118. This activation of ER by either EGF, insulin-like growth factor or activated ras, affects the N-terminal transactivation domain AF-1 and is not inhibited by tamoxifen, but is inhibited by pure anti-estrogen ICI 182.780 which inhibits both the N- and C terminal transactivation domains of ER, AF-1 and AF-2 respectively (Kato et al., 1995; Bunone et al., 1996). These various mechanisms of estrogen independent, tamoxifen-insensitive activation of ER may be responsible for failure of tamoxifen treatment of breast cancer that occurs in approximately one third of ER positive breast cancer cases treated with this anti-estrogen. A prediction from these studies would be that breast tumors that are positive for ER and show overexpression of cyclin D1 would not respond to tamoxifen treatment.

8. CONCLUSIONS

Activation of ER results in induction of cyclin D1 expression, whereas excessive amounts of cyclin D1 activate and stimulate ER mediated events. This co-stimulatory action provides a mechanism for estradiol mediated proliferation of mammary gland cells in mice, since disruption of one of these co-stimuli in cyclin D1 knock-out mice prevented estradiol-mediated outgrowth of mammary gland epithelium (Sicinsky et al., 1995; Fantl et al., 1995). Moreover, over-activation of cyclin D1 in cyclin D1-transgenic mice resulted in development of hyperplasia in the mammary gland, and eventually of mammary carcinomas (Wang et al., 1994). Additional oncogenic alterations apparently contributed to the latter development of carcinoma. These animal model systems provide strong evidence for an interactivity between cyclin D1 and ER and for an involvement of deregulation of cyclin D1 in breast cancer development. The mouse model system also clearly demonstrated that additional genetic alterations determine together with deregulation of cyclin D1 the fate of tumor progression. This applies most likely also to human breast cancer. In combination with such alterations, overexpression of cyclin D1 may well provide a marker to identify ER-positive breast cancer and HN-SCC patients for a better treatment. Most promisingly, the interaction between cyclin D1 and ER may become a direct target for improved therapy in ER positive breast cancer.

ACKNOWLEDGMENTS

This work is supported by grants of the Dutch Cancer Society. I thank Daan Muris for critical reading of the manuscript.

REFERENCES

Altucci, L, Addeo, R, Cicatiello, L, Dauvois, S, Parker, MG, Truss, M, Beato, M, Sica, V, Bresciani, F and Weisz, A. (1996). 17B-Estradiol induces cyclin D1 gene transcription, p36D1-p34cdk4 complex activation and p105 Rb phosphorylation during mitogenic stimulation of G1-arrested human breast cancer cells. Oncogene 12, 2315–2324.

Aronica, SM, Kraus, WL and Katzenellenbogen, BS. (1994). Estrogen action via the cAMP signaling pathway: stimulation of adenylate cyclase and cAMP-regulated gene transcription. Proc Natl Acad Sci USA 91, 8517–8521.

Assoian, RK. (1997) Anchorage-dependent cell cycle progression. J Cell Biol 136, 1, 1–4.

Baldin, V, Lukas, J, Mrcote, M-J, Pagano, M, Draetta, G. (1993). Cyclin D1 is a nuclear protein required for cell cycle progression in G1. Genes & Developm 7, 812–821.

Barbareschi, M, Pelosio, P, Caffo, O, Buttita, F, Pellegrini, S, Barbazza, R, Palma, PD, Bevilacqua, G, Marcetti, A. (1997). Cyclin-D1 gene amplification and expression in breast carcinoma: relation with clinicopathologic characteristics and with retinoblastoma gene product, p53 and p21 WAF1 immunohistochemical expression. Int J Cancer (Pred Oncol) 74, 171–174.

Bartek, J, Bartkova, J and Lukas, J. (1996). The retinoblastoma protein pathway and the restriction point. Curr Opin in Cell Biol 8, 805–814.

Bates, S, Parry, D, Bonetta, L, Vousden, K, Dickson, C and Peters, G. (1994). Absence of cyclin D/cdk complexes in cells lacking functional retinoblastoma protein. Oncogene 9, 1633–1640.

Bates, S, and Peters, G. (1995). Cyclin D1 as a cellular proto-oncogene. Cancer Biol, 6, 73–82.

Bellacosa, A, Almadori, G, Cavallo, S, Cadoni, G, Galli, J, Ferrandina, G, Scambia, G, Neri, G. (1996). Cyclin D1 gene amplification in human laryngeal squamous cell carcinomas: prognostic significance and clinical implications. Clin Cancer Res 2, 175–180.

Berry, M, Metzger, D, and Chambon, P (1990). Role of the two activating domains of the oestrogen receptor in the cell-type and promoter-context dependent agonistic activity of the anti-oestrogen 4-hydroxytamoxifen. EMBO J. 9, 2811–2818.

Boer de, CJ, Loyson, S, Kluin, PM, Kluin-Nelemans, HC, Schuuring, E, Krieken van, JHJM. (1993). Multiple breakpoints within the BCL-1 locus in B cell lymphoma: rearrangements of the cyclin D1 gene. Cancer Res 53, 4148–4152.

Bonapace, IM, Addeo, R, Altucci, L, Cicatiello, L, Bifulco, M, Laezza, C, Salzano, S, Sica, V, Bresciani, F, and Weisz, A. (1996). 17β-Estradiol overcomes a G1 block induced by HMG-CoA reductase inhibitors and fosters cell cycle progression without inducing ERK-1 and -2 MAP kinase activation. Oncogene 12, 753–763.

Bunone, G, Briand, PA, Miksicek, RJ, Picard, D. (1996). Activation of the unliganded estrogen receptor by EGF involves the MAP kinase pathway and direct phosphorylation. EMBO J. 15, 9, 2174–2183.

Catzavelos, C, Bhattacharya, N, Ung, YC, Wilson, JA, Roncari, L, Sandhu, C, et al., (1997). Decreased levels of cell-cycle inhibitor $p27^{Kip1}$ protein: Prognostic implications in primary breast cancer. Nature (Medicine) 3, 227–230.

Cobrink, D, Lee, MH, Hannon, G, Mulligan, G, Bronson, RT, Dyson, N, Harlow, E, Beach, D, Weinberg, RA, Jacks, T. (1996). Shared role of the pRB-related p130 and p107 proteins in limb development. Genes Dev 10, 1633–1644.

DeFriend, DJ, Howell, A, Nicholson, RI, Anderson, E, Dowsett, M, Mansel, RE, Blamey, RW, Bundred, NJ, Robertson, JF, Saunders, C, et al., . (1994). Investigation of a new pure anti-estrogen (ICI 182780) in women with primary breast cancer. Cancer Res. 54, 408–414.

Ewen, ME, Sluss, HK, Sherr, CJ, Matsushime, H, Kato, J, and Livingstone, DM. (1993). Functional interaction of the retinoblastoma protein with mammalian D-type cyclins. Cell 73, 487–497.

Fang, F, Orend, G, Watanabe, N, Hunter, T, Ruoslahti, E. (1996). Dependence of cyclin E-CDK2 kinase activity on cell anchorage. Science, 271, 499–501.

Fantl, V, Stamp, G, Andrews, A, Rosewell, I, and Dickson, C (1995). Mice lacking cyclin D1 are small and show defects in eye and mammary gland development. Genes & Development 9, 2364–2372.

Foekens, JA, Portengen, H, Look, MP, van Putten, WL, Thirion, B, Bontenbal, M, and Klijn, JG. (1994). Relationship of pS2 with response to tamoxifen therapy in patients with recurrent breast cancer. Br J Cancer 70, 1217–1223.

Fracchiolla, NS, Pruneri, G, Pignataro, L, Carboni, N, Capaccio, P, Boletini, A, Buffa, R, Neri, A. (1997). Molecular and immunohistochemical analysis of the bcl-1/cyclin D1 gene in laryngeal squamous cell carcinomas. Cancer 79, 6, 1114–1121.

Gillett, CE, Fantl, V, Smith, R, Fisher, C, Bartek, J, Dickson, C, Barnes, D and Peters, G (1994). Amplification and over-expression of cyclin D1 in breast cancer detected by immunohistochemical staining. Cancer Res. 54, 1812–1817.

Gillett, C, Smith, P, Gregory, W, Richards, M, Millis, R, Peters, G, Barnes, D. (1996). Cyclin D1 and prognosis in human breast cancer. Int J Cancer (Pred Oncol) 69, 92–99.

Guadagno, TM, Ohtsubo, M, Roberts, JM, Assoian, RK. (1993). A link between cyclin A expression and adhesion-dependent cell cycle progression. Science 262, 1572–1575.

Hall, M, and Peters, G. (1996). Genetic alterations of cyclins, cyclin-dependent kinases, and Cdk inhibitors in human cancer. Adv. Cancer Res 68, 67–108

Herber, B, Truss, M, Beato, M, Muller, R. (1994). Inducible regulatory elements in the human cyclin D1 promoter. Oncogene 9, 2105–2107.

Herrera, RE, Sah, VP, Williams, BO, Makela, TP, Weinberg, RA, Jacks, T. (1996). Altered cell cycle kinetics, gene expression, and G1 restriction point regulation in Rb-deficient fibroblasts. Mol Cell Biol 16, 2402–2407.

Hinds, PW, Dowdy, SF, Eaton, EN, Arnold, A, Weinberg, RA. (1994). Function of a human cyclin gene as an oncogene. Proc Natl Acad Sci USA 91, 709–713.

Hunter, T. (1997). Oncoprotein networks. Cell 88, 333–346.

Ignar-Trowbridge, DM, Teng, CT, Ross KA, Parker, MG, Korach, KS, McLachlan, JA . (1993). Peptide growth factors elicit estrogen receptor-dependent transcriptional activation of an estrogen-responsibe element. Mol. Endocrinol 7, 992–998.

Jordan, VC and Murphy, CS (1990). Endocrine pharmacology of anti-estrogens as antitumor agents. Endocr Rev 11, 578–610.

Kato, S, Endoh, H, Masuhiro, Y, Kitamoto, T, Uchiyama, S, Sasaki, H, Masushige, S, Gotoh, Y, Nishida, E, Kawashima, H, Metzger, D, and Chambon, P. (1995). Activation of the Estrogen Receptor Through Phosphorylation by Mitogen-Activated Protein Kinase. Science 270, 1491–1494.

Katzenellenbogen, BS, Miller, MA, Mullick, A, and Sheen, YY (1985). Anti-estrogen action in breast cancer cells: modulation of proliferation and protein synthesis, and interaction with estrogen receptors and additional anti-estrogen binding sites. Breast Cancer Res Treat 5, 231–243.

Kitagawa, M, Higashi, H, Jung, H-K, Suzuki-Takahashi, I, Ikeda, M, Tamai, K, Kato, J, Segawa, K, Yoshida, E, Nishimura, S and Tay, Y. (1996). The consensus motif for phosphorylation by cyclin D1-Cdk4 is different from that for phosphorylation by cyclin A/E-Cdk2. EMBO J 15, 7060–7069.

Klijn, JG, Berns, EM, and Foekens, JA (1993). Prognostic factors and response to therapy in breast cancer. Cancer Surv 18, 165–198.

Lukas, J, Bartkova, J, Bartek, J. (1996). Convergence of mitogenic signalling cascades from diverse classes of receptors on the cyclin D/cdk-pRb-controlled G1 checkpoint. Mol Cell Biol 16, 6917–6925.

Masuda, M, Hirakawa, N, Nakashima, T, Kuratomi, Y, Komiyama, S. (1996). Cyclin D1 overexpression in primary hypopharyngeal carcinomas. Cancer 78, 3, 390–395.

McIntosh GG, Anderson JJ, Milton I, Steward M, Parr AH, Thomas MD, Henry JA, Angus B, Lennard TW, Horne CH. (1995). Determination of the prognostic value of cyclin D1 overexpression in breast cancer. Oncogene 11, 885–891.

Meredith, SD, Levine, PA, Burns, JA, Gaffey, MJ, Boyd, JC, Weiss, LM, Erickson, NL, Williams, ME. (1995). Chromosome 11q13 amplification in head and neck squamous cell carcinoma. Arch Otolaryngol Head and Neck Surg 121, 790–794.

Metzger, D, Berry, M, Ali, S, and Chambon, P. (1995). Effect of antagonists on DNA binding properties of the human estrogen receptor in vitro and in vivo. Mol Endocrin 9, 579–591.

Michalides, R, van Veelen, N, Hart, A, Loftus, B, Wientjens, E, and Balm, A. (1995). Overexpression of cyclin D1 correlates with recurrence in a group of forty-seven opearble squamous cell carcinomas of the head and neck. Cancer Res 55, 975–978.

Michalides, R, Hageman Ph, Tinteren H van, Houben L, Wientjens E, Klompmaker R, Peterse J. (1996). A clinicopathological study on overexpression of cyclin D1 and of p53 in a series of 248 patients with operable breast cancer. Br J Cancer 73, 728–734.

Michalides, R, van Veelen, N, Kristel, P, Hart, A, Loftus B, Hilgers, F, and Balm, A.(1997). Overexpression of cyclin D1 indicates a poor prognosis in squamous cell carcinoma of the head and neck. Arch Oto Head and Neck Surg, in press.

Mushgrove, EA, Hamilton, JA, Lee, CS, Sweeney, KJ, Watts, CK, and Sutherland, RL (1993). Growth factor, steroid, and steroid antagonist regulation of cyclin gene expression associated with changes in T-47D human breast cancer cell cycle progression. Mol Cell Biol 13, 3577–3587.

Mushgrove EA, Lee CLE, Buckley MF, Sutherland RL. (1994).Cyclin D1 induction in breast cancer cells shortens the G1 and is sufficient for cells arrested in G1 to complete the cell cycle. Proc Natl Acad Sci USA 91, 8022–8026.

Ohtsubo, M, Theodoras, AM, Schumacher, J, Roberts, JM, Pagano, M. (1995). Human cyclin E, a nuclear protein essential for the G1-to-S- phase transition. Mol Cell Biol 15, 2612–2624.

Ohtani, K, DeGregori, J, Nevins, JR. (1995). Regulation of the cyclin E gene by transcripton factor E2F1. Proc Natl Acad Sci USA 92, 12146–12150.

Parry, D, Bates, S, Mann, DJ and Peter, G. (1995). Lack of cyclin D-Cdk complexes in Rb-negative cells correlates with high levels of p16INK4/MTS1 tumour suppressor gene product. EMBO J 14, 503–511.

Porter, PL, Malone, KE, Heagerty, PJ, Alexander, GM, Gatti, LA, Firpo, EJ, Daling, JR, Roberts, JM. (1997). Expression of cell-cycle regulator p27Kip1 and cyclin E, alone and in combination, correlate with survival in young breast cancer patients. Nature Med 3, 2, 222–234.

Reese, JC and Katzenellenbogen, BS (1991). Differential DNA-binding abilities of estrogen receptor occupied with two classes of anti-estrogens: studies using human estrogen receptor overexpressed in mammalian cells. Nucleic Acids Res 19, 6595–6602.

Resnitzky D, Gossen M, Bujard H, Reed SI. (1994). Acceleration of the G1/S phase transition by expression of cyclins D1 and E with an inducible system. Mol Cell Biol 14, 1669–1679.

Resnitzky, D, Reed, SI. (1995). Different roles for cyclins D1 and E in regulation of the G1-to-S transition. Mol Cell Biol 15, 3463–3469.

Rimokh, R, Berger, F, Bastard, Chr, Klein, B, French, M, Archimbaud, E, Rousault, JP, Santa Lucia, B, Duret, L, Vuillaume, M, Coiffier, B, Bryon, P.-A, Magaud, J-P. (1994). Rearrangement of CCND1 (BCLI/PRADI)3' untranslated region in mantle-cell lymphomas and t(11q13)-associated leukemias. Blood 83, 12, 3689–3696.

Rosenberg, CL, Kim, HG, Shows, TB, Kronenberg, HM, Arnold, A. (1991). Rearrangement and overexpression of D11S287E, a candidate oncogene on chromosome 11q13 in benign parathyroid tumors. Oncogene 6, 449–453.

Schulze, A, Zerfass, K, Spitkovsky, D, Middendorp, S, Berges, J, Helin, K, Jansen, Durr, P, Henglein, B. (1995). Cell cycle regulation of the cyclin A gene promoter is mediated by a variant E2F site. Proc Natl Acad Sci USA 92, 11264–11268.

Schulze, A, Zerfass-Thome, K, Berges, J, Middendorp, S, Jansen-Durr, P, Henglein, B. (1996). Anchorage-dependent transcription of the cyclin A gene. Mol Cell Biol 16, 9, 4632–4638.

Sherr, CJ. (1995). D type cyclins. TIBS 20, 187–190.

Sherr, CJ, and Roberts, JM. (1995). Inhibitors of mammalian G1 cyclin-dependent kinases. Genes Dev 9, 1149–1163.

Sherr, CJ. (1996). Cancer cell cycles. Science 274, 1672–1677.

Sicinski, P, Donaher, JL, Parker, SB, Li, T, Fazeli, A, Gardner, H, Haslam, SZ, Bronson, RT, Elledge, SJ, and Weinberg, RA. (1995). Cyclin D1 provides a link between development and oncogenesis in the retina and breast. Cell 82, 621–630.

Smith, CL, Conneely, OM and O'Malley, BW. (1993). Modulation of ligand-indepdent activation of the human estrogen receptor by hormone and antihormone. Proc Natl Acad Sci USA 90, 6120–6124.

Sutherland, RL, Watts, CKW, Musgrove, EA. (1993). Cyclin gene expression and growth control in normal and neoplastic human breast epithelium. J Steroid Biochem Molec Biol 47, 99–106.

Sweeney, KJ, Sarcevic, B, Sutherland, RL and Musgrove, EA. (1996). Cyclin D2 activates Cdk2 in preference to Cdk4 in human breast epithelial cells. Oncogene 14, 1329–1340.

Swerdlow, SH, Yang, W-I, Zukerberg, LR, Harris, NL, Arnold, A, Williams, ME. (1995). Expression of cyclin D1 protein in centrocytic/mantle cell lymphomas with and without rearrangement of the BCL1/cyclin D1 gene. Hum Pathol 26, 9, 999–1004.

Thompson, EW, Katz, D, Shima, TB, Wakeling, AE, Lippman, ME, and Dickson, RB. (1989). ICI 164–384 a pure antagonist of estrogen-stimulated MCF-7 cell proliferation and invasiveness. Cancer Res 49, 6929–6934.

Wang, TC, Cardiff, RD, Zukerberg, L, Lees, E, Arnold, A, and Schmidt, EV. (1994). Mammary hyperplasia and carcinoma in MMTV-cyclin D1 transgenic mice. Nature 369, 669–671.

Watts, CKW, Brady, A, Sarcevic, B, DeFazio, A, Musgrove, EA, Sutherland, RL. (1995). Anti-estrogen inhibition of cell cycle progression in breast cancer cells is associated with inhibition of cyclin-dependent kinase activity and decreased retinoblastoma protein phosphorylation. Mol Endocr 9, 1804–1813.

Weinberg, RA. (1995). The retinoblastoma protein and cell cycle control. Cell 81, 323–330.

Weinstat-Saslow, D, Merino, MJ, Manrow, RE, Lawrence, JA, Bluth, RF, Wittenbel, KD, Simpson, JF, Page, DL and Steeg, PS. (1995). Overexpression of cyclin D mRNA distinguishes invasive and in situ breast carcinomas from non-malignant lesions. Nature Med 1, 12, 1257–1260.

Weisz, A and Bresciani, F. (1988). Estrogen induces expression of c-fos and c-myc protooncogenes in rat uterus. Mol Endocrinol 2, 816–824.

Weisz, A and Rosales, R. (1990). Identification of an estrogen response element upstream of the human c-fos gene that binds the estrogen receptor and the AP-1 transcription factor. Nucleic Acids Res 18, 5097–5106.

Won, K-A, and Reed, SI. (1996). Activation of cyclin E/Cdk2 is coupled to site-specific autophosphorylation and ubiquitin-dependent degradation of cyclin E. EMBO J 15, 16, 4182–4193.

Zhang, T, Nanney, LB, Luongo, C, Lamps, L, Heppner, KJ, DuBois, RN and Beauchamp, RD. (1997). Concurrent overexpression of cyclin D1 and cyclin-dependent kinase 4 (Cdk4) in intestinal adenomas from multiple intestinal neoplasia (min) mice and human familial adenomatous polyposis patients. Cancer Res 57, 169–175.

Zwijsen, R, Klompmaker, R, Wientjens, E, Kristel, P, van der Burg, B, and Michalides, R. (1996). Cyclin D1 triggers autonomous growth of breast cancer cells by governing the cell cycle exit. Mol Cell Biol 16, 2554–2560.

Zwijsen, RML, Wientjens, E, Klompmaker, R, Sman, J van der, Bernards, R, Michalides, RJAM. Ligand-independent activation of ER-responsive gene transcription by cyclin D1. (1997) Cell 88, 405–415.

DISCUSSION

Helin: The implication from your study is that in breast tumors with overexpression of cyclin D, you would see an increased association between cyclin D1 and the estrogen receptor. Is that the case?

Michalides: That would be the implication, provided that cyclin D1 is present in a very high amount. If you can remember from the first picture I showed you, that when you transfect cyclin D1 together with cdk-4 you do not see an effect on estrogen receptor mediated readout. So, cdk-4 binding to cyclin D1 is preferred over binding of cyclin D1 to estrogen receptor. We also know that from mammalian to hybrid studies, where we could determine the preference of cyclin D1 to bind either to the estrogen receptor or to cdk-4. In these studies we came to a roughly three to four times preference of cyclin D1 for cdk-4, as compared to estrogen receptor. So, we need overexpression of cyclin D1, not to be bound to cdk-4 in order to activate an estrogen receptor. Those situations are hard to analyze actually *in vivo* so you have to devise a method to see free cyclin D1, which we have not really come up yet. In breast tumors there are some cases where we see high amounts but never incredibly high amounts of cyclin D1, they go up to levels of about ten to fifteen fold compared to normal tissue. So your idea is right, that with an overexpression of cyclin D1 you would see an association with an estrogen receptor, however, to demonstrate that *in vivo* is not that easy.

Mihich: Do you have any evidence, or has anybody any evidence that the same phenomenon occurs in cells which are estrogen receptor negative, both in mammalian breast and non-breast systems, which are subject to growth factors?

Michalides: Then you would expect an interaction between cyclin D1 and other growth factor receptors. We have not found any such interaction.

Livingston: Following up on what Kristian Helin asked you, if you take a breast carcinoma cell line that makes very high intracellular concentration of D1, and neutralize D1, let us say, with something like dominant negative cdk-4 or monoclonal antibody, can you demonstrate that the expression of an ER dependent gene changes?

Michalides: That is what the prediction would be. We have not done that experiment yet. In the MCF7 cell system we can increase cyclin D1 expression, but to a limited extent. In the tetracycline system we were never able to get more than ten fold overexpression but in that cell line we see an increased ERE dependent readout. In steroid-free conditions. That would suggest that the estrogen receptor mediated transcription is activated with that level of overexpression of cyclin D1. We also have preliminary data which indicate that growth of MCF7 cells with a normal level of cyclin D1 can be inhibited by tamoxifen, whereas overexpression of cyclin D1 reduces tamoxifen induced growth inhibition. These data indicate an interaction between ER and cyclin D1 in these cells.

Livingston: Weinstein reported on D1 amplified esophageal cancer lines some years ago. If you put an ER into those cells, does it fire in the absence of estrogen?

Michalides: We have not done those experiments, but you would expect so. When you put an estrogen receptor in those cells, or in other cell lines such as in some squamous cell carcinoma lines with high levels of cyclin D1 expression, you would expect that estrogen receptor would become activated in that system. We have not done that experiment yet.

Livingston: It would be interesting to see if the biology of such cells changes once one introduces ER and whether the ER is activated under those conditions.

Michalides: You wonder whether they react to other growth factors as well, and how the interplay between growth factors and the estrogen receptor would be. These cells have so many signaling pathways that it is hard to dissect the effects of the other signaling pathways from that of the estrogen receptor. But, I fully agree that it would be a very nice experiment to do and that is what we were planning also.

Zanker: Cyclin D1 is crystallographed?

Michalides: No.

Comoglio: As you know, the cancer pharmacologists are struggling for a cytostatic drug that will not kill stem cells. Would you make a guess, or what is your feeling, will it be feasible to develop an inhibitor of cyclin D1 or of any other cyclin, that will affect cancer cells overexpressing a cyclin but leave intact normal cells including bone marrow stem cells?

Michalides: That is a dream. As far as I know, bone marrow cells do not express cyclin D1 but express cyclin D3. However, in breast tumor cells there is an overexpression of cyclin D1 and when the cyclin D1 indeed activates an estrogen receptor and that activation is independent of tamoxifen, one may now look for other components that interfere with cyclin D1 estrogen receptor binding and examine other anti-estrogens that interfere with estrogen receptor mediated transcription. In that way one might for estrogen receptor positive, cyclin D1 overexpressing cells come up with specific anti-estrogens that could work in that way. These may work better than tamoxifen which only works in two-thirds of the ER positive breast cancer cases.

Pierotti: If I remember correctly, it was your experiment that showed that there is competition between CDK4 and cyclin D1 for binding to estrogen receptor. When you introduced CDK4 you had a reduced binding to the estrogen receptor. In other words, it is the same domain on cyclin D1 involved in binding to both?

Michalides: Whenever CDK4 is bound to cyclin D1 then the binding of cyclin D1 to the estrogen receptor is inhibited. Whether that means that also the same area in cyclin D1 is involved in binding is not clear.

Livingston: Do you know if you make an AR/ER chimera bearing an ER ligand binding domain. If so, does it bind E_2 and fire?

Michalides: You would expect it to fire because the interaction involves the ligand binding domain of ER. This was clear from immunoprecipitation studies, the binding study and it actually did so in the mammalian two hybrid assay. So you would also expect that to fire, yes.

Livingston: In the old days when oophorectomy was common as anti-estrogen therapy for women with advanced breast cancer, I suspect there must have been some patients whose primary tumors were ER+/D1 non-amplified. When some of these women relapsed, did their relapse show D1 amplified?

Michalides: We have not done those studies, but we are doing now the retrospective studies in terms of tamoxifen treatment and ask whether tamoxifen resistant cases are associated with overexpression of cyclin D1. However, you have to be careful in those studies. There might be various ways to reach tamoxifen resistance: Other kinases may also activate estrogen receptor, Rb mutations will probably make the cells insensitive to estradiol, so you have to do a complex study in order to make sure that overexpression of cyclin D1 may be associated with tamoxifen resistance.

Livingston: But if it were a primary driving force for autonomous replication, one might expect that eliminating normal estrogen access would result in the over production of D1 so that the circuit could still run.

Michalides: That might well be the case.

Zanker: You mentioned that you omitted in your medium phenol red. It is well known that phenol red can activate estrogen receptor. Did you also use charcoal treated serum because you do not want to contaminate your experiments with exogenous estrogens in the serum?

Michalides: No, in that medium we did not use any serum at all, we used phenol red medium and a limited amount of insulin and transferrin to keep the cells alive.

Visentin: This is more a comment than a question. Tamoxifen might possibly be not the most adequate drug for selecting and dissecting an ER-mediated mechanism, since it is not a pure antiestrogen, and might work also via modulation of other growth factors and receptors thereof. Indeed, Tamoxifen resistance is still an enigma, and nobody could rule out non ER-mediated sensitivity or resistance. Accordingly, it might be more fruitful to investigate in this context using as probes pure antiestrogens or relatively more pure than Tamoxifen itself.

Michalides: Well there have been some clinical studies done where the question was whether tamoxifen resistant breast tumors would still be sensitive to pure anti-estrogens. The answer is yes, but these studies have not been completely finished up yet.

13

DYSREGULATION OF PROGRAMMED CELL DEATH IN CANCER TOWARD A MOLECULAR UNDERSTANDING OF Bcl-2

John C. Reed

The Burnham Institute
10901 N. Torrey Pines Road
La Jolla, California 92037

The discovery of *BCL-2* over twelve years ago represents a milestone in tumor biology (1), because *BCL-2* is the first proto-oncogene identified which was found to contribute to neoplastic cell growth not by accelerating cell division but rather by halting cell turnover caused by programmed cell death (2). In all tissues in which cell division occurs, there exists a need for mechanisms that remove or eliminate cells. Otherwise, cells will accumulate in astounding numbers, given that in the course of a typical year each of us produces and eradicates a mass of cells equivalent to almost our entire body weight. This need for cell turnover is met in vivo by programmed cell death (PCD). PCD represents a mechanism by which cells actively commit suicide, culminating typically (but not always) in a constellation of morphological changes known as apoptosis, that include chromatin condensation and nuclear fragmentation (pyknosis), cell shrinkage, plasma membrane blebbing, and extrusion of cellular particles (apoptotic bodies). Unlike necrosis where cell lysis occurs followed by an inflammatory response, apoptotic cells and apoptotic bodies are rapidly cleared through phagocytosis by neighboring viable cells without accompanying inflammation.

PCD is controlled by the counteracting influences of a number of genes that either promote or block cell death. Many of the genes involved in the cell death pathway are genetically conserved from simple multicellular organisms such as the nematode *Caenorhabditis elegans* and the fly *Drosophila melanogaster* to mammals, including humans, thus implying distant evolutionary origins (3, 4). For example, the nematode contains a homolog of Bcl-2 called CED-9 which is essential for cell viability. Defects in CED-9 can be largely complemented by the human Bcl-2 protein (5, 6), thus illustrating the great extent to which the function of Bcl-2 is conserved across broad evolutionary distances.

The Biology of Tumors, edited by Mihich and Croce
Plenum Press, New York, 1998.

1. THE Bcl-2 PROTEIN FAMILY

Thus far, 15 cellular homologs of Bcl-2 have been reported. Many of the members of this family function similar to Bcl-2 as blockers of cell death, including the anti-apoptotic proteins Bcl-X_L, Mcl-1, A1/Bfl-1, Bcl-W, Nr-13 (avian), and CED-9 (nematode). However, others do just the opposite and promote cell death. The known pro-apoptotic members of the Bcl-2 family include Bax, Bcl-X_S, Bad, Bak, Bik, Bid, and Hrk (6–20). An additional human homolog BRAG-1, which was discovered because of its overexpression in gliomas, and two *Xenopus* homologs have also been described whose effects on cell life and death have not been assessed. However, they most likely are suppressors of apoptosis (21, 22). Of potential relevance to some types of cancer, five homologs of Bcl-2 have also been discovered in viruses: E1b-19 kD (adenovirus), BHRF-1 (Epstein Barr Virus), LMH-5W (African Swine Fever Virus), ORF-16 (Herpes Saimiri Virus), and KSbcl-2 (Herpes Virus 8) (23–27). The relative ratios of these pro- and anti-apoptotic members of the Bcl-2 family expressed in cells determine to a large extent how easily an individual cell will activate the program for apoptotic cell death when confronted with an appropriate stimulus.

Many of the Bcl-2 family proteins can physically interact with each other, forming a complex network of homo- and heterodimers (Figure 1). The various members of the Bcl-2 family can be grouped into one of four categories based on the impact they have on cell death and their dimerization properties. One group consists of pro-apoptotic proteins such as Bax and Bak, which can homodimerize with themselves and heterodimerize with anti-apoptotic proteins such as Bcl-2 and Bcl-X_L. A second group is comprised of anti-apoptotic proteins such as Bcl-2, Bcl-X_L, and Mcl-1. These proteins suppress cell death, and can both homodimerize with themselves and heterodimerize with pro-apoptotic proteins such as Bax and Bak. The third type of Bcl-2 family proteins is presented by Bcl-X_S, Bad, and Bik which have more limited dimerization capabilities. These death-promoting members of the Bcl-2 family appear to function essentially as trans-dominant inhibitors of the anti-apoptotic proteins such as Bcl-2 and Bcl-X_L. They can heterodimerize with anti-apoptotic members of the family presumably sequestering them so that they cannot suppress cell death, but appear to be incapable of either homodimerizing with themselves or of heterodimerize with pro-apoptotic proteins such as Bax and Bak (18, 28, 29). Although func-

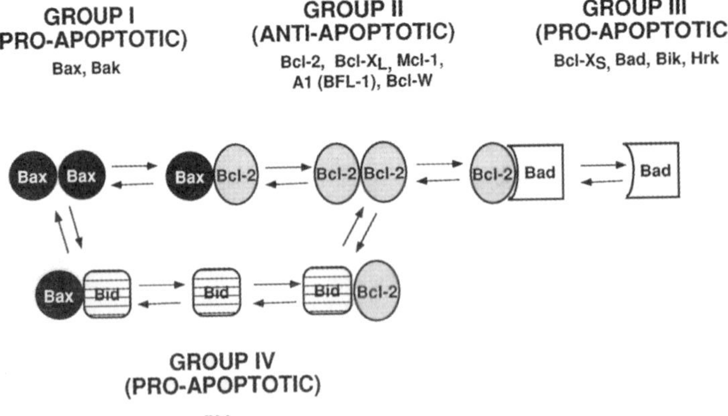

Figure 1. Bcl-2 family protein network of homo- and hetero-dimers. A model is presented showing some of the dimerization events that can occur among Bcl-2 family proteins. Four subgroups of proteins have been identified.

tioning as pro-apoptotic proteins, this third group of Bcl-2 family proteins appears to engage the cell death pathway indirectly though suppression of anti-apoptotic proteins such as Bcl-2 and Bcl-X_L, thus distinguishing themselves from pro-apoptotic proteins like Bax and Bak which seem to autonomously promote cell death irrespective of whether Bcl-2 or Bcl-X_L is present (28, 30–32) (Figure 1). A recently discovered member of the Bcl-2 family, Bid, may constitute a fourth group of proteins. Bid is a pro-apoptotic protein that probably does not directly promote cell death, analogous to the group III proteins such as Bcl-X_S and Bad, but which (unlike Bcl-X_S and Bad) can both dimerize with anti-apoptotic proteins (Bcl-2 and Bcl-X_L) and with pro-apoptotic proteins (Bax). Furthermore, unlike the group I proteins such as Bax and Bak that are suspected to directly promote cell death, Bid cannot homodimerize with itself (19). Thus, the dimerization properties of Bid are unique among the Bcl-2 family proteins identified thus far.

2. STRUCTURAL AND BIOLOGICAL ASSESSMENTS OF DIMERIZATION AMONG Bcl-2 FAMILY PROTEINS

The domains within Bcl-2 family proteins required for dimerization have been determined by deletional and mutational analysis, and the results recently corroborated by x-ray crystallographic and NMR-based structural studies (18–20, 30, 33–39). In this regard, the 3-dimensional structure of the Bcl-X_L protein consists of seven α-helices joined by flexible loops of variable length (38). Amino acid sequence alignments of Bcl-2 family proteins have demonstrated up to four evolutionarily conserved domains, termed Bcl-2 Homology (BH) domains: BH1, BH2, BH3, and BH4 (Figure 2). The BH4 and BH3 domains correspond to the first and second amphipathic α-helices in these proteins, as predicted from the 3 dimensional structure of Bcl-X_L (38). The BH1 domain coincides with a loop located upstream of the fifth α-helix in Bcl-X_L and extends partially into this α-helix,

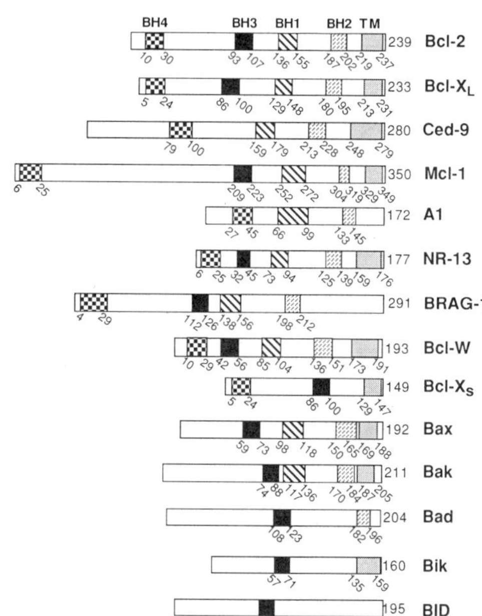

Figure 2. Bcl-2 protein family. The structures of the Bcl-2 family proteins are depicted, showing the locations of the BH1, BH2, BH3, BH4 and transmembrane (TM) domains.

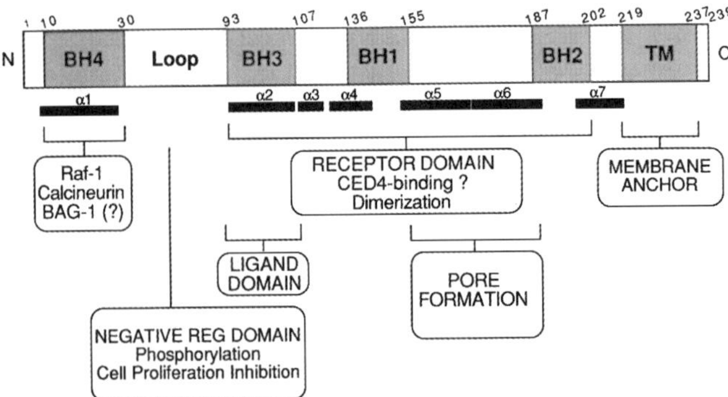

Figure 3. Topology of Bcl-2 protein. The topology of the Bcl-2 protein is depicted, showing the relative locations of the seven a-helices relative to the BH1, BH2, BH3, BH4 and transmembrane (TM) domains. The purported functions or interactions with other proteins associated with each domain are indicated below the diagram of Bcl-2.

whereas the BH2 domain has been assigned to the latter portion of the 6th α-helix and loop which follows thereafter.

The BH1, BH2, and BH3 domains in combination form the borders of a hydrophobic pocket located on the surface of the Bcl-X_L protein. Certain mutations that affect residues lining this pocket, such as Gly145Ala in the BH1 domain of human (hu) Bcl-2 protein and Gly159 Ala in the BH1 domain of hu-Bcl-X_L have been shown to abrogate their ability to dimerize with Bax (33, 37). Thus, this surface pocket appears to function analogous to a receptor, binding epitopes located on partner proteins that dimerize with Bcl-2 and Bcl-X_L. Through a combination of deletional and mutagenesis studies (18–20, 31, 35, 36), peptide competition assays (40), and NMR-based structural analyses (39), it has been determined that the BH3 domain represents the counter-structure on dimerizing Bcl-2 family proteins which inserts similar to a peptide ligand into the surface pocket created by the combination of BH1, BH2, and BH3.

This finding implies that at least some Bcl-2 family proteins exist in two conformations: one in which the protein creates a receptor-like pocket and the other in which the amphipathic α-helix that comprises BH3 rotates outward to expose its hydrophobic surface so that it can bury this side of the α-helix into the receptor-like pocket on dimerization partner proteins (39). Interestingly, some pro-apoptotic members of the Bcl-2 family, including Bik, Hrk, and Bid, contain only a BH3 domain (18–20), lacking the other BH-domains found in many other members of the Bcl-2 family of proteins (Figure 3). Deletion of the BH3 domains from these proteins uniformly abolishes their functions as inducers of cell death and antagonists of anti-apoptotic proteins such as Bcl-2 and Bcl-X_L. Moreover, over-expression of fragments of the pro-apoptotic protein Bak which include only the BH3 domain but not the other BH-domains is sufficient for inducing apoptosis in mammalian cells (36). In addition, substituting the BH3 domain of Bcl-2 for that of Bax has been reported to convert Bcl-2 from a protector to a killer protein (41).

Despite the advances that have been made in understanding the structural aspects of dimerization among Bcl-2 family proteins, the functional significance of many of these protein-protein interactions remains controversial. For example, mutagenesis studies of the anti-apoptotic proteins Bcl-2 and Bcl-X_L initially suggested that heterodimerization with the pro-apoptotic protein Bax was critical for their cell survival activities (33, 37). Subsequently, however, alanine substitution mutants of Bcl-X_L were reported that failed to dimerize with Bax but which retained the ability to promote cell survival (42). Whether such heterodimerization-defective mutants of Bcl-X_L could prevent cell death specifically induced by over-expression of Bax was not addressed, raising questions about whether heterodimerization might nevertheless be required for the mutual antagonism displayed by anti-apoptotic (Bcl-2/Bcl-X_L) and pro-apoptotic (Bax) Bcl-2 family proteins.

The role of homo- and hetero-dimerization in Bax- and Bak-induced apoptosis has also generated controversy. For example, deletion of the BH3 domain from the Bax or Bak proteins as well as removal of four amino acids from within the BH3 of Bax comprising a well conserved IGDE motif was reported to abolish Bax/Bak-mediated cell death in mammalian cells (31, 36). These same BH3 domain mutations also abrogated the ability of Bax and Bak to induce cell death in yeast and negated homo- and heterodimerization (31, 35, 36, 43). However, when tested in another mammalian cell line, Bax (ΔBH3) was reported to accelerate apoptosis induced by chemotherapeutic drugs as effectively as wild-type Bax protein and to abrogate the cytoprotective effects of Bcl-X_L over-expression, despite its inability to dimerize with either Bax or Bcl-X_L (44, 45). On balance, the data available thus far strongly suggest that it is not necessary for Bcl-2 or Bcl-X_L to bind Bax for them to suppress cell death. However, the role of homodimerization of Bax in its role as an inducer of apoptosis remains to be clarified.

3. ALTERED EXPRESSION OF Bcl-2 FAMILY PROTEINS IN CANCER

In ~85% of follicular non-Hodgkin's lymphomas (NHLs), as well as ~one-third of other types of NHLs (particularly diffuse large NHL), t(14;18) translocations involving the *BCL-2* gene are found (46). Follicular lymphoma (FL) is the most common of the NHLs with ~ 20,000 new cases annually in the United State alone and a prevalence in the U.S. population of over 100,000 cases. Molecular determination of the anatomy of t(14;18) chromosomal translocations has shown that the *BCL-2*, which normally resides on chromosome 18 at band q21, becomes juxtaposed with the immunoglobulin heavy-chain (IgH) locus located at 14q32 (1, 47, 48). The placing of *BCL-2* into a cis-configuration with the IgH locus brings this anti-apoptotic gene under the influence of strong enhancer elements whose normal job is to drive high levels Ig gene expression in B-cells. The result is loss of the normal regulation of *BCL-2* gene, and the production of inappropriately high levels of Bcl-2 protein in B-lymphocytes.

Though first discovered because of its involvement in the t(14;18) chromosomal translocations found in follicular B-cell lymphomas, expression of the *BCL-2* gene occurs in a wide variety of human cancers. Among the cancers that have been reported to contain high levels of Bcl-2 protein or that display abnormal patterns of Bcl-2 protein production are significant proportions of adenocarcinomas of the prostate, colon, breast, and stomach, squamous cell lung cancers, small cell lung cancers, renal carcinomas, neuroblastomas, melanomas, CLLs, ALLs, and CMLs (reviewed in 49, 50). Since structural alterations have only rarely been detected in the *BCL-2* genes of these other types of malignancies, presumably trans- rather than cis-regulatory mechanisms probably are responsible of high levels or aberrant patterns of *BCL-2* expression in these cancers. In some cases, it is unclear whether the high levels of Bcl-2 protein production reflect a tumor-specific alteration in the mechanisms that control *BCL-2* gene expression as opposed to changes in the overall differentiative state of tumor cells or origination of tumor from cellular subpopulations in which it is normal to have high levels of *BCL-2* expression. Examples of this include mantle cell NHL and B-cell CLL, where the malignant B-cells may be representative of the long-lived recirculating and memory B-cells populations of lymphocytes that normally contain high levels of Bcl-2 protein (51–53). Similarly, in many complex epithelial tissues, high levels of Bcl-2 protein are found in the stem cells that line the basement membrane but not in the more differentiated cells as they advance towards the body or lumen surface and die by programmed cell death. Since these stem cells represent a population of long-lived cells that typically have high proliferative rates, these cells may accumulate genetic errors over time that promote their malignant transformation. Indeed the presence of Bcl-2 in these stem cells may facilitate their transformation, by allowing these cells to tolerate genetic alterations that might otherwise, in the absence of Bcl-2, cause apoptosis. Among these changes that tend to induce apoptosis and which Bcl-2 can provide protection against are the expression of some types of oncogenes (such as *c-myc*), loss of integrin attachments to extracellular matrix proteins (a step important for invasion), and development of genetic instability due to loss of p53 or perhaps DNA repair enzymes (54–58).

Clearly, however, in some types of solid tumors, it does appear that expression of the *BCL-2* gene specifically becomes dysregulated during tumor progression. Perhaps one of the best examples of this is prostate cancer. Though variable results have been reported, in general only a small percentage of primary prostate cancers probably express Bcl-2 (59, 60). Consistent with this idea, most of these primary tumors also respond well to anti-androgen therapy, a stimulus that is known to trigger apoptosis in the androgen-dependent epithelial cells of prostate (61). In contrast, high levels of Bcl-2 proteins are commonly found in prostate tumors that have progressed to a metastatic and hormone-independent state (60, 62, 63). Thus, deregulated *BCL-2* expression appears to represent a relatively late event in the progression of prostate cancers.

3.1. Elevated Bcl-X_L Expression

Comparatively less is known about the expression of other *BCL-2* family genes in human tumors. Elevated levels of the anti-apoptotic protein, Bcl-X_L however have been found in a high proportion of colorectal carcinomas, particularly less differentiated tumors (64). Interestingly, Bcl-X_L and Bcl-2 expression tend to be reciprocally regulated in colorectal cancers, with Bcl-2 immunostaining being more prominent in well-differentiated tumors and Bcl-X immunostaining elevated (compared to normal colonic epithelium) in less differentiated cancers (64). Similarly, aggressive breast cancers have been reported to express Bcl-X_L at high levels, whereas the earlier-stage tumors that often contain high levels of Bcl-2 are often Bcl-X_L negative (65). In prostate cancers, both Bcl-X_L and Bcl-2 immunostaining tend to increase during histological progression as defined by Gleason grade (66). Thus, Bcl-X_L and Bcl-2 are not invariably regulated in a reciprocal fashion. Elevated expression of Bcl-X_L has also been reported in acute myelogenous leukemias and some other types of hematopoietic malignancies.

3.2. Bax Gene Inactivation

Single nucleotide mutations which inactive one or both copies of the *BAX* gene have been reported in colon cancers and some types of leukemia (67, 68). Interestingly, in colon cancers associated with the microsatellite instability phenotype, a homopolymeric stretch of 8 guanosines in one of the coding exons of *BAX* can commonly suffer additions or deletions that produce frame-shift mutations, thus preventing the resulting mRNA from encoding a stable Bax polypeptide (67). Thus, *BAX* appears to represent a bone fide tumor suppressor

Using immunohistochemical techniques and antibodies specific for Bax, we have observed that ~one-third of adenocarcinomas of the breast have markedly reduced or absence expression of Bax, in contrast to normal mammary epithelium which is Bax-immunopositive (69). Though the mechanisms responsible for reduced Bax in breast cancers remain undetermined, not only Bax protein but also mRNA is typically reduced (70). Moreover, gene transfer-mediated increases in Bax protein levels in breast cancer cell lines restored sensitivity to apoptotic stimuli such as serum withdrawal and anti-Fas/APO1 antibodies (71). Interestingly, reduced Bax immunostaining correlated with reduced Bcl-2 immunostaining, suggesting that these two genes may be co-regulated to some extent in breast cancers.

This correlation between Bcl-2 and Bax in breast cancers suggests a potential explanation for the seemingly paradoxical observation that Bcl-2 expression is often associated with favorable tumor characteristics and better clinical outcome in women with adenocarcinomas of the breast (72–75). Namely, since it is the ratio of anti-apoptotic and pro-apoptotic proteins that controls the ultimate sensitivity to cell death stimuli, presumably Bax-immunonegative tumors can tolerate the loss of Bcl-2. These findings, coupled with the observation that tumors with reduced Bcl-2 also generally have p53 mutations (69, 72), also suggest that a generalized deregulation of apoptotic control mechanisms may occur during the progression of breast cancers. In this regard, early-stage disease is typically ER-positive and Bcl-2-immunopositive, probably because *BCL-2* gene expression is under control of estrogens (76–78). Later, tumors can progress to an ER-negative state characterized almost uniformly by reduced *BCL-2* expression and very often associated with reduced *BAX* expression and p53 mutations (69).

3.3. Reduced Bak Expression

In addition to reduced or absent expression of Bax in some types of malignancies, >90% of colorectal adenocarcinomas and nearly as many gastric cancers have reduced immunostaining for Bak, another of the pro-apoptotic member of the Bcl-2 protein family (64). This suggests that reductions in either of these pro-apoptotic proteins can represent a step in the progression of epithelial cancers.

4. THE CELL DEATH PATHWAY

Though overly simplistic, the process of PCD can be divided into 3 fundamental steps: Initiation; Commitment; and Execution. A myriad of stimuli and insults can serve as initiators of the cell death pathway, including essentially all chemotherapeutic drugs, UV- and gamma-irradiation, cytokines such as TNF, Lymphotoxin-α, Lymphotoxin-β, Fas-ligand, and TGF-β, steroid hormones such as glucocorticoids and retinoids, cytolytic T-cells, growth factor deprivation, some types of viruses, oxidants, and agents that alter Ca^{2+} homeostasis. Because over-expression of Bcl-2 can protect cells at least to some extent from apoptosis induced by all of these stimuli, presumably the cellular pathways affected by these initiators of apoptosis ultimately converge onto what may represent a final common pathway that is governed by Bcl-2 and its homologs (Figure 4). This final common pathway culminates in the activation of a family of cysteine proteases with specificity for cleavage of their protein substrates at aspartic acid residues (termed "caspases"). The

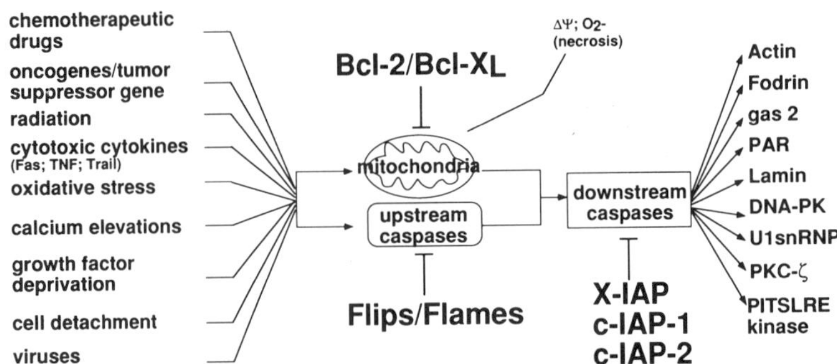

Figure 4. Cell death pathway. A simplistic model of the pathway for apoptotic cell death is presented. A wide variety of signals and insults can initiate the cell death process. Many of these converge on a mitochondria-dependent step, but some such as TNF-family death receptors engage a group of upstream caspases and may be less dependent on mitochondria for their actions. Both the release of protease-activators from mitochondria (such as cytochrome c [not shown]) and the proximal caspases can lead to activation of several downstream caspases such as caspases-2, 3, and 7. These distal caspases then cleave a variety of protein substrates, causing cell death. Though depicted as separate pathways, cross talk between the mitochondrial and upstream caspase pathways is probably extensive. If caspase activation is prevented, the changes in mitochondria can lead to cell death through mechanisms more closely resembling necrosis and involving generation of reactive oxygen species.

Among the inhibitors of the cell death pathway are the FLIP/Flame family proteins, which bind to DED domains in adaptor proteins such as Fadd/mort and in upstream caspase such as caspases-8 and -10. The Bcl-2 family protein act at the level of mitochondria but also have effects at other intracellular membrane compartments such as the ER/nuclear envelope. Some of the IAPs function as direct inhibitors of downstream caspases.

caspases exist as inactive zymogens in cells but become activated by proteolysis to generate functional proteases that consist of two subunits which assemble into a heterotetramer (reviewed in 79, 80). Though the mechanisms for proteolytic activation of caspases are still being defined, some can probably autocatalytically process themselves when enticed to do so by interactions with other proteins. However, the best documented mechanism involves activation by proteolysis, with many of the caspases having the ability to cleave and activate each other, thus suggesting the possibility of a cascade not unlike the complement or coagulation protease systems. These intracellular proteases represent the effector molecules of the cell death pathway, cleaving a variety of cytosolic, cytoskeletal, nuclear, and other proteins in cells and thereby producing either directly or indirectly most of the morphological and biochemical events that are commonly associated with apoptosis.

Bcl-2 family proteins appear to control the commitment step of the cell death pathway, determining whether certain caspases will or will not become processed and activated. Anti-apoptotic members of the Bcl-2 family such as Bcl-2 and Bcl-X_L tend to make it more difficult for apoptotic signals to trigger activation of these terminal effector proteases, whereas pro-apoptotic members such as Bax and Bak facilitate caspases activation.

However, not all cell death stimuli are blockable by over-expression of Bcl-2 or Bcl-X_L in all cellular contexts. Regular exceptions to the rule include certain members of the TNF family of cytokine receptors, such as Fas/APO-1 (CD95). In many cellular contexts, CD95-mediated death is poorly blocked by Bcl-2, implying that pathways for bypassing Bcl-2 exist. CD95 directly engages the caspases via an adapter protein Fadd/Mort that links this plasma membrane receptor to caspases-8 and -10 (81–84). Thus, opportunities may exist for side-stepping Bcl-2-mediated blocks to apoptosis through exploitation of the protease cascade initiated by CD95 and some other members of the TNF receptor family.

Of interest in this regard are reports that chemotherapy sometimes induces apoptosis in tumor cells through a process that involves induction of expression of the genes encoding CD95 and its ligand (Fas-ligand), resulting in an autocrine destruction of the cell. Unfortunately, examples of proteins which can interfere with the CD95-pathway for apoptosis have also been found in tumors. Among these blockades to CD95-induced apoptosis are FAP-1, a protein tyrosine kinase that binds to the C-terminus of CD95 and somehow thwarts its ability to induce apoptosis (85). Another such CD95 inhibitor is FLIP (also known as FLAME, CASH, I-FLICE, and Caspaser), a pseudo-Caspase protein which lacks enzymatic activity but shares striking amino-acid sequence homology with caspases-8 and -10 (reviewed in 86). FLIP binds to Fadd/Mort as well as caspases-8 and -10. Over-expression of FLIP has been reported to prevent CD95-induced apoptosis in at least some types of cells.

Interestingly, in those cellular contexts where Bcl-2 or Bcl-X_L have been shown to prevent CD95-induced apoptosis, upstream caspases can still become activated and some target proteins cleaved, yet apoptosis does not ensue (87). Thus, Bcl-2 and Bcl-X_L can dissociate some caspase activation events from cell death, presumably by interfering with a downstream event.

5. MITOCHONDRIAL PERMEABILITY TRANSITION

The molecular details of what constitutes the decision of the cell to either commit to apoptotic cell death or to survive despite exposure to apoptotic stimuli remain enigmatic. However, recently, it has been proposed that the phenomenon of mitochondrial permeability transition may play a central role as an integrator of the myriad of upstream signals that can potentially trigger apoptosis (reviewed in 88). Mitochondrial permeability transition refers to a phenomenon whereby the permeability of the inner membrane of mitochondria increases resulting in loss of the proton gradient across this membrane which creates the electrochemical gradient ($\Delta\Psi$) necessary for oxidative phosphorylation (reviewed in 89). Permeability transition (PT) has been associated with the opening of large pores ("megachannels") in mitochondria, which appear to be concentrated at the contact sites where the inner and outer mitochondrial membranes come into juxtaposition. Mitochondrial PT is induced very early during apoptosis. Moreover, pharmacological agents which directly induce mitochondrial PT also uniformly induce apoptosis in a variety of types of cells. Conversely, agents that prevent mitochondrial PT also suppress apoptosis induction by many apoptotic stimuli.

At least three cell death-inducing consequences of mitochondrial permeability transition are known to occur. First, loss of the $\Delta\Psi$ results in the shunting of electrons away from oxygen consumption and into free-radical production, resulting in damage to cellular membranes and other structures. Cell death occurring via this mechanism may bear more resemblance to necrosis than apoptosis. Second, loss of the $\Delta\Psi$ also causes Ca^{2+} which is normally stored in the matrix of mitochondria to be dumped into the cytosol, potentially activating Ca^{2+}-dependent proteases, kinases and phosphatases that can promote cell death. Third, swelling of the mitochondrial matrix after induction of PT indirectly causes rupture of the outer mitochondrial membrane, releasing apoptogenic proteins into the cytosol which are normally sequestered in the intermembrane space between the inner and outer mitochondrial membranes. These apoptogenic proteins include cytochrome *c* and a 50-kDa protein termed AIF for Apoptosis Inducing Factor which are both potently capable of activating the caspases through unknown mechanisms (90, 91).

A wide variety of factors influence the relative propensity of the megapore that controls mitochondrial PT to assume either an open or closed conformation. Thus, Ca^{2+}, loss of the $\Delta\Psi$, and oxidants all promote megapore opening, which in turn then leads to further Ca^{2+} release, greater dissipation of the electrochemical gradient, and more reactive oxygen species production in a positive feedback loop that sends the cell spirally towards death. Bcl-2 and most of its homologs are integral membrane proteins and have been localized at least in part to the outer mitochondrial membrane (92). Moreover, over-expression of Bcl-2 in cells provides protection against various apoptotic stimuli that induce mitochondrial PT (93, 94). Mitochondria isolated from Bcl-2 over-expressing cells are also more resistant to PT inducers in vitro (91, 95). Conversely, over-expression of Bax in cells can induce mitochondrial PT and apoptosis (96). Taken together, therefore, these findings have suggested the possibility that Bcl-2 and its homologs may directly or indirectly regulate the mitochondrial megapore, thus influencing whether this structure is likely to assume an open or a closed conformation. At present the molecular components of the megapore remain poorly defined, but appear to include the adenine nucleotide translocator and a voltage-dependent anion channel (reviewed in 88, 89).

While mitochondrial PT may serve the role of a central integrator that controls a commitment step in the cell death pathway, it should be noted that mitochondria may not be required for all apoptotic cell deaths (97) and that Bcl-2 may be capable of preventing cell death even without localization to mitochondria (98). Thus, there may also exist other mechanisms which can commit cells to triggering the cell death pathway. Moreover, it has recently been suggested that cytochrome *c* release from mitochondria can proceed the loss of $\Delta\Psi$ that accompanies permeability transition, thus pathways for extrusion of caspase-activating cytochrome *c* from mitochondria may exist which do not require antecedent opening of the megapore (99, 100). Taken together, while the bulk of evidence suggests that mitochondria can play an essential role in apoptosis under many circumstances and while most Bcl-2 family proteins are associated with mitochondrial membranes, the details of how mitochondria may participate in the commitment step of apoptosis await further elucidation.

6. Bcl-2 FAMILY PROTEINS AS ION-CHANNELS

A milestone in our understanding of Bcl-2 family protein function has come from the 3-dimensional structure of the Bcl-X_L protein, which has revealed striking structural similarity with the pore-forming domains of the bacterial toxins such as diphtheria toxin (DT) and the colicins (38). Some members of the Bcl-2 family including Bcl-2, Bax, and Bak can be modeled on the same Bcl-X_L crystallographic coordinates, suggesting that they also share this structural similarity to pore-forming domains. Other members of the Bcl-2 family however such as the BH3-only proteins Bik, Hrk, and Bid cannot, implying that they do not function as pore proteins and thus are relegated to the role of trans-dominant inhibitors or modulators of the other members of the family (unpublished observations). Consistent with these ideas, recombinant Bcl-2, Bcl-X_L, and Bax proteins have been shown to form ion-conducting channels in liposomes and planar bilayers in vitro (101–103), whereas Bid does not (unpublished data).

By analogy to the bacterial toxins, it has been speculated that the process of pore formation by Bcl-2 family proteins involves the perpendicular insertion into the membrane of two α-helices that are normally buried in the interior of the compact folded protein when in aqueous environments. Indeed, deletion of these two core α-helices (α5 and

$\alpha 6$) abolishes channel formation by Bcl-2 in vitro and destroys its ability to block cell death induced by Bax in mammalian cells and yeast (102). Bcl-2 has been reported to prevent the formation of channels by Bax in vitro (103), but how it accomplishes this remains unexplored.

It has been speculated that a minimum of two molecules of Bcl-2 are required to form a channel in membranes (102). This idea derives from studies of other α-helical-type ion-channels which have shown that a minimum of 4 amphipathic α-helices are required to form an ion-conducting ring in the membrane with the hydrophilic faces of the participating α-helices oriented towards the aqueous lumen of the channel and their hydrophobic surfaces contacting the membrane (104). Since Bcl-2, Bcl-X_L, Bax, and Bak are predicted to contain only two α-helices ($\alpha 5$ and $\alpha 6$) of sufficient length to span the lipid bilayer, it is envisioned that at least two molecules must interact after integration into the membrane, each contributing a pair of α-helices to the channel.

Given that both anti-apoptotic (Bcl-2; Bcl-X_L) and pro-apoptotic (Bax) proteins are capable of forming channels in membranes, controversy abounds as to how this pore-forming activity relates to the bioactivities of these proteins (reviewed in 105). One idea is that proteins such as Bcl-2 (death blocker) and Bax (death promoter) have different selectivities for the ions they transport (cations vs anions for example). Alternatively, it could be that both Bcl-2 and Bax have similar transport functions, possibly allowing transport of deleterious molecules that induce cell death. In this case, it might be speculated that heterodimerization of Bcl-2 and Bax plugs the pores, so that the critical issue is not whether Bcl-2 or Bax is present individually but rather the ratio of these proteins. This model would be consistent with data suggesting the excessive amounts of Bcl-2 can paradoxically promote cell death rather than block it in some circumstances (106) and with the finding that Bax and Bak can unexpectedly exhibit cytoprotective rather than cytotoxic functions in some cellular contexts (15, 107). Finally, it is entirely possible that heterodimerization of death-blockers such as Bcl-2 and death-promoters such as Bax creates chimeric channels with transport or regulatory properties are different from either Bcl-2 or Bax channels alone.

In addition, many other issues remain unresolved about the Bcl-2 and Bax protein channels, such as the diameter of the channels and how many Bcl-2, Bcl-X_L, or Bax proteins it takes to create an aqueous channel in membranes. In the case of DT, the channels are evidently large enough to transport a protein, since the primary function of DT is thought to be for allowing transport of the ADP-ribosylation factor subunit of the toxin from lysosomes and endosomes into the cytosol (108). In contrast, the bacterial colicins transport ions (109, 110). Though many details remain to be clarified, the structural and electrophysiological evidence of pore formation could fit nicely with some of the cellular phenomena that Bcl-2 family proteins have been reported to control, such as mitochondrial PT, and transporting either ions (Ca^{2+}) or proteins (Cytochrome *c* and AIF).

7. Bcl-2 FAMILY PROTEINS AS ADAPTER/DOCKING PROTEINS

Though capable of forming channels, Bcl-2 also has other ways of modulating cellular processes relevant to cell death. For example, Bcl-2 (or Bcl-X_L) has been reported to bind several other non-homologous proteins (Table II), including the protein kinase Raf-1, the protein phosphatase Calcineurin, the GTPases R-Ras and H-Ras, the p53-binding protein p53-BP2, the Prion protein Pr-1, and several novel proteins including BAG-1, CED-4, Nip-1, Nip-2, and Nip-3 (111–121). In most cases, neither the functional significance of

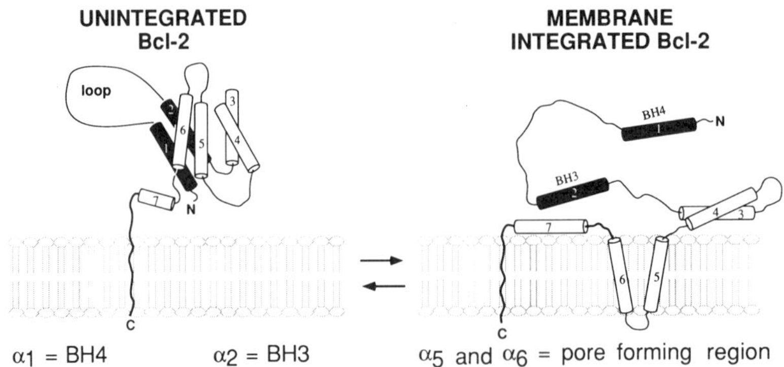

Figure 5. Integration of Bcl-2 into membranes. By analogy to the bacterial toxins, Bcl-2 and some of its related proteins may insert into membranes though a process where the fifth and sixth a-helices penetrate the lipid bilayer. The surrounding amphipathic a-helices presumably fold up against the membrane or may be freed to interact with other proteins in the case of the first a-helix (BH4 domain) and second a-helix (BH3 domain).

these interactions with Bcl-2 nor the regions on the Bcl-2 protein required for binding to these proteins has been explored.

Recently, however, it has been reported that Raf-1, Calcineurin, and possibly BAG-1 bind to Bcl-2 in a manner which depends upon the BH4 domain. Interestingly, unlike Bcl-2 and the other anti-apoptotic proteins, the pro-apoptotic members of the Bcl-2 family lack the BH4 domain, suggesting that this stretch of ~16 aminoacids located near the NH_2-terminus of these proteins may be uniquely involved in the suppression rather than the induction of cell death (Figure 3). Moreover, the Raf-1 and BAG-1 proteins can collaborate with Bcl-2 in the suppression of cell death (19, 111, 117, 122). As mentioned above, the BH4 domain of Bcl-2 corresponds to the first α-helix in the structure of Bcl-X_L. This amphipathic helix is separated from the rest of the protein by a flexible ~50 aminoacid long loop region, suggesting that the BH4 domain may be able to swing away from the body of the protein for the purpose of interacting with other proteins such as BAG-1, Raf-1, and calcineurin. In the crystal structure of Bcl-X_L, the BH4 domain is shown inserted on the back-side of the protein relative to the hydrophobic pocket that binds BH3 domains, with its hydrophobic face interacting with the core hydrophobic helices that are presumably responsible for penetration of the lipid bilayer. Upon insertion into the membrane or adopting an open conformation that resembles the membrane integrated state, therefore, the BH4 domain would be anticipated to swing free creating the opportunity to bind to other proteins (Figure 5).

How does the BH4-dependent interaction of Bcl-2 with Raf-1 and BAG-1 enhance protection from apoptosis? Recently it has been shown that BAG-1 not only binds to Bcl-2 but also can specifically interact with Raf-1. Moreover, BAG-1 can activate this kinase, presumably through a protein-protein interaction that involves binding of BAG-1 to the catalytic domain of Raf-1 (122). This activation of Raf-1 occurs in a Ras-independent fashion. Thus, the concept emerges of Bcl-2 as a docking or adapter protein around which "signal-transduction" events occur. In the case of Raf-1 and BAG-1, dimers of Bcl-2 could be viewed as a docking site that allows Raf-1 and BAG-1 to meet each other in the cells, interact, and result in transient reversible activation of the kinase Raf-1 locally in the vicinity of Bcl-2 on the surface of mitochondria, ER, or nuclear envelope membranes. A further implication is that a similar mechanism may be employed to bring a serine/threonine phosphatase such as calcineurin into the vicinity of Bcl-2, since a mechanism for reversing the effects of Raf-1 would also be needed. Once activated locally in the vicinity of Bcl-2, the kinase Raf-1 appears to mediate directly or indirectly phosphorylation of the pro-apoptotic protein BAD (19).

The unphosphorylated BAD protein can bind to anti-apoptotic proteins such as Bcl-2 and Bcl-X_L, preventing them from dimerizing with Bax and abrogating their function as blockers of cell death (13). Once phosphorylated, BAD no longer binds to Bcl-2 and Bcl-X_L (123). Moreover, the two serines that become phosphorylated reside within consensus binding sites for 14–3–3, an abundant cytosolic protein that sequesters the phosphorylated BAD protein in the cytosol where it cannot interact with Bcl-2 or Bcl-X_L. In this regard, it is important to note that BAD represents one of the few members of the Bcl-2 family that does not possess a C-terminal membrane anchoring domain and thus it is free to float between cytosol and membrane locations. Thus, the interaction of Bcl-2 with Raf-1 appears to provide a mechanism for eradicating the suppression of BAD. While perhaps not central to the purported pore forming functions of Bcl-2 family proteins, this role of Bcl-2 as adapter protein may nevertheless provide a means of modulating the activity of Bcl-2 family proteins for the purposes of either maintaining cell survival or promoting cell death.

Additional examples of this adapter protein function may explain the observation that Bcl-2 over-expression can prevent p53 translocation from cytosol to nucleus in some types of cells but not others (124, 125). In this regard, the p53-BP2 interactions with Bcl-2 could potentially provide a mechanisms for tethering Bcl-2 to p53. Though highly speculative, since Bcl-2 appears to be concentrated at nuclear pore complexes (92), p53-BP2-mediated bridging of Bcl-2 to p53 could hypothetically provide a convenient means for blocking the entry of p53 into the nucleus which would be cell-type dependent in the sense that only in those cells which express p53-BP2 would this phenomenon of Bcl-2-dependent blockage of p53 translocation be exhibited. Though the domains within Bcl-2 needed for binding to p53-BP2 have yet to be explored, a reasonable expectation is that the proline-rich loop region in Bcl-2 will be involved, given that p53-BP2 contains a SH3 domain and that such domains are known to bind to proline-rich sequences (126). Moreover, inasmuch as Bax and the other pro-apoptotic members of the Bcl-2 family lack this long proline-rich loop, this may explain why p53-BP2 binds to Bcl-2 but not to Bax. Finally, the ability of Bcl-2 to bind the Calcineurin, sequestering it at membranes and preventing its translocation with NF-AT into the nucleus, may also be at the heart of the mechanism by which Bcl-2 prevents apoptosis induced by this protein phosphatase (127).

8. THE CED-4 CONNECTION

Consistent with its role as a docking or adapter protein, Bcl-X_L has recently been reported to bind to CED-4. The CED-4 protein was originally discovered through genetic approaches in the free-living nematode, *C. elegans.* In the worm, CED-4 is essential for all 131 programmed cell deaths that occur in this organism during development, thus establishing CED-4 is a pro-apoptotic protein (128). Both human Bcl-X_L and the worm equivalent of Bcl-2 (CED-9) are capable of binding to CED-4. Moreover, it has been shown that CED-4 is normally found in the cytosol, but when co-expressed with Bcl-X_L or CED-9, the CED-4 protein is then sequestered at mitochondria and other intracellular membrane sites where Bcl-2 family proteins typically reside. Finally, mutants of Bcl-X_L and CED-9 that fail to block cell death also uniformly fail to bind CED-4 (119–121). The human homolog of CED-4 has only recently been cloned (129). Though it remains to be determined whether it can bind to Bcl-2 or Bcl-X_L, the tendency for evolutionary conservation of the cell death machinery makes it highly likely.

CED-4 provides Bcl-2 family proteins with a conduit to the cell death proteases. In *C. elegans*, the worm homolog of the caspases is CED-3, a cysteine protease whose enzymatic function highly resembles many of the mammalian caspases which are involved in apoptosis. CED-4 has been shown to bind CED-3 (119). The implication is that CED-4 functions as a pro-apoptotic protein, at least in part by binding to and somehow promoting the activation of CED-3. Consistent with this idea, genetic studies in the worm indicate that CED-4 functions upstream of CED-3 in the programmed cell death pathway. Moreover, CED-4 appears to be able to bind simultaneously to CED-9 or Bcl-X_L and CED-3 or mammalian caspases, such as ICE and FLICE (119). Thus, CED-4 bridges the anti-apoptotic members of the Bcl-2 family to the cell death proteases, presumably allowing Bcl-2 and its related apoptosis-blocking proteins to sequester certain proteases so that they cannot become activated. Interestingly, the pro-apoptotic members of the Bcl-2 family such as Bax, Bak, and Bik fail to bind CED-4, and instead displace CED-4 from its binding sites on Bcl-X_L upon heterodimerizing with Bcl-X_L. Thus, pro-apoptotic proteins such as Bax free CED-4 to travel into the cytosol where it presumably must reside in order to activate caspases. Though many details remain unexplored, it seems highly likely that the interaction of anti-apoptotic Bcl-2 family proteins with the Caspases via bridging CED-4 homologs will prove to be one of the major mechanisms by which Bcl-2, Bcl-X_L, and their related proteins suppress apoptosis.

9. POST-TRANSLATIONAL MODIFICATIONS OF Bcl-2 FAMILY PROTEINS

The loop region of Bcl-2 constitutes a negative regulatory domain in Bcl-2 and Bcl-X_L (130). At least for Bcl-2, it also appears to contain a phosphorylation site(s) or is necessary for binding of a kinase that induces phosphorylation of Bcl-2. Phosphorylation may possibly be a mechanism for inactivating the Bcl-2 protein, and has been correlated with impaired heterodimerize with Bax (131). A direct cause-and-effect relation between specific sites of phosphorylation and inhibition of Bcl-2 however has yet to be demonstrated. The kinases and phosphatases that regulate phosphorylation of Bcl-2 are currently unknown. It has been shown that anti-microtubule drugs such as taxol, vincristine, and nocodazole can induce phosphorylation of Bcl-2, suggesting the involvement of either a microtubule associated kinase (MAP kinase) or a cyclin-dependent kinase that becomes active when cells are pharmacologically arrested in M-phase (131, 132). The only known Bcl-2 associated kinase, Raf-1, appears not to be the relevant protein kinase (133).

Interestingly, mutations have been identified in the loop region of Bcl-2 in lymphomas containing t(14;18) chromosomal translocations involving the *BCL-2* gene, sometimes occurring in association with transformation from low-grade follicular lymphoma to aggressive diffuse large-cell non-Hodgkin's lymphoma (134, 135). Though little has been done so far to explore the function of Bcl-2 family proteins that contain loop mutations, alterations at one common mutation site (proline 59) have been associated with reduced Bcl-2 phosphorylation in response to taxol (133) and enhanced cell proliferation (136); and our unpublished data). In this regard, evidence that Bcl-2 can entry of quiescent cells into the cell cycle has been obtained for several types of cells (137–140). Thus, the enhanced proliferation seen in cells transfected with a Bcl-2 (pro 59) mutant relative to the wild-type Bcl-2 protein potentially may reflect a loss of interactions of Bcl-2 with other proteins responsible for this cell cycle slowing phenomenon seen in some types of cells that have been engineered to over-express Bcl-2. Recently, a laboratory-created mutant of Bcl-2 (Y328F) has been described which dissociates the cell proliferation inhibitory effects of Bcl-2 from its ability to protect against cell death, implying that either phosphorylation at this tyrosine or interaction of an unknown protein that requires this tyrosine is required for interference with cell cycle entry (141).

Candidate proteins that can potentially speed cell cycle rates and whose function may be inhibited by over-expression of Bcl-2 include the transcription factors NF-kB, and NF-AT. Inhibition of entry into the nucleus and/or successful trans-activation of target genes by NF-kB and NF-AT has been reported in Bcl-2 over-expressing cells (140, 142). In the case of NF-AT, a transcription factor first identified because of it ability to stimulate expression of lymphokine and lymphokine receptor genes in lymphocytes, it has been shown that dephosphorylation of NF-AT by calcineurin is required for NF-AT translocation from the cytosol into the nucleus (143). Thus, the ability of Bcl-2 to bind calcineurin, sequestering it at membranes where it cannot dephosphorylated NF-AT in the cytosol, may provide an explanation for at least one mechanism by which Bcl-2 contributes to cell cycle arrest (112). It should be noted however that the wild-type Bcl-2 protein does not cause a cell cycle delay in all types of cells, and indeed has been shown to enhance rather than inhibit cell proliferation in some cellular contexts (144, 145).

Nevertheless, the potential for Bcl-2 to modulate the cell proliferation by inhibiting entry of quiescent cells into the cell cycle may represent a protective function that prevent cells with damaged DNA for example from attempting to replicate. It may also be relevant to the biology of low-grade NHLs which contain t(14;18) translocations and which have low mitotic rates. Transformation to more aggressive lymphomas with higher mitotic rates, therefore, may require a breakdown in these mechanisms by which Bcl-2 potentially slows cell cycle entry, either through changes in the expression or function of other Bcl-2 interacting proteins or through mutations in the Bcl-2 protein that prevent it from interacting with those proteins.

10. STRATEGIES FOR INHIBITING Bcl-2 FUNCTION IN CANCERS

Since all anticancer drugs presently available rely upon the programmed cell death pathway for ultimately killing tumor cells, over-expression of Bcl-2 or, more specifically, alterations in the ratios of anti-apoptotic Bcl-2 family proteins relative to pro-apoptotic Bcl-2 family proteins can play an important role as a chemoresistance mechanism (reviewed in 49, 146, 147). Finding strategies for overcoming the cytoprotective effects of Bcl-2 and its related proteins such as Bcl-X_L therefore represents a major goal of researchers and pharmaceutical companies interested in this area.

Already, several strategies can be envisioned for abrogating the effects of Bcl-2 in human cancers and efforts are underway to implement them. These include small molecule blockers of Bcl-2 homo- or heterodimerization, as well as blockers of the interactions of Bcl-2 with CED4 or proteins such as BAG-1 and Raf-1 that promote cell survival in collaboration with Bcl-2. In addition, as the pore-like functions of Bcl-2 family proteins undergo further elucidation, and in particular their biological relevance becomes better understood, this may also create opportunities for screening for small molecules that alter the channel activities of these proteins by analogy to current cardiovascular drugs.

Another approach focuses on the idea that phosphorylation of Bcl-2 could be a mechanism for inactivating the protein. Clearly, therefore, if more were known about the kinases and phosphatases that control phosphorylation of Bcl-2, then this could suggest additional strategies for counteracting Bcl-2 in cancer cells. In this regard, while taxol and other anti-microtubule drugs can induce phosphorylation of Bcl-2, typically less than half of the Bcl-2 protein molecules become phosphorylated and, moreover, the phosphorylation is dependent on cells cycling through M-phase -- a potential limitation given the generally low proliferative fractions of most solid tumors. A search for drugs targeted against the kinase that phosphorylates BAD can also be contemplated as a means for inducing BAD-mediated suppression of Bcl-2 in cancer cells that express this pro-apoptotic member of the Bcl-2 family.

Antisense oligonucleotides targeted against Bcl-2 have also been employed both in vitro and in animal models for inducing apoptosis of malignant cells or sensitizing them to conventional chemotherapeutic drugs (148–152). In this regard, a recent Phase-I trial of a phosphorothioate *BCL-2* antisense oligonucleotide in patients with relapsed and refractory B-cell NHL provided encouraging results, suggesting that this approach warrants further investigations (153).

Finally, the recognition that expression of Bcl-2 family genes can be regulated by various biological response modifiers such as retinoids, estrogens, cytokines and growth factors, suggests opportunities for modulating the ratios of anti- and pro-apoptotic members of the Bcl-2 protein family in ways that might sensitize cancer cells to conventional cytotoxic drugs.

An important question when considering the idea of targeting Bcl-2 for cancer therapy is that of toxic side-effects. Though Bcl-2 and its homologs are expressed in a wide variety of normal tissues, tumor cells may be more dependent on the survival promoting functions of members of this family of apoptosis-regulating proteins. First, tumor cells typically contain alterations in the expression of *c-myc* and other cell-cycle genes that promote apoptosis. This deregulated proto-oncogene expression not only creates an incessant drive for cell proliferation but also sets up a clash in signals that would normally induce apoptosis, were it not for the actions of anti-apoptotic genes such as *bcl-2* (54, 55, 154). Given that normal cells do not have these contradictory signals caused by oncogene activation, drugs that target Bcl-2 or its homologs should be relatively specific for cancer cells, since cancer cells will presumably be more dependent on proteins such as Bcl-2 for their survival. Second, normal epithelial cells are highly dependent on cues from their surrounding environment for their survival. At least in part, these cues come from cell contact with the extracellular matrix via integrins, cell-cell interactions, and local production of growth factors (56, 155). Tumor cells are able to invade locally and metastasize to distant sites partly because they are less dependent on these environmental cues, probably as the result of deregulated expression of gene such as *bcl-2*. Hence, because of their localization to foreign environments, the survival of metastatic cells potentially may be severely comprised by drugs that disrupt the function of Bcl-2 or Bcl-X_L, whereas the viability of normal cell will not be. Third, the genetic instability that occurs in may tumors creates signals for apoptosis that would normally kill cells by triggering the endogenous program for cell death, were it not for the actions of anti-apoptotic proteins such as Bcl-2. Indeed, disturbances in the balance of pro- and anti-apoptotic protein that result in an anti-apoptotic state may be a prerequisite for the accumulation of genetic defects (67). Finally, the reason why existing chemotherapeutic drugs display some selectivity for tumor cells over normal cells can be attributed at least in part to the loss of cell cycle check point controls in cancer cells (156). These checkpoint controls normally cause cells with damaged DNA, for example, to arrest until repair mechanisms can correct drug-induced injury before resuming attempts at DNA-replication. The lack of cell cycle checkpoint control mechanisms in tumor cells drives these malignant cells to cycle despite the presence of drug-induced damage, thus creating signals that induce apoptosis. The problem with all too many tumors however is that these apoptotic signals are blocked from fulfilling their intended mission, presumably because of the actions of Bcl-2, Bcl-X_L or other anti-apoptotic proteins. Agents or strategies that counteract the effects of Bcl-2 and Bcl-X_L, therefore, should release these barriers to apoptosis in drug-damaged tumor cells and promote synergistic tumor responses to conventional chemotherapeutic drugs.

SUMMARY

After over 10 years of effort, we are beginning to understand the biochemical functions of Bcl-2 family proteins and to develop strategies for attempting to modulate their activity for the treatment of cancer. Bcl-2 and several of its homologs seem likely to be multifunctional proteins, that not only form channels in membranes but that also have important interactions with other types of proteins such as CED-4. Further knowledge about the structural and functional details of these two facets of Bcl-2 will contribute additional insights that can be applied in efforts to promote apoptosis selectively in tumors.

ACKNOWLEDGMENTS

I thank H. Gallant for manuscript preparation, and H-G. Wang and H. Zha for assistance with figure preparation. I also wish to thank the National Cancer Institute, American Cancer Society, Leukemia Society of America, U.S. Army, State of California, CaP-CURE, and the Susan B. Komen Foundation for their generous support.

REFERENCES

1. Tsujimoto, Y., J. Cossman, E. Jaffe, and C. Croce: Involvement of the bcl-2 gene in human follicular lymphoma. Science 1985, 228: 1440–1443.
2. Vaux, D. L., S. Cory, and J. M. Adams: Bcl-2 gene promotes haemopoietic cell survival and cooperates with c-myc to immortalize pre-B cells. Nature 1988, 335: 440–442.
3. Hengartner, M. O., and H. R. Horvitz: Programmed cell death in Caenorhabditis elegans. Curr Opin Gnet Dev 1994, 4: 581–586.
4. Steller, H.: Mechanisms and genes of cellular suicide. Science 1995, 267: 1445–1449.
5. Vaux, D. L., I. L. Weissman, and S. K. Kim: Prevention of programmed cell death in Caenorhabditis elegans by human bcl-2. Science 1992, 258: 1955–1957.
6. Hengartner, M. O., and H. R. Horvitz: C. elegans cell survival gene ced-9 encodes a functional homolog of the mammlian proto-oncogene bcl-2. Cell 1994, 76: 665–676.
7. Boise, L. H., M. Gonzalez-Garcia, C. E. Postema, L. Ding, T. Lindsten, L. A. Turka, X. Mao, G. Nunez, and C. B. Thompson: bcl-x, a bcl-2-related gene that functions as a dominant regulator of apoptotic cell death. Cell 1993, 74: 597–608.
8. Oltvai, Z., C. Milliman, and S. J. Korsmeyer: Bcl-2 heterodimerizes in vivo with a conserved homolog, Bax, that accelerates programmed cell death. Cell 1993, 74: 609–619.
9. Kozopas, K. M., T. Yang, H. L. Buchan, P. Zhou, and R. Craig: Mcl-1, a gene expressed in programmed myeloid cell differentiation, has sequence similarity to bcl-2. Proc.Natl.Acad.Sci.USA 1993, 90: 3516–3520.
10. Lin, E. Y., A. Orlofsky, M. S. Berger, and M. B. Prystowsky: Characterization of A1, a novel hemopoietic-specific early-response gene with sequence similarity to bcl-2. J.Immunol. 1993, 151: 1979–1988.
11. Gibson, L., S. P. Holmgreen, D. C. S. Huang, O. Bernard, N. G. Copeland, N. A. Jenkins, G. R. Sutherland, E. Baker, J. M. Adams, and S. Cory: bcl-w, a novel member of the bcl-2 family, promotes cell survival. Oncogene 1996, 13: 665–675.

12. Gillet, G., M. Guerin, A. Trembleau, and G. Brun: A Bcl-2-related gene is activated in avian cells transformed by the Rous sarcoma virus. EMBO J. 1995, 14: 1372–1381.
13. Yang, E., J. Zha, J. Jockel, L. H. Boise, C. B. Thompson, and S. J. Korsmeyer: Bad: a heterodimeric partner for Bcl-XL and Bcl-2, displaces bax and promotes cell death. Cell 1995, 80: 285–291.
14. Chittenden, T., E. A. Harrington, R. O'Connor, C. Flemington, R. J. Lutz, G. I. Evan, and B. C. Guild: Induction of apoptosis by the Bcl-2 homologue Bak. Nature 1995, 374: 733–736.
15. Kiefer, M. C., M. J. Brauer, V. C. Powers, J. J. Wu, s. R. Ubansky, L. D. Tomei, and P. J. Barr: Modulation of apoptosis by the widely distributed Bcl-2 homologue Bak. Nature 1995, 374: 736–739.
16. Choi, S. S., I.-C. Park, J. W. Yun, Y. C. Sung, S. Hong, and H. Shin: A novel Bcl-2 related gene, Bfl-1, is overexpressed in stomach cancer and preferentially expressed in bone marrow. Oncogene 1995, 11: 1693–1698.
17. Farrow, S. N., J. H. M. White, I. Martinou, T. Raven, K.-T. Pun, C. J. Grinham, J.-C. Martinou, and R. Brown: Cloning of a bcl-2 homologue by interaction with adenovirus E1B 19K. Nature 1995, 374: 731–733.
18. Boyd, J. M., G. J. Gallo, B. Elangovan, A. B. Houghton, S. Malstrom, B. J. Avery, R. G. Ebb, T. Subramanian, T. Chittenden, R. J. Lutz, and G. Chinnadurai: Bik, a novel death-inducing protein shares a distinct sequence motif with Bcl-2 family proteins and interacts with viral and cellular survival-promoting proteins. Oncogene 1995, 11: 1921–1928.
19. Wang, K., W.-M. Yin, D. T. Chao, C. L. Milliman, and S. J. Korsmeyer: BID: a novel BH3 domain-only death agonist. Genes & Development 1996, 10: 2859–2869.
20. Inohara, N., L. Ding, S. Chen, and G. Nunez: *Harakiri,* a novel regulator of cell death, encodes a protein that activates apoptosis and interacts selectively with survival-promoting proteins Bcl-2 and Bcl-X_L. EMBO J 1997, 16: 1686–1694.
21. Das, R., E. P. Reddy, D. Chatterjee, and D. W. Adrews: Identification of a novel Bcl-2 related gene, BRAG-1, in human glioma. Oncogene 1996, 12: 947–951.
22. Cruz-Reyes, J., and J. R. Tata: Cloning, characterization and expression of two Xenopus bcl-2-like cell-survival genes. Gene 1995, 158: 171–179.
23. Rao, L., M. Debbas, P. Sabbatini, D. Hockenbery, S. Korsmeyer, and E. White: The adenovirus E1A proteins induce apoptosis, which is inhibited by the E1B 19-kDa and bcl-2 proteins. Proc.Natl.Acad.Sci.USA 1992, 89: 7742–7746.
24. Henderson, S., D. Huen, M. Rowe, C. Dawson, G. Johnson, and A. Rickinson: Epstein-Barr virus-coded BHRF1 protein, a viral homologue of Bcl-2, protects human B cells from programmed cell death. Proc.Natl.Acad.Sci.USA 1993, 90: 8479–8483.
25. Neilan, J. G., Z. Lu, C. L. Afonso, G. F. Kutish, M. D. sussman, and D. L. Rock: An African swine fever virus gene with similarity to the proto-oncogene bcl-2 and the Epstein-Barr virus gene BHRF1. J.Virol. 1993, 67: 4391–4394.
26. Cheng, E. H.-Y., J. Nicholas, D. S. Bellows, G. S. Hayward, H.-G. Guo, M. S. Reitz, and J. M. Hardwick: A Bcl-2 homolog encoded by Kaposi sarcoma-associated virus, human herpesvirus 8, inhibits apoptosis but does not heterodimerize with Bax or Bak. Proc. Natl. Acad. Sci. USA 1997, 94: 690–694.
27. Smith, C. A.: A novel viral homolgue of Bcl-2 and Ced-9. Trends Cell Biol. 1995, 5: 344–0.
28. Sato, T., M. Hanada, S. Bodrug, S. Irie, N. Iwama, L. H. Boise, C. B. Thompson, E. Golemis, L. Fong, H.-G. Wang, and J. C. Reed: Interactions among members of the bcl-2 protein family analyzed with a yeast two-hybrid system. Proc.Natl.Acad.Sci.USA 1994, 91: 9238–9242.
29. Yang, E., J. Jockel, J. Zha, and S. Korsmeyer: Bad, a new bcl-2 family member, heterodimerizes with bcl-2 and bcl-XL in vivo, and promotes cell death. Blood 1994, 84 (Suppl. 1): 373a-0.
30. Hanada, M., C. Aimé-Sempé, T. Sato, and J. C. Reed: Structure-function analysis of Bcl-2 protein: identification of conserved domains important for homodimerization with Bcl-2 and heterodimerization with Bax. J.Biol.Chem. 1995, 270: 11962–11968.
31. Zha, H., H. A. Fisk, M. P. Yaffe, N. Mahajan, B. Herman, and J. C. Reed: Structure-function comparisons of the proapoptotic protein Bax in yeast and mammalian cells. Mol Cell Biol 1996, 16: 6494–508.
32. Jürgensmeier, J. M., S. Krajewski, R. Armstrong, G. M. Wilson, T. Oltersdorf, L. C. Fritz, J. C. Reed, and S. Ottilie: Bax- and Bak-induced cell death in the fission yeast *Schizosaccharomyces pombe*. Mol. Biol. Cell 1997, 8: 325–229.
33. Yin, X. M., Z. N. Oltvai, and S. J. Korsmeyer: BH1 and BH2 domains of bcl-2 are required for inhibition of apoptosis and heterodimerization with bax. Nature 1994, 369: 321–333.
34. Bodrug, S. E., C. Aimé-Sempé, T. Sato, S. Krajewski, M. Hanada, and J. C. Reed: Biochemical and functional comparisons of Mcl-1 and Bcl-2 proteins: evidence for a novel mechanism of regulating Bcl-2 family protein function. Cell Death Differ. 1995, 2: 173–182.

35. Zha, H., C. Aime-Sempe, T. Sato, and J. C. Reed: Pro-apoptotic protein Bax heterodimerizes with Bcl-2 and homodimerizes with Bax via a novel domain (BH3) distinct from BH1 and BH2. J.Biol.Chem. 1996, 271: 7440–7444.
36. Chittenden, T., C. Flemington, A. B. Houghton, R. G. Ebb, G. J. Gallo, B. Elangovan, G. Chinnadurai, and R. J. Lutz: A conserved domain in Bak, distinct from BH1 and BH2, mediates cell death and protein binding functions. EMBO J. 1995, 14: 5589–5596.
37. Sedlak, T. W., Z. N. Oltvai, E. Yang, K. Wang, L. H. Boise, C. B. Thompson, and S. J. Korsmeyer: Multiple Bcl-2 family members demonstrate selective dimerizations with Bax. Proc.Natl.Acad.Sci.USA 1995, 92: 7834–7838.
38. Muchmore, S. W., M. Sattler, H. Liang, R. P. Meadows, J. E. Harlan, H. S. Yoon, D. Nettesheim, B. S. Changs, C. B. Thompson, S. Wong, S. Ng, and S. W. Fesik: X-ray and NMR structure of human Bcl-XL, an inhibitor of programmed cell death. Nature 1996, 381: 335–341.
39. Sattler, M., H. Liang, D. Nettesheim, R. P. Meadows, J. E. Harlan, M. Eberstadt, H. S. Yoon, S. B. Shuker, B. S. Chang, A. J. Minn, C. B. Thompson, and S. W. Fesik: Structure of Bcl-xL-Bak peptide complex: recognition between regulators of apoptosis. Science 1997, 275: 983–986.
40. Diaz, J.-L., T. Oltersdorf, W. Horne, M. McConnell, G. Wilson, S. Weeks, T. Garcia, and L. C. Fritz: A common binding site mediates heterodimerization and homodimerization of Bcl-2 family members. J. Biol. Chem. 1997, 272: 11350–11355.
41. Hunter, J. J., and T. G. Parslow: A peptide sequence from Bax that converts Bcl-2 into an activator of apoptosis. J.Biol.Chem. 1996, 271: 8521–8524.
42. Cheng, E. H.-Y., B. Levine, L. H. Boise, C. B. Thompson, and J. M. Hardwick: Bax-independent inhibition of apoptosis by Bcl-XL. Nature 1996, 379: 554–556.
43. Ink, B., M. Zornig, B. Baum, N. Hajibagheri, C. James, T. Chittenden, and G. Evan: Human bak induces cell death in *Schizosaccharomyces pombe* with morphological changes similiar to those with apoptosis in mammalian cells. Mol Cell Biol 1997, 17: 2468–2474.
44. Simonian, P. L., D. A. M. Grillot, R. Merino, and G. Nunez: Bax can antagonize Bcl-XL during etoposide and cisplatin-induced cell death independently of its heterodimerization with Bcl-XL. J Biol Chem 1996, 271: 22764–72.
45. Simonian, P. L., D. A. Grillot, D. W. Andrews, B. Leber, and G. Nunez: Bax homodimerization is not required for Bax to accelerate chemotherapy-induced cell death. J Biol Chem 1996, 271: 32073–32077.
46. Weiss, L. M., R. A. Warnke, J. Sklar, and M. L. Cleary: Molecular analysis of the t(14;18) chromosomal translocation in malignant lymphomas. N.Engl.J.Med. 1987, 317: 1185–1189.
47. Cleary, M. L., S. D. Smith, and J. Sklar: Cloning and structural analysis of cDNAs for bcl-2 and a hybrid bcl-2/immunoglobulin transcript resulting from the t(14;18) translocation. Cell 1986, 47: 19–28.
48. Bakhshi, A., J. P. Jensen, P. Goldman, J. J. Wright, O. W. McBride, A. L. Epstein, and S. J. Korsmeyer: Cloning the chromosomal breakpoint of t(14;18) human lymphomas: clustering around JH on chromosome 14 and near a transcriptional unit on 18. Cell 1985, 41: 899–906.
49. Reed, J. C.: Bcl-2: Prevention of apoptosis as a mechanism of drug resistance. Hematology/Oncology Clinics of North America 1995, 9: 451–474.
50. Reed, J. C.: Regulation of apoptosis by bcl-2 family proteins and its role in cancer and chemoresistance. Curr Opin Oncol 1995, 7: 541–6.
51. Pezzella, F., A. G. D. Tse, J. L. Cordell, K. A. F. Pulford, K. C. Gatter, and D. Y. Mason: Expression of the bcl-2 oncogene protein is not specific for the 14;18 chromosomal translocation. Am.J.Pathol. 1990, 137: 225.
52. Hockenbery, D. M., M. Zutter, W. Hickey, M. Nahm, and S. J. Korsmeyer: Bcl-2 protein is topographically restricted in tissues characterized by apoptotic cell death. Proc.Natl.Acad.Sci.USA 1991, 88: 6961–6965.
53. Hanada, M., D. Delia, A. Aiello, E. Stadtmauer, and J. C. Reed: bcl-2 gene hypomethylation and high-level expression in B-cell chronic lymphocytic leukemia. Blood 1993, 82: 1820–1828.
54. Bissonnette, R. P., F. Exheverri, A. Mahboubi, and D. R. Green: Apoptotic cell death induced by c-myc is inhibited by bcl-2. Nature 1992, 359: 552–554.
55. Fanidi, A., E. A. Harrington, and G. I. Evan: Cooperative interaction between c-myc and bcl-2 proto-oncogenes. Nature 1992, 359: 554–556.
56. Frisch, S. M., and H. Francis: Disruption of epithelial cell-matrix interactions induces apoptosis. J.Cell Biol. 1994, 124: 619–626.
57. Zhang, Z., K. Vuori, J. C. Reed, and E. Ruoslahti: The alpha 5 beta 1 integrin supports survival of cells on fibronectin and up-regulates Bcl-2 expression. Proc.Natl.Acad.Sci.USA 1995, 92: 6161–6165.
58. Wang, Y., L. Szekely, I. Okan, G. Klein, and K. G. Wiman: Wild-type p53-triggered apoptosis is inhibited by bcl-2 in a v-myc-induced T-cell lymphoma line. Oncogene 1993, 8: 3427–3431.
59. Shabaik, A. S., S. Krajewski, A. Burgan, M. Krajewska, and J. C. Reed: bcl-2 proto-oncogene expression in normal, hyperplastic, and neoplastic prostate tissue. J.Urol.Pathol. 1994, 3: 17–27.

60. Furuya, Y., S. Krajewski, J. I. Epstein, J. C. Reed, and J. T. Isaacs: Enhanced expression of Bcl-2 and the progression of human and rodent prostatic cancers. Clin Cancer Res 1996, 2: 389–398.
61. Kyprianou, N., and J. T. Isaacs: Activation of programmed cell death in the rat ventral prostate after castration. Endocrinology 1988, 122: 552–562.
62. Colombel, M., F. Symmans, S. Gil, K. M. O'Toole, D. Chopin, M. Benson, C. A. Olsson, S. Korsmeyer, and R. Buttyan: Detection of the apoptosis-suppressing oncoprotein bcl-2 in hormone-refractory human prostate cancers. Am.J.Pathol. 1993, 143: 390–400.
63. McDonnell, T. J., P. Troncoso, S. M. Brisbay, C. Logothetis, L. W. K. Chung, J.-T. Hsieh, S.-M. Tu, and M. L. Campbell: Expression of the proto-oncogene bcl-2 in the prostate and its association with emergence of androgen-independent prostate cancer. Cancer Res. 1992, 52: 6940–6944.
64. Krajewska, M., S. Moss, S. Krajewski, K. Song, P. Holt, and J. C. Reed: Elevated expression of Bcl-X and reduced Bak in primary colorectal adenocarcinomas. Cancer Res. 1996, 56: 2422–2427.
65. Kumar, R., M. Mandal, A. Lipton, H. Harvey, and C. B. Thompson: Overexpression of HER2 modulates Bcl-2, Bcl-XL, and Tamoxifen-induced apoptosis in human MCF-7 breast cancer cells. Clin.Cancer Res. 1996, 2: 1215–1219.
66. Krajewska, M., S. Krajewski, J. I. Epstein, A. Shabaik, J. Sauvageot, K. Song, S. Kitada, and J. C. Reed: Immunohistochemical analysis of bcl-2, bax, bcl-X, and mcl-1 expression in prostate cancers. Am J Pathol 1996, 148: 1567–76.
67. Rampino, N., H. Yamamoto, Y. Ionov, Y. Li, H. Sawai, J. C. Reed, and M. Perucho: Frequent framshift somatic mutations in the pro-apoptotic gene bax in colon cancer of the microsatellite mutator phenotype. Science 1997, 275: 967–969.
68. Meiijerink, J. P. P., T. F. C. M. Smetsers, A. W. Slöetjes, E. H. P. Linders, and E. J. B. M. Mensink: Bax mutations in cell lines derived from hematological malignancies. Leukemia 1995, 9: 1828–1832.
69. Krajewski, S., C. Blomvqvist, K. Franssila, M. Krajewska, V.-M. Wasenius, E. Niskanen, and J. C. Reed: Reduced expression of pro-apoptotic gene Bax is associated with poor response rates to combination chemotherapy and shorter survival in women with metastatic breast adenocarcinoma. Cancer Res. 1995, 55: 4471–4478.
70. Bargou, R. C., P. T. Daniel, M. Y. Mapara, K. Bommert, K. Wagenert, C. Wagener, B. Kallinich, H. D. Royer, and B. Dörken: Expression of the bcl-2 gene family in normal and malignant breast tissue: low bax-a expression in tumor cells correlates with resistance towards apoptosis. Int.J.Cancer 1995, 60: 854–859.
71. Bargou, R. C., C. Wagener, K. Bommert, M. Y. Mapara, P. T. Daniel, W. Arnold, M. Dietel, H. Guski, A. Feller, H. D. Royer, and B. Dorken: Overexpression of the death-promoting gene bax-α which is down-regulated in breast cancer restores sensitivity to different apoptotic stimuli and reduces tumor growth in SCID mice. J Clin Invest 1996, 97: 2651–2659.
72. Silvestrini, R., S. Veneroni, M. G. Daidone, E. Benini, P. Boracchi, M. Mezzetti, G. Di Fronzo, F. Rilke, and U. Veronesi: The bcl-2 protein: a prognostic indicator strongly related to p53 protein in lymph node-negative breast cancer patients. J.Natl.Cancer Inst. 1994, 86: 499–504.
73. Joensuu, H., L. Pylkkänen, and S. Toikkanen: Bcl-2 protein expression and long-term survival in breast cancer. Am.J.Pathol. 1994, 145: 1191–1198.
74. Gasparini, G., M. Barbareschi, C. Doglioni, P. D. Palma, F. A. Mauri, P. Boracchi, P. Bevilacqua, O. Caffo, L. Morelli, P. Verderio, F. Pezzella, and A. L. Harris: Expression of bcl-2 protein predicts efficacy of adjuvant treatments in operable node-positive breast cancer. Clin.Cancer Res. 1995, 1: 189–198.
75. Hellemans, P., P. A. van Dam, J. Weyler, A. T. van Oosterom, P. Buytaert, and E. Van Marck: Prognostic value of bcl-2 expression in invasive breast cancer. Br.J.Cancer 1995, 72: 354–360.
76. Teixeira, C., J. C. Reed, and M. A. C. Pratt: Estrogen promotes chemotherapeutic drug resistance by a mechanism involving Bcl-2 protooncogene expression in human breast cancer cells. Cancer Res. 1995, 55: 3902–3907.
77. Wang, T. T. Y., and J. M. Phang: Effects of estrogen on apoptotic pathways in human breast cancer cell line MCF-7. Cancer Res. 1995, 55: 2487–2489.
78. Johnston, S. R. D., K. A. MacLennan, N. P. M. Sacks, J. Salter, I. E. Smith, and M. Dowsett: Modulation of Bcl-2 and Ki-67 expression in oestrogen receptor-positive human breast cancer by tamoxifen. Eur.J.Cancer 1994, 30A: 1663–1669.
79. Martin, S. J., and D. R. Green: Protease activation during apopotosis: death by a thousand cuts? Cell 1995, 82: 349–352.
80. Patel, T., G. J. Gores, and S. H. Kaufmann: The role of proteases during apoptosis. FASEB Journal 1996, 10: 587–597.
81. Chinnaiyan, A. M., K. O'Rourke, M. Tewari, and V. M. Dixit: FADD, a novel death domain-containing protein, interacts with the death domain of Fas and initiates apoptosis. Cell 1995, 81: 505–512.

82. Muzio, M., A. M. Chinnaiyan, F. C. Kischkel, K. O'Rourke, A. Shevchenko, J. Ni, C. Scaffidi, J. D. Bretz, M. Zhang, R. Gentz, M. Mann, P. H. Krammer, M. E. Peter, and V. M. Dixit: Flice, a novel FADD-homologous ICE/CED-3-like protease, is recruited to the CD95 (Fas/APO-1) death--induceing signaling complex. Cell 1996, 85: 817–827.
83. Boldin, M. P., E. E. Varfolomeev, Z. Pancer, I. L. Mett, J. H. Camonis, and D. Wallach: A novel protein that interacts with the death domain of Fas/APO1 contains a sequence motif related to the death domain. J.Biol.Chem. 1995, 270: 7795–7798.
84. Boldin, M. P., T. M. Goncharov, Y. V. Goltsev, and D. Wallach: Involvement of MACH, a novel MORT1/FADD-interacting protease, in Fas/APO-1- and TNF receptor-induced cell death. Cell 1996, 85: 803–815.
85. Sato, T., S. Irie, S. Kitada, and J. C. Reed: FAP-1: A protein tyrosine phosphatase that associates with Fas. Science 1995, 268: 411–415.
86. Wallach, D.: Placing death under control. Nature 1997, 388: 123–126.
87. Boise, L. H., and C. B. Thompson: Bcl-x(L) can inhibit apoptosis in cells that have undergone Fas-induced protease activation. Proc Natl Acad Sci U S A 1997, 94: 3759–64.
88. Petit, P. X., S.-A. Susin, N. Zamzami, B. Mignotte, and G. Kroemer: Mitochondria and programmed cell death: back to the future. FEBS Lett 1996, 396: 7–13.
89. Bernardi, P., K. M. Broekemeier, and D. R. Pfeiffer: Recent progress on regulation of the mitochondrial permeability transition pore; a cyclosporin-sensitive pore in the inner mitochondrial membrane. Journal of Bioenergetics and Biomembranes 1994, 26: 509–517.
90. Liu, X., C. N. Kim, J. Yang, R. Jemmerson, and X. Wang: Induction of apoptotic program in cell-free extracts: requirement for dATP and Cytochrome C. Cell 1996, 86: 147–157.
91. Susin, S. A., N. Zamzami, M. Castedo, T. Hirsch, P. Marchetti, A. Macho, E. Daugas, M. Geuskens, and G. Kroemer: Bcl-2 inhibits the mitochondrial release of an apoptogenic protease. J. Exp. Med. 1996, 184: 1331–1342.
92. Krajewski, S., S. Tanaka, S. Takayama, M. J. Schibler, W. Fenton, and J. C. Reed: Investigation of the subcellular distribution of the bcl-2 oncoprotein: residence in the nuclear envelope, endoplasmic reticulum, and outer mitochondrial membranes. Cancer Res 1993, 53: 4701–14.
93. Zamzami, N., P. Marchetti, M. Castedo, T. Hirsch, S. A. Susin, B. Masse, and G. Kroemer: Inhibitors of permeability transition interfere with the disruption of the mitochondrial transmembrane potential during apoptosis. FEBS Lett. 1996, 384: 53–57.
94. Marchetti, P., M. Castedo, S. A. Susin, N. Zamzami, T. Hirsch, A. Macho, A. Haeffner, F. Hirsch, M. Geuskens, and G. Kroemer: Mitochondrial permeability transition is a central coordinating event of apoptosis. J Exp Med 1996, 184: 1155–1160.
95. Zamzami, N., A. Susin, P. Marchetti, T. Hirsch, I. Gomez-Monterrey, M. Castedo, and G. Kroemer: Mitochondrial control of nuclear apoptosis. J.Exp.Med. 1996, 183: 1533–1544.
96. Xiang, J., D. T. Chao, and S. J. Korsmeyer: BAX-induced cell death may not require interleukin 1β-converting enzyme-like proteases. Proc Natl Acad Sci USA 1996, 93: 14559–14563.
97. Muzio, M., G. S. Salvesen, and V. M. Dixit: FLICE induced apoptosis in a cell-free system. J Biol Chem 1997, 272: 2952–2956.
98. Zhu, W., A. Cowie, L. Wasfy, B. Leber, and D. Andrews: Bcl-2 mutants with restricted subcellular location reveal spatially distinct pathways for apoptosis in different cell types. The EMBO Journal 1996, 15: 4130–4141.
99. Yang, J., X. Liu, K. Bhalla, C. N. Kim, A. M. Ibrado, J. Cai, I.-I. Peng, D. P. Jones, and X. Wang: Prevention of apoptosis by Bcl-2: release of cytochrome c from mitochondria blocked. Science 1997, 275: 1129–1132.
100. Kluck, R. M., E. Bossy-Wetzel, D. R. Green, and D. D. Newmeyer: The release of cytochrome c from mitochondria: a primary site for Bcl-2 regulation of apoptosis. Science 1997, 275: 1132–1136.
101. Minn, A. J., P. Velez, S. L. Schendel, H. Liang, S. W. Muchmore, S. W. Fesik, M. Fill, and C. B. Thompson: Bcl-x_L forms an ion channel in synthetic lipid membranes. Nature 1997, 385: 353–357.
102. Schendel, S. L., Z. Xie, M. O. Montal, S. Matsuyama, M. Montal, and J. C. Reed: Channel formation by antiapoptotic protein Bcl-2. Proc Natl Acad Sci U S A 1997, 94: 5113–8.
103. Antonsson, B., F. Conti, A. Ciavatta, S. Montessuit, S. Lewis, I. Martinou, L. Bernasconi, A. Bernard, J.-J. Mermod, G. Mazzei, K. Maundrell, F. Gambale, R. Sadoul, and J.-C. Martinou: Inhibition of Bax channel-forming activity by Bcl-2. Science 1997, 277: 370–2.
104. Montal: Design of molecular function: channels of communication. Annu Rev Biophys Biomol Struct 1995, 24: 31–57.
105. Reed, J. C.: Double identity for proteins of the Bcl-2 family. Nature 1997, 387: 773–776.
106. Chen, J., J. G. Flannery, M. M. LaVail, R. H. Steinberg, J. Xu, and M. I. Simon: bcl-2 overexpression reduces apoptotic photoreceptor cell death in three different retinal degenerations. Proc Natl Acad Sci USA 1996, 93: 7042–7047.

107. Middleton, G., G. Nunez, and A. M. Davies: Bax promotes neuronal survival and antagonises the survival effects of neurotrophic factors. Development 1996, 122: 695–701.
108. Donovan, J. J., M. I. Simon, R. K. Draper, and M. Montal: Diphtheria toxin forms transmembrane channels in planar lipid bilayers. Proc Natl Acad Sci U S A 1981, 78: 172–176.
109. Cramer, W. A., Y.-L. Zhang, S. Schendel, A. R. merrill, H. Y. Song, C. V. Stauffacher, and F. S. Cohen: Dynamic properties of the colicin E1 ion channel. FEMS Microbiology Immunol. 1992, 105: 71–82.
110. Konisky, J.: Colicins and other bacteriocins with established modes of action. Annu. Rev. Microbiol. 1982, 36: 125–144.
111. Wang, H.-G., T. Miyashita, S. Takayama, T. Sato, T. Torigoe, S. Krajewski, S. Tanaka, I. Hovey, L., J. Troppmair, U. R. Rapp, and J. C. Reed: Apoptosis regulation by interaction of bcl-2 protein and Raf-1 kinase. Oncogene 1994, 9: 2751–2756.
112. Shibasaki, F., E. Kondo, T. Akagi, and F. McKeon: Suppression of signalling through NF-AT by interactions between calcineurin and Bcl-2. Nature 1997, 386: 728–731.
113. Fernandez-Sarbia, M. J., and J. R. Bischoff: Bcl-2 associates with the ras-related protein R-ras p23. Nature 1993, 366: 274–275.
114. Chen, C.-Y., and D. V. Faller: Phosphorylation of Bcl-2 protein and association with p21Ras in Ras-induced apoptosis. J Biol Chem 1996, 271: 2376–2379.
115. Naumovski, L., and M. L. Cleary: The p53-binding protein 53BP2 also interacts with Bcl-2 and impedes cell cycle progression at G2/M. Mol.Cell.Biol. 1996, 16: 3884–3892.
116. Kurschner, C., and J. I. Morgan: The cellular prion protein (PrP) selectively binds to Bcl-2 in the yeast two-hybrid system. Mol.Brain Res. 1995, 30: 165–168.
117. Takayama, S., T. Sato, S. Krajewski, K. Kochel, S. Irie, J. A. Millan, and J. C. Reed: Cloning and functional analysis of BAG-1: a novel Bcl-2 binding protein with anti-cell death activity. Cell 1995, 80: 279–284.
118. Boyd, J. M., S. Malstrom, T. Subramanian, L. K. Venkatesch, U. Schaeper, B. Elangovan, C. D'Sa-Epper, and G. Chinnadurai: Adenovirus E1B 19 kDa and bcl-2 proteins interact with a common set of cellular proteins. Cell 1993, 79: 341–351.
119. Chinnaiyan, A. M., K. O'Rourke, B. R. Lane, and V. M. Dixit: Interaction of CED-4 with CED-3 and CED-9: a molecular framework for cell death. Science 1997, 275: 1122–1126.
120. Wu, D., H. D. Wallen, and G. Nunez: Interaction and regulation of subcellular localization of CED-4 by CED-9. Science 1997, 275: 1126–1129.
121. Spector, M. S., S. Desnoyers, D. J. Heoppner, and M. O. Hengartner: Interaction between the *C. elegans* cell-death regulators CED-9 and CED-4. Nature 1997, 385: 653–656.
122. Wang, H.-G., S. Takayama, U. R. Rapp, and J. C. Reed: Bcl-2 interacting protein, BAG-1, binds to and activates the kinase Raf-1. Proc Natl Acad Sci U S A 1996, 93: 7063–7068.
123. Zha, J., H. Harada, E. Yang, J. Jockel, and S. J. Korsmeyer: Serine phosphorylation of death agonist BAD in response to survival factor results in binding to 14–3–3 not BCL-X_L. Cell 1996, 87: 619–628.
124. Ryan, J. J., E. Prochownik, C. A. Gottlieb, I. J. Apel, R. Merino, G. Nuñez, and M. F. Clarke: c-myc and bcl-2 modulates p53 function by altering p53 subcellular trafficking during the cell cycle. Proc.Natl.Acad.Sci.USA 1994, 91: 5878–5882.
125. Marin, C. M., A. Fernandez, R. J. Bick, S. Brisbay, L. M. Buja, M. Snuggs, D. J. McConkey, A. C. von Eschenbach, M. J. Keating, and T. J. McDonnell: Apoptosis suppression by bcl-2 is correlated with the regulation of nuclear and cytosolic Ca^{2+}. Oncogene 1996, 12: 2259–2266.
126. Gorina, S., and N. P. Pavletich: Structure of the p53 tumor suppressor bound to the ankyrin and SH3 domains of 53BP2 [see comments]. Science 1996, 274: 1001–5.
127. Shibasaki, F., and F. McKeon: Calcineurin functions in Ca^{2+} -activated cell death in mammalian cells. J. Cell Biol. 1995, 131: 735–743.
128. Horvitz, H. R., S. Shaham, and M. O. Hengartner: The genetics of programmed cell death in the nematode Caenorhabditis elegans. Cold Spring Harb Symp Quant Biol 1994, 59: 377–385.
129. Zou, H., W. J. Henzel, X. Liu, A. Lutschg, and X. Wang: Apaf-1, a human protein homologous to C. elegans CED-4, particpates in cytochrome c-dependent activation of caspase-3. Cell 1997, 90: 405–413.
130. Chang, B. S., A. J. Minn, S. W. Muchmore, S. W. Fesik, and C. B. Thompson: Identification of a novel regulatory domain in Bcl-x_L and Bcl-2. EMBO J 1997, 16: 968–977.
131. Haldar, S., J. Chintapallli, and C. M. Croce: Taxol induces bcl-2 phosphorylation and death of prostate cancer cells. Cancer Res. 1996, 56: 1253–1255.
132. Blagosklonny, M., T. Schulte, P. Nguyen, J. Trepel, and L. Neckers: Taxol-induced apoptosis and phosphorylation of Bcl-2 protein involves c-Raf-1 and represents a novel c-Raf-1 transduction pathway. Cancer Res. 1996, 56: 1851–1854.
133. Aimé-Sempé, C., S. Kitada, and J. C. Reed: Investigations of Taxol®-mediated phosphorylation of bcl-2. Blood 1996, 88: 106.

134. Tanaka, S., D. C. Louie, J. A. Kant, and J. C. Reed: Frequent somatic mutations in translocated BCL2 genes of non-Hodgkin's lymphomas patients. Blood 1992, 79: 229.
135. Matolcsy, A., P. Casali, R. A. Warnke, and D. M. Knowles: Morphologic transformation of follicular lymphoma is associated with somatic mutation of the translocated Bcl-2 gene. Blood 1996, 88: 3937–3944.
136. Reed, J. C., and S. T. Tanaka: Somatic Point mutations in translocated bcl-2 alleles of non-Hodgkin's lymphomas and lymphocytic leukemias: Implications for mechanisms of tumor progression. Leuk.Lymphoma 1993, 10: 157–163.
137. Pietenpol, J. A., N. Papadopoulos, S. Markowitz, K. V. Willson, K. W. Kinzler, and B. Vogelstein: Paradoxical inhibition of solid tumor cell growth by bcl-2. Cancer Res. 1994, 54: 3714–3717.
138. Borner, C.: Diminished cell proliferation associated with the death-protective activity of Bcl-2. J.Biol.Chem. 1996, 271: 12695–12698.
139. Mazel, S., D. Burtrum, and H. T. Petrie: Regulation of cell division cycle progression by bcl-2 expression: a potential mechanism for inhibition of programmed cell death. J.Exp.Med. 1996, 183: 2219–2226.
140. Linette, G. P., Y. Li, K. Roth, and S. J. Korsmeyer: Cross talk between cell death and cell cycle progression: BCL-2 regulates NFAT-medicated activation. Proc. Natl. Acad. Sci. USA 1996, 93: 9545–9552.
141. Huang, D. C. S., L. A. O'Reilly, A. Strasser, and S. Cory: The anti-apoptosis function of Bcl-2 can be genetically separated from its inhibitory effect on cell cycle entry. EMBO J 1997, 16: 4628–4638.
142. Grimm, S., M. K. A. Bauer, P. Baeuerle, and K. Schulze-Osthoff: Bcl-2 down-regulates the activity of transcription factor NF-κB induced apoptosis. J Cell Biol 1996, 134: 13–23.
143. Shibasaki, F., E. R. Price, D. Milan, and F. McKeon: Role of kinases and the phosphatase calcineurin in the nuclear shuttling of transcription factor NF-AT4. Nature 1996, 382: 370–3.
144. Reed, J. C., H. S. Talwar, U. R. Rapp, M. Cuddy, and G. T. Fisher: Mitochondrial protein p26 bcl-2 reduces growth factor requirements of NIH3T3 fibroblasts. Exp.Cell Res. 1991, 195: 277–283.
145. Miyazaki, T., Z.-J. Liu, A. Kawahara, Y. Minami, K. Yamada, Y. Tsujimoto, E. L. Barsoumian, R. M. Perlmutter, and T. Taniguchi: Three distinct IL-2 signaling pathways mediated by bcl-2, c-myc, and lck cooperate in hematopoietic cell proliferation. Cell 1995, 81: 223–231.
146. Reed, J. C.: Regulation of apoptosis by Bcl-2 family proteins and its role in cancer and chemoresistance. Current Opinion in Oncology 1995, 7: 541–546.
147. Reed, J. C.: Bcl-2 family proteins: regulators of chemoresistance in cancer. Toxicol.Lett. 1995, 82/83: 155–158.
148. Kitada, S., S. Takayama, K. DeRiel, S. Tanaka, and J. C. Reed: Reversal of chemoresistance of lymphoma cells by antisense-mediated reduction of bcl-2 gene expression. Antisense Res.Dev. 1994, 4: 71–79.
149. Campos, L., O. Sabido, J.-P. Rouault, and D. Guyotat: Effects of Bcl-2 antisense oligodeoxynucleotides on in vitro proliferation and survival of normal marrow progenitors and leukemic cells. Blood 1994, 84: 595–600.
150. Berchem, G. J., M. Bosseler, L. Y. Sugars, H. J. Voeller, S. Zeitlin, and E. P. Gelmann: Androgens induce resistance to bcl-2-mediated apoptosis in LNCaP prostate cancer cells. Cancer Res. 1995, 55: 735–738.
151. Cotter, F. E., P. Johnson, P. Hall, C. Pocock, N. A. Mahdi, J. K. Cowell, and G. Morgan: Antisense oligonucleotides suppress B-cell lymphoma growth in a SCID-hu mouse model. Oncogene 1994, 9: 3049–3055.
152. Ziegler, A., G. H. Luedke, D. Fabbro, K.-H. Altmann, R. A. Stahel, and U. Zangemeister-Wittke: Induction of apoptosis in small-cell lung cancer cells by an antisense oligodeoxynucleotide targeting the Bcl-2 coding sequence. J Natl Cancer Inst 1997, 89: 1027–36.
153. Webb, A., D. Cunningham, F. Cotter, P. A. Clarke, F. di Stefano, P. Ross, M. Corbo, and Z. Dziewanowska: BCL-2 antisense therapy in patients with non-Hodgkin lymphoma. Lancet 1997, 349: 1137–41.
154. Kranenburg, O., A. J. van der Eb, and A. Zantema: Cyclin D1 is an essential mediator of apoptotic neuronal cell death. EMBO J. 1996, 15: 46–54.
155. Raff, M. C., B. A. Barres, J. F. Burne, H. S. Coles, Y. Ishizaki, and M. D. Jacobson: Programmed cell death and the control of cell survival: lessons from the nervous system. Science 1993, 262: 695–700.
156. Elledge, S. J.: Cell cycle checkpoints: preventing an identity crisis. Science 1996, 274: 1664–1667.

DISCUSSION

Lenaz: If I am correct, from your scheme, you were assuming that either Bax or the Bcl-2 dimers forms the channel for the release of cytochrome c from mitochondria. If this is so, maybe the channel is too small for allowing cytochrome c to pass through.

Reed: Yes, the channels we are measuring *in-vitro* would probably be too small to transport proteins. The largest of conductance we have measured in collaboration with Maurice Montal implies a tetrameric structure as I showed you. I am told by Seung Choe, who did the crystal structure on diphtheria toxin, and that such a tetrameter would be large enough to allow an unfolded polypeptide chain to go through a membrane. Thus, if you had chaperones to assist, it might be possible to get an unfolded polypeptide through. But, I think at least what we are measuring *in-vitro* is certainly more consistent with iron channels. However, there are two things to bear in mind. First, the channel activities we are measuring *in vitro* may not be what is occurring *in vivo*. Second, at least one group claims that they have been able to show quite large channels of Bax and they have some evidence that it may be a heptamer which presumably creates a large ring of α-helices in membranes. Clearly, this is very early days in terms of sorting all this out and I really present these ideas as hypotheses that must be tested. I do not strongly favor ion over protein channel because there is simply not enough data to support either hypothesis to the exclusion of the other.

Klausner: In the yeast system, you talked about high copy suppressors of Bax-induced lethality from mammalian cDNA libraries inhibitors, but what about just making use of the yeast genetics. Also, what is the mechanism of cell death that you are getting? Could you express Bax in yeast? It is cell death by apoptosis or necrosis?

Reed: The cell death induced by Bax in yeast does not look like apoptosis because yeast do not have caspases. It is very similar, however, to the cell death that occurs in mammalian cells when one expresses Bax but adds peptidyl inhibitors of the caspases. Under those circumstances where the proteases are inhibited, one still sees mitochondrial permanently transition and generation of reactive oxygen species. The associated morphology of mammalian cells dying under these circumstances by EM looks very much like the yeast when they are killed by Bax. Thus suggesting similar mechanisms. Regarding classical yeast genetics approaches, yes we have been using that to identify genes in yeast which are required for Bax-induced cell death. So far, we have determined that subunits of the mitochondrial FoFi-ATPase/proton-pump, when encoded is the nuclear genome, are required. We suspect the proton pump creates conditions conducive to Bax integrating into membranes and forming channels, but we cannot exclude the possibility that this ATPase participates as a downstream effector of Bax-induced death in yeast.

Klausner: Do you still see Bax-induced death in rho-zero cells?

Reed: Yes, rho zeros can still be killed by Bax. Thus, there is no requirement for oxidative phosphorylation of an intact respiratory chain. This highlights the uniqueness of the requirement seen for the proton pump in the inner membrane.

Livingston: What are rho-zero cells?

Klausner: Rho-zero are, in essence, mitochondrial DNA minus yeast. They actually have mitochondria but they do not respire. You can also derive them using mammalian cells.

Reed: You can create rho-zeros with mammalian cells. For example, you can grow the cells in ethidium bromide and they will lose their mitochondrial DNA. Several years ago, in collaboration with Martin Raff's group, we did that and showed Bcl-2 could still protect in the absence of oxidative phosphorylation in mammalian cells. Similarly, Bax will still kill yeast in the absence of oxidative phosphorylation. We are not entirely sure how to interpret the results of the genetic experiments indicating a requirement for the FoFi-ATPase proton pump, but it suggests there is some kind of a functional connection between the proton gradient and the function of Bax; we know that these channels are formed by Bcl2 family proteins, are modulated by PH and open more commonly at lower PH. It is interesting to think about the proton gradient across the inner membrane and whether there is a functional connection to the channel forming activity of Bax: it could be that in the absence of a proton gradient these channels do not form very readily.

Klausner: One other question: The Bcl-2 proline/serine 59 mutant that you identified in tumors, were these from tumors that were taxol resistant?

Reed: No, these tumors originated from patients before the days of taxol and they were all lymphomas that contained 14; 18 translocations. Analogous to V-genes is the immunoglobulin heavy chain locus, the translocated Bcl-2 gene can become peppered with mutations. Thus, once a Bcl-2 gene has moved into the heavy chain locus, it can start to accumulate somatic mutations. A number of years ago, we identified some mutations, and one that seemed to be recurring in a few tumors was proline 59. Another group did the same experiment and they saw a proline 59 mutant as well. We spent a long time trying to find some difference in the behavior of the wild type and mutant proteins with regards to prevention of cell death but could not detect anything. It was not until we looked at taxol that we saw a difference in the phenotype. Another thing I did not really have a chance to talk about is that Bcl-2 has been showed to have yet another activity. Namely, it can inhibit cell proliferation delaying the entry from G0 quiescence into the cell cycle. That phenotype seems to map generally to this loop region where mutants occur. We have evidence that some of those mutations also overcome that anti-proliferative effects of Bcl-2 so that what you are left with is a protein that still protects from apoptosis perfectly well but no longer has the anti-proliferative function. It may be a progression event involved in low grade lymphomas that start off as G0 B-cells with prolonged lifespans. About half of these neoplasms progress to more aggressive tumors that have high proliferative rates. The question is to what extent do these kinds of mutations occur in some of the more common cancers. Nobody has looked at this yet. It would be an interesting issue to ask whether there are mutations in other types of cancers, such as prostate, breast, and other solid tumors.

Zanker: May I refer to the pore forming capability of Bcl-2? Did you calculate from your NMR experiments lateral diffusion coefficients? In other words, is the pore forming capability a very rapid process, or it is a very slow process?

Reed: We have not done NMR to look at that. We have done planar bilayer experiments with Maurice Montal and we have done EPR with Wayne Hubbell. In our hands, it is a rapid process in the sense that the channels usually stay open only for short duration. I do remember the time constants, off the top of my head, but most of the spontaneous channel activity is of brief duration.

Livingston: I wonder if I can inquire once again about the yeast effect, that is the apparent death in yeast. Do you actually see a DNA ladder?

Reed: No, we do not get DNA ladders and that sort of thing. As a late event you get non-specific cleavage of DNA. Basically, a smear rather than ladders in gel-electrophoresis experiments. What I think yeast and mammalian cells have in common is the effects of Bax on mitochondria permeability transition. If you express Bax in yeast it can induce mitochondrial permeability transition, which is known to lead to generation of reactive oxygen species, so that part of the story is probably conserved between yeast and mammalian cells. I suspect this Bax-mediated effect reflects its direct proforming activity. We view yeast as a way of isolating the pore forming activity of Bax in the absence of all the heterodimerization events and effects on Bcl-2 interactions with CED4, Raf-1 or other proteins. Yeast do not seem to have caspases and so the downstream consequences of how a cell packages itself following cell death is different. The caspases, to a large extent, are really involved in the process of actually degrading and packaging the cell so it can be engulfed by other cells without bursting and spilling its contents and causing inflammation. There is no evidence that either yeast or plants, for that matter, have caspases, probably because plants do not need to remove their dead. In contrast, animals must to be mobile, they therefore cannot afford to have a lifetime's worth of dead cells on board but plants keep their dead as they grow. This may be the evolutionary difference in terms of caspases.

Comoglio: On one of your slides you mentioned Bag1 and the heat-shock proteins which appears to be a Bag1 adaptor, is this acquired for the interaction of Bcl2 with Bag1?

Reed: It appears to be yes. That is stuff I did not have a chance to get into but our current thinking about Bag1 is that it represents some sort of a novel component of the chaperone system that had not been previously appreciated. It binds very tightly to HSP70 and HSC70. *In vitro* using purified components we have been able to show that Bag1 and Bcl2 cannot et together in the absence of chaperones. So, we think that that is at the heart of what Bag1 does and that could then have a number of interesting repercussions: one is it could help induce confirmation changes in Bcl2 that are perhaps involved in disintegration into the membrane or changes that effect its binding to other proteins.

Arndt: My question is in relation to chemo-resistance, is there an imbalance between Bax and Bcl-2?

Reed: Yes, it has been documented in a number of tumor cell lines and through clinical correlative studies involving patients with some types of cancer. If one measures the levels of Bcl-2 or Bax proteins in various ways, elevated Bcl-2 or reduced Bax will often correlate with the more chemo-resistant phenotype. It depends on the specific type of tumor, however. Correlations can be complicated in that there is more than Bcl-2 and Bax. In fact, twelve cellular Bcl-2 homologues have been identified so far. Sometimes one can obtain misleading results if only Bcl-2 or Bax is examined. Nevertheless, there exists lots of precedence for participation of Bcl-2 family proteins in the modulation of chemo-responses. In gene transfer experiments, for example, once can restore Bax expression in cells that have lost it, and they show that they become sensitized to drugs. Conversely, if Bcl-2 is over-expressed by gene transfer, the cells become resistant to drugs.

Hanahan: There is a remarkable complexity developing here with the number of interactions possible for Bcl-2. It sounds as if there are going to be all sorts of different variants present in different states within the cell. So, how do you think this is all integrated?

Is there a threshold or, you would imagine that in one location in the cell there is a protein interaction sending a little of a death signal while in another place its protective.

Reed: Yes, I suspect there are threshold effects. The more we have learned, the more complicated parts of the Bcl-2 story have gotten. On the other hand, we now have two mechanisms that may functionally link Bcl-2 to the caspases. Thus, some aspects of Bcl-2 function are becoming clearer.

Hanahan: Not if there are dozens of them.

Lenaz: Has the mitochondrial protease activation, AIF, been cloned?

Reed: No, the protein has been described by Kroemer's group. He has purified it to homogeneity. It is a 50 KD protein. AIF is evidently encoded in the nuclear genome. It is presumably transported to the inter-membrane space of mitochondria. The purified protein can be mixed with purified pro-caspases and induces their cleavage and activation *in vitro*, so AIF is presumably a protease or something that allows the proteases to auto-process themselves, but it is not cloned yet.

14

CHARACTERIZATION OF ANTIGENS RECOGNIZED BY T CELLS ON HUMAN TUMORS

Pierre G. Coulie,[1,*] Benoît J. Van den Eynde,[2] Pierre van der Bruggen,[2] Aline Van Pel,[2] Etienne De Plaen,[2] and Thierry Boon[2]

[1]Cellular Genetics Unit
Catholic University of Louvain
74 avenue Hippocrate, UCL 7459, B-1200 Brussels, Belgium
[2]Ludwig Institute for Cancer Research, Brussels Branch
74 avenue Hippocrate, UCL 7459, B-1200 Brussels, Belgium

Much has been learned since the realization that anti-tumor cytotoxic T lymphocytes (CTL), which constitute the major effectors involved in tumor rejection, can be cultured *in vitro* and used as tools to identify the target antigens expressed by the tumor cells (reviewed by Boon et al.[1]). Tumor-specific CTL have been obtained in most mouse tumor models and in several human tumor types: mainly in melanoma, but also in renal carcinoma, head and neck carcinoma, or lung tumor. By using a genetic approach based on the transfection of genomic or cDNA libraries, we cloned the genes encoding a number of melanoma antigens. This led to the molecular identification of these antigens which consist of a peptide derived from an intracellular protein and presented to CTL by an HLA class I molecule. On the basis of their pattern of expression, these antigens can be classified in four groups. Antigens of the first group are encoded by genes that are expressed in the tumor cells but are silent in normal adult tissues except male germinal cells. The second group consists of differentiation antigens that are expressed in melanoma and in normal melanocytes. Antigens of the third group are unique to individual tumors and appear through tumor-specific mutations in genes expressed ubiquitously. Eventually there are antigens encoded by genes that are overexpressed in tumors.

---

* Address correspondence to P. G. Coulie, Cellular Genetics Unit, Catholic University of Louvain, 74 avenue Hippocrate - UCL 74.59, B. 1200 Brussels, Belgium. Tel. 32-2-764 75 81; Fax 32-2-764 75 90

The Biology of Tumors, edited by Mihich and Croce
Plenum Press, New York, 1998.

1. TUMOR SPECIFIC SHARED ANTIGENS

These antigens are encoded by genes like *MAGE*, *BAGE* or *GAGE*. Their specific expression in tumors and the fact that they are shared by a number of independent tumors of different histological origins make these antigens potentially useful for specific active immunotherapy of cancer. Remarkably, most CTL directed against these antigens were derived from the same melanoma patient, MZ2, a metastatic patient who enjoyed an extraordinarily favorable evolution. Blood samples from other patients with a tumor expressing some or all of these genes were tested and no such CTL were obtained by stimulating the lymphocytes with autologous tumor cells.

1.1. The *MAGE* Gene Family

The first antigen, named antigen MZ2-E, to be identified on a human melanoma was one of the several antigens that are recognized on the autologous melanoma cell line MZ2-MEL by CTL clones derived from the blood of patient MZ2. A cosmid library was prepared with DNA of the tumor cells. It was transfected, together with a plasmid conferring resistance to geneticin, into an antigen-loss variant of the tumor, obtained through an *in vitro* immunoselection with a tumor-specific CTL clone. Transfectants expressing the antigen were identified by their ability to stimulate the CTL clone to release tumor necrosis factor (TNF)[2]. From such a transfectant a cosmid was retrieved that transferred the expression of the antigen. This led to the identification of a new gene, that was called *MAGE-1*, which codes for the antigen[3].

Gene *MAGE-1* is about 5 kb long and comprises two short exons and a long third exon. An open reading frame coding for a protein of 309 amino acids is located in the third exon. The sequence of the *MAGE-1* gene of the melanoma cells is identical to that of a *MAGE-1* gene isolated from normal blood cells of patient MZ2[3]. *MAGE-1* is a member of a family of at least 12 closely related genes all located on chromosome Xq28[4]. A second family of *MAGE* genes, the *MAGE-B* genes, was recently located in Xp21. It contains four genes encoding proteins that have a predicted amino acid sequence showing 34–43% identity with those of MAGE. The function of the MAGE and MAGE-B proteins is unknown.

The expression of *MAGE* genes in various tumors and normal tissues was analyzed by reverse transcription and polymerase-chain-reaction amplification (RT-PCR) using primers specific for each *MAGE* gene. No expression of *MAGE-1* was found in normal tissues, except in testis. This expression in testis appears to occur in germ-line cells, more precisely spermatocytes and spermatogonia[5]. A similar observation has been made with the mouse equivalent of a *MAGE* gene by in situ hybridization[6]. Because these germline cells do not express MHC class I molecules, gene expression should not result in antigen expression. Forty-eight percent of metastatic melanoma samples were positive for *MAGE-1* (Table 1)[7]. Besides melanomas, a significant proportion of breast tumors[8], non small cell lung carcinomas[9], sarcomas, bladder carcinomas, and head and neck carcinomas express *MAGE-1* (Table 1).

Antigen MZ2-E consists of a nine amino-acid peptide derived from the MAGE-1 protein and presented by HLA-A1[10]. Cell lines expressing *MAGE-1* and HLA-A1 are recognized by the anti-MZ2-E CTL. Since about 26% of Caucasians express the HLA-A1 allele, and 48% of metastatic melanomas express *MAGE-1*, about 12% of them are expected to express antigen MZ2-E.

Several other anti-tumor CTL clones of patient MZ2 were found to recognize another antigen, which was found to be presented by HLA-Cw16. When COS-7 cells were cotransfected with HLA-Cw16 and with the *MAGE-1* cDNA, the transfectants stimulated

Table 1. Expression in tumor samples of genes encoding T cell antigens[a]

	Percentage of tumors expressing					
Histological type	*MAGE-1*	*MAGE-3*	*BAGE*	*GAGE-1,2*	*RAGE-1*	*GnTV*
Melanomas						48
Primary lesions	16	36	8	13	2	
Metastases	48	76	26	28	5	
Non small cell lung carcinomas	49	47	4	19	0	0
Head and neck tumors	28	49	8	19	2	0
Bladder carcinomas	22	36	15	12	5	0
Sarcomas	14	24	6	25	14	0
Mammary carcinomas	18	11	10	9	1	0
Prostatic carcinomas	15	15	0	10	0	0
Colorectal carcinomas	2	17	0	0	0	0
Renal cell carcinomas	0	0	0	0	2	0
Leukemias and lymphomas	0	0	0	1	0	0

[a]Expression was measured by RT-PCR on total RNA using primers specific for each gene.

the CTL to produce TNF[11]. The MAGE-1.Cw16 antigenic peptide is a nonamer that is different from the MAGE-1.A1 peptide (Table 2). Thus, two different peptides derived from the MAGE-1 protein can bind to different HLA class I molecules within the same cells and constitute two antigens recognized by different CTL.

Yet another anti-tumor CTL clone of patient MZ2 was analyzed. When tested on a panel of melanoma cell lines it was found to recognize tumors that expressed both HLA-A1 and *MAGE-3*[12]. Moreover, COS-7 cells cotransfected with HLA-A1 and *MAGE-3* stimulated the CTL. The MAGE-3.A1 antigenic peptide was identified as EVDPIGHLY (Table 2). The same peptide extended by one residue at its N-terminus is presented to CTL by HLA-B44[13]. Like *MAGE-1*, gene *MAGE-3* is silent in normal tissues except testis, and is expressed in a number of tumors of different histological types (Table 1). Since *MAGE-3* is more often expressed by melanomas than *MAGE-1*, it has a wider applicability for specific anti-tumor immunization of melanoma patients. Moreover, most of the *MAGE-1*-positive melanomas also express *MAGE-3*, allowing simultaneous immunization of these patients against both MAGE-1 and MAGE-3 antigens.

Another antigenic peptide derived from gene *MAGE-3* was recently identified by a very different approach: because the consensus motif for binding to HLA-A2 is well char-

Table 2. Tumor-specific antigens shared by different tumors

Gene	Normal expression	MHC	Peptide	Position
MAGE-1	testis	HLA-A1	EADPTGHSY	161–169
		HLA-Cw16	SAYGEPRKL	230–238
MAGE-3	testis	HLA-A1	EVDPIGHLY	168–176
		HLA-A2	FLWGPRALV	271–279
		HLA-B44	MEVDPIGHLY	167–176
BAGE	testis	HLA-Cw16	AARAVFLAL	2–10
GAGE-1/2	testis	HLA-Cw6	YRPRPRRY	9–16
GAGE-3-6	testis	HLA-A29	—	—
RAGE-1	retina	HLA-B7	—	—
GnTV (atypical transcript)	none	HLA-A2	VLPDVFIRC	38–64

acterized and because HLA-A2 is the most frequent HLA class I allele (49% of Caucasians), the MAGE-3 protein sequence was searched for peptides that fitted the A2-binding motif. Nine such peptides were localized and synthesized. They were tested *in vitro* for their capacity to bind to HLA-A2 and the three best binders were used for *in vitro* stimulation of blood lymphocytes from normal HLA-A2 individuals. The responder cells were cloned and CTL clones were obtained that specifically lysed HLA-A2+ cells incubated with peptide FLWGPRALV (Table 2). These CTL also recognized HLA-A2+ melanoma cells expressing gene *MAGE-3*[14].

1.2. Gene *BAGE*

Another gene was identified by the genetic approach and found to encode an antigen recognized by autologous CTL on MZ2-MEL cells. This new gene was named *BAGE*. It codes for a small protein of 43 amino acids that includes a peptide which is presented to CTL by HLA-Cw16 (Table 2)[15]. Like the *MAGE* genes, *BAGE* is not expressed in normal tissues except testis, and is expressed in a number of tumors of various histologies (Table 1).

1.3. The *GAGE* Gene Family

Another family of genes, named *GAGE*, was recently identified by a similar approach, using CTL derived from the same melanoma patient[16]. Six different cDNAs were identified, encoding proteins of 116–138 amino acids. *GAGE-1* and *GAGE-2* produce a peptide that is presented to CTL by HLA-Cw6, whereas *GAGE-3, -4, -5* and *-6* produce a peptide presented by HLA-A29 (Table 2). Again, these genes are silent in normal adult tissues besides testis, but are expressed in a significant proportion of tumors of various types (Table 1).

1.4. Gene *RAGE*

Even though they are expressed on various types of tumors, all these antigens were characterized on a single tumor type, which is melanoma. We recently used a similar approach to characterize antigens recognized by CTL on a kidney tumor. We derived CTL clones directed against an antigen expressed on several renal cell carcinoma lines and presented by HLA-B7. This antigen was found to be encoded by a previously unknown gene that we called *RAGE*[17]. This gene is silent in normal tissues except retina, and is expressed in a small proportion of tumors, mainly in sarcomas, bladder tumors and melanomas (Table 2). Since most retinal cells do not express MHC class I molecules, this antigen is probably tumor-specific, although the formal proof of this will require the identification of the retinal cell type that expresses *RAGE*.

1.5. Antigen NA17-A

The recent characterization of antigen NA17-A revealed a very peculiar mechanism responsible for the expression of this antigen on melanomas. Here it seems that a gene that is ubiquitously expressed, namely N-acetyl-glucosaminyltransferase V, contains an intron which itself appears to carry near its end a promoter that is activated only in melanoma cells[18]. This atypical activation occurs in more than 50% of melanomas. This produces a message containing a new open reading frame, which codes for the antigenic peptide in its intronic part. This peptide is presented by HLA-A2 to melanoma-specific CTL (Table 2).

2. DIFFERENTIATION ANTIGENS

The observation that autologous CTL can be generated readily against differentiation antigens present on normal melanocytes as well as melanoma cells was unexpected. Four genes encoding melanoma differentiation antigens have been identified: tyrosinase, Melan-A/MART-1, gp100 and gp75[19–24]. Most of the identified antigenic peptides are presented by HLA-A2, but other HLA-peptide combinations have been found[23–32]. One tyrosinase peptide is presented by HLA-DR4 to CD4 T cells[28].

The pattern of CTL precursors directed against these differentiation antigens appears to be very different from that observed with the MAGE-like antigens. Here, most melanoma patients have CTL precursors that can be readily restimulated in vitro with autologous tumor cells[32;33]. Populations of tumor infiltrating lymphocytes (TIL) also contain these CTL[27]. How these findings affect the immunotherapy potential of these antigens is unclear. The fact that many patients carry CTL precursors against these antigens implies that active immunization resulting in an increase in the number of these CTL should be possible. On the other hand, the fact that many of these patients have progressive disease suggests that these CTL are not very effective.

There is concern for the potential side-effects of active or passive immunization against melanoma differentiation antigens. Not so much for the skin, where vitiligo due to the destruction of melanocytes might occur, but for the retina where melanocytes are present in the choroid layer. However, vitiligo has been associated with good prognoses in melanoma and also with adoptive transfer of TILs, without noticeable eye lesions[25;31;34;35]. Carefully devised immunotherapy trials based on these antigens therefore seem permissible.

3. UNIQUE ANTIGENS

Several years ago, the study of immunogenic variants of mouse tumors obtained by mutagenesis showed that point mutations in genes expressed ubiquitously could create antigenic epitopes recognized by tumor-specific CTL[36;37]. Subsequent work on mouse tumors provided two interesting examples of tumor antigens resulting from point mutations[38;39]. This mechanism also accounts for the expression of antigens by human tumors. As was seen with the mouse antigens induced by mutagens, the mutations are located in the region coding for the antigenic peptide, enabling it to bind to the MHC molecule or generating a new epitope. This first example is a point mutation in a previously unknown gene, that produces a new antigenic peptide which, remarkably, is partially encoded by the 5′ end of an intron[40]. Another example is a point mutation of the cyclin-dependent kinase 4 gene, which alters the regulation of the activity of this protein and may therefore contribute to oncogenesis[41]. Yet another antigenic peptide is produced by a mutation in the β-catenin gene. This mutation creates an anchor residue enabling the peptide to bind to HLA-A24[42]. Point mutations may also create tumor antigens by directly altering an HLA molecule: we recently observed that autologous CTL directed against a human renal cell carcinoma recognized an HLA-A2 molecule that was altered as a result of a point mutation changing one amino acid in the alpha-2 helix[43].

The most recently identified antigen resulting from a mutation is encoded by a gene involved in the induction of apoptosis[44]. From blood lymphocytes of patient BB49, we have previously isolated a CTL clone which specifically lysed the autologous head and neck squamous cell carcinoma line. To clone the gene conferring recognition by the CTL, we prepared a cDNA library from RNA of the tumor cells and transfected this library into COS-7 cells together with a plasmid coding for the HLA-B35 presenting molecule. Having isolated cDNA clone which transferred the expression of the antigen, we then identified an antigenic peptide that sensitized autologous EBV-transformed B cells to lysis by the CTL.

The sequence of cDNA clone was nearly identical to that of recently discovered gene *CASP-8*. This gene, which has also been named *FLICE* and *MACH*, codes for protease caspase-8, which is required for induction of apoptosis through the Fas and TNFR-1 receptors. The comparison of the sequence of the cDNA with that of the reported *FLICE* and *MACH* sequences in the region encoding the antigenic peptide revealed a single difference. This region was therefore amplified by PCR on DNA extracted from normal and tumor cells of patient BB49, and the PCR product was sequenced. Both the tumor cell line and the original tumor sample contained a wild-type and a mutated *CASP-8* allele, and this mutated allele was not found in the normal cells of the patient.

The point mutation suppresses the stop codon and adds an Alu repeat to the coding region, thereby lengthening the protein by 88 amino acids. The five last amino acids of the antigenic peptide are encoded by the extension of the reading frame caused by the mutation. Another consequence of this mutation is an alteration of the function of the protein. As observed with *ICE* family members, when *CASP-8* is transiently transfected and therefore overexpressed, apoptosis is triggered. We therefore examined whether the mutation in cDNA 668 affected the ability of the abnormal caspase-8 protein to induce apoptosis. The ability of the abnormal protein to trigger apoptosis was reduced but not abolished.

The antigens generated by point mutations ought to be absolutely specific for the tumor cells, and the CTL precursors directed against these antigens should not have undergone any of the depletion or anergy that accompany natural tolerance. On the other hand, they are expected to be unique for an individual tumor or restricted to very few. This should make it difficult to develop cancer therapeutic vaccines based on these antigens. But one should not exclude the possibility that technological progress may one day make the identification of such antigens so easy that strictly individual immunogens will become a realistic possibility.

4. ANTIGENS RESULTING FROM OVEREXPRESSION OF GENES IN TUMOR CELLS

A class of semi-specific tumor antigens are proteins encoded by genes that are overexpressed in tumors. Lymphocytes infiltrating some HLA-A2$^+$ ovarian carcinomas have been found to recognize peptides derived from HER-2/neu, an oncogene expressed in normal tissues at a low level and overexpressed in 30% of breast and ovarian carcinomas[45].

We recently made an observation that explains how CTL recognizing antigens that are not strictly tumor specific may not be harmful to normal cells[46]. We have studied the patterns of antigens recognized by autologous CTL on two melanoma clonal lines, MEL.A and MEL.B. These were derived from metastases removed several years apart from patient LB33. The MEL.A cells were obtained after surgery in 1988 and a large number of CTL clones directed against these cells were obtained from blood lymphocytes of the patient, collected in 1990. *In vitro* selection of melanoma cells which were resistant to these CTL clones indicated that at least six different antigens were recognized on the MEL.A cells by autologous CTL. Four of these antigens were found to be presented by HLA-A28, B13, B44 and Cw6, respectively.

The patient remained disease-free until 1993 when a metastasis was detected and used to obtain the MEL.B cell line, which proved resistant to lysis by all the CTL clones directed against MEL.A. The MEL.B cells showed no expression of HLA class I molecules except for HLA-A24. These results suggested that the melanoma cells in patient LB33 lost the expression of several HLA molecules under the selective pressure of an anti-tumor CTL response.

By stimulating autologous blood lymphocytes collected in 1994 with MEL.B cells, we obtained CTL clone 17 which lysed these cells. The gene coding for the target antigen, presented by HLA-A24, was identified by transfection of a cDNA library. This new gene, which was named *PRAME* (PReferentially expressed Antigen of MElanoma), is expressed in a large proportion of tumors of different histological origins (91% of melanomas, 78% of lung squamous cell carcinomas, 39% of sarcomas). Gene *PRAME* is also expressed by 33% of acute leukemia samples which do not express any of the *MAGE*, *BAGE*, *GAGE* and *RAGE* genes. When the expression of gene *PRAME* was studied on Northern Blots, no band was obtained with the RNA of normal adult tissues except testis. With reverse transcription and PCR amplification, however, some expression was also found in endometrium, ovary, and adrenal samples. These tissues express gene *PRAME* at a low level, corresponding to 2–5% of that found in the LB33 melanoma cells, with the exception of some endometrium samples where expression reaches 30% of that found in the tumor cells.

Surprisingly, the anti-PRAME CTL 17 lysed the MEL.B cells but failed to lyse the MEL.A cells, even though these cells expressed *PRAME* and HLA-A24 at the same level as MEL.B. A major difference between the MEL.A and MEL.B cells is that the former also express all the other HLA class I alleles of patient LB33: A28, B13, B44, Cw6 and Cw7. Our observation was therefore reminiscent of the inhibition of the cytolytic activity of NK cells, and of a fraction of T cells, by MHC class I molecules expressed on target cells. We tested the lytic activity of CTL 17 against MEL.A cells in the presence of a monoclonal antibody recognizing HLA-B and HLA-C molecules. In these conditions, MEL.A cells were efficiently lysed by the CTL. To identify the inhibitory HLA-B or C molecules, the lytic activity of CTL 17 was tested on MEL.A-1.1.1, an HLA-loss variant derived from MEL.A cells by several rounds of selection with autologous CTL clones restricted by HLA-A28 and HLA-B44 molecules. MEL.A-1.1.1 had lost an entire HLA class I haplotype: HLA-A28, B44, Cw7. These cells were efficiently lysed by CTL 17. No significant inhibition of lysis was observed after transfection of MEL.A-1.1.1 cells with an HLA-B44 construct, whereas transfection with a cDNA clone encoding HLA-Cw7 protected these cells from lysis by the CTL. These results suggested that an inhibitory receptor binding to HLA-Cw7 prevented lysis of MEL.A by CTL 17.

CTL 17 was labelled with a monoclonal antibody, GL183, which recognizes the Natural Killer (NK) inhibitory receptor p58.2. This receptor interacts with a subset of HLA-C molecules including HLA-Cw1, 3, 7 and 8. Addition of increasing concentrations of antibody GL183, which blocks the interaction between p58.2 and HLA-Cw7 molecules, restored lysis of the MEL.A cells by CTL 17. Taken together, these results indicate that the MEL.A cells express the *PRAME*-encoded antigen, but are not lysed by the anti-PRAME CTL 17 because they bear HLA-Cw7 molecules. The presence of these molecules inhibits the lytic activity of the CTL upon engagement of the NK inhibitory receptor p58.2.

CTL 17 may be representative of a new category of anti-tumor T lymphocytes, situated between tumor specific CTL and NK cells, a category which shows specificity for tumor cells which have lost expression of some, but not all, HLA class I molecules. Such partial HLA-losses are frequently observed in tumors and may often result from selection by anti-tumor T-cell responses. The anti-tumor CTL with inhibitory receptors may therefore play a role after the classical CTL, which recognize tumor specific antigens presented by HLA class I molecules, and before NK cells, which usually express multiple inhibitory receptors and therefore lyse only those cells that have lost all their HLA molecules.

5. CLINICAL PROSPECTS

Specific active immunotherapy as an adjuvant treatment of cancer has become a rational experimental approach since these molecular identifications of tumor antigens recognized by CTL. The most promising class of antigens are the shared tumor-specific antigens (Table 1).

The knowledge of the molecular nature of these antigens first allows the selection of patients whose tumor actually expresses a given antigen. Eligible tumors should express the relevant gene along with the appropriate HLA class I specificity, and this can be tested readily by RT-PCR on RNA extracted from a small tumor sample. In Caucasians, about 60% of melanomas express at least one of the tumor-specific antigens thus far defined.

The definition of the molecular nature of tumor antigens allows the rational design of highly specific vaccine preparations. These could consist of engineered cells expressing the antigens or of antigenic peptides mixed with appropriate adjuvants. The availability of the genes encoding the antigens also allows the preparation of recombinant proteins that can be combined with adjuvants, or the preparation of recombinant defective viruses such as adenoviruses or poxviruses carrying either the sequence coding for the entire protein or one or several "minigenes" coding for antigenic peptides. An interesting alternative will be to collect dendritic cells from patients, pulse them with the antigen and reinfuse them to the patient.

T cell immunization and tumor rejection responses obtained in murine models have helped to sort out several promising approaches to anti-tumor vaccination. Immunization of mice with a recombinant adenovirus containing the P1A antigen of the mouse tumor P815 has been shown to induce a strong anti-P1A CTL response, provided that the mice have not been primed previously with adenovirus[47]. Peptide-pulsed dendritic cells have also been shown to induce strong CTL responses in vivo[48]. Interleukin-12, a potent stimulus of the Th1 response, could lead to strong anti-tumor responses when injected with a mutated p53 peptide in adjuvant[49].

A small number of patients with advanced disease have received several subcutaneous injections of an antigenic peptide encoded by MAGE-3 and presented by HLA-A1, in the absence of adjuvant. Tumor regressions have been observed in five out of 17 melanoma tumor-bearing patients[50]. No CTL response against the immunizing peptide could be detected in the blood of these patients. It is difficult, though, to imagine that a peptide could induce a tumor regression response without the involvement of T cells. More sensitive methods of detection of specific T cells will be required to understand the mechanisms of these tumor regressions.

Since a number of tumors appear to express several antigens, patients bearing these tumors could be immunized simultaneously with several distinct defined antigens. This ought to eliminate the tumor cells more effectively. It should also reduce the emergence of antigen-loss variants arising by loss of antigen expression, as it is unlikely that the same variant would simultaneously loose distinct nominal antigens.

REFERENCES

1. Boon, T., Cerottini, J.-C., Van den Eynde, B., van der Bruggen, P. and Van Pel, A. (1994) *Annu. Rev. Immunol.* 12, 337–365
2. Traversari, C., van der Bruggen, P., Van den Eynde, B. *et al.* (1992b) *Immunogenetics* 35, 145–152
3. van der Bruggen, P., Traversari, C., Chomez, P. *et al.* (1991) *Science* 254, 1643–1647

4. De Plaen, E., Arden, K., Traversari, C. *et al.* (1994) *Immunogenetics* 40, 360–369
5. Takahashi, K., Shichijo, S., Noguchi, M., Hirohata, M. and Itoh, K. (1995) *Cancer Res.* 55, 3478–3482
6. Chomez, P., Williams, R., De Backer, O., Boon, T. and Vennström, B. (1995) *Immunogenetics* 43, 97–100
7. Brasseur, F., Rimoldi, D., Liénard, D. *et al.* (1995) *Int. J. Cancer* 63, 375–380
8. Brasseur, F., Marchand, M., Vanwijck, R. *et al.* (1992) *Int. J. Cancer* 52, 839–841
9. Weynants, P., Lethé, B., Brasseur, F., Marchand, M. and Boon, T. (1994) *Int. J. Cancer* 56, 826–829
10. Traversari, C., van der Bruggen, P., Luescher, I.F. *et al.* (1992) *J. Exp. Med.* 176, 1453–1457
11. van der Bruggen, P., Szikora, J.-P., Boël, P. *et al.* (1994) *Eur. J. Immunol.* 24, 2134–2140
12. Gaugler, B., Van den Eynde, B., van der Bruggen, P. *et al.* (1994) *J. Exp. Med.* 179, 921–930
13. Herman, J., van der Bruggen, P., Luescher, I. *et al.* (1996) *Immunogenetics* 43, 377–383
14. van der Bruggen, P., Bastin, J., Gajewski, T. *et al.* (1994) *Eur. J. Immunol.* 24, 3038–3043
15. Boël, P., Wildmann, C., Sensi, M.-L. *et al.* (1995) *Immunity* 2, 167–175
16. Van den Eynde, B., Peeters, O., De Backer, O. *et al.* (1995) *J. Exp. Med.* 182, 689–698
17. Gaugler, B., Brouwenstijn, N., Vantomme, V. *et al.* (1996) *Immunogenetics* 44, 323–330
18. Guilloux, Y., Lucas, S., Brichard, V.G. *et al.* (1996) *J. Exp. Med.* 183, 1173–1183
19. Brichard, V., Van Pel, A., Wölfel, T. *et al.* (1993) *J. Exp. Med.* 178, 489–495
20. Coulie, P.G., Brichard, V., Van Pel, A. *et al.* (1994) *J. Exp. Med.* 180, 35–42
21. Kawakami, Y., Eliyahu, S., Delgado, C.H. *et al.* (1994) *Proc. Natl. Acad. Sci. USA* 91, 3515–3519
22. Bakker, A.B.H., Schreurs, M.W.J., de Boer, A.J. *et al.* (1994) *J. Exp. Med.* 179, 1005–1009
23. Cox, A.L., Skipper, J., Chen, Y. *et al.* (1994) *Science* 264, 716–719
24. Wang, R.-F., Robbins, P.F., Kawakami, Y., Kang, X.-Q. and Rosenberg, S.A. (1995) *J. Exp. Med.* 181, 799–804
25. Kawakami, Y., Eliyahu, S., Delgado, C.H. *et al.* (1994) *Proc. Natl. Acad. Sci. USA* 91, 6458–6462
26. Kawakami, Y., Eliyahu, S., Sakaguchi, K. *et al.* (1994) *J. Exp. Med.* 180, 347–352
27. Robbins, P.F., El-Gamil, M., Kawakami, Y. and Rosenberg, S.A. (1994) *Cancer Res.* 54, 3124–3126
28. Topalian, S.L., Rivoltini, L., Mancini, M. *et al.* (1994) *Proc. Natl. Acad. Sci. USA* 91, 9461–9465
29. Wölfel, T., Van Pel, A., Brichard, V. *et al.* (1994) *Eur. J. Immunol.* 24, 759–764
30. Castelli, C., Storkus, W.J., Maeurer, M.J. *et al.* (1995) *J. Exp. Med.* 181, 363–368
31. Kawakami, Y., Eliyahu, S., Jennings, C. *et al.* (1995) *J. Immunol.* 154, 3961–3968
32. Brichard, V.G., Herman, J., Van Pel, A. *et al.* (1996) *Eur. J. Immunol.* 26, 224–230
33. Sensi, M.L., Traversari, C., Radrizzani, M. *et al.* (1995) *Proc. Natl. Acad. Sci. USA* 92, 5674–5678
34. Bystryn, J.-C., Darrell, R., Friedman, R.J. and Kopf, A. (1987) *Arch. Dermatol.* 123, 1053–1055
35. Richards, J.M., Mehta, N., Ramming, K. and Skosey, P. (1992) *J. Clin. Oncol.* 10, 1338–1343
36. Boon, T. (1992) *Adv. Cancer Res.* 58, 177–210
37. Boon, T., De Plaen, E., Lurquin, C. *et al.* (1992) *Cancer Surveys* 13, 23–37
38. Mandelboim, O., Berke, G., Fridkin, M. *et al.* (1994) *Nature* 369, 67–71
39. Monach, P.A., Meredith, S.C., Siegel, C.T. and Schreiber, H. (1995) *Immunity* 2, 45–59
40. Coulie, P.G., Lehmann, F., Lethé, B. *et al.* (1995) *Proc. Natl. Acad. Sci. USA* 92, 7976–7980
41. Wölfel, T., Hauer, M., Schneider, J. *et al.* (1995) *Science* 269, 1281–1284
42. Robbins, P.F., El-Gamil, M., Li, Y.F. *et al.* (1996) *J. Exp. Med.* 183, 1185–1192
43. Brändle, D., Brasseur, F., Weynants, P., Boon, T. and Van den Eynde, B. (1996) *J. Exp. Med.* 183, 2501–2508
44. Mandruzzato, S., Brasseur, F., Andry, G., Boon, T. and van der Bruggen, P. (1997) *Immunity* Submitted.
45. Fisk, B., Blevins, T.L., Wharton, J.T. and Ioannides, C.G. (1995) *J. Exp. Med.* 181, 2109–2117
46. Ikeda, H., Lethé, B., Lehmann, F. *et al.* (1997) *Immunity* 6, 199–208
47. Warnier, G., Duffour, M.-T., Uyttenhove, C. *et al.* (1996) *Int. J. Cancer* 67, 303–310
48. Zitvogel, L., Mayordomo, J.I., Tjandrawan, T. *et al.* (1996) *J. Exp. Med.* 183, 87–97
49. Noguchi, Y., Richards, E.C., Chen, Y.-T. and Old, L.J. (1995) *Proc. Natl. Acad. Sci. USA* 92, 2219–2223
50. Marchand, M., Weynants, P., Rankin, E. *et al.* (1995) *Int. J. Cancer* 63, 883–885

DISCUSSION

Allison: I like this story about the selectivity arising about loss of MHC in some of the tumors, but I thought it was my recollection that the inhibitory receptors (KIRs) were not expressed on naive T-cells but were expressed only after some stimulation.

Coulie: That is correct. Maybe Lorenzo you could comment on that. Apparently, indeed the fraction of T-lymphocytes that express the NK receptors are activated T-lymphocytes (memory T-lymphocytes) so it does not seem that we have to start with two distinct populations of T-cells, the normal one and the ones expressing the inhibitory receptors. We are still trying to work that out in this particular patient, but one possible scenario would be that she had an anti-PRAME CTL response before which we could not detect. And then, because these CTL were activated for a long time, or for another type of reason, the inhibitory receptors were expressed and then these CTL were extremely effective against this new metastasis which had lost the inhibitory molecule.

Parmiani: The reconstruction of the tumor progression is a fascinating story, of course. I wonder whether in the last step of tumor growth, the one that you showed us, actually the complete absence of MHC class I, is a very rare event. In fact, we have looked at approximately 60 melanoma lines and only in five cases did we really find the complete absence of all the possible alleles.

Coulie: I fully agree. But there is some discrepancy there. People doing histochemistry claim that very often their slides show tumor cells that are class I-negative. There are some studies in which eighty percent of melanoma metastasis are class I-negative. Then, indeed, as you say, when we try to obtain a class I-negative cell line, it is extremely difficult. So, maybe there is another problem associated with the complete loss of class I, and a consequence could be that the MHC class I-negative cells do not grow very well.

Moretta: Firstly, I would like to comment on Giorgio's question. I mean the point that you have only five out of sixty MHC class I-negative cell lines. It is a low percentage for a complete class I loss. But there is not a substantial difference in terms of the ability of NK cells to deal with melanoma cells, with a complete or a partial loss of class I alleles because these inhibitory NK receptors are allele-specific and are clonally distributed. Frequently, NK cells express only one receptor for a given allele, so those cells will sense even a single allelic loss; operationally, NK will lyze both types of melanomas. Secondly, James Allison's question was about the pre-existence of T cells which express KIRs as a kind of a separate category of T cells. We have now clear evidence that, *in vitro,* in response to both alloantigens or super antigens, you can, for example in the presence of IL15, induce T cells to express these receptors. So it is something that T-cells can acquire under special conditions.

Melief: I was interested in your comment that perhaps NK surveillance against MHC class I loss variants of tumors could play a role *in vivo*. If I recall properly the Rosenberg experiments with IL2 LAK therapy, the observation was that this therapy was only effective on immunogenic tumors. So that would suggest that all of the cells that played a role in those models were actually tumor specific CTL's. Now in order to know whether NK cells could play a role, it would be interesting to note whether or not there are specific expansions of either NK cells or CTL's with NK inhibitory receptors in patients that have allele specific MHC class I losses. Has anybody looked at this in patients with melanoma?

Coulie: Not to my knowledge. But since IL2 amplifies both NK and tumor specific T cells, it might be quite difficult to sort things out.

Melief: Is it possible that A24-restricted CTL's would exist in your patient without the p58.2 receptors? Maybe you could stimulate those cells following a stimulation *in vitro* of PBLs with PRAME-encoded peptides.

Coulie: We have tried that and we did not succeed.

Boon: But, Cornelis, could one essentially reverse your argument and say that, if IL2 therapy is effective only on a certain category of cases with those immunogenic tumors, it is because only those patients with immunogenic tumors generate a special kind of MHC-loss variants because they have had to escape to a first immune response and that these cells are then sensitive to a certain type of NK that is itself stimulated by IL2. It seems to me that the argument could go both ways.

Melief: No, I am just referring to the murine studies of Rosenberg where, if he depleted his LAK cells of NK cells, all of the activity was gone, now the situation in human melanoma may be different because you go through a long period of selection.

15

IDENTIFICATION OF HUMAN TUMOR ANTIGENS USING THE B-CELL REPERTOIRE

Michael Pfreundschuh,* Özlem Türeci, and Ugur Sahin

Medizinische Klinik I
Saarland University Medical School

1. SUMMARY

Specific vaccines for the immunotherapy of human neoplasms require specific human tumor antigens. While efforts to identify such antigens by the analysis of the T-cell repertoire yielded only few antigens, the application of SEREX, the serological identification of antigens by recombinant expression cloning, has brought a multitude of new antigens. Several specific antigens have been identified in each tumor tested, suggesting that many, if not all human tumors elicit multiple immune responses in the autologous host. The frequency of human tumor antigens, which can be readily defined at the molecular level, facilitates the identification of T-cell dependent antigens and provides a basis for peptide and genetherapeutic vaccine strategies.

2. INTRODUCTION

A prerequisite for the successful application of tumor vaccines for the treatment of cancer is the recognition of tumor-specific or tumor-associated antigens, i.e. of molecules that are either specifically expressed or overexpressed in tumor cells when compared to normal cells, by the patient's immune system. In contrast to animal models, the existence of antigens that are specifically expressed in neoplastic, but not normal tissues was not unequivocally demonstrated in humans until 1991, when Thierry Boon's group described the antigen MAGE-1 in maligant melanoma[1]. Since then, extensive efforts have been made to identify tumor-specific antigens in humans by the analysis of the T-cell repertoire against tumors using two strategies. The first strategy, which was pioneered by T. Boon's group[2],

---

* Correspondence: Michael Pfreundschuh, Medizinische Klinik I, Universität des Saarlandes, D-66421 Homburg. Tel. 49-6841163002; Fax: 49-6841163101

The Biology of Tumors, edited by Mihich and Croce
Plenum Press, New York, 1998.

makes use of antigen-loss tumor cell variants which are transfected with recombinant DNA libraries or cDNA libraries prepared from antigen-positive tumor cell lines. The second strategy is based on a biochemical approach[3]. It uses the acid elution of antigenic peptides bound to major histocompatibility complex class I molecules from tumor cells[4,5]. The peptides are then tested for recognition by tumor-specific CTL clones[6]. Using these approaches, several human tumor antigens have been defined at the molecular level, most notably in malignant melanoma[1,2,7–11]. Whether and to what extent the cytotoxicity of T cells against tumor cells which has been used to detect human tumor antigens *in vitro*, is also operative *in vivo*, remains to be shown. Moreover, we do not know, how many antigens are expressed by a given tumor and how often such antigens elicit an immune response in the tumor-bearing patient.

3. SEROLOGICAL APPROACHES TO THE DEFINITION OF HUMAN TUMOR ANTIGENS

"Autologous typing", originally established by Lloyd Old's group is probably the most stringent approach to the serological identification of human tumor antigens pursued in the 1970s and 1980s[12]. Using autologous serum and tumor cell lines from cancer patients, and extensive absorption analyses for the definition of the tumor specificity and the expression spectrum of the respective antigens, only few antigens could be identified and further defined biochemically[13–15]. Nevertheles, the observation that sera from some tumor patients specifically recognize structures selectively expressed on autologous tumor cells provided a strong evidence for the existence of tumor-specific antigens and for the the specificity of the antibody response in cancer patients[12]. Tumor serology of the 80s was dominated by the definition of human tumor antigens with murine monoclonal antibodies[16]. Even though some of the antigens primarily identified by murine monoclonal antibodies were later shown to elicit an immune response in cancer patients, e.g. gp 75[17,18], gp100[19,20], tyrosinase[21,22] and restin[23,24], the recognition of tumor-associated molecules by the murine immune system does not provide any clue as to the immunogenicity of the respective structures in humans.

With the the recent advances of recombinant technology several groups set out to investigate humoral immune responses against molecularly predefined structures. Indeed, several oncogenes and signal transduction molecules could thus be demonstrated to elicit humoral immune responses in cancer patietns. Her2/neu[25,26], p53[27], ras[28], E6 and E7[29], c-myc[30], c-myb[31] and MUC1[32] are only some examples of a rapidly growing list of such antigens.

4. SEREX

By implementing molecular cloning techniques into the original strategy of autologous typing, our group developed SEREX, an acronym that stands for the serological analysis of tumor antigens by recombinant cDNA expression cloning[24]. In the SEREX approach, a cDNA library is constructed from fresh tumor specimens, packaged into lambda-phage vectors and expressed recombinantly in Escherichia coli. Recombinant proteins are transferred onto nitrocellulose membranes, and identified as antigens by their reactivity with high-titered IgG antibodies using an enzyme-conjugated secondary antibody specific for human IgG. Positive clones are subcloned to monoclonality, and the nucleotide sequence of the inserted cDNA is determined.

The SEREX approach is characterized by several features: The use of fresh tumor specimens restricts the analysis to genes that are expressed by the tumor *in vivo* and circumvents *in vitro* artifacts. The expression spectrum of an antigen can be determined by Northern blots and reverse transcriptase polymerase chain reaction (RT-PCR). While SEREX allows an unbiased search for antigenic proteins, epitopes that undergo conformational changes when expressed in bacteria or glycosylated epitopes will escape detection.

We first analyzed the cDNA-libraries of a human melanoma, renal cell cancer, astrocytoma and tumor tissue from Hodgkin's disease. Of a total of 5×10^6 clones screened, 109 clones representing 24 different transcripts were found to be reactive with IgG antibodies. This indicated that some inserts were expressed by multiple clones in the cDNA library of a given tumor. At least four different antigens were expressed in each tumor demonstrating that many human tumors express multiple antigens which elicit an immune response in the autologous host. Meanwhile we have extended our analysis to other types of human neoplasms and have identified more than 100 different human tumor antigens. That among these are known tumor antigens such as the melanoma antigens MAGE-1, MAGE-4a and tyrosinase[24,33], demonstrates that at least some of the serologically identified antigens are also targets for cytotoxic T cells (CTL). A second group of antigens is comprised of transcripts that are either identical or highly homologous to known genes which have not been known to elicit immune responses in humans, e.g. restin which had originally been described by a murine monoclonal antibody specific for Hodgkin and Reed-Sternberg cells[23,24]. The third and by number the largest group of serologically defined antigens consists of previously unknown genes, such as HOM-RCC-3.1.3, a new carbonic anhydrase detected in a renal cell cancer [manuscript in preparation], and HOM-HD-21, a new galectin from a tissue affected by Hodgkin's disease[34].

5. SPECIFICITY OF SEREX ANTIGENS

Sequence data as well as the results of Northern blot and RT-PCR analyses revealed different types of antigen specificity (Table 1). *"Shared" tumor antigens* such as MAGE-1, MAGE-3, MAGE-4a, which had already been defined by T cellular approaches, and the new antigens HOM-MEL-40[33] and NY-ESO-1[35] are selectively expressed in a variety of neoplams, but not in normal tissues except for testis. *Differentiation antigens* show a lineage-specific expression in tumors, but also in normal cells of the same origin at a given stage of differentition. The prototype of such an antigen is the melanocyte-specific antigen tyrosinase, which was shown to be antigenic in malignant melanoma. *Antigens encoded by mutated genes* have been demonstrated only rarely by the serological approach, with mutated p53 being one example [M. Scanlan personal communication]. However, as antibody responses due to mutated molecules may be directed against non-mutated regions[36] and/or may thus crossreact with transcripts derived from unmutated alleles, mutations of serologically defined antigens are difficult to exclude. A *virus-encoded antigen* that elicits an autologous antibody response is the env protein of the human endogenous retrovirus HERV-K10, which was found in a renal cell cancer[37]. Many antigens detected by SEREX, however, are not tumor-specific in the strict sense of the word. Rather, these antigens show overexpression in tumors compared to normal tissues. *Overexpressed genes* can elicit immune responses by overriding thresholds critical for the maintenance of tolerance[38]. An example of such an overexpressed SEREX antigen is is HOM-RCC-3.1.3, a new carbonic anhydrase. *Gene amplification* was the underlying mechanism for overexpression in some cases, for example for the translation initiation factor eIF-4γ in a squamous cell lung cancer[39]. *Splice*

Table 1. Specificity of tumor antigens detected by SEREX

Class	Antigen	Homology/Identity	Source
Shared tumor antigens	HOM-MEL-40	SSX-2	melanoma
Differentiation antigens	HOM-MEL-55	tyrosinase	melanoma
Mutated genes	NY-COL-2	p53	colon cancer
Splice variants	HOM-HD-397	restin	Hodgkin
Overexpressed genes	HOM-HD-21	galectin-9	Hodgkin
Amplified genes	HOM-NSCLC-11	eIF-4γ	lung cancer
Viral genes	HOM-RCC-1.14	HERV-K10 (env)	renal cell cancer
Cancer related autoantigens	HOM-MEL-2.4	CEBPγ	melanoma
Cancer independent autoantigen	HOM-TES-11	PCM-1	testis

variants as the likely immunogenic trigger have been shown for the Hodgkin's disease-associated restin. Autoantigens are immunogenic molecules which are expressed ubiquitously and at a similar level in neoplastic and normal tissues. *Cancer-related autoantigens* such as HOM-MEL-2.4 which represents the CCAAT enhancer binding protein elicit antibody responses only in cancer patients. Their immunogenicity might result from tumor-associated post-translational modifications or from changes in antigen processing and/or presentation in tumor cells[40,41]. *Cancer-independent autoantigens* have also been identified, for example HOM-RCC-10 which represents mitochondrial DNA or HOM-TES-11 which is identical to pericentriol material-1 (PCM-1). As these cancer-independent autoantigens do elicit humoral immune responses both in patients with and without cancer, their immunogenicity is obviously not related to neoplastic disease. The detection by SEREX of such autoantigens supports the notion that anti-tumor immunity and autoimmunity are two faces of the same medal[42,43]. The spectrum of tumor antigen specificities suggests that the immunogenicity of a given molecule depends more on the context in which it is presented (e.g. "danger") than on its more or less restricted expression in certain tissues[44,45].

6. CANCER-TESTIS-ANTIGENS

The MAGE[46,47], BAGE[48] and GAGE[49] family of genes all of which had originally been identified by CTL, as well as the serologically defined HOM-MEL-40[34] and NY-ESO-1[35] are members of an expanding family of tumor antigens that are expressed selectively in a varying proportion of different cancers as well as in testis. This has led to the designation of cancer/testis antigens (CTA)[35]. The realization of the existence of a superfamily of CT antigens prompted us to apply SEREX with allogenic sera from tumor patients to a testis library which was enriched for testis-specific transcripts by subtracting total cDNAs from other normal tissues. This approach has led to the rapid identification of several new CT antigens, among others HOM-TES-14 which is expressed at a high frequency in gliomas and breast cancer. HOM-TES-14 is the first CTA with a known function. It codes for SCP-1, an essential part of the synaptonemal complex which aligns the homologous chromosomes during meiosis I. Further investigations have to elucidate the role of this meiosis-specific protein in non-meiotic tumor cells. The molecular mechanisms for the selective expression pattern of the CT antigens may rely on the genome-wide hypomethylation of cancer cells and testis. Methylation of promoter regions is an important regulatory mechanism for silencing gene expressions. Although the functional significance of the selective activation of CT antigens in cancer and testis is unknown and is the scope of ongoing research, one clue comes from the finding of a higher frequency of MAGE and BAGE expression in metastatic melanoma compared to primary melanoma[48]. Hypomethylation is constitutive in testis and also occurs in tumors[50,51], where it accounts for the genomic instability, and switches on genes which are normally silenced.

7. SIGNIFICANCE OF ANTIBODIES AGAINST HUMAN TUMOR ANTIGENS

At the functional level, antibodies to intracellular proteins would not be expected to affect normal cell function; however, antibodies against surface molecules, such as the HOM-HD-21 antigen in Hodgkin's disease, may interfere with its adhesive properties and influence disease progression.

The clinical significance of anti-tumor antibodies in cancer patients is largely unknown. While the presence of p53 antibodies has been reported to be associated with a poor prognosis[52,53], the clinical significance of anti-HER-2/neu antibodies and antibodies to mutated ras oncogenes remains to be determined[54]. An initial survey of sera from patients with various malignant diseases and healthy controls showed that antibodies against many SEREX antigens can be detected at varying rates only in the sera of patients with the same type of tumor[24]. While long-term follow-up studies of sera during the course of the disease are not yet available for SEREX antigens, it is our preliminary impression that the presence of an antigen-expressing tumor is a prerequisite for the production of tumor antigen-specific antibodies. Investigations of obvious clinical parameters, such as stage or tumor burden in melanoma patients did not identify any significant correlations between these parameters and the presence of antibodies to HOM-MEL-40[33]. More patients must be analyzed to determine whether the development of antibodies to tumor antigens is associated with clinically relevant features and might be used for the serodiagnosis of cancer or as a prognostic marker. In addition, the role of the MHC class I and II molecules, and the integrity of the antigen processing machinery of the tumor cells for the induction of an antibody response reamain to be determined.

8. REVERSE T-CELL IMMUNOLOGY

The fact that SEREX antigens induce high-titered IgG responses implies the coexistence of cognate T-cell help. Indeed, during our initial screening, two antigens were detected (MAGE-1 and tyrosinase) which had originally been identified by cytotoxic T-cell responses. These results, together with recent reports on a $CD4^+$ response to tyrosinase[55,56], suggest an integrated immune response against human tumor antigens, which is mediated by the concerted action of specific $CD4^+$ and $CD8^+$ T-cells, and B-cells. To address the question whether a T-cell reactivity can also be demonstrated for other SEREX antigens, we follow three different strategies. For the "peptide approach" the sequences of serologically identified antigens are screened for peptides containing the binding motifs for defined MHC molecules, which are then tested for MHC binding and used for the expansion of specific T cells. In the "whole protein approach", professional antigen presenting cells are transfected with cDNA encoding SEREX antigens and are used for priming and establishment of $CD4^+$ and $CD8^+$ T-cell clones. To ensure that naturally processed peptides are recognized by these T cell clones antigen-expressing tumor cell lines are tested for specific recognition by these T cell clones. Both approaches were successful in establishing antigen-specific CTLs.

A third strategy is similar to an approach used by Steve Rosenberg's group[57] and employs "cross analysis of SEREX antigens with preestablished T-cell lines". A multitude of CTLs have been established in laboratories from all over the world from peripheral blood lymphocytes and tumor infiltrating lymphocytes of cancer patients, which react specifically with antigens presented by the tumor. However, very few of the involved tumor antigens could be molecularly defined. To check whether such CTLs recognize SEREX antigens, we tested a panel of nine T-cell lines derived from tumor-infiltrating lymphocytes of patients with renal cell cancer for reactivity against HLA-2$^+$ CHO-cell lines which had been transfected with 31 different cDNA clones identified by SEREX. Preliminary data indicate that one T-cell line recognized two of the renal-cancer derived cDNA clones, If confirmed, thsi observation would again support the notion that an integrated cellular and humoral immune response against tumor antigens is maintained by the autologous host's immune system.

9. REVERSE TUMOR CYTOGENETICS

Amplified regions of chromosomes from malignant tumors are commonly believed to contain genes which are important for tumor growth. However, such amplified regions, besides the genes of interest (the target genes of amplification) contain a majority of bystander genes. To identify such target genes, amplified regions of defined chromosomes can be labelled by FISH and dissected under the microscope. Submitting a cDNA prepared from the microdissected region of chromosome 3q from a squamous cell lung cancer to SEREX with the autologous patient's serum, we identified the translation initiation factor eIFγ4 as a prime candidate for a target gene in the amplified region in the respective tumor[39].

10. BIOLOGICAL SIGNIFICANCE OF SEREX ANTIGENS

While the function of most SEREX tumor antigens remains to be determined, the function of some serologically defined tumor antigens can be derived from sequence analysis. HOM-NSCLC-1 codes for the translation initiation factor eIFγ4[39]; the Hodgkin's disease derived antigen HOM-HD-21 is a new human galectin (galectin-9), which may function as an adhesion molecule and might play a role in the interactions between the minority of neoplastic Hodgkin- and Reed-Sternberg cells and their surrounding reactive (mostly CD4$^+$ T cells[34]). The melanoma antigen HOM-MEL-40, encoded by the SSX-2 gene, is involved in the t(X;18) translocation that occurs in >80% of human synovial sarcomas and might function as a regulated transcription factor and putative oncogene[33].

11. CONCLUSIONS AND PERSPECTIVES

The modest yield brought about by the extensive efforts from many tumor immunologist around the world to define specific antigens in human cancer by the analysis of the T-cell repertoire has lead to the common belief among tumor immunologists (and even more among cancer researchers from other fields) that specific immune responses to tumor antigens are the exception rather than the rule in humans. The multitude of tumor-specific antigens identified by SEREX has changed this view dramatically. With the identification and molecular definition of multiple antigens expressed by a given tumor, that elicit an immune response in the autologous host, it has now become evident that the recognition of antigens that are specifically expressed or significantly overexpressed in the tumor is not the basic problem in tumor immunity. Rather, it seems to be the effector arm of the immune system that is responsible for the failure to control cancer by immunological means. The availability of molecularly defined genes which are specifically expressed or overexpressed in many (and possibly all) human cancers now provides a tool for the redirection, e.g. via peptide or genetically engineered vaccines, of the effector arm of the immune response towards an efficient cytotoxic response against malignant cells.

The abundance of human tumor antigens enables us to leave therapeutic avenues that were doomed to fail from the outset: cancer vaccines using tumor cells transfected with one or more costimuli, such as B7.1 and/or cytokines, might only be successful if the antigen presented by the tumor cell is an immunodominant peptide. This is unlikely to be the case in a human cancer cell, where tumor-specific antigens represent a minority among the overwhelming majority of normal differentiation antigens. Instead of using whole tumor-cell vaccines which consist of a majority of tolerizing "normal self proteins", we can now proceed to the development of mono- or even polyvalent vaccines for a wide spectrum of human cancers using pure preparations of molecularly defined antigens or antigenic peptide fragments[58,59]. That tumor vaccines can be successful, if used as a pure preparation without contaminating tolerogens, has recently been demonstrated by Alex Knuth's group, who obtained complete remissions in melanoma patients by intradermal vaccination with tumor-specific peptide fragments and the systemic application of GM-CSF[60]. That some tumors defied this therapy by the outgrowth of antigen-loss variants[59], underlines even more the need for the molecular definition of additional tumor antigens for the development of specific multivalent vaccines.

ACKNOWLEDGMENT

We thank Dr. L.J. Old for continuing discussion and encouragement. This work was supported in part by Deutsche Forschungsgemeinschaft, Deutsche Krebshilfe and SFB 399.

REFERENCES

1. van der Bruggen, P. et al.: A gene encoding an antigen recognized by cytolytic T-lymphocytes on a human melanoma. *Science* 1991, **254**: 1643–1647
2. Boon, T. et al.: Tumor antigens recognized by T-lymphocytes. *Annu Rev Immunol* 1994, **12**:337–366
3. Mandelboim, O., et al.: CTL induction by a tumour-associated antigen octapeptide derived from a murine lung carcinoma. *Nature* 1994, **369**: 67–71
4. Falk, K. et al.: Allele-specific motifs revealed by sequencing of self-peptides eluted from MHC molecules. *Nature* 1991, **351**: 290–296
5. van Bleek, G., and Nathenson, S.: Isolation of an endogenously processed immunodominant vial peptide from the class I H-$2K^b$ molecule. *Nature* 1990, **348**: 213–216
6. Cox, A.L. Identification of a peptide recognized by five melanoma-specific human cytotoxic T-cell lines. *Science* 1994, **264**: 716–719
7. Brichard, V. et al.: The tyrosinase gene codes for an antigen recognized by autologous cytolytic T-lymphocytes on HLA-A2 melanomas. J. Exp. Med. 1993, **178**: 489–495

8. Coulie, P., et al.: A new gene coding for a differentiation antigen recognized by autologous cytolytic T-lymphocytes on HLA-A2 melanomas. *J. Exp. Med.* 1994, **180**: 35–42
9. Kawakami, Y., et al.: Recognition of tyrosinase by tumor-infiltrating lymphocytes from a patient responding to immunotherapy. *Proc. Natl. Acad. Sci. USA* 1994, **91**: 3515–3519
10. Kawakami, Y., et al.: Identification of the immunodominant peptide of the MART-1 human melanoma antigen recognized by the majority of HLA-A2 restricted tumor infiltrating lymphocytes. *J Exp Med* 1994, **180**: 347–352
11. Wölfel, T. et al.: A p16INK4a-insensitive CDK4 mutant targeted by cytolytic T lymphocytes in a human melanoma. *Science* 1995, **269**: 1281–1284
12. Old, L.J.: Cancer Immunology: The search for specificity - G.H.A. Clowes Memorial Lecture. *Cancer Res.* 1981, **47**: 261–275
13. Shiku, H., et al.: Cell surface antigens of human malignant melanoma. II. Serological typing with immune adherence assays and definition of two new surface antigens. *J Exp Med* 1976, **144**: 873–881
14. Pfreundschuh, M. et al.: Serological analysis of cell surface antigens of malignant human brain tumors. *Proc Natl Acad Sci USA* 1978, 75: 5122–6
15. Ueda, R., et al.: Cell surface antigens of human renal cancer defined by autologous typing. *J Exp Med* 1979, **150**: 564–79
16. Neville AM: Detection of tumor antigens with monoclonal antibodies: immunopathology and immunodiagnosis. *Current Opin Immunol* 1991, **1**: 674–8
17. Vijayasaradhi, S., et al: The melanoma antigen gp75 is the human homolog of the mouse b (brown) locus gene product. *J Exp Med* 1990, **171**: 1375–1380
18. Wang, R.F., et al.: Identification of a gene encoding a melanoma tumor antigen recognized by HLA-A31-restricted tumor infiltrating lymphocytes. *J Exp Med* 1995, **181**: 799–804
19. Adema, G.J., et al.: Molecular characterization of the melanocyte lineage specific antigen gp100. *J Biol Chem* 1994, **269**: 20126–20133
20. Kawakami, Y., et al: Identification of a human melanoma antigen recognized by tumor infiltrating lymphocytes associated with in vivo tumor rejection. *Proc Natl Acad Sci USA* 1994, **91**: 6458–6462
21. Kwon, B.S. et al.: Isolation and sequence of a cDNA clone for human tyrosinase that maps at the mouse c-albino locus. *Proc Natl Acad Sci USA* 1988, **85**: 6352–6359.
22. Brichard, V. et al.: The tyrosinase gene codes for an antigen recognized by autologous cytolytic T lymphocytes on HLA-A2 melanomas. *J Exp Med* 1993, **178**: 489–495.
23. Bilbe, G., et al.: Restin: a novel intermediate filament-associated protein highly expressed in the Reed-Sternberg cells of Hodgkin`s disease. *EMBO J* 1992, **11**: 2103–2113.
24. Sahin, U. et al.: Human neoplasms elicit multiple immune responses in the autologous host. *Proc Natl Acad Sci USA* 1995, **92**: 1180–11813.
25. Pupa, S.M., et al.: Antibody response against c-erbB-2 oncoprotein in breast carcinoma petients. *Cancer Res* 1993, **53**: 5864–5866.
26. Disis, M.L., et al.: Existent T-cell and antibody immunity to HER-2/neu protein in patients with breast cancer. *Cancer Res* 1993, **54**: 16–20.
27. Crawford, L.V. et al.: Detection of antibodies against the cellular protein p53 in sera from patients with breast cancer. *Int J Cancer* 1982, **30**: 403–408.
28. Takahashi, M., et al.: Antibody responses to ras protein in patients with colon cancer. *Clin Canc Res* 1995, **1**: 1071–1077.
29. Galloway, D.A., and Jenison, S.A.: Characterization of the humoral immune response to genital papillomaviruses. Mol Biol Med 1990, 7: 59–72.
30. Ben-Mahrez. K. et al.: Circulating antibodies against c-myc oncogene product in sera of colorectal cancer patients. *Int J Cancer* 1990, **46**: 35–38.
31. Sorokine, I. et al.: Presence of ciculating anti c-myb oncogene product antibodies in human sera. *Int J Cancer* 1991, **47**: 665–669.
32. Kotera, Y., et al.: Humoral immunity against a tandem repeat epitope of human mucin MUC-1 in sera from breast, pancreatic and colon cancer patients. *Cancer Res* 1994, **54**: 2856–2860.
33. Türeci, Ö. et al.: The SSX2 gene, which is involved in the t(X,18) translocation of synovial sarcomas, codes for the human tumor antigen HOM-Mel-40. *Cancer Res* 1996, **56**: 4766–4772.
34. Türeci, Ö., et al.: Molecular definition of a novel human galectin which is immunogenic in patients with Hodgkin's disease. *J Biol Chem* 1997, **272**: 6416–6422.
35. Chen, T. et al.: A testicular antigen aberrantly expressed in human cancers detected by autologous antibody screening. *Proc Natl Acad Sci USA* 1997, **94**: 1914–1918.
36. Schlichtholz, B., et al.: The immune response to p53 in breast cancer patients is directed against immunodominant epitopes unrelated to the mutational hotspot. *Cancer Res* 1992, **52**: 6380–6384.

37. Ono, M., et al.: Nucleotide sequence of human endogenous retrovirus genome related to the mouse mammary tumor virus genome. *J Virol* 1986, 60: 589–98
38. Viola, A., and Lanzavecchia, A.: T cell activation determined by T cell receptor number and tunable thresholds. *Science* 1996, **273**: 104–6
39. Brass, N. et al.: Translation initiation factor eIF-4gamma is encoded by an amplified gene and induces an immune response in squamous cell lung carcinoma. *Hum Mol Genet* 1997, **6**: 33–39.
40. Skipper, J.C.A., et al.: An HLA-A2-restricted tyrosinase antigen on melanoma cells results from posttranslational modification and suggests a novel pathway for processing of membrane proteins. *J Exp Med* 1996, **183**: 527–534.
41. Chen, W., et al.: CTL recognition of an altered peptide associated with asparagine bond rearrangement. Implications for immunity and vaccine design. *J Immunol* 1996, **157**: 1000–5
42. Schönberger, S.P., and Sercarz, E.E.: Harnessing self reactivity in cancer immunotherapy. *Semin Immunol* 1996, **8**: 303–309.
43. Nanda, N.K., and Sercarz, E.E.: Induction of anti-self immunity to cure cancer. *Cell* 1995, **82**: 13–17.
44. Matzinger, P.: Tolerance, Danger and the extended family. *Annu Rev Immunol* 1994, **12**: 991–1045.
45. Fuchs, E.J., and Matzinger, P.: Is cancer dangerous to the immune system? *Semin Immunol* 1996, **8**: 271–80.
46. van der Bruggen, P. et al.: A gene encoding an antigen recognized by cytolytic T lymphocytes on a human melanoma. *Science* 1991, **254**: 1643–1647.
47. Gaugler, B. et al.: Human Gene MAGE-3 codes for an antigen recognized on a melanoma by autologous cytolytic T lymphocytes. *J Exp Med* 1994, 179: 921 930.
48. Boel, P. et al.: BAGE: a new gene encoding an antigen recognized on human melanomas by cytolytic T lymphocytes. *Immunity* 1995, **2**: 167–175.
49. van den Eynde, P, et al.: A new family of genes coding for an antigen recognized by autologous cytolytic T lymphocytes on a human melanoma. *J Exp Med* 1995, **182**: 689–698.
50. De Smet, C. et al.: The activation of human gene MAGE-1 in tumor cells is correlated with genome-wide demethylation. *Proc Natl Acad Sci USA*, 1996, **93**: 7149–7153.
51. Erickson, R.P., et al.: Gene expression, X-inactivation, and methylation during spermatogenesis: The case of Zfa, Zfx, and Zfy in mice. *Mol Reprod Dev* 1993, **35**: 114–120.
52. Houbiers, J.G.A., et al.: Antibodies against p53 are associated with poor prognosis of colorectal cancer. *Br J Cancer* 1995, **72**: 632–641.
53. Coomber, D., et al.: Characterisation and clinicopathological correlates of serum anti-p53 antibodies in breast and colon cancer. *J Cancer Res Clin Oncol* 1996, **122**: 757–62.
54. Disis, M.L., and Cheever, M.A.: Oncogenic proteins as tumor antigens. *Curr Opin Immunol* 1996, **8**: 637–642.
55. Topalian, S.L., et al.: Human CD4 T cells specifically recognize a shared melanoma-associated antigen encoded by the tyrosinase gene. *Proc Natl Acad Sci USA* 1994, **91**: 9461–5.
56. Topalian, S.L., et al.: Melanoma-specific CD4+ T cells recognize nonmutated HLA-DR restricted tyrosinase epitopes. *J Exp Med* 1996, **183**: 1965–71.
57. Bakker, A.B, et al.: Melanocyte lineage-specific antigen gp100 is recognized by melanoma-derived tumor-infiltrating lymphocytes. *J. Exp Med.* 1994, **179**: 1005–1009 (1994).
58. Melief, C.J., et al.: Peptide-based cancer vaccines. *Curr Opin Immunol* 1996, **8**: 651–7.
59. Jäger, E., et al.: Inverse releationship of melanocyte differentiation antigen expression in melanoma tissues and CD8+ cytotoxic-T-cell responses: evidence for immunoselection of antigen-loss variants in vivo. *Int J Cancer* 1996, **66:** 470–476.
60. Jäger, E., et al.: Granulocyte-macrophage-colony-stimulating factor enhances immune responses to melanoma-associated peptides in vivo. *Int J Cancer* **67**: 54–62 1996

DISCUSSION

Mihich: Using antibodies to pick up unique expression products which may be internal to the cells and may not be on the cell surface implies that they internalize. Do your antibodies in fact internalize and how do you take care of the possibility that some of these antibodies will not internalize?

Pfreundschuh: We do not know whether they internalize and I think that most of the antibodies are produced only after presentation of and processing of the antigens by antigen presenting cells because most of the antigens that we detect by these antibodies are not as surface membrane antigens on the tumor cells, but they are cytoplasmatic or nuclear antigens. So, within the question of immune response against tumors, I do not think that they play a crucial role, I think that they show us that antigens are recognized by the immune system and then these antigens are used to demonstrate that T cell responses exist against the same antigens.

Berns: How many of these tumor antigens might be recognized by normal human T cells because, obviously, there are many auto-immune diseases as well, so the question is, to what extent are they really specific for patients with tumors?

Pfreundschuh: I showed you the one table, the antigens which are over expressed in tumors, usually we find antibody responses only in tumor patients, while in the auto-antigens we can distinguish between what we call cancer related auto-antigens and non-cancer related antigens. Some auto-antigens you find antibody responses to even in normal controls and in patients with auto-immune diseases, while against other auto-antigens where we had no mutations, no different expression between tumors compared to normal tissues and we found anti-body responses only in tumor patients. This may relate to post-translational differences in, or changes of the antigens which we cannot detect by our approach because we detect only proteins and we would not detect epitopes that have been changed by glycosylation. It might have to do with different antigen processing in tumor cells, we do not know the reason yet. But there is a wide spectrum of specifities, both of the antibody response and the tumor antigens.

Zanker: It is really so surprising that you find antibodies? Because cells always make antibodies but it is the question of their affinity, if you really find high affinity antibodies I think that has a biological meaning. So, my question is, did you see any influence on the cause of the disease in those patients with antibodies?

Pfreundschuh: We have too few patients to really answer this question. But these antibodies are high affinity antibodies because we restrict our analysis to these antibodies by the initial screening. We dilute the serum 1:1,000 and these are IgG antibodies. We definitely excluded IgM antibodies which have low affinity and can cross react with everything. And I think the fact that we detect antigens, at least some antigens which are really specifically expressed, really says that the approach was right and that it has some biological meaning.

Parmiani: Can you see different antibodies against different antigens in the same patient?

Pfreundschuh: It seems to be a polyclonal response in most patients and in one case where we did an analysis of a patient with ovarian cancer we also detected IgA antibody responses against the same antigen which was originally identified by IgG screening and what happened to be the MAGE-4.

DePinho: Could you comment on why the search for tumor antigens tends to yield normal proteins that are often expressed during embryogenesis and serve normal functions, rather than antigens that represent somatically mutated products such as activated Ras and mutant p53.

Pfreundschuh: I think from the serological approach we are biased because as has been shown with the p53 you might elicit an antibody response by a mutated antigen but the antibody response is directed against the non-mutated antigen, too. So I had only the p53 as an example but I, as I said, it is very hard to really prove that behind the antigens that we think are non-mutated antigens there might not be a mutated clone in one of ten clones that we pick. So, that is one reason. I think the second, what comes out from our analysis is that I think the presentation or the conditions under which an antigen is presented, for example, in the context of danger, is more important to break the tolerance and induce an antibody response than the more or less limited expression of a given molecule in certain tissues.

Boon: Let me make a quick comment on that. It is certainly striking that, up to now, anybody who tried to obtain CTL by stimulating autologous blood cells with tumor cells did not come up with CTL against the most obvious mutated candidates such as p53 and Ras. Maybe it means that it is not enough to have a mutation, you need to have the peptide processed and you need to have the peptide bind to the particular HLA molecule this person is carrying. When you add up all those restrictions, and the fact that after all there are many mutations in cancer cells, maybe you end up with the other ones. Perhaps mutated p53 and Ras are minor, immunologically minor, compared to the rest of the available antigens. But it is certainly very striking that all those antigens one was waiting for logically have not come up. But to me, it means the immune system has a different way of looking at this than what we imagined.

DePinho: This seems very counter intuitive. Many of the tumor epitopes that have been identified are likely to be expressed ubiquitously during embryogenesis. That is certainly the case for many "tissue specific" genes expressed in the mouse. For many genes expressed in a tissue-specific manner postnatally, there is ubiquitous expression during embryogenesis followed by tight-tissue specific expression in mature organ systems.

Boon: But it could be early embryogenesis before tolerance is installed. And you see there is a big bonus for antigens generated by specific expression versus mutation. If you are expressing a protein that normal other cells do not express, there is a lot of room there for the immune system to select a peptide that is agreeable to your immune system.

Visentin: I agree that, generally speaking, we should expect responses once the antigens have been processed and presented. Secondly, it may be worthwhile to link immunogenicity with the multiplicity of antigens, that finds some correspondence in certain ongoing trials of antibodies cocktails, especially for immunotherapy. However, the major problem, in my view, resides in the portion of tumor cells non-expressing antigens. What is an antigenic heterogeneity? Heterogeneity of tumors both at the same stage, and in the process of progression? This may well relate to possible intervention--perhaps not exclusively or predominantly in malignant melanoma, but also in other solid tumors. Immmunotherapy is assumed to be more effective in the adjuvant setting, possibly implying a relationship of antigenic heterogeneity, as heterogeneity in general, to tumor bulk.

Pfreundschuh: We do not know much about the heterogeneity of these antigens in these tumors because there are very few monoclonal antibodies available with which we can do immunohistology, for example. I think most attempts to establish monoclonal antibodies against the MAGE family have failed. I can only tell you for antigens that we have antibodies against, for example, the Restin antigen which is expressed in a hundred percent of the Reed-Sternberg cells that we have investigated; for the other antigens, we have to wait for the availability of the respective monoclonal antibodies. Perhaps Thierry Boon could comment. I think there is a monoclonal antibody which works fine in immunohistology for the MART-1/Melan-A. Is that right?

Boon: I do not know.

Visentin: What is the percentage of the Reed-Sternberg (RS) cells?

Pfreundschuh: Reed-Sternberg cells are a totally different exception from the rule because they represent less than one percent of the whole cell population; but all Reed-Sternberg cells express Restin.

Melief: Yes, again, in comment to Ron DePinho's question about why the mutant sequences are not readily recognized. I think indeed, as I will also discuss this afternoon, if you immunize, for example, p53 knockout mice with mutant p53, then lots of epitopes come up but not the mutant p53. Mutant p53 epitopes are just not processed in the case of many different p53 mutations. As Thierry Boon said, you have to be extremely lucky for a given mutant peptide to adhere to all the stringent requirements for processing and presentation and, on the other hand, the tolerance mechanisms are quite sloppy, as we now know. For example, the ease with which you can induce experimental allergic encephalomyelitis with myelin basic proteins. An autologous protein, has been known for years. There is a repertoire for all of these auto-antigens, the question is how to arouse the repertoire and really get tumoricidal T cells out. So, in that regard, I have a question for you, since you said there were ITT antibodies, have you looked at T-cell recognition of these antigens and, if so, is there a T-cell response which would be compatible with anti-tumor activity like IL-2, interferon gamma type, T helper 1 responsiveness?

Pfreundschuh: No, we did not have the chance yet. The problem with some of our antigens is that they have been identified in fresh tumor and for some of the antigens we have problems finding tumor cell lines which express the antigens in sufficient amounts. So, the only response is that we could verify so far is that against the HOM-MEL-40 antigen, the melanoma antigen with CD8 cells.

Melief: Because in the case of p53, there is good evidence for proliferative helper cells against self sequences.

Pfreundschuh: We implied that there must be T cell help because of the high titered IgA response.

Pierotti: My question is related to the point of tumor specific antigens. The most dramatic, I guess, tumor specific antigen is the fusion of different genes that you have in some leukemias. Did you check these type of tumors for antibodies?

Pfreundschuh: Well, the melanoma antigen HOM-MEL-40 happens to be coded by the SSX2 gene and the SSX2 gene is involved with the t(X;18) translocation and probably plays a role as a regulator transcription factor. We had very few sarcomas to test, I think three or five, and we had an antibody response in SSX2 positive sarcomas only in one out of three or five cases.

Bankert: I have a comment about the biological significance of what you found. Two things: One is, in the SCID mouse model where we co-engrafted the tumor and the TIL or the PBL, we find lots of tumor reactive antibodies in western blots. Yet if we deplete from these cells, it has no effect on tumor progression as opposed to depleting the $CD8^{+}T$ cell. Perhaps the antibodies indicate a failure to generate the preferred TH_1 response. It would suggest that what you are looking at in patients that are responding and producing tumor reactive antibodies may have little to do with tumor rejection.

Pfreundschuh: I completely agree with you. I do not think that these antibodies have anything to do with tumor rejection. For us, they are just indicators that the tumor is recognized by the immune system, and give us a chance to identify the antigen and then it is the big question, which antigen do we use clinically and how do we manipulate the immune system so that the effector arm of the immune system really helps us to destroy the tumor?

Bankert: But they actually may have a biological relevance in a negative way; it could be an indicator of something that has gone wrong, and you want to reverse the TH_2 response to a TH_1 response.

DePinho: This is, again, coming from a non-immunologist, but as the structural basis for peptide recognition has been more or less worked out, you know, you have your pocket, there is a limited number of amino acids and so on, has anyone gone back and looked at all of these peptides that apparently get recognized, these antigens, and to see if there is any theme that is emerging on a structural level and whether or not that information could be used to rationally design vaccines or peptides that might be tested experimentally *in vivo*? Anyone?

Boon: There is no theme, except for the fact that when a peptide binds to certain HLA molecules it usually has the right anchoring residue. I mean that there are some common anchoring residues. Because otherwise there is no theme.

DePinho: So can one use that information to at least assign in some sort of algorithmic way a minimum requirement for an antigen in a given context and potentially use that more or less targeted type of information to design something?

Boon: The minimum requirement is that the peptide has to bind. But many peptides that bind are not well recognized by T cells for reasons that, I believe, escape most of us.

Melief: The general requirements are that there is no prediction at this moment, but the hope is that the rules of proteasome processing will become known because the proteasome has now been crystallized, at least the Arche bacterium proteasome, and so people are working on the mammalian proteasome structure and, apparently, there are distal effects of amino acid substitution, so there is no simple prediction of proteasome cleavage but this is a very important issue. Once we learn to predict proteasome cleavage, as well as peptide transport, in addition to MHC binding stability, then eventually, perhaps, some sort of prediction will arise, but at the moment it is not there.

Boon: When we look at all these sequences, except for the anchoring residue, we do not see anything in common.

Melief: But because of the spacial effects, there may be rules that are hard to discern from linear amino acid sequences which will immediately become apparent if you know the crystal structure.

Boon: Yes, but we first have to start knowing the crystal structure of all of the potential antigens, which is going to be a long venture.

Anderson: I am curious about your strategy where you are screening an expression library. It seems like it would give you some indication of the abundance of transcripts associated with antigens that are being recognized, and I am just wondering what are the thresholds for the degree of expression? In other words, when you are screening these libraries, are you finding the same clones over and over, and never find unique clones? Or is it a linear response; can you pick up those altered proteins that are expressed at very low levels or transcribed at low levels?

Pfreundschuh: Some of the antigens show up over and over again, for example, the NY-ESO-1, the one cancer tested antigen which was originally described in esophagus carcinoma, we picked it up in screening an ovarian carcinoma and we also picked it up when screening a testis-enriched library. And then perhaps another question relating to that is, it is really striking when you compare different tumors. The melanoma is really the one tumor where the positive clones really pop up. It is a whole difference compared to other tumors. And then once in a while you find some patients who obviously lack an antibody response, maybe one or two out of ten patients do not have antibody responses. Why that is, I do not know.

Boon: In this regard, may I ask you, have you looked whether you tend to get antibody responses only at late stages of the disease or also at early stages?

Pfreundschuh: No, we just do not have enough experience with serial analyses of patients. We tried some of these antigens, to test them in Western blot, but that is problematic because it is very difficult to really get rid of E. coli proteins and often you get specific bands in the western blot and it cannot be confirmed in the phage assay and to check for an antibody response in phage assay you have to absorb the serum and that is very tedious, it takes time, it is like making a good wine. To absorb all the cross reactivities out of the serum it takes about two weeks and that makes serial analysis of these sera very, very tedious.

16

T CELL TOLERANCE VERSUS TUMOR IMMUNITY OR AUTOIMMUNITY

Pamela S. Ohashi and Daniel E. Speiser

Ontario Cancer Institute
Departments of Medical Biophysics and Immunology
610 University Avenue, Toronto, ON M5G 2M9, Canada

1. INTRODUCTION

With the recent identification of tumor associated antigens[1–4] together with the strong correlation between certain viruses and cancer[5,6], efforts have become more focussed upon developing tumor immunotherapy as an alternative modality for cancer treatment. One concern regarding the efficacy of tumor immunotherapy is the possibility that T cells may be tolerant or develop tolerance towards the tumor 'self' antigens. A significant amount of research has been directed towards understanding the mechanisms of thymic tolerance and peripheral T cell tolerance and will be outlined in this review. Apsects of T cell tolerance are relevant for the development of effective tumor vaccines.

2. THYMIC TOLERANCE

2.1. Clonal Deletion

Clonal deletion is the major mechanism for the induction of T cell tolerance to self antigens expressed at reasonable levels in the thymus. Direct evidence for this process was reported in 1987[7]. Using monoclonal antibodies specific for particular T cell receptor V gene segments, Kappler, Marrack and coworkers showed T cells expressing certain Vβ+ TCR were absent in strains of mice expressing a particular self molecule. Using a variety of TCR transgenic and nontransgenic models, further studies have shown that clonal deletion occurs in the presence of various superantigens and conventional antigens[8–13] (for review see[14,15]).

Experiments involving the administration of peptides in vivo or cocultivation in vitro were also able to demonstrate thymocyte deletion[16,17]. Since peptides are being extensively used in many cancer vaccines, considerations of the affects this may have on developing thymocytes and peripheral T cells may be considered. Tolerance studies have

The Biology of Tumors, edited by Mihich and Croce
Plenum Press, New York, 1998.

suggested that the concentration of peptide required to induce thymocyte clonal deletion was less than the concentration required to activate T cell clones[18,19]. Studies using TCR transgenic mice have suggested that the concentration of peptide antigen required to induce thymic deletion is the same concentration that is required for the activation of naive T cells[20–22]. These studies suggest that thymocyte clonal deletion is at least, if not more sensitive than T cell activation.

The sensitivity of thymocyte clonal deletion has also been investigated using peptide antigen variants that contain alterations in amino acid residues. Using TCR transgenic mice, several studies have suggested that specific thymcocyte deletion may occur without activation of T cells expressing the same TCR[18,22,23]. Other studies have also suggested that the ligands that cannot induce T cell activation, are still sufficient for clonal deletion[18,24,25].

Because thymocyte deletion is efficient and sensitive, it is important that the peptide used for immunotherapy is not derived from a protein that is expressed abundantly in the thymus.

2.2. Unresponsiveness, Anergy, or Tuning?

Evidence suggests that thymocytes may become tolerant to self ligands present in the thymus by clonal unresponsiveness or anergy[26–28](reviewed in[29]). Unresponsiveness is a term that generally refers to the presence of T cells that are unable to make a functional response to a defined ligand. Anergy also implies the presence of non-functional T cells, however, the unresponsiveness may be reversed by the addition of IL-2[30].

Conceptually, the presence of non-functional T cells in the mature repertoire is not sensible. Why would the immune system want to maintain useless Tcells? Theoretical models were postulated by Grossman and colleagues, which predicted that T cells were able to tune or modify activation thresholds[31,32]. In this way, T cells could be rendered unresponsive to self ligands, but still be reactive to encounter with higher affinity ligands. Experimental support for this model has come from several studies using a defined TCR transgenic population. The results demonstrate that thymocyte maturation in the presence of a defined ligand, leads to unresponsiveness to that ligand, but reactivity to other related or unrelated antigens[33,34]. Therefore, it is possible that unresponsiveness to thymic antigens is a consequence of T cell modification or alteration of the resting thresholds (for review see[35]). As a result, the T cells are tuned so that they do not respond to self ligands, but react with higher affinity antigens.

3. PERIPHERAL TOLERANCE

Several models have been developed to investigate the mechanisms of tolerance to extrathymic ligands. Studies suggest that self specific T cells may either undergo clonal deletion, become unresponsive or in many cases remain unaware of the particular self ligand expressed in a tissue specific location. These 'ignorant' T cells persist in the T cell repertoire, remain naive and are potentially fully reactive to self ligands. The fate of the self reactive T cell largely depends on the expression level of the tolerizing ligand or whether the T cell routinely encounters a particular self ligand.

3.1. Clonal Deletion

After systemic challenge with high amounts of superantigens or peptide antigens, specific T cells undergo a rapid expansion, where they acquire effector cell function, fol-

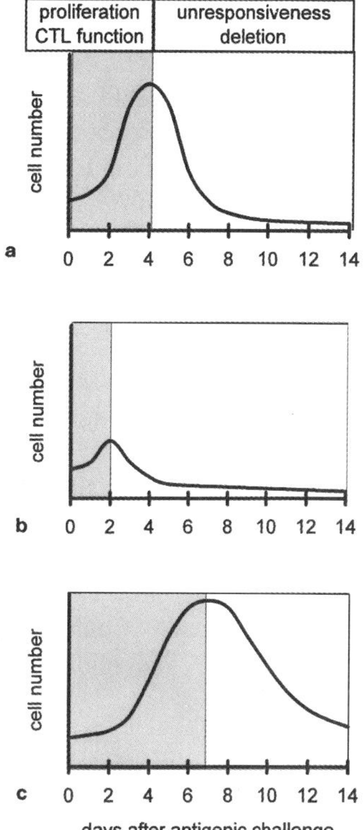

Figure 1. T cell responses to tolerizing and activating ligands. a) Efficient T cell induction leading to T cell tolerance. Antigenic challenge occurs on day 0. T cells generally respond by proliferation and the aquisition of effector function (shown in shaded regions). This is followed by T cell unresponsiveness and a decline in cell number. b) Suboptimal T cell induction. This may occur when the T cell encounters antigen in the absence of a CD28 mediated signal, or if a suboptimal peptide ligand is present. Less T cell expansion is seen, and a narrow window of T cell function is detected. c) Induction of immunity. T cell activation for full effector function generally is extended over a longer period of time. Unresponsive T cells are not detectable and instead functional memory occurs. The days shown on the scale are approximate and vary slightly between model systems.

lowed by deletion (Figure 1a)[21,36–38](reviewed in[15]). These and many other findings provided the basis for clonal deletion or exhaustion as a mechanism of mature T cell tolerance. It may also be possible that peptide immunization for tumor therapy be biased towards tolerance rather than activation, depending on the peptide and protocol used for administration[39,40].

Recent experiments using mice expressing a defined model transgene in a tissue specific manner demonstrated that clonal deletion may occur to natural self ligands[41–43]. Studies suggest that self antigens may be presented by class I[42] or class II[43] molecules in the local draining lymph nodes, and induce tolerance at these sites. These experiments suggest that tumor associated antigens, which would be more abundant as the tumor mass grows, may have the potential to induce peripheral tolerance of tumor specific T cells. Using this model, it will be important to identify the type of cell that is presenting the self ligand in the regional lymph node, and understand why these interactions lead to tolerance by deletion as opposed to longterm functional T cell activation.

3.2. Unresponsiveness and Anergy

In many models, antigen specific T cells have been generated that do not respond to further antigenic challenge in vitro and in vivo. The kinetics and mechanisms of various states of unresponsiveness may be different and are often not distinguished in the literature.

Unresponsiveness has been demonstrated as a mechanism of T cell tolerance in many reports[36–38,44,45]. After challenge with superantigens or peptide antigens, specific T cells have been shown to expand, transiently secrete cytokines or demonstrate cytotoxic effector function, followed by a transient state of unresponsiveness and subsequent deletion (Figure 1a). In these models, the T cells interact with antigen presented by antigen presenting cells (APC), and therefore also receive costimulatory signals in vivo.

A different type of immunological unresponsiveness has also been described in several transgenic models using H-2K^b as the self molecule. By using different promoters to express H-2K^b, Arnold and colleagues have demonstrated that unresponsiveness to the peripheral H-2K^b antigen may be induced by downregulating the TCR specific for H-2K^b, or decreasing the density of CD8[46–49](reviewed in 6601).

Several models have been proposed to explain how T cell interactions lead to T cell activation or tolerance. The traditional two signal model for T cell activation predicts that efficient T cell activation only occurs when the T cell receives a specific stimuli through the TCR, together with a second costimulatory signal[50,51]. If the T cell receives a TCR mediated signal 1, in the absence of signal 2, the T cell is predicted to become anergic. The major costimulatory signal is believed to be generated through interactions with CD28 and the B7 family of molecules expressed by the APC or IL-2[52]. However, there is very little evidence to support the prediction that signal 1 alone leads to a longlasting state of T cell tolerance where the T cells are present but are unable to respond to further stimuli[53–55].

Transgenic models or gene deficient mice have suggested that the absence of costimulatory molecules does not lead to the induction of anergy as a longlasting mechanism of T cell unresponsiveness. Studies have shown that either CD28 deficient mice were able to mount an efficient response to viruses or allogeneic stimuli[56–60]. Using TCR transgenic mice specific for the lymphocytic choriomeningitis virus (LCMV), that do not express CD28 (TCR+CD28–/–), experiments have shown that anergy may be induced by in vivo administration of peptides. However in these studies peptide challenge led to reduced proliferation, early cytotoxic function, followed by anergy and deletion (Figure 1b)[59]. This sequence of events are identical to those observed after superantigen or peptide administration in CD28+ mice (as outlined in Figure 1a). Therefore, this study suggests that the role of CD28 is primarily to enhance T cell activation. Other studies have demonstrated that CD28 also enhances the response to subdominant tumor epitopes and suboptimal ligands in vitro and in vivo[61,62]. In addition, several experiments have demonstrated that the expression of B7–1 by tumors or normal tissues now confer immunogenicity in vivo[63–68]. Collectively, these reports suggest that the primary function of costimulatory molecules such as CD28 and B7–1 are to augment T cell activation. The absence of these signals, in the presence of TCR mediated signals does not lead to long term T cell anergy. These findings are relevant for tumor immunotherapy, since tumors are generally not APCs and generally do not express the B7 family of molecules. Therefore, it is likely that the tumor does not induce anergy to tumor reactive T cells in vivo.

3.3. Ignorance

Many experiments have demonstrated that self reactive T cells exist to peripheral antigens, and that these T cells remain naive, but may be induced through appropriate stimuli to acquire full effector function[69–76]. Studies that were able to examine both a defined self antigen together with self reactive T cells were based upon transgenic models expressing the LCMV glycoprotein (GP) in the β-islets of the pancreas using the rat insulin promoter (RIP)[72,73,77]. By breeding TCR transgenic mice specific for LCMV-GP with RIP-GP animals, double transgenic mice RIP-GP/TCR were generated to follow the fate of the self reactive LCMV-GP specific T cells. Studies clearly demonstrated that the

LCMV-GP specific T cells were not tolerant to the LCMV-GP self antigen in both RIP-GP/TCR double transgenic mice or RIP-GP single transgenic mice, which have a normal frequency of LCMV-GP specific T cells. Infection with LCMV lead to the activation of virus specific T cells, specific infiltration of the pancreatic islets expressing LCMV-GP and autoimmunity. At this point, we considered whether it was feasible to examine whether this LCMV-GP self antigen would also serve as a tumor associated antigen and began to develop this model (see next section).

4. T CELLS ARE NOT TOLERANT TO TUMOR ASSOCIATED ANTIGENS IN VIVO

Tumor specific T cell clones together with various strategies have led to the identification of a variety of tumor associated antigens[1–4,78–83]. Tumor associated antigens may be tumor specific and may arise from mutant oncogenes. Other studies have characterized tumor antigens to be differentiation antigens or tissue specific proteins that are expressed at low levels or in limited tissues. These findings have led to the proposal that tumor immunotherapy may be directed against normal self antigens[84,85]. Studies have shown that T cells are not tolerant to many of these tissue specific antigens[86,87], which parallel predictions made by earlier studies that focussed on peripheral T cell tolerance.

5. TUMOR IMMUNOTHERAPY

In the 1980's, a series of studies examined the induction of an anti-tumor immune response using viral vaccines and demonstrated that the immune response could control and eliminate tumors in vivo[88–90] . Recombinant vaccinia viruses were generated that contained proteins from the tumorigenic polyoma (PY) virus[88]. Rats were given syngeneic PY-transformed rat 3T3 cells. Rats previously immunized subcutaneously with recombinant vaccinia viruses expressing the PY proteins were able to reject the tumors, whereas non-immunized controls were unable to eliminate the tumors. Further studies identified the role of CD8 cytotoxic T cells as playing a major role in the elimination of tumors in vivo[91,92].

More recent studies have addressed the potential for various vaccines to be used to induce immunotherapy (for reviews see[6,93]) and reports from clinical trials have been optomistic[94,95]. Peptides derived from tumor associated antigens have also been considered in conjunction with various strategies that would maximize the immune response[96–101]. In addition, recent experiments have also focussed on identifying the relative efficiency of priming an anti-tumor response using dendritic cells expressing tumor associated antigens[102–105].

We have developed a model to examine the effectiveness of tumor immunity where the tumor naturally develops in vivo. This is an important distinction from other models because we do not rely on cell lines that may have acquired alterations in vitro which may enhance immunogenicity. The tumors should be normally vascularized and we may more confidently address questions including tumor recurrance and T cell responsiveness at different stages of tumor growth, regression, and regrowth. Transgenic mice (RIP-Tag2) have been generated that express the SV40 large T antigen in the β islets cells of the pancreas using the rat insulin promoter. These mice develop islet tumors at approximately 8—16 weeks of age[106,107]. An increased tumor mass correlates with increased insulin production and therefore decreased blood glucose levels. Another transgenic line expressing the

LCMV-GP in the islets cells have also been described[72]. The LCMV-GP was used as a model tumor associated antigen. The LCMV-GP is an ideal antigen since LCMV is a natural mouse pathogen that induces an efficient cytotoxic response. In the spleen of acutely infected C57Bl/6 mice approximately 1/4—1/10 CD8+ spleen cells are specific for LCMV[13]. Therefore LCMV infection should lead to an effective therapy against a tumor expressing the LCMV-GP.

Islet hyperplasia and tumor growth from single RIP-Tag2 mice and double transgenic RIP(GP-Tag2) mice were investigated[108]. Animals with a significant tumor mass were immunized with LCMV. An anti-tumor response was monitored by blood glucose levels, histological analysis and functional analysis. LCMV infection prolonged the survival of the animals by 4–5 weeks indicating the success of the tumor immunotherapy. Noteably, the tumors recurred in all cases. The tumors were still expressing immunologically detectable levels of the tumor associated antigen LCMV-GP and functional memory LCMV-GP specific CTL were detectable. This raises several important points. T cell tolerance to tumor associated antigens did not develop in vivo, even in the presence of a high tumor burden with recurring tumors. It appears that the islet tumor was not sufficiently immunogenic, even to maintain a functional memory CTL response in vivo. In addition, memory CTL did not provide sufficient protection against tumor regeneration. Clinical studies have also found evidence for memory tumor specific T cells in patients with cervical carcinoma[109]. Together, this suggests that multiple stimulation against tumor associated antigens may be required to maintain tumor immunosurveillance in vivo.

6. TUMOR IMMUNITY VERSUS AUTOIMMUNITY?

One major concern if tumor associated antigens are used as targets for immunotherapy is that the immunizations may lead to a chronic autoimmune response[84,110]. It is also possible that as tumors are lysed by tumor specific CTL, that other self antigens are released, picked up by APC which then are subsequently capable of inducing a cascade of autoimmune responses against a variety of self ligands. It is interesting to note that in the studies with the RIP(GP-Tag2) animals, although immunization led to a reduction in tumor mass, hyperglycemia rarely occurred. In contrast, diabetes was always detected in RIP-GP single transgenic control animals[108].

Several studies have previously suggested that the relative risk for the induction of autoimmunity is low despite the fact that the target tumor associated antigen is a self protein[111–116]. Greenberg and his colleagues have examined a mouse model using Friend murine leukemia virus (FMuLV). Transgenic mice were generated that expressed the envelope protein of FMuLV in lymphoid cells. This led to T cell tolerance of env specific T cells in the host. However, adoptive transfer experiments with env specific T cells demonstrated that autoimmunity against the lymphoid compartment could not be detected. Nevertheless, the adoptive transfer of env specific T cells were capable of promoting survivial of mice given the Friend virus induced leukemic cell line, FBL.

7. CONCLUDING REMARKS

Tumor associated antigens serve as potential targets for tumor immunotherapy since T cells are not naturally tolerized to many of these antigens in vivo. Experimental models also suggest that T cell tolerance does not develop even in the presence of a large tumor burden or recurring tumor growth in vivo. This is probably because tumors as well as normal tissues are generally poorly immunogenic and poorly tolerogenic in vivo. Efficient strategies for the activation but not tolerance induction of tumor specific T cells must be developed to provide effective therapies in vivo.

Many models predict that the generation of memory cells may be sufficient to provide immune protection against tumor growth. However, this should be more rigorously tested in in vivo models. Our model suggests that this will not be sufficient to confer protection to tumors. Therefore, the success for future therapies should also consider the effectiveness of memory T cells and the requirement for multiple challenges with a variety of tumor associated antigens in vivo.

REFERENCES

1. C. L. Slingluff,Jr., D. F. Hunt and V. H. Engelhard, Direct analysis of tumor-associated peptide antigens. *Curr. Opin. Immunol.* 6, 733–740 (1994).
2. S. Ostrand-Rosenberg, Tumor immunotherapy: the tumor cell as an antigen-presenting cell. *Curr. Opin. Immunol.* 6, 722–727 (1994).
3. A. Van Pel, P. van der Bruggen, P. G. Coulie, V. G. Brichard, B. Lethe, B. Van den Eynde, C. Uyttenhove, J.-C. Renauld and T. Boon, Genes coding for tumor antigens recognized by cytolytic T lymphocytes. *Immunol. Rev.* 145, 229–250 (1995).
4. M. Disis and M. Cheever, Oncogenic proteins as tumor antigens. *Curr. Opin. Immunol.* 8, 637–642 (1996).
5. R. Tindle, Human papillomavirus vaccines for cervical cancer. *Curr. Opin. Immunol.* 8, 643–650 (1996).
6. C. Melief, R. Offringa, R. Toes and W. Kast, Peptide-based cancer vaccines. *Curr. Opin. Immunol.* 8, 651–657 (1996).
7. J. W. Kappler, N. Roehm and P. Marrack, T cell tolerance by clonal elimination in the thymus. *Cell* 49, 273–280 (1987).
8. H. R. MacDonald, R. Schneider, R. K. Lees, R. C. Howe, H. Acha-Orbea, H. Festenstein, R. M. Zinkernagel and H. Hengartner, T cell receptor Vβ use predicts reactivity and tolerance to Mls^a - encoded antigens. *Nature* 332, 40–45 (1988).
9. J. W. Kappler, U. D. Staerz, J. White and P. Marrack, Self tolerance eliminates T cells specific for Mls-modified products of the major histocompatibility complex. *Nature* 332, 35–40 (1988).
10. A. M. Pullen, P. Marrack and J. W. Kappler, The T cell repertoire is heavily influenced by tolerance to polymorphic self-antigens. *Nature* 335, 796–801 (1988).
11. P. Kisielow, H. Blüthmann, U. D. Staerz, M. Steinmetz and H. von Boehmer, Tolerance in T cell receptor transgenic mice involves deletion of nonmature $CD4^+8^+$ thymocytes. *Nature* 333, 742–746 (1988).
12. J. White, A. Herman, A. M. Pullen, R. Kubo, J. W. Kappler and P. Marrack, The V_β-specific superantigen staphylococcal enterotoxin B: stimulation of mature T cells and clonal deletion in neonatal mice. *Cell* 56, 27–35 (1989).
13. H. Pircher, K. Bürki, R. Lang, H. Hengartner and R. Zinkernagel, Tolerance induction in double specific T-cell receptor transgenic mice varies with antigen. *Nature* 342, 559–561 (1989).
14. G. V. Nossal, Negative selection of lymphocytes. *Cell* 76, 229–239 (1994).
15. J. Sprent and S. R. Webb, Intrathymic and extrathymic clonal deletion of T cells. *Curr. Opin. Immunol.* 7, 196–205 (1995).
16. K. M. Murphy, A. B. Heimberger and D. Y. Loh, Induction by antigen of intrathymic apoptosis of CD4+CD8+TCR/lo thymocytes in vivo. *Science* 250, 1720–1723 (1990).
17. L. M. Spain and L. J. Berg, Developmental regulation of thymocyte susceptibility to deletion by "self"-peptide. *J. Exp. Med.* 176, 213–223 (1992).
18. D. M. Page, J. Alexander, K. Snoke, E. Appella, A. Sette, S. M. Hedrick and H. M. Grey, Negative selection of CD4+CD8+ thymocytes by T-cell receptor peptide antagonists. *Proc. Natl. Acad. Sci. USA* 91, 4057–4061 (1994).
19. N. Vasquez, J. Kaye and S. M. Hedrick, In vivo and in vitro clonal deletion of double-positive thymocytes. *J. Exp. Med.* 175, 1307–1316 (1992).
20. C. Mamalaki, T. Norton, Y. Tanaka, A. Townsend, P. Chandler, E. Simpson and D. Kioussis, Thymic depletion and peripheral activation of class I major histocompatibility complex-restricted T cells by soluble peptide in T-cell receptor transgenic mice. *Proc. Natl. Acad. Sci. USA* 89, 11342–11346 (1992).

21. R. S. Liblau, R. Tisch, K. Shokat, X. Yang, N. Dumont, C. C. Goodnow and H. O. McDevitt, Intravenous injection of soluable antigen induces thymic and peirpheral T-cell apoptosis. *Proc. Natl. Acad. Sci. USA* 93, 3031–3036 (1996).
22. M. Bachmann, A. Oxenius, D. Speiser, S. Mariathasan, H. Hengartner, R. Zinkernagel and P. Ohashi, Peptide induced TCR-down regulation on naive T cells predicts agonist/partial agonist properties and strictly correlates with T cell activation. *Eur. J. Immunol.* (1997).
23. H. Pircher, U. Hoffmann Rohrer, D. Moskophidis, R. M. Zinkernagel and H. Hengartner, Lower receptor avidity required for thymic clonal deletion than for effector T cell function. *Nature* 351, 482–485 (1991).
24. J. Bill, O. Kanagawa, D. L. Woodland and E. Palmer, The MHC molecule I-E is necessary but not sufficient for the clonal deletion of Vβ11-bearing T cells. *J. Exp. Med.* 169, 1405–1419 (1989).
25. J. Yagi and C. A. Janeway, Ligand thresholds at different stages of T cell development. *Int. Immunol.* 2, 83–89 (1990).
26. F. Ramsdell, T. Lantz and B. J. Fowlkes, A nondeletional mechanism of thymic self tolerance. *Science* 246, 1038–1041 (1989).
27. D. E. Speiser, Y. Chvatchko, R. M. Zinkernagel and H. R. MacDonald, Distinct fates of self specific T cells developing in irradiation bone marrow chimeras: Clonal deletion, clonal anergy or in vitro responsiveness to self Mls-1a controlled by hemopoietic cells in the thymus. *J. Exp. Med.* 172, 1305–1314 (1990).
28. M. A. Blackman, H. Gerhard-Burgert, D. L. Woodland, E. Palmer, J. W. Kappler and P. Marrack, A role for clonal inactivation in T cell tolerance to Mls-1a. *Nature* 345, 540–542 (1990).
29. F. Ramsdell and B. J. Fowlkes, Clonal deletion versus clonal anergy: the role of the thymus in inducing self tolerance. *Science* 248, 1342–1348 (1990).
30. R. H. Schwartz, A cell culture model for T lymphocyte clonal anergy. *Science* 248, 1349–1356 (1990).
31. Z. Grossman and W. E. Paul, Adaptive cellular interactions in the immune system: the tunable activation threshold and the significance of subthreshold responses. *Proc. Natl. Acad. Sci. USA* 89, 10365–10369 (1992).
32. Z. Grossman, Cellular tolerance as a dynamic state of the adaptable lymphocyte. *Immunol. Rev.* 133, 45–73 (1993).
33. K. Kawai and P. S. Ohashi, Immunological function of a defined T-cell population tolerized to low-affinity self antigens. *Nature* 374, 68–69 (1995).
34. E. Sebzda, T. M. Kündig, C. T. Thomson, K. Aoki, S. Y. Mak, J. Mayer, T. M. Zamborelli, S. Nathenson and P. S. Ohashi, Mature T cell reactivity altered by a peptide agonist that induces positive selection. *J. Exp. Med.* 183, 1093–1104 (1996).
35. P. S. Ohashi, T cell selection and autoimmunity: flexibility and tuning. *Curr. Opin. Immunol.* 8, 808–814 (1996).
36. S. Webb, C. Morris and J. Sprent, Extrathymic tolerance of mature T cells: Clonal elimination as a consequence of immunity. *Cell* 63, 1249–1256 (1990).
37. D. Kyburz, P. Aichele, D. E. Speiser, H. Hengartner, R. M. Zinkernagel and H. P. Pircher, T cell immunity after a viral infection versus T cell tolerance induced by soluble viral peptides. *Eur. J. Immunol.* 23, 1956–1962 (1993).
38. E. R. Kearney, K. A. Pape, D. Y. Loh and M. K. Jenkins, Visualization of peptide-specific T cell immunity and peripheral tolerance induction in vivo. *Immunity* 1, 327–339 (1994).
39. R. E. M. Toes, R. Offringa, R. J. J. Blom, C. J. M. Melief and W. M. Kast, Peptide vaccination can lead to enhanced tumor growth through specific T-cell tolerance induction. *Proc. Natl. Acad. Sci. USA* 93, 7855–7860 (1996).
40. P. Aichele, K. Brduscha-Riem, R. M. Zinkernagel, H. Hengartner and H. P. Pircher, T cell priming versus T cell tolerance induced by synthetic peptides. *J. Exp. Med.* 182, 261–266 (1995).
41. P. Bertolino, W. R. Heath, C. L. Hardy, G. Morahan and J. F. A. P. Miller, Peripheral deletion of autoreactive $CD8^+$ T cells in transgenic mice expressing $H\text{-}2K^b$ in the liver. *Eur. J. Immunol.* 25, 1932–1942 (1995).
42. C. Kurts, W. R. Heath, F. R. Carbone, J. Allison, J. F. A. P. Miller and H. Kosaka, Constitutive class I-restricted exogenous presentation of self antigens in vivo. *J. Exp. Med.* 184, 923–930 (1996).
43. I. Forster and I. Lieberam, Peripheral tolerance of CD4 T cells following local activation in adolescent mice. *Eur. J. Immunol.* 26, 3194–3202 (1996).
44. H.-G. Rammensee, R. Kroschewski and B. Frangoulis, Clonal anergy induced in mature Vβ6+ T lymphocytes on immunizing $Mls\text{-}1^b$ mice with $Mls\text{-}1^a$ expressing cells. *Nature* 339, 541–544 (1989).
45. Y. Kawabe and A. Ochi, Programmed cell death and extrathymic reduction of Vβ8+ CD4+ T cells in mice tolerant to Staphylococcus aureus enterotoxin B. *Nature* 349, 245–248 (1991).
46. G. Schoenrich, U. Kalinke, F. Momburg, M. Malissen, A.-M. Schmitt-Verhulst, B. Malissen, G. J. Hämmerling and B. Arnold, Downregulation of T cell receptors on self-reactive T cells as a novel mechanism for extrathymic tolerance induction. *Cell* 65, 293–304 (1991).

47. G. Schoenrich, F. Momburg, M. Malissen, A.-M. Schmitt-Verhulst, B. Malissen, G. Hammerling and B. Arnold, Distinct mechanisms of extrathymic T cell tolerance due to differential expression of self antigen. *Int. Immunol.* 4, 581–590 (1992).
48. I. Ferber, G. Schonrich, J. Schenkel, A. L. Mellor, G. J. Hammerling and B. Arnold, Levels of peripheral T cell tolerance induced by different doses of tolerogen. *Science* 263, 674–676 (1994).
49. A. Tafuri, J. Alferink, P. Moller, G. J. Hammerling and B. Arnold, T cell awareness of paternal alloantigens during pregnancy. *Science* 270, 630–633 (1995).
50. P. Bretscher and M. Cohn, A theory of self-nonself discrimination. *Science* 169, 1042(1970).
51. C. H. June, J. A. Bluestone, L. M. Nadler and C. B. Thompson, The B7 and CD28 receptor families. *Immunol. Today* 15, 321–331 (1994).
52. D. J. Lenschow, T. L. Walunas and J. A. Bluestone, CD28/B7 system of T cell costimulation. *Annu. Rev. Immunol.* 14, 233–258 (1996).
53. D. Lo, L. C. Burkly, G. Widera, C. Cowing, R. A. Flavell, R. D. Palmiter and R. L. Brinster, Diabetes and tolerance in transgenic mice expressing class II MHC molecules in pancreatic beta cells. *Cell* 53, 159–168 (1988).
54. D. Lo, L. C. Burkly, R. A. Flavell, R. D. Palmiter and R. L. Brinster, Tolerance in transgenic mice expressing class II major histocompatibility complex on pancreatic acinar cells. *J. Exp. Med.* 170, 87–104 (1989).
55. L. C. Burkly, D. Lo, O. Kanagawa, R. L. Brinster and R. A. Flavell, T-cell tolerance by clonal anergy in transgenic mice with nonlymphoid expression of MHC class II I-E. *Nature* 342, 564–566 (1989).
56. A. Shahinian, K. Pfeffer, K. P. Lee, T. M. Kündig, K. Kishihara, A. Wakeham, K. Kawai, P. S. Ohashi, C. B. Thompson and T. W. Mak, Differential T cell costimulatory requirements in CD28-deficient mice. *Science* 261, 609–612 (1993).
57. J. M. Green, P. J. Noel, A. I. Sperling, T. L. Walunas, G. S. Gray, J. A. Bluestone and C. B. Thompson, Absence of B7-dependent responses in CD28-deficient mice. *Immunity* 1, 501–508 (1994).
58. K. Kawai, A. Shahinian, T. Mak and P. Ohashi, Skin allograft rejection in CD28-deficient mice. *Transplantation* 61, 352–355 (1996).
59. T. M. Kündig, A. Shahinian, K. Kawai, H. W. Mittruecker, E. Sebzda, M. F. Bachmann, T. W. Mak and P. S. Ohashi, Duration of TCR stimulation determines costimulatory requirements of T cells. *Immunity* 5, 41–52 (1996).
60. D. Speiser, M. Bachmann, A. Shahinian, T. Mak and P. Ohashi, Acute graft-versus-host disease without costimulation via CD28[1]. *Transplantation* 63, (1997).
61. J. V. Johnston, A. R. Malacko, M. T. Mizuno, P. McGowan, I. Hellstrom, K. E. Hellstrom, H. Marquardt and L. Chen, B7-CD28 costimulation unveils the hierarchy of tumor epitopes recognized by major histocompatibility complex class I-restricted CD8+ cytolytic T lymphocytes. *J. Exp. Med* 183, 791–800 (1996).
62. M. F. Bachmann, E. Sebzda, T. M. Kündig, A. Shahinian, D. E. Speiser, T. W. Mak and P. S. Ohashi, T cell responses are governed by avidity and costimulatory thresholds. *Eur. J. Immunol.* 26, 2017–2022 (1996).
63. L. Chen, S. Ashe, W. A. Brady, I. Hellström, K. E. Hellström, J. A. Ledbetter, P. McGowan and P. S. Linsley, Costimulation of antitumor immunity by the B7 counterreceptor for the T lymphocyte molecules CD28 and CTLA-4. *Cell* 71, 1093–1102 (1992).
64. S. E. Townsend and J. P. Allison, Tumor rejection after direct costimulation of CD8+ T cells by B7-transfected melanoma cells. *Science* 259, 368–370 (1993).
65. L. Chen, P. McGowan, S. Ashe, J. Johnston, Y. Li, I. Hellstrom and K. E. Hellstrom, Tumor immunogenicity determines the effect of B7 costimulation on T cell-mediated tumor immunity. *J. Exp. Med.* 179, 523–532 (1994).
66. S. Guerder, J. Meyerhoff and R. Flavell, The role of the T cell costimulator B7–1 in autoimmunity and the induction and maintenance of tolerance to peripheral antigen. *Immunity* 1, 155–166 (1994).
67. D. M. Harlan, H. Hengartner, M. L. Huang, Y. H. Kang, R. Abe, R. W. Moreadith, H. Pircher, G. S. Gray, P. S. Ohashi, G. J. Freeman, L. M. Nadler, C. H. June and P. Aichele, Mice expressing both B7 and viral glycoprotien on pancreatic beta cells along with glycoprotein-specific transgenic T cell develop diabetes due to a breakdown of T lymphocyte unresponsiveness. *Proc. Natl. Acad. Sci. USA* 91, 3137–3141 (1994).
68. M. G. von Herrath, S. Guerder, H. Lewicki, R. A. Flavell and M. B. A. Oldstone, Coexpression of B7–1 and viral ("self") transgenes in pancreatic β cells can break peripheral ignorance and lead to spontaneous autoimmune diabetes. *Immunity* 3, 727–738 (1995).
69. G. Gammon and E. E. Sercarz, How some T cells escape tolerance induction. *Nature* 342, 183–185 (1989).
70. H. J. Schild, O. Rötzschke, H. Kalbacher and H.-G. Rammensee, Limit of T cell tolerance to self proteins by peptide presentation. *Science* 247, 1587–1589 (1990).
71. J. Boehme, K. Haskins, P. Stecha, W. van Ewijk, M. LeMeur, P. Gerlinger, C. Benoist and D. Mathis, Transgenic mice with I-A on islet cells are normoglycemic but immunologically intolerant. *Science* 244, 1179–1183 (1989).

72. P. S. Ohashi, S. Oehen, K. Bürki, H. Pircher, C. T. Ohashi, B. Odermatt, B. Malissen, R. Zinkernagel and H. Hengartner, Ablation of "tolerance" and induction of diabetes by virus infection in viral antigen transgenic mice. *Cell* 65, 305–317 (1991).
73. M. B. A. Oldstone, M. Nerenberg, P. Southern, J. Price and H. Lewicki, Virus infection triggers insulin-dependent diabetes mellitus in a transgenic model: role of anti-self (virus) immune response. *Cell* 65, 319–331 (1991).
74. J. D. Katz, B. Wang, K. Haskins, C. Benoist and D. Mathis, Following a diabetogenic T cell from genesis through pathogenesis. *Cell* 74, 1089–1100 (1993).
75. J. Goverman, A. Woods, L. Larson, L. P. Weiner, L. Hood and D. M. Zaller, Transgenic mice that express a myelin basic protein specific T cell receptor develop spontaneous autoimmunity. *Cell* 72, 551–560 (1993).
76. J. J. Lafaille, K. Nagashima, M. Katsuki and S. Tonegawa, High incidence of spontaneous autoimmune encephalomyelitis in immunodeficient anti-myelin basic protein T cell receptor transgenic mice. *Cell* 78, 399–408 (1994).
77. P. S. Ohashi, S. Oehen, P. Aichele, H. Pircher, B. Odermatt, P. Herrera, Y. Higuchi, K. Buerki, H. Hengartner and R. M. Zinkernagel, Induction of diabetes is influenced by the infectious virus and local expression of MHC class I and TNF-α. *J. Immunol.* 150, 5185–5194 (1993).
78. N. K. Nanda, K. K. Arzoo, H. M. Geysen, A. Sette and E. E. Sercarz, Recognition of multiple peptide cores by a single T cell receptor. *J. Exp. Med.* 182, 531–539 (1995).
79. C. Castelli, W. Storkus, M. Maeurer, D. Martin, E. Huang, B. Pramanik, T. Nagabhushan, G. Parmiani and M. Lotze, Mass spectrometric identification of a naturally processed melanoma peptide recognized by $CD8^+$ cytotoxic T lymphocytes. *J. Exp. Med.* 181, 363–368 (1995).
80. T. Wolfel, M. Haurer, J. Schneider, M. Serrano, C. Wolfel, E. Klehmann-Hieb, E. De Plaen, T. Hankeln, K. H. Meyer zum Buschenfelde and D. Beach, A $p16^{INK4a}$-insensitive CDK4 mutant targeted by cytolytic T lymphocytes in a human melanoma. *Science* 269, 1281–1284 (1995).
81. P. Robbins, M. El-Gamil, Y. Li, Y. Kawakami, D. Loftus, E. Appella and S. Rosenberg, A mutated β-catenin gene encodes a melanoma-specific antigen recognized by tumor infiltrating lymphocytes. *J. Exp. Med.* 183, 1185–1192 (1996).
82. J. Skipper, D. Kittlesen, R. Hendrickson, D. Deacon, N. Harthun, S. Wagner, D. Hunt, V. Engelhard and J. C. Slingluff, Shared epitopes for HLA-A3-restricted melanoma-reactive human CTL include a naturally processed epitope from Pmel-17/gp100[1]. *J. Immunol.* 157, 5027–5033 (1996).
83. A. Huang, P. Gulden, A. Woods, M. Thomas, C. Tong, W. Wang, V. Engelhard, G. Pasternack, R. Cotter, D. Hunt, D. Pardoll and E. Jaffee, The immunodominant major histocompatibility complex class I-restricted antigen of a murine colon tumor derives from an endogenous retroviral gene product. *Proc. Natl. Acad. Sci. USA* 93, 9730–9735 (1996).
84. N. K. Nanda and E. E. Sercarz, Induction of anti-self-immunity to cure cancer. *Cell* 82, 13–17 (1995).
85. A. Houghton, Cancer antigens: immune recognition of self and altered self. *J. Exp. Med.* 180, 1–4 (1994).
86. Y. Kawakami, S. Eliyahu, C. H. Delgado, P. F. Robbins, K. Sakaguchi, E. Appella, J. R. Yanelli, G. J. Adema, T. Miki and S. A. Rosenberg, Identification of a human melanoma antigen recognized by tumor-infiltrating lymphocytes associated with in vivo tumor rejection. *Proc. Natl. Acad. Sci. USA* 91, 6458–6462 (1994).
87. T. Nakamura, R. K. Lee, S. Y. Nam, B. K. Al-Ramadi, P. A. Koni, K. Bottomly, E. R. Podack and R. A. Flavell, Reciprocal regulation of CD30 expression on $CD4^+$ T cells by IL-4 and IFN-gamma. *J. Immunol.* 158, 2090–2098 (1997).
88. R. Lathe, M. P. Kieny, P. Gerlinger, P. Clertant, I. Guizani, F. Cuzin and P. Chambon, Tumor prevention and rejection with recombinant vaccinia. *Nature* 326, 878–880 (1987).
89. R. Bernards, A. Destree, S. McKenzie, E. Gordon, R. A. Weinberg and D. Panicali, Effective tumor immunotherapy directed against an oncogene-encoded product using a vaccinia virus vector. *Proc. Natl. Acad. Sci. USA* 84, 6854–6858 (1987).
90. C. D. Estin, U. S. Stevenson, G. D. Plowman, S. L. Hu, P. Sridhar, I. Hellström, J. P. Brown and K. E. Hellström, Recombinant vaccinia virus vaccine against the human melanoma antigen p97 for use in immunotherapy. *Proc. Natl. Acad. Sci. USA* 85, 1052–1056 (1988).
91. P. Greenberg, Adoptive T cell therapy of tumors: mechanisms operative in the recognition and elimination of tumor cells. *Adv. Immunol.* 49, 281–355 (1991).
92. W. M. Kast, R. Offringa, P. J. Peters, A. C. Voordouw, R. H. Meloen, A. J. van der Eb and C. J. M. Melief, Eradication of adenovirus E1-induced tumors by E1A specific cytotoxic T lymphocytes. *Cell* 59, 603–614 (1989).
93. N. Restifo, The new vaccines: building viruses that elicit antitumor immunity. *Curr. Opin. Immunol.* 8, 658–663 (1996).

94. X. Hu, N. G. Chakraborty, J. R. Sporn, S. H. Kurtzman, M. T. Ergin and B. Mukherji, Enhancement of cytolytic T lymphocyte precursor frequency in melanoma patients following immunization with the MAGE-1 peptide loaded antigen presenting cell-based vaccine. *Cancer Res.* 56, 2479–2483 (1996).
95. L. Borysiewicz, A. Fiander, M. Nimako, S. Man, G. Wilkinson, D. Westmoreland, A. Evans, M. Adams, S. Stacey, M. Bournsnell, E. Rutherford, J. Hickling and S. Inglis, A recombinant vaccinia virus encoding human papillomavirus types 16 and 18, E6 and E7 proteins as immunotherapy for cervical cancer. *The Lancet* 347, 1523–1527 (1996).
96. O. Mandelbolm, G. Berke, M. Fridkin, M. Feldman, M. Eisenstein and L. Eisenbach, CTL induction by a tumour-associated antigen octapeptide derived from a murine lung carcinoma. *Nature* 369, 67–71 (1994).
97. B. Minev, B. McFarland, P. Spiess, S. Rosenberg and N. Restifo, Insertion signal sequence fused to minimal peptides elicits specific $CD8^+$ T-cell responses and prolongs survival of thymoma-bearing mice. *Cancer Res.* 54, 4155–4161 (1994).
98. Y. Noguchi, E. Richards, Y. Chen and L. Old, Influence of interleukin 12 on p53 peptide vaccination against established Meth A sarcoma. *Proc. Natl. Acad. Sci. USA* 92, 2219–2223 (1995).
99. M. Wang, Bronte, V, P. Chen, L. Gritz, D. Panicali, S. Rosenberg and N. Restifo, Active immunotherapy of cancer with a nonreplicating recombinant fowlpox virus encoding a model tumor-associated antigen. *J. Immunol.* 154, 4685–4692 (1995).
100. K. Irvine, B. McCabe, S. Rosenberg and N. Restifo, Synthetic oligonucleotide expressed by a recombinant vaccinia virus elicits therapeutic CTL. *J. Immunol.* 154, 4651–4657 (1995).
101. K. Irvine, J. Rao, S. Rosenberg and N. Restifo, Cytokine enhancement of DNA immunization leads to effective treatment of established pulmonary metastases. *J. Immunol.* 156, 238–245 (1996).
102. M. Ossevoort, M. Feltkamp, K. van Veen, C. Melief and W. Kast, Dendritic cells as carriers for a cytotoxic T-lymphocyte epitope-based peptide vaccine in protection against a human papillomavirus type 16-induced tumor. *J. Immunotherapy* 18, 86–94 (1995).
103. J. Young and K. Inaba, Dendritic cells as adjuvants for class I major histocompatibility complex-restricted antitumor immunity. *J. Exp. Med.* 183, 7–11 (1996).
104. F. Hsu, C. Benike, F. Fagnoni, T. Liles, D. Czerwinski, B. Taidi, E. Engleman and R. Levy, Vaccination of patients with B-cell lymphoma using autologous antigen-pulsed dendritic cells. *Nature Medicine* 2, 52–58 (1996).
105. A. Porgador, D. Snyder and E. Gilboa, Induction of antitumor immunity using bone marrow-generated dendritic cells. *J. Immunol.* 156, 2918–2926 (1996).
106. D. Hanahan, Heritable formation of pancreatic β-cell tumours in transgenic mice expressing recombinant insulin/simian virus 40 oncogenes. *Nature* 315, 115–122 (1985).
107. T. E. Adams, S. Alpert and D. Hanahan, Non-tolerance and autoantibodies to a transgenic self antigen expressed in pancreatic β cells. *Nature* 325, 223–228 (1987).
108. D. Speiser, R. Miranda, A. Zakarian, M. Bachmann, K. McKall-Faienza, B. Odermatt, D. Hanahan, R. Zinkernagel and P. Ohashi, Tumor antigens of peripheral solid tumors do not activate naive or primed T cells: implications for immunotherapy and autoimmune disease. *submitted* (1997).
109. M. Ressing, W. van Driel, E. Celis, A. Sette, R. Brandt, M. Hartman, J. Anholts, G. Schreuder, W. ter Harmsel, G. Fleuren, B. Trimbos, W. Kast and C. Melief, Occasional memory cytotoxic T-cell responses of patients with human papillomavirus type 16-positive cervical lesions against a human leukocyte antigen-A 0201-restricted E7-encoded epitope. *Cancer Res.* 56, 582–588 (1996).
110. A. Houghton, On course for a cancer vaccine. *Lancet* 345, 1384–1385 (1995).
111. J. Hu, W. Kindsvogel, S. Busby, M. Bailey, Y. Shi and P. Greenberg, An evaluation of the potential to use tumor-associated antigens as targets for antitumor T cell therapy using transgenic mice expressing a retroviral tumor antigen in normal lymphoid tissues. *J. Exp. Med.* 177, 1681–1690 (1993).
112. Y. Kawakami, S. Eliyahu, C. Delgado, P. Robbins, K. Sakaguchi, E. Appella, J. Yannelli, G. Adema, T. Miki and S. Rosenberg, Identification of a human melanoma antigen recognized by tumor-infiltrating lymphocytes associated with *in vivo* tumor rejection. *Proc. Natl. Acad. Sci. USA* 91, 6458–6462 (1994).
113. Hara, I, Y. Takechi and A. Houghton, Implicating a role for immune recognition of self in tumor rejection: passive immunization against the *Brown* locus protein. *J. Exp. Med* 182, 1609–1614 (1995).
114. M. Visseren, A. van Elsas, E. van der Voort, M. Ressing, W. Kast, P. Schrier and C. Melief, CTL specific for the tyrosinase autoantigen can be induced from healthy donor blood to lyse melanoma cells. *J. Immunol.* 154, 3991–3998 (1995).
115. O. Mandelboim, E. Vadai, M. Fridkin, A. Katz-Hillel, M. Feldman, G. Berke and L. Eisenbach, Regression of established murine carcinoma metastases following vaccination with tumour-associated antigen peptides. *Nature Medicine* 1, 1179–1183 (1995).
116. S. Rosenberg and D. White, Vitiligo in patients with melanoma: normal tissue antigens can be targets for cancer immunotherapy. *J. Immunotherapy* 19, 81–84 (1996).

DISCUSSION

Melief: Have you tried to prolong the survival by additional treatment with interleukin-2?

Ohashi: No we have not done those experiments yet.

Melief: I would argue that in most systems where specific CD8 cells really have a strong anti-tumor effect *in vivo,* that you need interleukin-2 to really amplify and keep them going.

Pfreundschuh: Did you check for the zeta chain expression?

Ohashi: On the T cells? No we have not done that. That is probably going to be the answer to a lot of my questions. This is really recent work that we have not published and we are just getting to a lot of these kinds of questions and asking whether peptide therapy will work. We have also different recombinant that we would like to try to boost an anti-tumor response with vaccinia viruses.

Hanahan: Yes, I think it is very provocative. You seem to activate the immune system in a way that, for example, when the oncogene is the initiating antigen we do not see. But I am wondering in your recurrent tumors, in terms of this thought that I was raising about the access of the lymphocytes to the tumor, in your recurrent tumors at the end stage, either at the 137 days or the 168, if you look then do you see cellular infiltration or are they devoid, what are the characteristics of the tumors after recurrence at end stage?

Ohashi: Most mice survive approximately 50 days after LCMV infection. We have looked 30 days and we still see CD8 cells in the tumors.

Hanahan: The response is rather scattered, I mean, you do not see dense sort of focal infiltration; is it pretty well just scattered around? It is not just within the vasculature, it is really in the tumor?

Ohashi: Yes, I guess I can say, in response to your question as well about the CD3 zeta chains, these T cells still pretty much seem fully functional, especially when we re-challenge them *in vivo.* Our bias right now is basically that the anti-tumor response really follows the pattern that you see just against the normal LCMV infection, and that is when the virus is cleared, the CTL's kind of quieted down and that is when the tumor progresses as well. So, I think it has to do with virus load in the presence of the antigen.

Pfreundschuh: The infiltrations into the tumor by the T cells is a random event that the T cells just evade more than they used to do before the challenge with the virus, or is it specifically an invasion into the tumor? Did you cross label the T cells?

Ohashi: Well when we look at multiple sections and we ask if normal islets are infiltrated preferentially over the tumors or any sized tumors, we just see a specific but random infiltration all over the islets cells, as long as they express LCMVGP they are just in all islets basically.

Pfreundschuh: But they invade the tumor more than they do other organs?

Ohashi: There is no infiltration in other organs. In the islets, I would say the density of CD8 infiltration in the normal islet versus the tumor is the same; it is just that the tumor is bigger so overall there is more.

DePinho: Is there any merit in trying to construct a model that would recapitulate more what happens in cancer where antigens emerge somatically after the immune system tolerance has been executed? In other words, would it be useful to design a transgenic model in which you express a tumor antigen only after the immune response has gone through its learning process. You are expressing your antigen constitutively throughout development. My question is whether there would be any difference in immune response if an antigen is expressed during post-natal life only.

Ohashi: I am not so sure that there would be a difference because, on this particular model, we know that the expression of the antigen is throughout embryonic development and early adulthood and the mouse actually does not alter the T cell response; that is, we know that we have no T cell tolerance, so I do not know that you would actually get, say, an improved version of immunotherapy. So, just because we know there is no T cell tolerance to this antigen, I think it serves us as a fine tumor associated antigen.

Berns: For your adoptive transfer you use an argument that there is still expression of the LCMV glycoprotein but have you looked more directly, because I have found the effect of the adaptive transfer rather subtle. Could it be that the expression of the LCMV glycoprotein is significantly diminished or, alternatively, that you have selected only a small fraction of the cells that still express the LCMV?

Ohashi: Yes, absolutely that would be one concern of ours.

Berns: Could you not look directly by histochemistry to see how the expression is in the tumors?

Ohashi: Right, what we can do is RT/PCR and that is about the same because it is so sensitive. But we do not really have a good antibody right now against the glycoprotein so it would be really nice to assess that question, but unfortunately we cannot right now.

Bankert: Have you looked at the possible role of antibody in this? Does antibody to the glycoprotein go up during regression of the tumor or actually go down during regression of tumor?

Ohashi: Yes, we have not asked those questions, that would be nice.

Zanker: I am always puzzled by the question, why should lymphocytes have a greater access to tumors when the tumors are infected by virus proteins, or whatever?

Ohashi: Jim has a comment.

Allison: Another possibility is something that I will address in my talk. That is that one of the effects of the virus may be to infect a cell that can present the antigen in the context of additional co-stimulatory signals which are not there in the cells which are expressed in the GP, so the virus initiates the response by infecting dendritic cells, which then activate the T cells so that then the T cells get activated and then they can go find the antigen, wherever it is. But I will talk about this in more detail.

Zanker: But this means that you have to involve the lymphocytes of the tumor because there are so many other cells to give stop signals for migration.

Allison: I would say it is because the other cells do not have the start signals to get them to be able to migrate, so I would argue it the opposite way.

Ohashi: I would completely agree with Jim that in our particular system the activation of the virus specific T cells are on antigen presenting cells *per se*. So, because of this idea of co-stimulation, that would provide the effective immunization or priming of your GP specific T cells, we went on to also look at a triple transgenic model where we have got now B71 expressed on the beta islets cells of the pancreas. So, we have gotten the tumors with the GP antigen and B7 to ask the question, if we now provide the co-stimulation at this site, if you actually have a more anti-tumor response. But, the answer is no.

Zanker: I would partially agree if the priming is this local in the tumor or on the lymph node, that is the question.

Ohashi: Yes, and we would say the priming is on the lymph node.

Melief: Actually, a very puzzling question is why there is no tolerance in either normal or TCR transgenic mice because, presumably, the T cells can get there, so T cells are supposed to recirculate and see the antigen there, so why is there no tolerance?

Ohashi: I think a lot of different transgenic models have addressed those questions too, and they found similar answers to our experiments and they could show that, by and large, for a lot of tissue specific antigen, you actually do not have T cell tolerance and I think that could be due to a number of reasons, one, being a fairly low expression of the antigen. Also, most tissues actually have very low Class 1 MHC levels, so you can imagine that together with low MHC levels, they are actually about 100 to 1,000 fold lower than those on a normal T cell actually, especially in the islet cells because we have had quite a good look. So, you have gotten maybe approximately 1,000 MHC molecules on the cell surface, together with a relatively low expression of your particular antigen, you can guess that maybe you have only gotten ten epitopes on the cell surface. For me, I can imagine that you do not have tolerance against a self-epitope like that, together with the lack of co-stimulation.

Melief: But if you express the GP under other promoters, like in the skin, this is the experiment that Bernd Arnold did, so you see different levels of tolerance arising, depending on the amount of expression.

Ohashi: Yes, actually the group in Zurich has made quite extensive efforts to try and get the LCMV glycoprotein expressed in other tissues and, unfortunately, they have been unsuccessful in that.

MANIPULATION OF T CELL ACTIVATION IN THE ANTI-TUMOR IMMUNE RESPONSE

Arthur A. Hurwitz, Dana R. Leach, Andrea van Elsas, Sarah E. Townsend, and James P. Allison*

Cancer Research Laboratory and Howard Hughes Medical Institute
415 Life Sciences Addition
University of California
Berkeley, California 94720

Recent advances in the understanding of the regulation of T cell activation have allowed the rational design of new treatment protocols aimed at inducing or enhancing T cell responses against tumors. This review describes the progress our laboratory has made in applying these principles to several different murine tumor model systems. These findings are discussed in the context of developing an effective immunotherapeutic treatment for cancer.

1. T CELL ACTIVATION

Full activation of naive T cells requires two signals through different receptors (reviewed in[1,2]). The primary signal involves engagement of the T cell receptor (TcR) by its cognate ligand, an antigenic peptide bound to the restricting major histocompatibility complex (MHC) molecule. The second, antigen-independent costimulatory signal is provided through interaction of CD28 on the T cell with B7 on the antigen-presenting cell (APC). Concomitant ligation of the TcR and CD28 results in activation, including cytokine production, clonal expansion, and functional differentiation of the antigen-specific T cells. In this respect, it is important to consider that only 'professional' APCs (dendritic cells, B cells, and macrophages) express both the antigenic peptide/MHC complexes and costimulatory B7 molecules.

CTLA-4 is a second ligand for B7 that has a 20-fold higher avidity for binding B7 than does CD28[3]. Unlike the relatively high constitutive level of CD28 detected on T

* Address Correspondence to: Dr. James P. Allison, Howard Hughes Medical Institute, 401 Life Sciences Addition, University of California Berkeley, California 94720. (510) 643-6012; (510) 643-5692 (fax); jallison@uclink4.berkeley.edu

The Biology of Tumors, edited by Mihich and Croce
Plenum Press, New York, 1998.

cells, maximal CTLA-4 expression is reached 24–48 hours after complete activation[4,5]. Overwhelming evidence shows that CTLA-4 plays an important down-regulatory role in T cell responses[4,6–8]. These inhibitory effects of CTLA-4 may be involved in both initiation and termination of T cell responses. At the early stages of T cell activation, CTLA-4 may set a threshold of stimulatory signals needed for full activation of naive T cells by preferentially sequestering B7 molecules on the APC. Unless enough B7 is available to allow engagement of CD28, T cell activation may be inhibited. CTLA-4/B7 interactions may also serve to dampen or downregulate the antigen-specific T cell response after T cell activation, when CTLA-4 expression is at its peak. At both stages, blockade of CTLA-4/B7 interactions would serve to enhance T cell response.

2. ENHANCEMENT OF COSTIMULATORY INTERACTIONS IN TUMOR IMMUNOTHERAPY

The fact that B7 expression is restricted to professional APCs and that most tumors of epithelial origins do not express B7 suggested that the poor immunogenicity of many tumors is, in part, due to their lack of B7 expression. Therefore, we and others postulated that conferring B7 expression to tumors might enhance their immunogenicity[9,10]. Transduction of tumors of various tissue origins to express B7-1 was sufficient in many, but not all cases, to promote rejection of the tumor as well as provide immunity to rechallenge with the unmodified, 'parental' tumor (Table 1). Rejection correlated with the inherent immunogenicity of the tumor[11] and was principally dependent on CD8+ T cells. The latter finding suggested that tumors genetically modified to express B7 acted directly as APCs to prime CD8+ effector T cells. Subsequent studies indicated that in some tumor systems, genetic modification to express B7-2 also resulted in tumor rejection[12,13]. These findings laid the groundwork for future studies that examined enhancement of T cell activation in the anti-tumor response.

Several poorly immunogenic tumors were not rejected when transduced to express B7[11,14]. Therefore, we tested whether co-transduction of these tumors to express cytokines that would enhance the Th1 or pro-inflammatory T cell response would sufficiently enhance their immunogenicity. In our laboratory, we studied a Balb/C-derived, N-methyl-N-nitrosourea-induced mammary carcinoma, SM1[15]. Transfection with *B7-1* did not alter the tumorigenicity of the parental tumor. Transduction with *IFN-γ* slowed tumor growth, but this tumor line was still 100% tumorigenic in syngeneic mice. In contrast, SM1 tumors that expressed both B7-1 and IFN-γ were rejected by a T cell-dependent mechanism. Mice that rejected the double transductant were also immune to rechallenge with the parental SM1

Table 1. Rejection of tumors after transduction with B7-1

Tumor	Tumor type	Rejection	Protection against rechallenge with parental tumor
EL-4	lymphoma	+	+
TRAMP	prostatic adenocarcinoma	+	+
CM5153	fibrosarcoma	+	+
51BLim10	colorectal	+	+
K1735	melanoma	+	+/– (limited)
B16BL6	melanoma	–	–
SM1	mammary carcinoma	–	–

tumor as well as another BAlb/C-derived mammary tumor. Interestingly, synergy was not observed in SM1 tumors that expressed both B7 and granulocyte-macrophage colony-stimulating factor (GM-CSF). Given that both B7 and IFN-γ can work at the level of effector activation whereas GM-CSF is principally involved in host-derived APC differentiation and activation, these findings suggest that synergy is best attained when enhancing similar pathways of T cell activation. Thus, coexpression of both B7 and GM-CSF may 'confuse' the immune system by enhancing both tumor and host-derived APC function whereas coexpression of B7 and IFN-γ may focus enhancement of T cell activation by solely relying on the tumor as APC. Work from other laboratories using B7 and IL-12, a cytokine that induces IFN-γ and involved in Th1 development, supports this idea and also demonstrated that 'co-therapy' may be a feasible alternative for enhancing immunogenicity of tumors[16,17].

3. BLOCKADE OF INHIBITORY INTERACTIONS

As the inhibitory role of CTLA-4 on T cell activation became clear, we sought to determine whether blocking CTLA-4/B7 interactions would enhance immune responses to tumors. Previous studies showed that B7-transfected cells of the murine colorectal carcinoma 51BLim10 underwent substantial regression[14]. However, there was often an initial period of tumor growth prior to regression and not all tumors were rejected. Therefore, we tested whether an antibody that blocked CTLA-4/B7 interactions could enhance rejection of $B7^+$51BLim10. We demonstrated that administration of anti-CTLA-4 to mice bearing recently implanted $B7^+$51BLim10 resulted in rapid and complete rejection[18]. More impressive was our finding that treatment of mice implanted with a control vector-transduced 51BLim10 tumor (V51BLim10) with anti-CTLA-4 also resulted in tumor regression. This finding suggested that B7 expression was not required for rejection as a consequence of CTLA-4 blockade and that anti-CTLA-4 was acting at the level of host-derived APCs.

Subsequently, we tested the efficacy of CTLA-4 blockade in a variety of murine tumor models (Table 2). We found that many, but not all tumors were rejected following treatment with anti-CTLA-4. The pattern of effectiveness was similar to that observed with tumors transfected with B7. Some tumors, such as 51BLim10 and especially the fi-

Table 2. Rejection of various tumors as a consequence of CTLA-4 blockade

Tumor	Rejection (n)	Reduced growth	Comments
SaIN	73% (80)	+	
TRAMP	approx. 75%	+	More effective with TRAMPC2 (lower MHC expression)
V51BLim10	33% (75)	+	
Renca	72% (25)	+	
P815	+	+	
SM1	–	+	Occasional rejection at smaller tumor doses
K1735	–	+	
EL-4	–	+	Limited data
MOPC 315	–	–	
B16BL6	–	–	Consistent, but small effect

brosarcoma Sa1N, had significant rates of rejection, and those tumors that did develop grew much slower than in control antibody-treated animals. However, in several other tumor models, only a reduction in growth rates were observed and very few animals remained tumor-free. Finally, as with B7 transfection, CTLA-4 blockade had minimal or no appreciable effect against tumors that are generally considered to be poorly immunogenic (e.g., B16-BL6 and SM1).

Using more immunogenic tumor models, significant results were seen even when anti-CTLA-4 treatment was delayed until there was a substantial and measurable tumor burden. Treatment of the rapidly growing Sa1N tumor was very effective when begun seven days after tumor injection, but only marginally so at 14 days. The slower growing 51Blim10 could be successfully treated at 14, but not 21, days post inoculation and treatment of K1735 tumor-bearing mice was more effective if begun seven days after inoculation instead of at the time of tumor challenge. In general, we found that CTLA-4 blockade was most effective when delayed a few days after tumor cell injections, possibly due to some delay in the induction of local inflammation and/or the processing and presentation of tumor antigens by host APC.

The mechanisms by which CTLA-4 blockade induces tumor regression are not entirely understood. From initial *in vivo* lymphocyte depletion studies, we do know that both CD4+ and CD8+ T cells are necessary for rejection of Sa1N and 51BLim10 tumor cells. In support of this idea, mice deficient in T cells due to a germ-line deletion within the T cell receptor beta chain gene no longer reject TRAMPC2 prostate tumor cells after CTLA-4 blockade. Additionally, *in vitro* restimulation of T cells from Sa1N tumor-bearing mice treated with anti-CTLA-4 antibody consistently resulted in 2–3 fold more IFN-γ than T cells from control antibody-treated mice. Thus it is clear that T cells mediate tumor rejection induced by CTLA-4 blockade, although the precise effector mechanisms are not well understood. However, enhancement of tumor immunity is probably due to host-derived APCs since the tumors in these systems do not express either class II MHC that presents antigen to CD4+ cells or costimulatory molecules required for efficient T cell activation.

4. COMBINATION OF CTLA-4 BLOCKADE WITH TUMOR-DERIVED CYTOKINE EXPRESSION

As with tumors transduced to express B7, some tumors did not respond to anti-CTLA-4 immunotherapy, presumably due to inherently poor immunogenicity. We postulated that CTLA-4 blockade affected B7/CTLA-4 interactions during cross-priming of tumor-reactive T cells by host bone marrow-derived APCs. Therefore, we sought to enhance antigen presentation by host APCs by using an irradiated, cytokine-expressing tumor vaccine. We chose GM-CSF because of its ability to generate potent APCs[19–23]. Exposure of dendritic cells and macrophages to GM-CSF results in upregulation of both MHC and B7 expression. In addition, GM-CSF can promote growth and differentiation of stem cells (bone marrow- and umbilical cord blood-derived) into dendritic cells. Consistent with this idea, a study that tested 10 different cytokines and adhesion proteins demonstrated that GM-CSF was the most effective at enhancing the immunogenicity of a melanoma[24].

We used SM1 and B16-BL6 cells modified to express GM-CSF (GMSM1 and GMB16) in combination with CTLA-4 blockade. As described above, anti-CTLA-4 was ineffective at promoting regression of SM1 tumors (Table 2). In addition, treatment with GMSM1 at the time of tumor implantation had no effect on SM1 growth. However, treatment with both GMSM1 at the time of implantation and anti-CTLA-4 four days later resulted in regression of the parental SM1 tumor. Mice were immune to rechallenge with SM1 30 days later, supporting the idea that rejection was not solely due to transient effector mechanisms. Lymphocyte depletion studies indicated that both CD4+ and CD8+ T cells were required for rejection. These findings support the idea that GM-CSF secreted by tumors serves to recruit and activate host-derived APCs that could prime both CD4+ and CD8+ T cells capable of recognizing the MHC class II-negative SM1 cells. Therefore, synergy between tumor-derived GM-CSF expression and CTLA-4 blockade presumably occurs at the level of host APCs priming T cells and like the synergy observed between B7 and IFN-γ, is most effective because of enhancement of a similar axis of T cell priming.

The spontaneously-arising melanoma B16 and its selected variants are poorly immunogenic. The variant B16-BL6 is highly invasive and probably the least immunogenic. B16-BL6 expresses low levels of class I MHC that may allow escape from NK mediated lysis and renders tumor cells incapable of activating T cells. Treatment of B16-BL6 by CTLA-4 blockade alone did not have a significant effect on tumor growth *in vivo*. In contrast, vaccination with irradiated, GMB16 cells in combination with CTLA-4 blocking antibodies was found to induce rejection of the unmanipulated B16-BL6 tumor. These findings support the idea that T cell activation was hampered by poor antigen presentation by the tumor.

An interesting observation within this B16 system was the loss of skin and hair pigmentation following combination treatment. It has been hypothesized that autoantigens in human and murine melanomas will provide an antigen source for immunotherapy of these tumors. Such antigens remain immunologically silent in normal tissues but may serve a pivotal role in tumor rejection. Presumably, the GM-CSF-producing cellular vaccine expresses an immunogenic autoantigen, whose function is related to pigment synthesis. Only in cooperation with enhanced T cell activation after CTLA-4 blockade is a response against this self antigen generated, resulting in tumor rejection and autoimmunity.

5. CONCLUSIONS

The approaches to enhance activation of T cells to eradicate tumors reviewed in this paper have generated strong interest for possible application to human cancer. Transfection of B7 into tumors with some degree of immunogenicity was found to induce successful rejection in mouse tumor models. However, memory to rechallenge with the original tumor was limited and this strategy was not very successful against established tumors. In addition, *in vitro* transduction of tumors would be labor-intensive and targeting B7 expression to tumors *in vivo* could have undesirable consequences; inadvertent induction of B7 expression by normal tissues would pose a risk of developing autoimmune reactions. In contrast, application of CTLA-4 blocking antibodies *in vivo* has shown therapeutic potential against a wide variety of murine tumors including more established tumors. It can be used alone or in combination with GM-CSF-expressing, whole cell vaccination. No evident autoimmunity was induced in any of the tumor systems after CTLA-4 blockade, with the single exception of the vitiligo-like condition in mice implanted with the melanoma B16-BL6 and treated with GMB16 and anti-CTLA-4. Therefore, careful consideration of the target tissue would need to be considered when designing an immunotherapeutic strategy. In general, manipulation of T cell activation as described above shows promise for treating poorly or non-immunogenic tumors. Our current studies aimed at understanding the cellular mechanisms involved in anti-tumor immunity utilize a transgenic model of primary (autochthonous) tumor development[25]. This approach will provide a model system more similar to human cancer pathogenesis and assist in designing clinical trials using CTLA-4 blockade for tumor immunotherapy.

ACKNOWLEDGMENTS

The authors would like to thank Stan Grell, Tina Yu, and Jennifer Ziskin for their technical assistance. This work was performed in collaboration with Drs. Satyabrata Nandi and Rafael Guzman (UC Berkeley), Drs. Norman Greenberg and Barbara Foster (Baylor College of Medicine), and Drs. Maurice Burg and Eugene Kwon (NHLBI, NIH).

J.P.A. is an investigator of the Howard Hughes Medical Institute. A.A.H. is a fellow of the Department of Defense Breast Cancer Research Program. A.v.E. is a fellow of the Nederlandse Kankerbestrijding.

This work was supported in part by NIH CA 57986 (J.P.A.), NIH CA 71613 (D.R.L.), DOD DAM 9127-97-1-705 (A.A.H.), and the CaP Cure Foundation.

REFERENCES

1. Allison, J.P. 1994. CD28-B7 interactions in T cell activation. *Curr Opin Immunol* 6, no. 4:414–419.
2. June, C.H., J.A. Bluestone, L.M. Nadler, and C.B. Thompson. 1994. The B7 and CD28 receptor families. *Immunol Today* 15, no. 7:321–331.
3. Linsley, P.S., J.L. Greene, W. Brady, J. Bajorath, J.A. Ledbetter, and R. Peach. 1994. Human B7–1 (CD80) and B7–2 (CD86) bind with similar avidities but distinct kinetics to CD28 and CTLA-4 receptors. *Immunity* 1:793–801.
4. Krummel, M.F., and J.P. Allison. 1995. CD28 and CTLA-4 have opposing effects on the response of T cells to stimulation. *J Exp Med* 182, no. 2:459–65.
5. Linsley, P.S., J. Bradshaw, J. Greene, R. Peach, K.L. Bennett, and R.S. Mittler. 1996. Intracellular trafficking of CTLA-4 and focal localization towards sites of TCR engagement. *Immunity* 4, no. 6:535–43.
6. Walunas, T.L., D.J. Lenschow, C.Y. Bakker, P.S. Linsley, G.J. Freeman, J.M. Green, C.B. Thompson, and J.A. Bluestone. 1994. CTLA-4 can function as a negative regulator of T cell activation. *Immunity* 1, no. 5:405–13.
7. Krummel, M.F., and J.P. Allison. 1996. CTLA-4 engagement inhibits IL-2 accumulation and cell cycle progression upon activation of resting T cells. *J Exp Med*, no. in press.
8. Walunas, T.L., C.Y. Bakker, and J.A. Bluestone. 1996. CTLA-4 ligation blocks CD28-dependent T cell activation. *J Exp Med* 183, no. 6:2541–50.
9. Chen, P.W., and H.N. Ananthaswamy. 1993. Rejection of K1735 murine melanoma in syngeneic hosts requires expression of MHC class I antigens and either class II antigens or IL-2. *J Immunol* 151, no. 1:244–255.
10. Townsend, S., and J.P. Allison. 1993. Tumor rejection after direct costimulation of CD8+ T cells by B7-transfected melanoma cells. *Science* 259:368–370.
11. Chen, L., P. McGowan, S. Ashe, J. Johnston, Y. Li, I. Hellstrom, and K.E. Hellstrom. 1994. Tumor immunogenicity determines the effect of B7 costimulation on T cells-mediated tumor immunity. *J Exp Med* 179:523–532.
12. Hodge, J.W., S. Abrams, J. Schlom, and J.A. Kantor. 1994. Induction of antitumor immunity by recombinant vaccinia viruses expressing B7–1 or B7–2 costimulatory molecules. *Cancer Res* 54:5552–5555.
13. Yang, G., K.E. Hellstrom, I. Hellstrom, and L. Chen. 1995. Antitumor immunity elicited by tumor cells transfected with B7–2, a second ligand for CD28/CTLA-4 costimulatory molecules. *J immunol* 154:2794–2800.
14. Townsend, S.E., F.W. Su, J.M. Atherton, and J.P. Allison. 1994. Specificity and longevity of anti-tumor immune responses induced by B7-transfected tumors. *Cancer Res* 54:6477–6483.
15. Hurwitz, A.A., S.E. Townsend, T.F.-Y. Yu, J. Atherton, and J.P. Allison. 1998. Enhancement of the Anti-Tumor Immune Response Using a Combination of Interferon-γ and B7 Expression in an Experimental Mammary Carcinoma. *Int J Cancer*, no. in press.
16. Coughlin, C.M., M. Wysocka, H.L. Kurzawa, W.M. Lee, G. Trinchieri, and S.L. Eck. 1995. B7–1 and interleukin 12 synergistically induce effective antitumor immunity. *Cancer Research* 55, no. 21:4980–4987.

17. Zitvogel, L., P.D. Robbins, W.J. Storkus, M.R. Clarke, M.J. Maeurer, R.L. Campbell, C.G. Davis, H. Tahara, R.D. Schreiber, and M.T. Lotze. 1996. Interleukin-12 and B7.1 co-stimulation cooperate in the induction of effective antitumor immunity and therapy of established tumors. *Eur J Immunol* 26, no. 6:1335–1341.
18. Leach, D.R., M.F. Krummel, and J.P. Allison. 1996. Enhancement of antitumor immunity by CTLA-4 blockade. *Science* 271:1734–1736.
19. Heufler, C., F. Koch, and G. Schuler. 1988. Granulocyte/macrophage colony-stimulating factor and interleukin 1 mediate the maturation of murine epidermal Langerhans cells into potent immunostimulatory dendritic cells. *J Exp Med* 167, no. 2:700–705.
20. Inaba, K., M. Inaba, N. Roomani, H. Aya, M. Deguchi, S. Ikehara, S. Muramatsu, and R.M. Steinman. 1992. Generation of large numbers of dendritic cells from mouse bone marrow cultures supplemented with granulocyte/macrphage colony-stimulating factor. *J Exp Med* 176, no. 6:1693–1702.
21. Caux, C., C. Dezutter-Dambuyant, D. Schmitt, and J. Banchereau. 1992. GM-CSF and TNF-alpha cooperate in the generation of dendritic Langerhans cells. *Nature* 360:258–261.
22. Larsen, C.P., S.C. Ritchie, R. Hedrix, P.S. Linsley, K.S. Hathcock, R.J. Hodes, R.P. Lowry, and T.C. Pearson. 1994. Regulation of immunostimulatory function and costimulatory molecule (B7–1 and B7–2) expression on murine dendritic cells. *J Immunol* 152, no. 11:5208–5219.
23. Inaba, K., M. Witmer-Pack, M. Inaba, K.S. Hathcock, H. Sakuta, M. Azuma, H. Yagita, K. Okumura, P.S. Linsley, S. Ikehara, S. Muramatsu, R.J. Hodes, and R.M. Steinman. 1994. The tissue distribution of the B7–2 costimulator in mice: abundant expression on dendritic cells *in situ* and during maturation *in vitro*. *J Exp Med* 180:1849–1860.
24. Dranoff, G., E. Jaffee, A. Lazenby, P. Golumbek, H. Levitsky, K. Brose, V. Jackson, H. Hamada, D. Pardoll, and R.C. Mulligan. 1993. Vaccination with irradiated tumor cells engineered to secrete murine granulocyte-macrophage colony-stimulating factor stimulates potent, specific, long-lasting anti-tumor immunity. *Proc Nat'l Acad Sci* 90:3539–3543.
25. Greenberg, N.M., F. DeMayo, M.J. Finegold, D. Medina, W.D. Tilley, J.O. Aspinall, G.R. Cunha, A.A. Donjacour, R.J. Matusik, and J.M. Rosen. 1995. Prostate cancer in a transgenic mouse. *Proc Nat'l Acad Sci* 92:3439–3443.

DISCUSSION

Mihich: Jim, I think you know of our previous work on adriamycin and we find that adriamycin has a number of immunomodulating affects, the important one in connection with your talk is an induction of the production of IL2, which seems to be occurring at the transcriptional level. Now, if you combine adriamycin with IL2, with tolerated doses of IL2, in a syngeneic system with a tumor which has been made resistant to adriamycin, therefore there is no tumor necrosis by the drug, you can have complete cures of a large percentage of animals and memory that lasts for two and a half to three years, both in the peripheral tissue and interestingly enough, in the thymus. The reason for my comment is that in the case of your platinum experiment, for instance, you may not need to cause tumor necrosis because platinum also has some immunomodulating effects. So, is that due a priming because of cells killed by low doses of drug or is it due to a mechanism related, perhaps, to the expression of your CTLA-4 at the transcriptional level? And I wonder whether you had the chance, or whether you would be enthused to try adriamycin in your system because we really do not know yet exactly the molecular mechanism but we are interested in the issue of the CTLA-4 versus B7 balance. An unrelated question that I would have is, in the case of your antibody that allows stimulation of T-cells proliferation, do you have any evidence that that treatment would facilitate the development of leukemias?

Allison: No, we have not. How would you do that? Look in AKR mice or some other strain prone to leukemia and treat them for a long period of time? Anyway we do not have any data, but it probably is something that we should look at. We have been concentrating mostly on more short-term aspects of auto-immunity than lymphoma.

Rauscher: Your anti-CTLA-4 administration does not at all mimic the knockout, but is that because you are treating adults? If you treat neonates with the anti-CTLA-4, do you make a large lymph node?

Allison: We have not treated neonates yet. But I think that that really is the key to the difference and that experiment is being done soon, I hope. Because the main difference in the knockouts is that when the cells are first maturing there is no CTLA-4 around to modulate responses; whereas, in normal mice treated with anti CTLA-4, we are dealing with T cells that have already gone through that period, become adjusted to the presence of CTLA-4.

Rauscher: What happens with the truncated CTLA-4 molecule, so it has no extra-cellular domain? Is it a constitutive signaler?

Allison: I do not know. Nobody has done that experiment. We have made truncated the cytoplasmic domain and it does not work anymore.

Rauscher: Will a full length CTLA-4 molecule engage this negative regulatory program in a naive cell, like a fibroblast?

Allison: I do not know whether it would function in a fibroblast that obviously does not express the T cell antigen receptor. However, we have shown that, at least with the antibodies, the program can work on a naive T cell and just shut off progression to activation.

Bankert: Have you, or are you, going to address the question of whether environmental antigens are driving the T cell proliferation in your knockout mice? In other words, if you grow them in gnotobiotic conditions, are they not proliferating as greatly?

Allison: We have not done that yet. Other labs may be a little bit further along. What we have been trying to do is to cross the knockout mice with TCR transgenic mice so that you have, presumably a clonal population of T cells with defined specificity. You have to do it in a rag background just to look at what happens in the absence of a normal antigen. I can tell you that you can completely prevent the lymphoadenopathy by treating the mice with soluble CTLA-4 fusion protein. But again, that is CD28 dependent: if you cover it up, it does not happen.

Bankert: I have one more question. In most of your experiments where you are using anti-CTLA-4, particularly in the EAE or the auto-immune disease you induced, you see exacerbation, but then you see a tailing off. Do you think there is another damping effect?

Allison: The fas/fas ligand pathway still works, so I think that what you are seeing in a down response is just a normal thing that FAS is doing to the bulk of the T cells.

Hanahan: Were all these tumors that you tested done subcutaneously? I am wondering whether you tried other sites, like under the kidney capsule or IP?

Allison: We are beginning to look at metastatic carcinoma, but we have not yet done much there. We are in the second phase of work now, we are beginning to look at metastases as well as primary tumors in several transgenic models as well.

Hanahan: Metastasis will also be found.

Allison: Right. We had hoped to do the metastases right away but this particular sub line of that BLM-10 that we have does not metastasize anymore.

Melief: Have you tried mice of different backgrounds? Including those that have a natural tendency towards auto-immunity, like the MLR, LPR or NZB/W mice? Is auto-immunity accelerated by this treatment? That is question number one. The second question is, can you think of any strategy by which you would inactivate CTLA-4 only locally at the site of vaccination and get rid of the systemic effects, would this make it radically much safer?

Allison: In my lab we have done a lot of work in normal strains of mice and not really seen any evidence of activation. We have started a collaboration with Brian Kotzin in Denver to look in the autoimmune-prone strains of mice such as NZB and NZW. I do not know what the results are, although he did tell me he has not seen anything really terrifically overt happening yet in antibody treated mice. There are subtle effects but I do not know what those are, this was just a few weeks ago. We are starting some experiments to see whether we accelerate diabetes in that model. As far as local administration, I think that would be something very advantageous. Although I think the action, again, is in the lymph nodes so you are going to get it, at least, not just where the tumor is; you would want to get it into draining lymph nodes.

Croce: Jim, can you have a mutation, a deletion of CTLA-4 in acute T-cell leukemia in human?

Allison: I do not know of any example of that. There are, though, two recent papers that have reported RLFP's and in one case a marker that is in the leader sequence of CTLA-4 associated respectively with susceptibility of Grave's Disease and in the other case susceptibility to diabetes.

Croce: Where is the gene in human?

Allison: Chromosome 1.

Croce: Which band?

Allison: I do not know. I have got it. I can tell you sometime.

Zanker: Your CTL-4 mice obviously break tolerance, do you have any idea what kind of thymocytes are released into the periphery?

Allison: That is an interesting thing and I have the slides that I did not put in for time. One of the things that was noticed in initial reports of the mice was that if you look at them after a couple of weeks, what you see are a lot of single positives and diminished numbers of double positives. We looked at that very closely, and there is also an increase in double negatives, but one of the first things we noticed were the double negatives actually had B220 and IgM on their surface. So what is happening is that the apparent perturbation in the normal phenotypic display of CD-4 and CD-8 is actually a result of the growth of the parathymic lymph nodes into the thymus; if you dissect those away very carefully and look at the residual thymus there is no detectable difference in ratios of CD-4 to CD-8 or in the V beta repertoire or any activation marker you want to look at. If you pulse the cells with BUDR and look at proliferation there is no evidence of abnormal proliferation of any of the thymic subsets, until the very latest stages of the disease when the mice are about to die anyway. We have not yet done the sort of detailed analysis that we need to do, that is, to breed a number of different TCR transgenes to the CTLA-4 knockouts and ask questions about positive and negative selection. But my own bias, based on that phenotypic analysis, is that CTLA-4 does not have anything to do with selection at all. All it does is to regulate the threshold of signals a T cell needs for activation. That is a property that is acquired post selection, associated with the final maturation of a T-cell. The syndrome is initiated really in the periphery. Now that kind of begs the question as to whether it is auto-immune or not, it becomes semantic after a time, but I do not think that what is happening is that there are cells escaping negative selection. Again, we need to do a little bit more work. But I think rather that the threshold for activation is just generally lowered in these cells.

Parmiani: Since the tumor antigen expression appears to be a crucial factor in your system, have you done any direct comparison using, in the same tumor transfected with a specific tumor antigen to see whether you can really dissect this effect?

Allison: No, we are doing that now. In fact, we are doing a couple of things. One is to use a model marker along the lines of what Pam described. The other thing we are doing that I think is going to be interesting too, is taking advantage of EL4, which is a lousy tumor to work with. EL4 is so laden with antigens that it is practically xenogeneic in mice and it is normally not a good model for tumor immunotherapy. But in another way, it is a great one because it is loaded with different recombinant retroviruses and there are a lot of different potential antigens. Although when you immunize with EL4 alone the response is directed to just one peptide. One of the things that we are trying to address now is, if we immunize with and without anti CTLA-4 will we get responses to more peptides? Will we get a stronger height of the response? But will we also get a wider breadth of the response? Because if we are correct about this threshold argument, what we may do is to reveal subdominant epitopes and we may get a more vigorous response by getting T cells to see things that they would not see otherwise. But it is a very important experiment. I should stress we really have not done very much mechanistic work in this system yet. Again, I am perplexed by the result with the MHC negative tumor, so there may be a lot going on that we are not aware of at this point.

Ohashi: I have some comments to some of the questions that are extensions of the discussion. In respect to the TCR transgenic mice in the CTLA-4 knockout background, we have done some of the studies in collaboration with Tak Mack so we can tell you that the HY TCR transgenic and the CTLA-4 knockout background has normal positive and negative selection. And the same with the LCMV specific T cell specific receptor transgenes. So I agree with Jim, I think the thymus is completely normal in terms of its positive selection and deletion. And also getting back to your question, what are the antigens that are triggering this vast expansion of the T cells in the absence of CTLA-4? Paul Waterhouse has done the experiment also, putting almost all T cell transgenics in a rag knockout background together with the CTLA-4 knockout. Huge breeding experiments, and they find the mice are fine up to six months.

GENERAL DISCUSSION

Boon: I am eager to know what ideas the community of tumor geneticists has regarding clinical trials utilizing their findings. Is there something going on, or is there something being prepared?

White: Just to initiate the discussion, I indicated in my presentation two days ago that there are ongoing clinical trials for polyp therapy with a number of different non-steroidal anti-inflammatories. So that pathway which has been identified genetically has yielded a target which has proven effective both in mouse and in human. That target has been verified in Japan through knockouts of the target protein and that is as effective as the use of the non-steroidals in a mouse model system.

Hanahan: Just because it comes from the perspective of animal models, I think one of the issues there is to define, in terms of the biology of tumors, what are the key events in tumor development. We have heard a lot about genomic instability: Well, what are the key parameters that represent key capabilities that a tumor has to acquire, ranging from the ability to escape the immune system, to the ability to resist apoptosis, to the ability to induce angiogenesis? I think the philosophy that I see is, for instance, taking the observation that angiogenesis is turned on during tumor development, is to play with angiogenesis inhibitors and to ask not only can you cure established tumors, but in the longer run could you prevent developing tumors. And, as for our capabilities to detect early different kinds of tumors in their development, one can imagine going on with strategies like this. For example, an interesting contrast with Pamela's story, that although in our model we get this activated immune response against the oncogene that is causing the tumors, we do not get rejection. We are now trying to use combinations, either classic chemotherapeutic agents plus agents activating the immune system, or anti-angiogenesis therapy along with immunotherapies. The question is how can you play, particularly in animal models, as a relevant precedent to trying to look forward and to design how you would do this in an actual clinical setting. But I think the issue is, at least from my perspective, if we could really define more precisely the key mechanistic features in the biology of tumors that could become valid parameters to consider. So, basically, above and beyond cell proliferation for classical chemotherapeutic agents, for immunology two key parameters would be apoptosis and angiogenesis. Those are going to be valid targets and I envision them being combined with classical chemotherapeutic approaches targeting the cell cycle along with immunotherapies.

Boon: And for apoptosis? What do you envision that could be used for?

Hanahan: I think that, as John Reed was also alluding, as we understand more and more about the regulation of apoptosis it is going to be possible to design drugs that activate the apoptotic pathway. Because every tumor retains, I believe, the capability to undergo apoptosis, whether or not it has developed p53 mutations, and if we can understand the fundamental regulation of apoptosis, it is going to be possible to design drugs that will induce apoptosis in tumor cells. So I think that with angiogenesis there are already compounds available and clinical trials and that is a bit further along. I envision that apoptosis is going to become a fertile area for drug development and that none of these may stand alone but that if you can attack the tumor via the immune system, by inhibitors of cell proliferation, by anti-angiogenesis, by apoptosis inducers, maybe collectively one will really produce cancer therapies.

Boon: I certainly agree with that concept but there is not yet any drug being seen for apoptosis induction or are there already some proposals?

Livingston: Oh yes. At the present time I know of several major pharmaceutical companies that have initiated multimillion dollar projects searching for small molecule leads that effect one or another step in the apoptotic process. This is a growth industry at the present time.

Boon: And they have some of these compounds already tested in mice?

Livingston: Yes. In this respect, I think Doug Hanahan also makes a really important point. We are in the middle of explosive growth, I would say explosive growth, in the identification of cancer therapy molecular targets. There are going to be many targets to test. Some will likely prove interesting and valid. A really interesting question is: How to choose the optimal targets? Answering this question accurately is going to be a significant problem largely because of the immense cost associated with drug development. In this regard, there is an enormous need to reduce the number of false negatives that come through pre-clinical testing and fail at the development level where the costs are so enormous. I think there will be no shortage of brilliant ideas. But, with the existing methods of candidate drug evaluation, there is a shortage of funds.

Mihich: I agree with David, there are a lot of agents being developed in that direction. But the proof of principle could be obtained also with agents that are available already, at least in experimental systems because several of the anti-cancer drugs, anti-proliferative agents that we have end up inducing apoptosis and so the principle could be verified with available agents. There is another type of compound and that is TNF. TNF is a very potent inducer of apoptosis and, in fact, it is also a mediator of apoptosis for the reaction of several agents. The problem with TNF is the non-apoptotic related toxicity which prevents clinical application. However, there are now some peptides which are derived from some segments of the TNF molecule which are capable of inducing apoptosis and we do not know yet whether those have the untoward toxicity of TNF, but that would be another way to go—to take fragments of a cytokine like TNF which is inducing apoptosis.

Gray: I wanted to come back to the point that David was making a minute ago and try to bring in this concept of the arrays that surfaced earlier in the meeting. One of the things that makes these clinical trials so expensive right now is that, in a sense, you have to treat all cancers as a homogeneous center entity and I think that the arrays or expression pattern or whatever, offer the possibility of fingerprinting these things genetically so that you could identify subset of tumors that you would expect to respond to your particular therapeutic agent and then test only those and I think this would lead us to much smaller clinical trials than we now have.

Livingston: I think that is extremely astute. Once again, in his own modest way, I think what Joe Gray is pointing out is that, in the future, genetic fingerprinting of all malignant tumors will be necessary and there will have to be smart doctors to understand the fingerprints, smart enough to be able to translate a fingerprint into non-action or, ideally, appropriate therapeutic action. A simple case in point; there will be almost 350,000 cases of prostrate cancer in the United States in the next twelve months, the vast majority of them diagnosed on the basis of an abnormal blood test, the PSA, followed by an abnormal prostatic biopsy. Now, comes the question: whom do you treat? The death rate from prostate cancer has fallen in the past five years, not risen, but the apparent incidence has increased three-fold, all because of increased detection. Given the relatively low death rate, only a small fraction needs to be treated. Right now, it is difficult to discern who many of these men are. I am hoping that, in the future, tumor genotype/clinical phenotype outcome analysis will be part of the solution to the problem of whom to treat.

Visentin: To continue this observation, I think it is of fundamental importance to emphasize that risk-adjusted treatment has been, so far, essentially based on patients' characteristics, while there is still an obvious difficulty in adapting therapeutic strategies to the very features of the heterogeneous tumors themselves. This, however, should be taken into consideration; in other words, I believe that a major challenge we must meet is mapping individual neoplasms in order to properly select therapeutic options. A first step forward might possible be recognizing the varying determinants of chemo- and/or hormone resistance. This could assist distinction of tumors on the basis of, for instance, variable intensities or mechanisms of drug resistance, in order to try different ways of reverting, circumventing, or anyway modifying resistance. For instance, despite the recognized fundamental role of apoptosis in sensitivity and resistance, nobody knows as yet whether clinical resistance due to apoptotic blockade (e.g., from *bcl-2* overexpression, p53 mutations/inactivation, etc.) may possibly be reverted by intensified treatment, like it is more or less assumed for conventional mechanisms of resistance mediated by transport proteins like Pgp-170 or MRP. For this purpose, only preliminary indications are available, but the point should be further clarified. As a matter of fact, the role of intensified regimens cannot be assessed over 50,000 patients, without discriminating between the actual risk categories, including of course, resistance mechanisms as well as other variables. Such a categorization is required not only to save money, but to confer significance on clinical trials.

Hanahan: Yes but, not to belabor the point, I think many of us are also hopeful that we have on the horizon the ability to really develop increasingly accurate animal models of human cancers which rather than looking in the environment of established tumors transplanted with tumor cell lines would be based on looking in the proper micro-environment of tumors that may indeed be able to mimic different aspects of the heterogeneity that one sees in the human population; this will allow a more rapid discrimination because, as David pointed out, you cannot do everything in clinical trials, it is economically impossible; but the question is, can we discriminate more accurately using better animal models? I think that real opportunities will be offered by a knowledge of mechanism. For example, I think that CTLA-4 is another target to develop better treatments than the historical immunotherapies. I mean, now we have a molecular target which can be refined in terms of activating an immune system in a specific way and I think this is yet another example of the progress of cancer research; we can develop new treatments more from a position of strength and I think that if we build more models and develop more knowledge about mechanism, then really we may have exciting opportunities for therapeutic exploitation.

Livingston: Is there now a process that will keep track of the results obtained with the myriad of experimental cancer vaccines now being tested? If so, will this process be able to provide unifying insights that will simplify the relevant clinical research?

Draetta: In line with what Dr. Livingston was just saying, I would like to emphasize that given the many genetic alterations and, therefore, the many potential drug targets likely to be identified, there will be quite a burden on pharmaceutical companies to prioritize their efforts in drug discovery. First, there will be the issue of target validation, i.e., of all those preclinical studies needed to demonstrate the biological effects generated by altering a given pathway. Second, the need of coordinated efforts to determine the frequency of a given genetic alteration will emerge. I believe it should be the responsibility of International Cancer Agencies to coordinate these efforts. Given the "fragmented" cancer market of the future, it is unlikely that any pharmaceutical company will take over this work on its own.

Boon: I would be delighted to answer your questions as best I can for immunotherapy but, since this is a switch from genetics to immunotherapy, I cannot resist asking a last question about the use of tumor genetics and then I will try to give you the best answer I can. Regarding genetics, I was impressed by what Ray White proposed because that seems to be a fairly innocuous therapy. Regarding the other ideas based on oncogenes and things of the kind, especially based on your remark that some classical anti-cancer agent, in a way, act at that level, I would like to ask, is it your gut feeling, and you already have some mouse experiment to believe, that it will be possible to find inhibitors of oncogenes or activators of tumor suppressor genes that will not be too toxic? It is a very important question, and one that is extremely unclear to me. I would like to know your gut feeling about that.

Parmiani: I think this is a major issue; I mean we are having now a lot of molecular targets and more will come in the next few weeks or months, but still we have a very weak situation concerning the way we have to use the different agents to target such molecules. This remains the area of major concern in any therapeutic approach, that is the reason why I like more the immunological approach because selectivity and targeting are exactly the features of the immune system. Therefore, I am concerned on how to reach the targets once you have drugs or reagents that can really see only the molecules that are already available today.

Mihich: I think that we may and I say MAY, have an answer in this regard to your question in a couple of years, IF there are not sufficient alternatives to the telomerases. There we do seem to have relative uniqueness for tumor cell as compared to normal tissues, except for the germinal tissues. However, this is why I was very interested in the second presentation of Dr. DePinho about the telomer lengths essentiality. But, if there are alternatives to telomerase in assuring the so-called immortality, then we are in trouble. But, if there is a sufficient number of cells where the lengths of the telomers is required for immortality, and if it is true that, in addition to tumors, this is so only in some normal hematopoietic and germ cells, then we may get closer to an answer to your question, in terms of antitumor specificity of treatments. As far as the inhibition of transcription of several genes that are involved in several aspects or regulation of the cancer cell, there is a big question there, because we do not know whether we are just getting to a more elegant and sophisticated way to get additional cytotoxic agents which may, or may not, be very specific and that remains to be seen. We are thriving towards specificity, but maybe the telomerase will give an early answer.

DePinho: A comment on the telomerase issue since this topic has emerged several times in the course of sideline discussions regarding telomerase inhibitors and the potential for independent mechanisms, telomerase independent mechanisms that would serve to rescue tumor cells from critical shortening and telomerase deficiency. If we can extrapolate from the lessons learned in yeast, elimination of the essential RNA priming component ultimately leads to genomic instability and crisis. Rare survivor cells emerge in which telomeres are reconstituted through a recombination mechanism. A key point to keep in mind is that the vast majority of these telomerase-deficient recombination-competent cells, 99.9% do, in fact, die. Thus, telomere maintenance by telomerase is a very essential process that serves to enhance the viability of the cells. Given such circumstances, telomerase-inhibitors may have the potential to eliminate many cells and, in combination with other therapeutic modalities (angiogenesis inhibitors, immunotherapy, radiation and surgery), would provide for a powerful armentarium against tumor growth.

Hanahan: To follow up on that, I think this is another obvious area for consideration because if telomerase is important, then any therapy that knocks down the tumors that requires further expansion and additional proliferation is going to drive you toward that telomere length limit and whether there is increased apoptosis or immunotherapies to balance the proliferation. So, again, you can really see the framework for new rational therapies which, right now, is a rarity, right? But it seems to me that you can see mechanistically how these things can start playing into each other.

DePinho: I think one of the other issues that may need to be considered more fully is the general physiology of the tumor and how its special properties, in terms of blood flow, lymphatics, and endothelial biology impact upon the delivery of drugs to this bulky structure. I was impressed with the sequestration of these tumor cells away from the immune response and whether or not we really understand enough about the tumor "organ" properties and whether these tumor-specific features may not allow the proper delivery of drugs to all cells in the tumor.

Hanahan: Rakesh Jain is going to talk directly to that issue tomorrow.

Visentin: The question is also that resistance may well imply concurrent blockade of several potential antitumor mechanisms. It depends, among other things, on tumor topography and geometry. Of course, what holds true for hematological neoplasms cannot be directly applied to solid tumors and the tumor bulk is a fundamental variable, like primary versus metastatic disease—all of these being known components of tumor heterogeneity. However, I stick to the old idea that even if, theoretically, you could kill 99% of tumor cells, the problem is whether the remaining cells will represent an actual minimal residual disease (MRD), cloning and reproducing the tumor, and hence a mechanism of regrowth resistance and/or systemic recurrence. A very conservative but not unreasonable statement may be that combined cell kill approach is more or less mandatory in order to control at least the MRD. Indeed, MRD eradication remains to be documented even with high-dose chemotherapy, that however can kill most of cancer cells, as was approximately quantitated, not in solid tumors, but in aggressive hematological malignancies. Thus, the problem is how to eradicate MRD--a classical example are chronic, low-grade, indolent hematological malignancies, where it was speculated that biotherapy might control MRD via cell-differentiating and apoptogenic, rather than cytocidal/cytotoxic mechanisms. This could possibly be extrapolated to other settings, including antimetastatic treatments (essentially for adjuvant control of micrometastases) in solid neoplasms.

Anderson: The point I was trying to make, when David Livingston was talking about prostate cancer, is that I think that genetics has a useful role in diagnostics as well as therapy. David pointed out how with prostate cancer you are able to detect large numbers of cases of early stage disease, but you have this problem of identifying who is going to progress and who actually needs treatment. The same issue exists in early stage breast cancer. The notion of whole genome fingerprinting on what genes are involved in cancer is attractive but very cumbersome. A much simpler approach that I thought was really beautiful was work which Manuel Perucho put on the Internet last December, on the Symposium on Genomic Aberrations in Cancer Detected by DNA Fingerprinting, where he is using a technique analogous to the inter-SSR PCR technique we are using. Perucho has utilized arbitrarily primed PCR to measure overall genomic damage in a retrospective study of very early stage breast cancer. Those very early stage tumors which had a lot of genomic damage, indicative of genomic instability, were much more likely to progress rapidly. In other words, simply by looking at how scrambled the genome is, in a very early stage tumor, you get an excellent indicator of the prognosis. Some of these simpler approaches are available here and now, and it would be nice to see them in use.

Mc Namee: There may be a new target which is the drug regulatory authorities. Would anyone comment on how such authorities are going to cope with what may be the basis of these new scientific therapies?

Hanahan: Just to follow up on Jim's talk, do you see CTLA-4 blockade then as a basic event for retreating tumors and how do you see the path forward here, in terms of making a case for that? Do you think that this is worth considering?

Allison: Well, there are antibodies to human CTLA-4, but they have not been characterized completely. Unfortunately, this is an area where the patent situation has led to a proprietal aspect of some of the antibodies slowing progress toward the clinic. What we are doing is making antibodies to human CTLA-4 and characterizing them with respect to *in vitro* functional effects. As soon as those are available with some sponsorship from a pharmaceutical company but largely with the help of the NCI antibody facility at Frederick, we will try to get enough antibody made to then distribute to investigators and have ongoing trials that have built into them some objective measure of enhancement of T cell responses and determine the effect of CTLA-4 blockade in those patients. So, that is our plan, to get out that way. Meanwhile, there are some companies that are working on small molecules, but that has been one of the debates actually worth talking about too, is that some of the companies that are trying to move on CTLA-4 have got a very conservative approach and are not going to do anything until they get their small molecule made. But, I think if we can get an antibody that has at least the affinity of the antibody that is successful in mice, then I think a real proof of principle would come from a small scale trial along the lines described and then we can at least decide whether to reject it or to keep it going.

Mihich: One dimension that we may keep in mind for further study which was mentioned before during this discussion, is the problem of residual cells and presumably residual resistant cells. Way back, we reported that some resistant tumor cell lines are markedly more immunogenic than the corresponding parent line and, later, Bonmassar in Italy expanded this using DCTC as the agent to elicit this phenomenon. But that is something to consider also, whether one could elicit a response either by virtue of an innate great immunogenicity or by induced increased responses against resistant cells.

Boon: To answer your questions regarding defined immunotherapy and where we are going, I will tell you about our experience with the defined antigens. We have immunized 23 melanoma tumor-bearing patients with the MAGE-3 peptide presented by HLA-A1. Six tumor regressions have been observed, some of these have been complete and long-lasting. Paradoxically, we have failed to detect in the blood of these patients any increase in the frequency of anti-MAGE CTL precursors.

Berns: You refer to our trial; in the meanwhile, 28 patients have been treated by autologous vaccines expressing GMCSF, and I think it is difficult to draw strong conclusions at this moment except that there are some anecdotal data. You see responses which might indicate that something is happening. What it indicates is that you want to look further into what is actually happening to those patients, because, again, there is not a significant effect on CTL precursor frequency. Clearly, this is not a parameter which is significantly altered and related to what is seen in terms of responses. So I think what one learns from these trials is that if something does not happen, if there is not immediately a significant result, that one should not give up. But at least try to use the information which is obtained to see if it can be a starting point for next set of trials. For example, if you see the effects with GMCSF, and you see the effects of CTLA-4 then it would make a lot of sense to test what that combination would do. And so you would not lose all the expertise and information you gathered in GMCSF trial, but would amplify the effects rather than having all independent attempts which you cannot compare. I could imagine that it makes much more sense if immunologists who are doing peptide vaccinations would agree on using one or a few peptides and try to figure out how that works, rather than using 20 different ones and nobody knows finally what it means. So I think that it is a matter of organization, that you are willing to give up your favorite peptide rather than trying to hang on to it, because you might become famous with your peptide. You really want to establish mechanisms and I think you can only do that by reaching consensus and then limit the number of things you are going to try out.

Boon: You know there may be about fifty concurrent studies with different peptides at the present time. May I make one more comment that relates to your presentation, Dr. Ohashi. I was most interested to learn that adding B7 to your transgenics did not lead to rejection of the tumor cells. I would be most interested to know if adding an additional one, two, three, four antigens would solve the problem, because that, to us, would be an enormous indication for what we have to give priority to in terms of peptide immunization. We have a double peptide immunization on the way, but maybe what we need is triple or quadruple peptide because maybe the immune system has some co-operativity to it. So I would be thrilled to know the result in the context of your experiment.

Ohashi: That is exactly what we would like to do next, as well. We are in the process of making vaccinia virus expressing different peptides that we know are expressed on beta islet cells, for example, so we are going to go to self-peptides completely and see if together, or with three, you can get much more efficient responses. But I also have a question for you, when you say you looked in the blood for CTL's, when was that?

Boon: We tried about two weeks after the third immunization.

Ohashi: Because a lot of studies by several groups have shown that the CTL's, of course, like to home where the antigen is. And you either have to check the local draining lymph nodes or very close to the site of the tumor.

Boon: But that is not easy to do.

Ohashi: I can imagine.

Melief: I think in dealing with peptides we are facing a number of problems. One is the delivery of peptides. I will show the data this afternoon, but what we have seen is that peptides are such powerful biologicals that, depending on how you deliver them, and this is highly individual peptide specific, you either induce immunity or you can even cause tolerance and may make matters worse for the cancer patient. That is problem number one. The second problem is that although proof of principle with peptides would be fantastic to achieve, as these results suggested, if you can correlate it with specific CTL responses, the problem we experience in the clinic is that if you only treat HLA-A2 positive patients, then you have to say to the other patients, "Sorry, but you have the wrong HLA type, we cannot treat you". And that is where I think the protein comes in as a much better option if we learn how to deliver the protein properly. In fact, in our HPV tumor model, and, again, I will show you the data this afternoon, protein works as well to induce protective CTL responses as peptides do and, in addition, protein has the great advantage of offering all potential helper epitopes in addition to all potential CTL epitopes for each tissue type. I think in tumor immunology the issue of specific help has really been underestimated, as I will also show this afternoon in a Rauscher virus tumor model where really by far the best immunization was achieved by a combination of the specific helper peptide and a specific class I peptide and get ninety percent protection where each peptide alone could only achieve thirty percent.

Moretta: Of course, all these peptides are designed because they bind very well to class I molecules and CTL's. But my question was really whether it is not useful to concentrate also on combining helper peptides. I mean peptides seen by helper cells not only by cytolytic cells. In relation to this, I would also suggest perhaps a cell frequency analysis be done not only for CTLs but also for cells which really provide help in order to see whether efficient help is also generated in a given patient.

Boon: To close the discussion, the cytokine brings us to another comment that was made. Clearly now we would like to combine the efficacy of recombinant virus cytokines, peptides, proteins, whatever, and then once you want to make such a combination and do a trial with it, you have to negotiate with three companies and I can tell you, already negotiating with one, is quite something, but to negotiate with three and trying to induce them to collaborate together, you can spend your life doing that.

18

CYTOKINES AND TUMOR IMMUNOGENICITY

Toward an Appropriate Cancer Vaccine

Federica Cavallo, Katia Boggio, Mirella Giovarelli, and Guido Forni

Dipartimento di Scienze Cliniche e Biologiche
University of Turin
10043 Orbassano, Italy

CANCER VACCINES INTO THE BREACH

Molecular biology and genetics are currently providing a definition of tumor-associated antigens (TAA). This important issue enables the question of immune recognition of tumors be stated in defined terms. Immunological investigation of T lymphocyte receptor, costimulatory molecules, signal transduction and cytokines has progressively led to a much more exact description of the requirements for the induction of an immune response. Refinement of cell genetic engineering is making it almost daily easier to use molecular and genetic information to construct new cancer vaccines. The convergence of these issues is once again placing tumor immunology at the cutting edge of biological research[1,2].

A recent survey by Science[3] indicates that most scientists believe antitumor vaccination will be an established therapeutical option in the near future. When compared with conventional cancer management, vaccination is a *"soft"*, noninvasive treatment free from particular distress and iatrogenic side-effects. Antitumor vaccines can be expected to have a considerable social impact, since they will significantly improve the quality of life of cancer patients, while there are many with minimal residual disease after surgical and antiblastic drug management whose life expectancy could be extended in this way.

Tumor vaccines, however, go hand in hand with a peculiar immunological situation. By contrast with microbial vaccines where individuals are immunized prior to encountering the pathogenic microorganism, cancer patients have to be immunized when they already bear a tumor, since it is not yet possible to predict which gene mutations will give rise to cancer. The common clinical setting, therefore, is elicitation of a systemic immune response in a patient already bearing a tumor, rather than prior to tumor development[4].

The very concept of vaccine is distorted since it has moved from being preventive to therapeutic. *"Therapeutic vaccination"* has not had much success in the handling of infectious diseases. Its use against the progression of an established tumor, minimal residual

The Biology of Tumors, edited by Mihich and Croce
Plenum Press, New York, 1998.

disease or incipient metastases is very challenging, since it must secure an effective immune response capable of getting the better of a well-established, proliferating tumor actively interacting with its microenvironment.

The path is also being opened towards antitumor preventive vaccination. Genetic studies are leading to the identification of gene mutations that predispose to cancer and hence are holding out the possibility of identifying not-yet patients or *"unpatients"* with a defined genetic prognosis[5]. This probing of the human genome is raising new ethical, psychological and cultural issues at the same time as it is pushing back the medical horizon. The identification of the gene at risk and its mutated or amplified products is also providing a heaven-sent opportunity to vaccinate susceptible patients before the development of cancer and thus avoid most of the difficulties in breaking the peripheral tolerance that arises after prolonged TAA expression[6]. Determination of the altered gene products that are predictability destined to become TAA will enable the proteins or peptides to be used for vaccination to be selected. This situation prospects the use of specific cancer vaccines in an unprecedented setting, where they may prove as strikingly effective as those employed against infectious diseases. Vaccines and specific immunity will eventually be used in what medicine has shown to be the right context to prevent the onset or inhibit the initial growth of tumors in *"unpatients"* with a high risk of cancer. Pasteur, one feels, would be overjoyed.

STRATEGIES FOR BUILDING ANTITUMOR VACCINES

The ability to elude an immune response, proceed without eliciting any response at all or even exploit a response to achieve better growth is the hallmark of transformed cells that become tumorigenic. The evasive immunogenicity of tumor cells stems from many features among which the limping processing and presentation of TAA peptide fragments by the few glycoproteins of the major histocompatibility complex (MHC) expressed on their membrane, the absence of adhesion molecules and costimulatory signals appears to be the most important. Moreover tumors hinder the recruitment of host antigen presenting cells (APC), thus making TAA indirect presentation inefficient. Tumor cells themselves modulate the immune response by expressing on their membrane the glycoprotein of Fas ligand family that may induce apoptosis of activated lymphocytes[7]. They release characteristic repertoires of cytokines and other soluble factors by which they recruit and suppress reactive leukocytes, deviate the kind of immune response, and modulate the activity of endothelial and stromal cells. A tumor can be characterized in terms of the cytokines and factors it produces and the cytokine receptors it expresses (reviewed in[8]). The result is that any antigenic signal tumor cells may present to the immune system is both ignored by natural immune mechanisms and insufficient to bring about those of specific and adaptative immunity[9].

Many ways to build vaccines able to overcome tumor-borne suppression and anergy are being explored. The strategies employed range from distinct technological and intellectual approaches to the enhancement of TAA immunogenicity, and are targeted to influence individual checkpoints set up during the establishment of an immune response. The most rational way to make tumors more immunogenic is gene engineering. This flexible technology allows the selective insertion of new genes into the genome of a tumor cell, thus forcing it to express the molecules on which its immune recognition will depend. Which gene should be transferred and with what rationale, and what kind of strategies should be designed for enhancing tumor immunogenicity are current challenging questions.

TUMORS CELLS TRANSFORMED INTO FRANKENSTEINIAN APC

Enhancement of tumor immunogenicity has been sought by coupling foreign antigenic determinants to the tumor cell membrane. Elicitation of a delayed-type hypersensitivity to the foreign determinant is associated with better T lymphocyte recognition of antigen-bearing tumor cells[10,11]. Today, however, a fuller understanding of the mechanisms of immune recognition is being combined with genetic engineering to transform tumor cells themselves into Frankensteinian cells that directly present TAA to T lymphocytes[9].

Whether or not a tumor cell directly leads to a T lymphocyte response depends on serial triggering of its receptors (TCR) by TAA peptides associated with the glycoproteins of the major histocompatibility complex (MHC) on the tumor cell membrane. A series of elegant works by the Lanzavecchia group has shown that irrespective of the nature of the triggering ligand, T lymphocytes appear to count the number of triggered TCR, and respond when a threshold of approximately 8000 events is reached[12–14]. The signals delivered by the simultaneous interaction between costimulatory molecules and their receptors reduce this number to 1500[12]. The efficiency of TCR scanning, however, depends on the stability of the interactions between the T lymphocyte and the tumor cell. Scarcity or absence of MHC glycoproteins or adhesion molecules on the tumor cell decreases the duration and extent of TCR triggering, and may extinguish the signaling and prevent T lymphocyte activation. Once activated, effector T lymphocytes recognise their target and exert their function in the absence of a costimulus[13].

Tumors Engineered with Costimulatory and Adhesion Molecules

B7-1 and B7-2 are prototypes of costimulatory molecules. Besides decreasing the T lymphocyte's response threshold, their interaction with its CD28 counter-receptors both increases IL-2 production and prevents the cell itself from becoming anergic by allowing it to progress through the cycle and differentiate into CTL[15,16].

This need for multiple membrane signals is one reason why tumors that do not display enough MHC glycoproteins to present TAA peptides lead to marginal T lymphocyte activation. The introduction of costimulatory molecules enables the direct presentation of TAA to specific T lymphocytes by tumor cells. Indeed, many tumor cells transfected with genes encoding B7-1 costimulatory molecules become effective cellular vaccines against their wild-type tumor. The improved immunity conferred by B7-1 expression is dependent on newly induced tumor-specific $CD8^+$ and/or $CD4^+$ T lymphocytes, but can only be achieved with tumors that naturally express a suitable number of TAA on their membrane (intrinsically immunogenic tumors), or those rendered immunogenic by transfection with viral or human onco-related genes[9,17–19]. By contrast, B7-1 expression is not enough to allow the immune recognition of nonimmunogenic tumors[20–21]. It is still debated whether the costimulatory effects of B7-2, too, are confined to immunogenic tumors. Many comparative data obtained with such tumors show that it is as efficient as B7-1, whereas our findings suggest that poorly immunogenic tumors engineered to express B7-2 induce a better protective and curative immunity than those expressing B7-1[21] Although the inflammatory responses triggered by tumor cells engineered to express B7-1 or B7-2 appear macroscopically similar, they may induce a qualitatively or quantitatively different set of cytokines. It has been recently reported that B7-1 or B7-2 costimulation can drive the differentiation of $CD4^+$ T lymphocytes towards a polarized Th-1 or Th-2 phenotype, and the possibility that this is equally true of antitumor $CD8^+$ effector T-lymphocytes is an intriguing prospect. Most probably, both the density of costimulatory molecules and the expression of accessory molecules, such as ICAM-1, serve to determine the outcome of the costimulatory signal delivered by B7-1 and B7-2[9,22].

In effect, the poor or non-immunogenicity of a tumor may also depend on the lessening of adhesion molecules. We have found that the transduction of both B7-1 and ICAM-1 is required to elicit T lymphocyte recognition and establishment of an effective memory in a series of intrinsically immunogenic and other tumors[21].

The costimulation story, however, is more complex. In effect, while interactions between CD28 and members of the B7 family costimulate and enhance T lymphocyte response, interaction of B7 with CTLA-4 (a CD28 homolog) has the opposite effect[23,24]. Administration of antibodies blocking CTLA-4 results in both the rejection of otherwise lethal tumors, including those that are pre-established, and immunity to a secondary tumor challenge. Triggering of CTLA-4 appears to bring the response of a T lymphocyte to an end by restricting its IL-2-dependent transition from the G1 to the S phase or inducing its apoptosis[25]. Blockade of the inhibitory effects of CTLA-4 allows and potentiates an effective immune response.

In our studies, B7 and ICAM-1 co-expression on tumor cell surface of poorly immunogenic tumors positively correlates with memory T lymphocyte induction. However, modulation of the TCR activation threshold may be not the whole story. B7 transduction into tumor cells, in fact, surprisingly activates a potent inflammatory response. Both lymphocytes and polymorphonuclear leukocytes are rapidly recruited to the transduced tumor cell inoculation site. Depletion experiments have confirmed that $CD8^+$ T lymphocytes, NK cells and polymorphonuclear leukocytes activated by the B7-expressing vaccines are critically involved in tumor rejection[9,20–21]. If this means that professional APC are recruited at the tumor site, costimulation with B7 may also result in the indirect presentation of TAA to T lymphocytes. The involvement of host APC and TAA indirect presentation is, however, a diverse issue.

Tumors Engineered with MHC Glycoproteins

Many tumors display a defective MHC glycoprotein expression. Transduction of their cells with syngeneic MHC class I genes offers one way of improving TAA presentation (reviewed in detail[9]).

Moreover, Ostrand-Rosenberg has shown that mouse tumor cells transfected with syngeneic MHC class II genes are highly immunogenic in the autologous host, and induce a potent tumor-specific immunity against the wild-type tumor[26–28]. The immunity stimulated by simultaneous transduction of B7-1 plus MHC class II genes is stronger than that achieved with cells carrying either gene alone or mixtures of single-gene-transduced cells. It also lasts a very long time and involves both $CD4^+$ and $CD8^+$ T lymphocytes[28,29].

Allogeneic bone marrow chimeras have been used to determine whether $CD4^+$ activation rests on indirect presentation of TAA by host APC or their direct recognition on MHC class II engineered tumor cells. It would seem that these engineered tumor cells truly are able to directly present within one week of immunization[30].

Vaccination with tumor cells engineered to express allogeneic MHC membrane glycoproteins (allo-MHC) allows recognition of both TAA and allo-MHC on the same cell membrane (reviewed in[9]). However, effector T lymphocytes activated in this way have to react against wild-type tumors that may express different TAA associated with syngeneic-MHC. On most occasions, therefore, transduction of allo-MHC has just been used as a nonspecific stimulus to make allo-reactive T lymphocytes release cytokines that may favor the indirect presentation of TAA by attracting professional APCs and inducing the expansion of helper or cytolytic T-lymphocytes[31,32].

Direct enhancement of a tumor's ability to elicit a significant immune reaction through gene engineering is an attractive prospect, since the main membrane signals that transform its cells into an effective APC can be individually studied. How commonly direct TAA presentation takes place in vivo is, however, a debatable issue[17,20–21,33,34].

PROFESSIONAL APC DISPLAYING TAA

The specularly opposed way to prepare antitumor vaccines is to pulse normal or genetically engineered dendritic cells (DC) with TAA, or engineer DC to express TAA constitutively. Progressive identification of TAA expressed on mouse and human tumors is making it possible to select TAA peptides that best associate with the MHC grooves (reviewed in[35]). Transcription of the information gleaned from the molecular biology of the mutated gene is thus straightforward in immunologic terms. The crux of this elegant approach is discovery of the *"right"* peptide. Molecular identification of a TAA provides a defined target protein or peptide, but cannot show that it is the most important antigen. Various findings indicate that tumor cells can express more than one TAA[36]. Other as yet unknown molecules may be more convenient or less down-modulable targets for T lymphocyte activity, or more resistant to immune selection.

Many papers have shown the efficacy of this kind of vaccine. DC pulsed with synthetic tumor peptides[37] or nonidentified peptides[38] eluted from MHC class I molecules expressed by tumor cells, for example, have acted as a preventive vaccine and eradicated incipient mouse tumors; immunization with synthetic peptide-pulsed autologous APC has induced an antigen specific CTL response in human melanoma patients[39].

Why in vitro TAA-pulsed APC should be more effective than APC naturally exposed to TAA in vivo is not perfectly clear. Genetic modification of DC to make them more titillating for the immune system may be the way to overcome host tolerance to TAA.

Many variations of this basic approach are being pursued. For example, a non analytic but effective way to immunize is to use DC fused with tumor cells. In this way, multiple known and unknown TAA are presented[40].

The clinical application of peptide pulsed DC, the use of live as opposed to fixed APC, MHC restriction, single or multiple TAA are but a few of questions to be tackled[35]. The dramatic social and emotional problems raised by cancer and the incredible flexibility of technology, however, make these problems a secondary concern. Efficacy must be the only important issue. Comparative evaluation in multiple tumor settings of the immunogenic potential of professional APC vs Frankensteinian tumors will show which strategy should be pursued.

DNA VACCINES

Vaccination by inoculating DNA encoding for TAA is another fascinating endeavor that has already been rewarded with a variety of interesting results[41–43], though it would seem that TAA proteins elicit a more effective response[44].

Its intriguing advantage is the ease with which DNA could be modified to code for proteins with slightly varied sequences or additional immunogenic signals. The flexibility of this promising technique, indeed, has already been demonstrated by the increased response obtained when a DNA vaccine coding for a myeloma idiotype was supplemented with sequences coding for an IL-1β nonapeptide[44].

CYTOKINE BASED ANTITUMOR VACCINES

A Rationale

Activation of cytotoxic T lymphocytes (CTL) is commonly viewed as the key issue in antitumor immunity. However, the immune system's control mechanisms can be manipulated to induce more effective antitumor reactions in which CTL are only one of the players. We have shown that following TAA recognition a few T helper lymphocytes produce cytokines by which they recruit several reactive cell populations, initiate DTH reactions and assist the activation and expansion of tumor-specific CTL[11,45]. Could cytokines be used instead of lymphocytes ? Experiments in non-tumoral systems indicating that they offset defective antigen recognition[46] and overcome tolerance[47] have suggested their use to impair tolerance and activate effective and specific TAA immune recognition. A strong antitumor reaction was activated by repeated injections of a mouse tumor mass with very low doses of exogenous IL-1[48], IL-2[49], IL-4[50] and IFN-γ[51], and was followed in a few cases by specific, systemic and persistent immunity against nonimmunogenic tumors. A certain number of complete or partial clinical remissions have also been observed in humans[52]. Recombinant molecules, indeed, could be used instead of Th lymphocytes: local administration of low doses of cytokines can thus be viewed as a new way of looking at cytokine-mediated immunotherapy. Engineering of tumors with cytokine genes was the logical next step[53].

Debulking of TSA Cells Engineered to Release Cytokines

Tumor cells engineered with cytokine genes can be regarded as actively self-replicating minipumps that constitutively secrete increasing amounts of a particular cytokine. When this cytokine reaches a pharmacologic threshold in the microenvironment, it activates a local reaction that may affect the growth of the engineered cells themselves.

In our studies, the TSA-parental cell (TSA-pc) line established from the first in vivo transplant of an aggressive, metastatising and moderately differentiated mammary adenocarcinoma that arose spontaneously in a BALB/c (H-2^d) female mouse was mainly used. In common with most mouse and human breast carcinomas, TSA-pc constitutively produce significant amounts of granulocyte colony stimulating factor, granulocyte-macrophage colony stimulating factor and TGF-β[54].

Histologic and ultrastructural investigation of the events that follow the challenge of syngeneic mice with cytokine gene engineered TSA cells has shown that the features of the reaction and its efficacy are decided by the cytokine released. Similar reactions are elicited by locally injected cytokines. Those released from engineered cells, however, are more effective, probably because they are continuously secreted and their amount is governed by a feedback mechanism, increasing as the tumor expands, then decreasing in function of the intensity of the reaction they induce. Debulking of most engineered tumors depends on a host reaction mounted a few hours after the challenge[55,56]. Generally speaking, the greater the amount of cytokine, the quicker the rejection.

The repertoire of the leukocytes involved was unexpected. With TSA cells engineered to release IL-2 (TSA-IL2), killing appears to be mostly the work of neutrophils[57]. The reaction activated by TSA-IL4[58] results in efficient inhibition of tumor growth through the early intervention of eosinophils, that activated by TSA-IL7[55,56] is mainly mediated by lymphocytes, that activated by TSA-IL10 by macrophages, NK cells, neutrophils and T-lymphocytes[59], and those activated by TSA-TNFα[55,56] and TSA-IFNγ[60] by macrophages. Macrophages activated by TSA-IFNγ produce prolonged tumor dormancy, but takes are eventually observed in most mice. The reaction elicited by TSA-IFNα depends on $CD8^+$ lymphocytes and polymorphonuclear leukocytes[61]. The reaction elicited by TSA-IL5, TSA-IL6 and TSA-GM CSF cells is not strong enough to inhibit their growth[55,56].

The mounting of these reactions requires a few days, which means that a specific T-lmphocyte-mediated reactivity is unlikely to make a significant contribution. Later, however, the killer activity of the inflammatory cells appears to be sustained through their intense crosstalk with T lymphocytes[62]. The dominant ultrastructural hallmark of this stage is the membrane contacts between polymorphonuclear leukocytes, macrophages, fibroblasts and lymphocytes. T lymphocytes are still too few to cause any significant tumor cell destruction on their own. Their continuous contact with the inflammatory cells, on the other hand, suggests that they act as guides, probably through an intense release of the secondary cytokines revealed by in situ hybridization experiments. The importance of this guidance is confirmed by the fact that in mice selectively deprived of T lymphocytes the antitumor reaction of inflammatory cells elicited by small amounts of IL-2, IL-4, IL-10 and G-CSF becomes marginal. Progressive killing of the engineered cells and replacement of the tumor itself by granulation tissue represent the final stage of the reaction. In many cases, the engineered cells are totally destroyed, whereas in others a few remain intermingled in the granulation tissue, and may eventually form a new tumor after a period of dormancy.

Tumor-draining nodes often enlarge progressively during the rejection process and display expanded cortical and paracortical areas, as well as numerous tingible-body macrophages and polymorphonuclear leukocytes. The morphologic and functional findings make it clear that rejection of engineered tumors is not the work of a single population, but the outcome of a complex reaction in which the weight of each leukocyte population varies from one cytokine to another, though polymorphonuclear leukocytes and $CD8^+$ T lymphocytes are constantly to the fore. $CD8^+$ T lymphocytes are a late and numerically small component, but they are always important[55,56,62]. Their accumulation in the tumor mass is mostly due to the local increase in the expression of adhesion molecules that follows polymorphonuclear leukocyte activation.

Induction of a Tumor-Specific Immune Memory

T lymphocytes seem to guide the activity of effector leukocytes as rejection comes to an end. This dialogue is eventually reversed. Inflammation and the presence of nonspecific leukocytes, in fact, are probably needed for the induction of a specific, long-lasting immune memory. This is of particular interest, because the special features of the rejection scenario enable tumors previously classed as poorly immunogenic or nonimmunogenic to generate an effective signal. Indeed, the ability of tumors engineered to release cytokines to elicit an efficient systemic immune memory against tumor parental cells forms the rationale for their use as vaccines. It was therefore of importance to compare the efficacy of the systemic immune memory against a subsequent lethal challenge by TSA-pc elicited by a single immunizing injection of irradiated or mitomycin C treated, non-proliferating or proliferating engineered TSA cells. Thirty days later, immunized mice were challenged in the right flank with a lethal dose of TSA-pc. The non-proliferating cells gave only slight protection, whereas the immunity elicited by the growth and rejection of proliferating cells was much stronger: protection was complete in mice that rejected TSA-IL4 and TSA-IL10 and in the few that rejected TSA-IFNγ; it ranged between 32–60% in mice that rejected TSA-IL2, TSA-IL7, TSA-IL12, or TSA-IFNα, but was only marginal in those that rejected TSA-TNFα[61,63]. Rejection of engineered tumor cells is clearly a new and efficient form of vaccination. And this is an important issue.

Following immunization with proliferating cells, intense tumor destruction, release of debris and phagocytosis by cytokine-activated macrophages and polymorphonuclear leukocytes and recruited DC create the antigen load and immunogenic environment required for effective indirect TAA presentation to T lymphocytes. This presentation may be supposed to be further favored by both the cytokines released by the engineered cells and secondary cytokines released by recruited leukocytes that up-modulate the expression of adhesion molecules and MHC glycoproteins on polymorphonuclear leukocytes, macrophages, DC, fibroblasts and endothelia[62]. Biomedical technology is ready to introduce cytokine genes into human cultured or freshly explanted tumor cells, and many phase I clinical trials are in progress[64].

Does a Cytokine-Induced Immunogenicity Really Exist ?

As we have seen, a tumor's ability to induce an immune memory is not greatly increased by non-proliferating engineered cells. Similar protection, indeed, can be conferred by admixing its cells with the conventional adjuvant *Corynebacterium parvum*[63].

Establishment of an antitumor immune memory seems to require:

1. loading of the immune system with sufficient tumor antigen
2. intervention of appropriate host APC
3. presence of the cytokine.

The existence of a true cytokine-induced tumor immunogenicity can be assessed by isolating a few of the variables involved. Cytokines might, for example, simply trigger a debulking reaction without altering a tumor's immunogenicity. Those that elicit a sound immune memory against a nonimmunogenic tumor could be merely causing the slow regression of engineered cells that have achieved a certain degree of growth. This would explain the poor immunogenicity of non-proliferating tumor cells. Alternatively, immunogenicity may stem from recruitment by released cytokines of particular repertoires of inflammatory cells, whose differing ability to act as APC and secrete secondary cytokines may shape both immunogenicity and deflection of the ensuing immune memory towards a Th1 or Th2 type of response. Thirdly, cytokines may deliver pivotal accessory signals to tumor-poised memory T lymphocytes.

This subject was investigated by transfecting TSA-pc with the cytosine deaminase (CD) suicide gene, which enables the nontoxic prodrug 5-fluorocytosine (5FC) to selectively destroy TSA-CD cells after an initial growth. This regression is very similar to that elicited by tumor-released cytokines[65] and leaves an equally efficient systemic memory giving immunity to a subsequent TSA-pc challenge. The higher immunogenicity of engineered cells evidently rests on the ability of the cytokine they release to make the tumor regress.

A Role for Local Cytokines: Selection of the Immune Mechanisms

Do cytokines injected or released locally simply lead to the regression of an incipient tumor? The results obtained with TSA-CD cells suggest that regression is itself a major immunogenic stimulus, no matter how it is brought about[65].

There are interesting differences in the reaction mechanisms elicited by debulking. The systemic immunity that follows local TSA-IL2 debulking mainly rests on CTL, and that elicited by TSA-IFNα rejection is mainly characterized by a CTL-mediated response[61], whereas that elicited by TSA-IL4 rejection rests on the interaction between non-cytolytic $CD8^+$ T lymphocytes, eosinophils, and IgG1, IgA and IgE anti-tumor antibodies, and CTL are absent[58]. The few mice that reject TSA/IFNγ display a marked CTL response along with a minor antibody production[60]. The debulking reaction elicited by TSA-IL10 is characterized by a strong CTL and IgG3 anti-TSA response and confers 100% protection against TSA-pc[59]. It is the cytokine released by an engineered tumor, therefore, that dictates the immune memory mechanisms by prompting TAA presentation by different sets of APC and inducing the release of distinct repertoires of secondary factors. Selection of the cytokine could thus be used to promote or inhibit a particular type of reaction.

Curative Potential of Cytokine-Gene Engineered Vaccines

Induction by engineered TSA cells of systemic immunity is not necessarily associated with their ability to cure established tumors[66]. Four twice-weekly s.c. injections of non-proliferating or proliferating engineered TSA cells were given to evaluate their curing of mice bearing 1- and 7-day tumors. Commencement of these injections 1 day after challenge means that the systemic immunity elicited has to deal with a fast-growing TSA-pc mass of about 5×10^{-4} mm^3. By day 7, this mass has formed a visible, vascularized subcutaneous tumor with many mitotic figures. Treatment begun after 1 day with proliferating TSA-IL2, TSA-IL7, and non-proliferating TSA-IFNγ cells cured a few mice only. Proliferating TSA-IL12 cured 30% of mice and were the most effective. Very few mice were cured when the treatment began after 7 days[63].

These data suggest that therapeutic vaccination with engineered cells is not effective against established tumors. Pessimism with regard to a possible clinical setting, however, must be tempered by the consideration that transplantable tumors grow extremely fast in mice. Very little time elapses between the initial vaccination and the growth of a tumor to a point where it can no longer be controlled by the host's immune reactions. The ineffectiveness of these vaccines in the mouse reflects their need for a long induction period and the inability of their destructive powers to overcome the kinetics of tumor growth. Since human tumors grow much more slowly, they would have enough time to become effective[63].

WHY IS THE INDUCTION OF SYSTEMIC IMMUNITY NOT CORRELATED WITH THERAPY

The immunologic scenario created by a tumor regressing as a consequence of the local presence of a cytokine is different from that generated by elicitation of a systemic immunity or the immune therapy of an established tumor. The players may be the same, but the importance of their roles varies as the plot progresses.

Preventive vaccination prompts immune mechanisms that efficiently protect against a subsequent challenge of monodispersed tumor cells. In addition to tumor kinetics, its efficacy is menaced by other difficulties. Established tumors build up an immune privileged site because of the positive pressure inside their mass, and the barrier imposed by their extracellular matrix, which prevents the penetration of leukocytes. Immune mechanisms leading to the rejection of non-neoplastic tissues are ineffective against an established tumor[67]. In addition, a tumor's own activities interfere with several steps of the immune reaction and form a strong obstacle. This, however, is not immunologically unbreakable. Most tumors suppressive and lethal in syngeneic mice are, in fact, recognised and rejected when injected in allogeneic hosts, even if their histoincompatibility only takes the form of a few minor antigens. Rejection also takes place when tumors form evident masses before the activation of the immune system, as happens when tumor cells are injected in sublethally irradiated allogeneic mice. They give rise to an established mass before the immune reactivity returns and leads to their eradication[68]. These experiments show the great possibilities of a correctly activated immune system.

WHAT CAN WE DO TO MAKE THERAPEUTIC VACCINES MORE EFFECTIVE?

A closer examination must be made of what happens when a local cytokine elicits the debulking of an established tumor. The case of IL-10 is particularly instructive. TSA-IL10 cells form a large mass that is rejected about one month after the challenge. The rejection involves many interacting cells and leads to a major vessel alteration that eventually results in ischemic necrosis of the tumor[59]. Exogenous IL-12, too, induces the regression of established tumors. Here again, multiple cell populations are involved and a major vessel alteration is evident[63]. In both cases, nonspecific immune reactions are of great importance in debulking. Both rejections, however, require a pivotal T lymphocyte guidance of the reaction. Tumor rejection results solely from the extensive crosstalk between these different immunologic worlds.

In our opinion, therefore, an effective therapeutic antitumor vaccine will be one that elicits memory lymphocytes able to orchestrate the crosstalk between specific and nonspecific immunity and vessel endothelial cells. A few new cytokines appear to be particularly promising in the induction of these memory T lymphocytes[59,63].

ACKNOWLEDGMENTS

We thank Dr. John Iliffe for careful review of the manuscript. This work was supported by grants from the Italian Association for Cancer Research (AIRC), the National Research Council (CNR PF-ACRO), and Istituto Superiore di Sanita' special project for gene therapy.

REFERENCES

1. Cancer Vaccines. Once more unto the breach, *The Economist* Dec: 78–81 (1994).
2. M.L. Disis, M.A. Cheever, Oncogenic proteins as tumor antigens. *Curr. Opin. Immunol.* 8: 637–642 (1966).
3. *Science*: Immunology Future, http://www.aas.org/science.immunology

4. A.N. Houghton and J.J. Lewis, Active Specific Immunotherapy In Humans, In:*Cytokine-Induced Tumor Immunogenicity. From exogenous molecules to gene therapy.* G. Forni, R. Foa, A. Santoni and L. Frati Eds., Academic Press, pages 37–54, London, (1994).
5. A.R. Jonsen, S.J. Durfy, W. Burke, and A.G. Motulsky, The advent of the 'unpatients' *Nature.* Med. 2: 622–624 (1966).
6. X. Ye, J.M. McCarrick, L. Jewett and B.B. Knowles, Timely immunization subverts the development of peripheral nonresponsiveness and suppress tumor development in Simian virus 40 tumor antigen-transgenic mice. *Proc. Natl. Acad. Sci. USA.* 91: 3916–3920 (1994).
7. P.R. Walker, P. Saas, P.Y. Dietrich, Role of Fas ligand (CD95L) in immune escape: the tumor cells strikes back. *J. Immunol.* 158: 4521–4524 (1997).
8. G. Forni and R. Foa, The role of cytokines in tumour rejection. In: *Tumour Immunology*, A. Dalgleish and R. Browning Eds., Cambridge University Press, Pag 199–218, Cambridge, (1996).
9. F. Cavallo, P. Nanni, P. Dellabona, P.L. Lollini, G. Casorati and G. Forni, Strategies For Enhancing Tumor Immunogenicity (or how to transform a tumor cell in a Frankenstenian APC). In T. Blankenstein and F. Hermann eds., *Gene Therapy*, Principles and Applications, Chapman & Hall, Weinhein, 1997 (in press).
10. G.M. Iverson, Ability of CBA mice to produce anti-idiotypic sera to 5563 myeloma protein. *Nature* 227: 273–275 (1970).
11. G. Forni, H. Fujiwara, F. Martino, T. Hamaoka, C. Jemma, P. Caretto and M. Giovarelli, Helper strategy in tumor immunology: expansion of helper lymphocytes and utilization of helper lymphokines for experimental and clinical immunotherapy. *Cancer Metastasis Rev.* 7: 289–309 (1988).
12. A. Viola, A. Lanzavecchia, T cell activation determined by T cell receptor number and tunable threshods. *Science* 273: 104–106 (1996).
13. S. Valitutti, S. Muller, M. Dessing and A. Lanzavecchia, Different responses are elicited in cytotoxic T lymphocyte by different levels of T cell receptor occupancy. *J. Exp. Med.* 183: 1917–1921 (1996).
14. S. Valitutti, S. Muller, M. Dessing and A. Lanzavecchia, Signal extinction and T cell repolarization in T helper cell-amtigen-presenting conjugates. *Eur. J. Immunol.* 26: 2012–2016 (1996).
15. D.W. Talmage, J.A. Woolnough, H. Hemmingsen, L. Lopez, K.J. Lafferty, Activation of cytotoxic T cells by nonstimulating tumor cells and spleen factor(s). *Proc. Natl. Acad. Sci. USA.* 90: 5687–5694 (1970).
16. L.K. Koulova, E.A. Clark, G. Shu, B. Dupont, The CD28 lignad B7/BB1 provides the costimulatory signal for alloactivation of CD4+ T cells. *J. Exp. Med.* 173: 759–764 (1991).
17. G. Forni, F. Cavallo, M. Consalvo, A. Allione, P. Dellabona, G. Casorati and M. Giovarelli, Molecular approaches to cancer immunotherapy, *Cytokines and Molecular Therapy.* 1: 225–248 (1995).
18. L. Chen, S. Ashe, W.A. Brady, I. Hellstrom, K.E. Hellstrom, J.A. Ledbetter, P. McGowan, and P.S. Lindsey, Costimulation of antitumor immunity by the B7 counterreceptor for T lymphocyte molecules CD28 and CTLA-4. *Cell.* 71: 1093–1099 (1992).
19. S.A. Townsend, J.A. Allison, Tumor rejection after direct costimulation of CD8+ T cells by B7-transfected melanoma cells. *Science (Wash DC)* 259: 368–371 (1993).
20. F. Cavallo, A. Martin-Fontecha, M. Bellone, S. Helthai, E. Gatti, P. Tornaghi, M. Freschi, G. Forni, P. Dellabona and G. Casorati, Coexpression of B7-1 and ICAM-1 on tumors is required for rejection and the establishment of a memory response. *Eur. J. Immunol.* 25: 1154–1162 (1995).
21. A. Martin-Fontecha, F. Cavallo, M. Bellone, S. Helthai, G. Iezzi, P. Tornaghi, Freschi, N. Nabavi, G. Forni, P. Dellabona and G. Casorati, Heterogenous effects of B7-1 and B7-2 in the induction of both protective and therapeutic anti tumor immunity against different mouse tumors. *Eur. J. Immunol.* 26: 1851–1859 (1996).
22. G. Forni and M.P. Colombo. *Cytokine based tumor immunotherapy. In: Molecular Approaches to Tumor Immunotherapy*, Y. Liu Ed. Word Scientific, New York, 1997 (in press).
23. D.R. Leach, M.F. Krummel, J.P. Allison, Enhancement of antitumor immunity by CTLA-4 blockade. *Science* 271: 1734–1736 (1996).
24. M.F. Krummel, T.J. Sullivan, J.P. Allison, Superantigen responses and co-stimulation: CD28 and CTLA-4 have opposing effects on T cell expansion in vitro and in vivo. *Int. Immunol.* 8: 519–523 (1996).
25. M.F. Krummel, J.P. Allison, CTLA-4 engagement inhibits IL-2 accumulation and cell cycle progression upon activation of resting T cells. *J. Exp. Med.* 183: 2533–2540 (1996).
26. S. Baskar, V.K. Clements, L.H. Glimcher, N. Nabavi, Ostrand-Rosenberg S Rejection of MHC class II-transfected tumor cells requires induction of tumor-encoded B7-1 and/or B7-2 costimulatory molecules. *J. Immunol.* 156: 3821–3827 (1996).
27. S. Ostrand-Rosenberg, S. Baskar, N. Patterson, V.K. Clements, Expression of MHC Class II and B7-1 and B7-2 costimulatory molecules accompanies tumor rejection and reduces the metastatic potential of tumor cells. *Tissue Antigens.* 47: 414–421 (1996).

28. S. Baskar, L. Glimcher, N. Nabavi, R.T. Jones, S. Ostrand-Rosenberg, Major histocompatibility complex class II+B7-1+ tumor cells are potent vaccines for stimulating tumor rejection in tumor-bearing mice. *Exp. Med.* 181: 619–629 (1995).
29. S. Baskar, V. Azarenko, E. Garcia Marshall, E. Huges, S. Ostrand-Rosenberg, MHC class II transfected tumor cells induce long-term tumor-specifici immunity in authologous mice. *Cell. Immunol.* 155: 123–133 (1994).
30. S. Ostrand-Rosenberg, V.K. Clements, S. Amstrong, S. Baskar and Pulasky B, Enhancing tumor immunity: improving the generation of tumor specifici T helper cells.In *Cellular Immunology and Immunotherapy of Cancer III. Keystione Symp Mol. Cell Biology*, Copper Mountain, page 10 (1997).
31. P.L. Lollini, C. De Giovanni, L. Landuzzi, G. Nicoletti, F. Frabetti, F. Cavallo, M. Giovarelli, G. Forni, A. Modica, A. Modesti, P. Musiani and P. Nanni, Transduction of genes coding for a histocompatibility (MHC) antigen and for its physiological inducer gamma-interferon in the same cell. Efficient MHC expression and inhibition of tumor and metastasis growth. *Hum. Gene. Ther.* 6: 743–52 (1995).
32. M. Giovarelli, A. Santoni and G. Forni, Alloantigen-activated lymphocytes from mice bearing a spontaneous "non-immunogenic" adenocarcinoma inhibit its grown by recruiting host immunoreactivity. *J. Immunol.* 133: 3596–3603 (1985).
33. Y.C. Huang, P. Golumbeck, M. Ahmadzadeh, E. Jaffee, D. Pardoll, H. Levitsky, Role of bone-marrow derived cells in presenting MHC class I-restricted tumor antigens. *Science (Washington, DC)* 264: 961–965 (1994).
34. S. Cayeux, G. Richter, G. Noffz, B. Dorken and T. Blankenstein, Influence of gene-modified (IL-7, IL-4, and B7) tumor cell vaccines on tumor antigen presentation. *J. Immunol.* 158: 2834–2841 (1997).
35. C.J.M. Melief, R. Offring, R.E.M. Toes and W. Marin, Peptide-based cancer vaccines. *Curr. Opin. Immunol.* 8: 651–657 (1966).
36. T. Boon, T.F. Gajewski and P.G. Coulie, From defined tumor antigens to effective immunization? *Immunol Today.* 16: 334–336 (1995).
37. J. Mayordomo, T. Zorma, W.J. Storkus, L. Zitvogel, C. Celluzi, L.D. Falo, C.J.M. Melief, S.T. Ildstad, W.M. Kast, A.B. Deleo, M.T. Lotze, Bone marrow-derived dendritic cells pulsed with synthetic tumor peptides elicit protective and therapeutic anti-tumor immunity. *Nat. Med.* 1: 1297–1302 (1995).
38. L. Zitvogel, J. Mayordomo, T. Tjandrawan, A.B. Deleo, M.R. Clarke, M.T. Lotze, W.T. Storkus, Therapy of murione tumors with peptide pulsed dendritic cells: Dependence on T-cells, B7 costimulation, and T helper cell1-associated cytokine. *J. Exp. Med.* 183: 87–97 (1996).
39. B. Mukherij, N.G. Chakraborty, S. Yamashi, T. Okino, H. Yamase, J.R. Sporn, S.K. Kurtzman, M.T. Egrin, J. Ozolos, J. Meehan, F. Mauri, Induction of antigen-specific cytolytic T cells in situ in human melanoma by immunization with synthetic peptide-pulsed autologous antigen presenting cells. *Proc. Natl. Acad. Sci. USA.* 92: 87078–8082 (1995).
40. J. Gong, D. Chen, M. Kashiwara, D. Kufe, Induction of antitumor activity by immunization with fusion of dendritic and carcinoma cells. *Nat. Med.* 3: 588–561 (1997).
41. D.M. Pardoll, A.M.L. Beckerleg, Exposing the immunology of naked DNA vaccines. *Immunity.* 3: 165–169 (1995).
42. DNA Vaccine Web. http://www.genweb.com/Dnavax/faq.html totaldna
43. A.Concetti, A. Amici, C. Petrelli, A. Tibaldi, M. Provinciali, F.M. Venanzi, Autoantibody to p185erbB2/neu oncoprotein by vacciantion with xenogenic DNA *Cancer Immunol.Immunother.* 43: 307–315 (1996).
44. I. Hakin, S. Levy and R. Levy, A nine amino acid peptide from IL-1b augment antitumor immune responses induced b protein and DNA vaccines. *J. Immunol.* 157: 5503–5510 (1996).
45. G. Forni and M. Giovarelli, In vitro reeducated T-helper cells from sarcoma bearing mice inhibit sarcoma growth in vivo. *J. Immunol.* 132: 527–533 (1984).
46. H. Kawamura, S.A. Rosenberg, J.A. Berzofsky, Immunization with antigen and interleukin 2 in vivo overcomes Ir gene low responsiveness. *J. Exp. Med.* 162: 381–389 (1985).
47. M. Malkowsky, P.M. Medawar, D.R. Thacher, J. Toy, L. Hunt, S. Rayfield, C. Dore`, Acquired immunological tolerance of foreign cells is impaired by recombinant interleukin 2 or vitamin A. *Proc. Natl. Acad. Sci. USA.* 82: 536–540 (1985).
48. G. Forni, T. Musso, C. Jemma, D. Boraschi, A. Tagliabue, and M. Giovarelli, Lymphokine activated tumor inhibition (LATI) in mice: ability of a nonapeptide of the human Interleukin-1 to recruit antitumor reactivity in recipient mice. *J. Immunol.* 142: 712–718 (1989).
49. G. Forni, A. Santoni, M. Giovarelli, Lymphokine acivated tumor inhibition in vivo. I. The local administration of Interleukin -2 triggers non reactive lymphocytes from tumor bearing mice to inhibit tumor growth. *J. Immunol.* 134: 1305–1311 (1985).

50. M.C. Bosco, M. Giovarelli, M. Forni, A. Modesti, S. Scarpa, L. Masuelli, G. Forni, Low doses of interleukin-4 injected perilymphatically in tumor bearing mice inhibit the growth of poorly and apparently nonimmunogenic tumors and induce a tumor specific immune memory. *J. Immunol.* 145: 3136–3143 (1990).
51. M. Giovarelli, F. Cofano, A. Vecchi, M. Forni, S. Landolfo and G. Forni, Interferon-Activated Tumor Inhibition in vivo: Small amounts of interferon-gamma inhibit tumor growth by eliciting host systemic immunoreactivity. *Int. J. Cancer.* 37: 141–148 (1986).
52. G. Cortesina, A. De Stefani, E. Galeazzi, G.P. Cavallo, F. Badellino, G. Margarino, C. Jemma and G. Forni, Temporary regression of recurrent squamous cell carcinoma of the head and neck achieved with low but not with high doses of recombinant interleukin-2 injected perilymphatically. Brit. *J. Cancer.* 69: 572–576 (1984).
53. *Cytokine-Induced Tumor Immunogenicity. From exogenous molecules to gene therapy,* G. Forni, R. Foà, A. Santoni, and L. Frati, eds., Academic Press, London (1994).
54. G. Nicoletti, C. De Giovanni, P.L. Lollini, G.P. Bagnara, K. Scotlandi, L. Landuzzi, B. del Re, G. Zauli, G. Prodi, P. Nanni, In vivo and in vitro production of haematopoietic colony-stimulating activity by murine cell lines of different origin: A frequent finding. *Eur. J. Cancer. Clin. Oncol.* 25: 1281–1286 (1989).
55. P. Musiani, A. Allione, A. Modica, P.L. Lollini, M. Giovarelli, F. Cavallo, F. Belardelli, G. Forni and A. Modesti, Role of neutrophils and lymphocytes in inhibition of a mouse mammary adenocarcinoma engineered to release IL-2, IL-4, IL-7, IL-10, IFN-alpha, IFN-gamma, and TNF-alpha. *Lab. Invest.* 74: 146–157 (1996).
56. P. Musiani, A. Modesti, M. Giovarelli, F. Cavallo, M.P. Colombo, P.L. Lollini and G. Forni, Cytokines, tumor cell death and immunogenicity: a question of choice. *Immunol. Today.* 18: 32–36 (1997).
57. F. Cavallo, M. Giovarelli, A. Gulino, A. Vacca, A. Stoppacciaro, A. Modesti and G. Forni, Role of neutrophils and CD4+ T lymphocytes in the primary and memory response to nonimmunogenic murine mammary adenocarcinoma made immunogenic by IL-2 gene transfection. *J. Immunol.* 149: 3627–3635 (1992).
58. F. Pericle, M. Giovarelli, M.P. Colombo, G. Ferrari, P. Musiani, A. Modesti, F. Cavallo, F. Novelli and G. Forni, An efficient Th-2-type memory follows CD8+ lymphocyte driven and eosinophil mediated rejection of a spontaneous mouse mammary adenocarcinoma engineered to release IL-4. *J. Immunol.* 153: 5659–5672 (1994).
59. M. Giovarelli, P. Musiani, A. Modesti, P. Dellabona, G. Casorati, A. Allione, M. Consalvo, F. Cavallo, F. Di Pierro, C. De Giovanni, T. Musso and G. Forni, The local release of IL-10 by transfected mouse mammary adenocarcinoma cells does not suppress but enhances antitumor reaction and elicits a strong cytotoxic lymphocyte and antibody dependent immune memory. *J. Immunol.* 155: 3112–3123 (1995).
60. P.L. Lollini, M.C. Bosco, F. Cavallo, C. De Giovanni, M. Giovarelli, L. Landuzzi, P. Musiani, A. Modesti, G. Nicoletti, G. Palmieri, A. Santoni, H.A. Young, G. Forni and P. Nanni, Inhibition of tumor growth and enhancement of metastasis after transfection of the interferon-gamma gene. *Int. J. Cancer.* 55: 320–329 (1993).
61. M. Ferrantini, M. Giovarelli, A. Modesti, P. Musiani, A. Modica, M. Venditti, E. Peretti, P.L. Lollini, P. Nanni, G. Forni and F. Belardelli, IFN-alpha gene expression into a metastatic murine adenocarcinoma (TS/A) results in CD8+ T cell-mediated tumor rejection and development of antitumor immunity. Comparative studies with IFN-gamma producing TS/A cells. *J. Immunol.* 153: 4604–4615 (1994).
62. M.P. Colombo, A. Modesti, G. Parmiani and G. Forni, Perspectives in Cancer Research: Local cytokine availability elicits tumor rejection and systemic immunity through granulocyte-T-lymphocyte cross-talk. *Cancer Res.* 52: 4853–4857 (1992).
63. F. Cavallo, P. Signorelli, M. Giovarelli, P. Musiani A. Modesti, M.J. Brunda, M.P. Colombo and G. Forni, Antitumor efficacy of adenocarcinoma cells engineered to produce IL-12 or other cytokines compared with exogenous IL-12. *J. Natl. Cancer Inst.* (in press) 1997
64. M.P. Colombo and G. Forni, Cytokine gene transfer in tumor inhibition and tentative tumor therapy: Where are we now ? *Immunol. Today.* 15: 48–51 (1994).
65. M. Consalvo, C. Mullen, A. Modesti, P. Musiani, A. Allione, F. Cavallo, M. Giovarelli and G. Forni, 5-Fluorocytosine induced eradication of murine adenocarcinomas engineered to express the cytosine deaminase suicide gene requires host immune competence and leaves an efficient memory. *J. Immunol.* 154: 5302–5312 (1995).
66. M.P. Colombo and G. Forni, Immunotherapy I: Cytokine gene transfer strategies. *Cancer Metastasis Rev.* 15: 317–328 (1996).
67. S. Singh, S.R. Ross, M. Acena, D.A. Rowley, H. Schreiber, Stroma is critical for preventing or permitting immunological destruction of antigenic cancer cells. *J. Exp. Med.* 175: 139–146 (1992)
68. N. Bellomo, L. Preziosi and G. Forni, Tumor immune system interactions: The kinetic cellular theory. In: *A Survey of Models for Tumor-Immune System Dynamics*, J.A. Adam and N. Bellomo Eds., Birkhauser, Boston, Pag. 135–180 (1997)

DISCUSSION

Mihich: In the model where you have the cytokine transfected tumor cells, do you see a difference with some of the cytokine for which you may have neutralizing antibody? By giving such antibody, are you completely excluding the cytokine participation? What happens if you have the tumor transfected and then at a certain time when the tumor is lysing you add antibodies against the cytokine that you used to transfect.

Forni: I am confused. Are you speaking about the debulking of the initial tumor or the immunity to a subsequent challenge?

Mihich: I am talking about the immunity that is developed as a result of the lysis of the original vaccine tumor.

Forni: If you obtain tumor rejection, or in particular tumor growth and then rejection, the presence of any kind of cytokine does not matter: you always induce a very strong immunity. However if a cytokine is present it privileges the induction of some kind of memory mechanisms. When the cells are releasing cytokines directly you deflect the immune memory mechanism towards Th1 or Th2 reactivity. You can obtain almost the same effect by injecting recombinant cytokines locally and repeatedly. So, during the rejection of the tumor the presence of a cytokine or another influences the mechanism of immune memory.

Mihich: There is no participation of the possible modification of the immunogenicity of the tumor by virtue of being transfected with the cytokine?

Forni: Well, It cannot be excluded that gene transduction alters cell immunogenicity. However, since we can reproduce almost the same results by using parental tumor cells and injecting recombinant cytokines locally we are quite convinced that the selective promotion of certain immune mechanisms is dependent on the antitumor reaction activated by the cytokine.

Hanahan: I have two questions on the IL-12 study in the *neu* mice. You started when the mice were one month of age, and you treated for five days in a row and then waited?

Forni: We treated *neu* transgenic mice for 5 days every month and then we kept them under observation until the 33rd week of age.

Hanahan: Gave them a rest, then five sequential days in the next month throughout the life of the animal?

Forni: Yes.

Hanahan: The other comment is that IL-12 is also, in addition to being an immune activator, a demonstrable in angiogenesis inhibitor.

Forni: We are convinced that the anti-angiogenic activity of IL-12 plays a major role in inhibiting spontaneous tumors.

Hanahan: So the question is, are you looking at vessel density, do you think this is a factor?

Forni: Yes, I have no data here, but vessel alteration appears to be the major effect of IL-12. However I don't know how much anti-angiogenic activity directly depends on IL-12 or on the immune response elicited by IL-12. The pathologists told us that the vessels are altered, and we have a scarce vascularization.

Hanahan: That is very exciting because it really is almost a common therapy in itself.

Pierotti: I think that your model with the transgenic mice is very provocative. In other words, if you have a transgenic model with p53, for example, you should provide with related peptides vaccination some answers to the story of the Li-Fraumeni syndromes in which we have no particular cure, and also in the case of breast carcinoma you have the same model with BRCA-1 or 2, and to see whether with the related peptides you can prevent the outcome of the tumor.

Forni: You know, it is a pain in the neck to obtain these transgenic mice, because we have to breed them ourselves. We are trying to make some arrangements to have them bred commercially, even though it is very expensive. So far our animal house is full of transgenic mice and you can imagine how difficult this situation is.

Moretta: I was actually rather surprised to see the effect of IL-10. In other words, I would like to have your comment about this finding and whether you have some information on the histology of the tumor since it is an immuno-suppressive cytokine?

Forni: IL-10 normally inhibits cytokines and factors production. We believe that in this tumor system as well as in other tumors we have a collapse of established tumors instead of a real tumor rejection. Tumors collapse because the tumor vascularization is very scarce. You have many areas of ischemic necrosis which attracts a lot of reactive polymorphonuclear cells. So IL-10 induces a cascade of events: the necrosis probably depends on the lack of angiogenetic factors that are normally released by parental wild-type tumor cells. I think that this cascade of events leads to provocative results since commonly IL-10 is believed to enhance tumor growth because it inhibits the immunity. When a high amount of IL-10 is released locally it first enhances tumor growth. Subsequently IL-10 action becomes a double edge sword. It kills the tumor, as without angiogenic factors it cannot survive. The lack of vessels induces necrosis and necrosis induces tumor rejection.

Bankert: You tried lots of cytokines but there is a growing awareness that a combination of cytokines and also the level and the timing of the dose of the cytokines have differential effects. With IL-2 and IL-12 there is work that has been reported, and we have seen this also, that a combination of IL-2 and IL-12 has a tremendously synergistic effect, particularly when IL-2 is at low dose and the IL-12 is at high dose. I wonder if you have done any combination of dose effects?

Forni: We tested several combinations of recombinant cytokines injected at tumor site. Later we tested a few combinations of IL genes. We obtained some synergisms when IL-12 was injected intraperitoneally in mice bearing tumor cells engineered with IL-2 gene. However the synergism was always very unpredictable and the possible combinations were too many. Therefore we decided it was too complicated to continue to test all possible combinations. Nevertheless I believe that some combinations could lead to interesting synergisms.

Bankert: I wonder whether the persistence of the cytokine can actually be inhibitory. Therefore, in using cytokine gene therapy, it may be preferable to achieve only a transient expression by using techniques that result in transfection of the cytokine gene without integration.

Forni: Yes. In fact engineered tumor cells are rejected after 2 \ 3 \ 6 days, depending on the type of cytokines they release. Only in the case of cells releasing interferon gamma or IL-10 you have a prolonged tumor growth. So engineered cells are continuously releasing cytokines, but they are destroyed in a few days: therefore we have only a transient release of cytokines.

Parmiani: You have shown that proliferating cells are much more immunogenic than non-proliferating cells. Since we have to inject patients, we cannot give proliferating cells; can you reach the same result by giving a higher number of non-proliferating cells or multiple injections of non-proliferating cells?

Forni: I believe so but I have never done this kind of experiment. We were interested in having the best vaccination system and not a system to be immediately applied with humans. So we stuck to proliferating cells. But I am sure that very similar results can be obtained through repeated injections of non proliferating tumor cells.

Hanahan: On the IL-10 then, have you done systemic treatment or just the engineered cells? I mean, much like you did the IL-12, have you done the same thing with IL-10?

Forni: Not yet. I saw a paper by Michael Lotze. In this paper the systemic injection of huge amounts of IL-10 was inducing tumor regression. So far we used IL-10 releasing cells only.

Hanahan: Just to clarify: Although you may have gotten into this from an immunotherapy perspective, at this point, your best guess from the histology is that these are acting as angiogenesis inhibitors, both IL-10 and IL-12. Would that be a fair interpretation?

Forni: Yes, right.

Hanahan: It would not explain the memory but it would explain the effect on the tumors which we have been discussing.

Parmiani: We may have both effects. In fact, we have injected melanoma patients with IL-12 and we see both lymphocyte stimulation and signs of inhibition of angiogenesis. So, it depends probably on the schedule of immunization on the amount of cytokine and so on. You may have a prevalence of one or the other effect but, certainly, both are there.

Zanker: Do you have any kinetic information on the release of the engineered cytokine from tumor cells and maybe on the amount? Because many people are pursuing this approach but mostly the cytokine release from these engineered cells is unpredictable.

Forni: Yes, it is unpredictable, but you can select the clones that release the amount you want in general. As you probably noticed in some slides the clones have a superscript number. This number shows the units or micrograms of cytokines released by 10^5 in 48 hours. So for each clone you have a defined idea of the amount of cytokine that is released. What is important for immunogenicity is that the greatest amount of cytokine released is not the most effective. For each cytokine you have to find the right amount that best stimulates the immune system.

Zanker: But how do you find the right clone that releases the right amount of cytokine?

Forni: In mice it is easy. You have to test various clones comparatively so you can get some idea about what is the most effective amount of cytokines.

Bankert: It has been shown by other people, including you, very nicely that the proliferating cells are more effective inducers of immunity. I just wondered if you could get approved perhaps a protocol that includes a suicide gene that is able to be triggered by a drug.

Forni: Yes. We did this experiment. We were very disappointed because using tumor cells engineered with a suicide gene we obtained exactly the same, or similar , immunity to the one obtained when the tumor cells are rejected because of the release of a cytokine. The most important signal is to have a tumor that grows and then is rejected. The local presence of a cytokine can only privilege some kinds of immune mechanism.

Bankert: It is possible that you were saying that you could not use that in a clinical setting and perhaps you could if you could guarantee that after proliferating the first two or three days that you could shut it off.

Parmiani: In such a case, the problem is to convince an ethical committee that you will destroy one hundred percent of cells.

Forni: I am positive I saw there are some clinical trials based on proliferating tumor cells in the USA.

Parmiani: I was aware of one, which has been stopped.

Hanahan: I do not remember what doses we did but, a couple of years ago, we tried IL-12 in the mice just as an angiogenesis strategy but the mice all died, which was about the time the patients were dying which is why the trial was stopped. So I will have to compare notes and see what you did differently, because obviously, you are not seeing death in the mice.

Forni: The animals are quite healthy. We started the experiments by injecting very low doses of IL-12. Then we increased the dose up to 0.1 micrograms. We used this dose when the animals were 10 weeks of age if I remember correctly. Before the dose was much lower.

19

T-CELL AND NK-MEDIATED SURVEILLANCE OF CANCER

A Delicate Balancing Act

M. J. W. Visseren, S. H. van der Burg, M. Vierboom, M. E. Ressing, R. Toes, R. Offringa, and C. J. M. Melief

Leiden University Medical Center
Department of Immunhematology and Bloodbank*
P.O. Box 9600
2300 RC Leiden, The Netherlands

1. THE CELLS OF THE IMMUNE SYSTEM THAT CAN LYSE CANCER CELLS

In humans the immune system protects the individual from death by infection with for instance bacteria and viruses. Both the humoral and cellular parts of the immune response cooperate to achieve this. For some time we know that certain members of the cellular immune system not only react against infectious agents, but can also destroy tumor cells. Monocytes, macrophages, natural killer cells (NK cells) and cytotoxic T lymphocytes (CTL) have been proven tumoricidal. This thesis deals with the tumoricidal potential of NK cells, and focuses on the anti-tumor actions of CTL.

1.1. Natural Killer Cells

These natural born killers recognize a cell by their receptors (NKR-P1 in the mouse) specific for certain cell surface molecules, the nature of which is still unknown, that are ubiquitously expressed on all cells (1). This receptor engagement results in lysis of the recognized cell, unless a message is received by the NK cell from its killing inhibitory receptors (KIR) (2), that recognize the $\alpha 1$-domain of MHC class I, or from other inhibitory receptors that recognize the $\alpha 1/\alpha 2$ domain of MHC class I (3,4). In the cytoplasmic tails of the inhibi-

* Tel.: 31-71-5263800; Fax: 31-71-5216751

The Biology of Tumors, edited by Mihich and Croce
Plenum Press, New York, 1998.

Table 1. NK receptors for MHC-class I antigens

NK receptor	Type	Species	Ligand	Sensitivity
p58.1	IgSF, KIR-2D	human	HLA-Cw2, -Cw4, -Cw5, -Cw6	high
p58.2	IgSF, KIR-2D	human	HLA-Cw1, -Cw3, -Cw7, -Cw8	high
p70	IgSF, KIR-3D	human	HLA-Bw4	high
p140	IgSF, KIR-3D	human	HLA-A3, -A11	high
CD94/NKG2-A	type II C-lectin	human	HLA-A, -B, -C	low
Ly49A	type II C-lectin	mouse	H-2^{d}, $^{-k}$, $^{-b}$, $^{-s}$	?
Ly49 C	type II C-lectin	mouse	H-2 K^{b}	?
Ly49 G2	type II C-lectin	mouse	H-2 D^{d}, L^{d}	?

Sources: Moretta et al., 1997, Immunological Reviews 155:105–117; Yu et al., 1996, Immunity 4:67–76; Lanier, 1997, Immunity 6:371–378.

tory receptors a conserved immunoreceptor tyrosine-based inhibiting motif (ITIM) is found. Upon cross-linking of the receptors the tyrosines are phosphorylated, and are subsequently linked to SHP-1 tyrosine phosphatase. SHP-1 is crucial for delivery of the negative signal that prevents NK cell mediated killing (5). Several types of MHC-specific inhibitory receptors have been identified in mouse and man (Table 1). Some receptors are type II membrane glycoproteins of the C-type lectin superfamily, whereas others belong to the immunoglobulin superfamily (IgSF). The IgSF-type receptors are named KIR, and have thus far only been found in human NK cells and T cells. The KIR can be grouped based on the number of immunoglobulin-like domains: either 2 (KIR-2D) or 3 (KIR-3D) domains. In man, the specificities of the receptors is the clearest. The KIRs only recognize specific MHC molecules presenting self peptides, whereas the CD94/NKG2A receptor has a broad specificity recognizing most HLA-A, -B, and -C alleles. The majority of human NK-cells in one individual express the less sensitive CD94/NKG2A receptor, accounting for recognition of cells that have lost (a proportion of) their HLA expression. Most nucleated body cells express MHC class I molecules, and hence are protected from lysis by NK cells. The body cells devoid of class I probably express ligands for other inhibitory receptors. However, tumor cells tend to downregulate their MHC class I molecules (see part 4) and thus become more sensitive to lysis by NK cells (6–8). In addition, altered biochemistry of tumor cells may result in presentation of a different set of peptides (see part 3). Such novel peptide-MHC complexes sometimes fail to trigger the more specific and sensitive NK receptors, i.e. p58, p70, and p140 (9,10), probably by masking the KIR-specific docking site at the α1-helix. The abrogation of this negative signal results in attack and destruction of tumor cells expressing novel peptide-MHC complexes.

1.2. Cytotoxic T Lymphocytes

CTL recognize small peptides of about nine aminoacids long that are captured in an MHC class I molecule and are thus presented to the T cell receptor of the CTL (11). Each CTL displays a unique specificity for a certain peptide-MHC complex, which is formed by the combination of the α and β chain of the T cell receptor. In the thymus both chains are engineered from gene segments, chosen out of a pool, that are linked together (12). Additional diversity is gained by random reduction and addition of nucleotides at certain linkage points (13). T cells with a functional T cell receptor are positively selected on the MHC class I molecules present in the thymus (14). However, CTL with a high affinity for the self-peptides expressed in the MHC class I molecules of the thymus are clonally deleted, and do not leave the thymus (15,16). It is estimated that about half to two-thirds of the positively selected T cells undergo subsequent negative selection (17).

Three regions of hypervariability in each TCR chain can be distinguished, called CDR1, CDR2, and CDR3 (CDR = complementarity determining region). Of those the CDR3 is the most diverse, as it is created by the process of random reduction and addition of nucleotides as mentioned above. In a properly folded TCR the CDR of both chains are positioned on the side that faces the MHC molecule which presents a peptide captured between its α-helices (18). Of the TCRα chain all three CDR contact the α-helices of the MHC molecule, and CDR1 and CDR3 contact the peptide. Of the β-chain only CDR3 contacts an α-helix of the MHC molecule, whereas again CDR1 and CDR3 of this chain contact the peptide (19).

Virgin CTL that leave the thymus can be activated into cell division and lysis of target cells upon proper stimulation. Dendritic cells and monocytes, the classical antigen presenting cells, are equipped for this task. Their expression of the costimulatory molecules B7.1 (CD80) and B7.2 (CD86) is crucial to activation of CTL, since engagement of the T cell receptor in the absence of a costimulatory signal can result in anergy or apoptotic cell death (20). Likewise, expression of B7.1 on murine tumor cells significantly enhances the anti-tumor CTL response (21). Once a CTL has been properly activated, recognition of a peptide-MHC complex results in lysis of the target cell.

It is of note that some CTL express KIR. Most likely this expression is induced in CTL that are chronically exposed to antigen in particular micro-environmental conditions, one example of which could be autoreactive CTL. The KIR on CTL may have a regulatory function in recognition of autoantigens, tumor cells, or virus-infected cells (22). Indeed HLA-A24-restricted CTL expressing the p58.2 KIR-2D have been described that recognized only melanoma cells that had lost expression of HLA-C molecules but not the HLA-A24 molecule. In the presence of HLA-C the negative signal of the p58.2 KIR-2D overruled the positive signal of the TCR, resulting in abrogation of lysis (23).

An overview of the antigens present on tumor cells that can be recognized by CTL is given in part 3.

2. MANIPULATING THE TUMORICIDAL CAPACITY OF NK CELLS AND CTL

2.1. Cytokine Activation

Both NK cells and CTL can be activated with cytokines, for instance with Interleukin-2 (IL-2). However, cytokines have a relatively short half-life and thus normally act in the vicinity of the cell that secretes them. When given systemically to cancer patients IL-2 proved to be quite toxic. At dosage levels allowing acceptable toxicity, weak anti-tumor effects have been reported (24).

When given only at the tumor site the IL-2-dosage can be raised to levels high enough to activate NK-cells. Due to its short half-life, the IL-2 should be delivered repetitively. An alternative way is to construct tumor cells that themselves continuously secrete IL-2. In mice IL-2 producing tumor cells have been shown to activate NK cells, resulting in a lower tumor take rate (25). Outgrowth of IL-2 producing tumor cells was not completely abrogated, illustrating the poor anti-tumor effect of NK cells in this model. When too many IL-2 producing tumor cells are introduced into the mouse, the fast growing tumor cells probably outnumber the activated NK cells.

In some IL-2 producing tumor models CTL-activation specific for the tumor was found, and in others it was absent (26,27) depending on the tumor model investigated. A more predictable strategy for CTL-induction is intraperitoneal vaccination with large amounts of non-dividing (irradiated) IL-2 producing tumor cells, which also in our model led to activation of CTL that specifically recognized and killed larger amounts of subcutaneously injected non IL-2 producing tumor cells. The exact mechanism whereby IL-2 produced by tumor cells induces tumor-specific CTL is not known. Most likely, the IL-2 activates several leukocytes that start to secrete other cytokines which on their turn attract and activate other leukocytes or have their effect on the tumor cells. Murine NK cells were shown to produce IFNγ upon stimulation with IL-2 (28), which upregulates MHC class I-expression on the tumor cells and boosts their protein processing machinery (see part 3). In this process (some of) the IL-2 producing tumor cells are lysed and picked up by antigen presenting cells. Subsequently these cells activate tumor-specific CTL that are able to kill parental tumor cells (29).

Since the tumor cells are scavenged and processed by the patients' antigen presenting cells, allowing presentation of tumor-specific peptides in the proper MHC molecules, the MHC type of the IL-2 producing tumor cells is not crucial. Thus, completely allogeneic tumor cells still can induce a tumor-specific CTL response (30). In theory therefore, only one IL-2 producing tumor cell line has to be engineered for each tumor type, provided that generally expressed tumor (type-specific) antigens exist as observed in melanoma (31). Application of non-autologous IL-2 producing melanoma cells as a vaccine in a phase I/II study on 34 advanced stage melanoma patients has yielded one complete remission, one partial remission, six stable diseases, and several mixed responses in which some metastases regress, whereas others grow out (personal communication from Dr. S. Osanto at the department of clinical oncology of the university hospital Leiden). These results were confirmed by another group that had immunized melanoma patients with IL-2-producing HLA-A2-compatible melanoma cells (32). Other strategies circumventing elaborate culturing and gene transduction procedures for each patient include injection of mixtures of autologous tumor cells and IL-2 producing allogeneic fibroblasts (33), or mixtures of autologous tumor cells and cytokine-containing microspheres (34).

Downmodulation of ζ-chain and $p56^{lck}$ in T-cells of mice bearing an IL-2 producing tumor was not observed, as opposed to T-cells of mice bearing a conventional tumor (35). It is hypothesized that the IL-2 produced by the tumor cells results in intracellular signal transduction events mimicking the costimulation signal (36), which is crucial for proper activation of CTL (20).

Anti-tumor effects of tumor cells producing other cytokines such as IL-4, IFNγ, TNF, IL-6, IL-7, GM-CSF, IL-3, MCP-1, and G-CSF have been reviewed (37). Some effects may be the result of direct CTL activation. However, the anti-tumor effect caused by GM-CSF is most likely mediated via activation of antigen presenting cells that subsequently trigger both T-helper cells and CTL (38).

2.2. Guided Activation with Tumor-Specific Peptides

Whereas cytokines stimulate all cells expressing the proper receptor, regardless of their specificity, peptides stimulate only the CTL that recognize that particular peptide-MHC complex. Some peptides are uniquely (or selectively) generated in tumor cells. CTL recognizing those peptides can therefore specifically lyse tumor cells without damaging normal cells. In part 3 an overview of tumor-antigens is given.

Vaccination of mice with suitable synthetic peptides results in activation of CTL specific for those peptides (39, Chapter 3). Furthermore, mice vaccinated with the appropriate synthetic peptides are protected against an otherwise lethal dose of tumor cells (40).

In the body specialized cells present antigenic peptides to CTL resulting in activation of these cells. Both dendritic cells and monocytes have been reported to be able to activate CTL. The antigen presenting capacity of activated B lymphocytes is still controversial. When using the proper antigen presenting cells, it is possible to activate CTL *in vitro* with synthetic peptides using the buffy coat cells isolated from blood (41, Chapter 4). The *in vitro* induced CTL seem to be similar to CTL that are induced in the body with respect to affinity, fine-specificity, and diversity in T cell receptor composition of the activated CTL. Such *ex vivo* stimulated and expanded specific CTL could be reinfused into the patient.

A more direct way would be to vaccinate the patients themselves, instead of expanding the desired CTL in the laboratory. Phase I studies are being performed to examine the feasibility of vaccinating humans with synthetic peptides. The first results show, as observed in mice, specific CTL activation (42–44). However, in some mouse models peptide vaccination can induce specific CTL tolerance resulting in enhanced tumor cell growth (45,46), which seems to depend on the intrinsic nature of the peptide. Interestingly, the same epitope expressed in tumor cells or in an adenovirus vaccine conferred immunity and prevented tumor outgrowth. It might therefore be more preferable to use vaccination strategies in which exposure to the naked peptide is avoided. Such strategies could include DNA vaccination (47), or engineered adenovirus (48,49) or vaccinia virus vectors (50). The presence of cytokines such as IL-12 (51) and GM-CSF (52) in the vaccine can amplify its effect. In addition, the presence of costimulatory signals enhances the anti-tumor effect of such vaccines (53). Proper costimulation is also ensured by transduction of tumor associated genes in antigen presenting cells such as dendritic cells (54), or transducing tumor cells with costimulatory molecules (E. Hooijberg, personal communication).

3. TUMOR ANTIGENS

Larger cellular proteins are degraded in the cell by the action of proteolytic enzymes such as the proteasome complex (55–57). The importance of the proteasome complex for the generation of CTL epitopes has recently been demonstrated (58). This multi-unit enzyme complex cleaves large linearized proteins into smaller peptides, which are actively transported into the endoplasmatic reticulum (ER) by the peptide transporter heterodimer TAP (59). The human TAP proteins efficiently transport peptides of 8–12 aa, but longer peptides can be transported (60,61). In the ER the newly synthesized MHC class I heavy chain molecules, that are associated with β2 microglobulin, interact with calreticulin, and subsequently bind to TAP with the help of tapasin (62). When the peptide fits precisely into the groove and pockets of that particular MHC class I molecule (63), a stable complex is formed from which TAP, tapasin and calreticulin dissociate. TAP-transported peptides can also bind to the ER-resident stress protein gp96 (64), that might act as a chaperonin for the peptide and aids subsequent MHC class I binding. Furthermore, peptides that are introduced into the ER by other mechanisms can also bind to MHC class I (65). Even gradual trimming of longer peptides to the optimal MHC class I binding length can take place in the ER (66,67). Upon transportation through the Golgi compartment the β2 microglobulin-MHC-peptide complex is expressed on the outside of the cell. In this manner all cellular proteins are sampled for recognition by CTL.

3.1. Tumor-Specific Antigens

In some instances, CTL against point mutated normal proteins were found (68–70). Most likely these point mutations are unique for those patients, resulting from the unbridled growth that ends in less severe control of the genetic integrity of the tumor cell. Other patient-specific antigens comprise the hyper-variable regions of the immunoglobulins of B-cell lymphomas. The uniqueness of those antigens implies extensive research for each patient. This is not only very costly, but also time consuming.

Ubiquitously expressed proteins are therefore more useful. However, few proteins are exclusively expressed in tumor cells, and hence can be named tumor-specific proteins. Viral proteins of virally induced tumors can be recognized by CTL, and are truly tumor-specific. In mice several viral oncoproteins are recognized by CTL, for example retroviral antigens (71,72), Adenovirus E1A and E1B (73,74), and simian virus large T (75). Also in man CTL specifically recognizing viral tumor proteins of Epstein-Barr virus (76,77), human T cell leukemia virus (78), and human papilloma virus (79,80) were isolated. However, most tumors are not virally induced.

Another group of tumor-specific antigens comprise the mutated oncogene products, that are exclusively present in tumor cells and are involved in the uncontrolled growth. CTL against mutated ras (81–83) were found in mice and man, and CTL against mutated p53 (84) have been isolated in mice. Those mutated p53-specific CTL were capable of diminishing established p53-expressing sarcomas in mice (85). The product of the tumor-specific Bcr-Abl fusion gene is also a candidate for CTL recognition (86). However, so far no CTL recognizing the breakpoint-spanning region that lyse chronic myelogenous leukemia cells containing the Bcr-Abl fusion gene have been found (87).

Although mutated oncogene-specific CTL can be effective against cancer, not all mutations are contained within MHC-binding CTL epitopes. Thus, such CTL cannot be elicited in all cancer patients.

3.2. Tumor-Associated Antigens

This group of antigens is mainly though not uniquely expressed in tumor cells. An example is the MAGE-family of antigens that are expressed in melanoma cells and other tumor cells, but also in testis and placenta (88–95). Expression of MAGE genes is caused by a genome wide demethylation process, that occurs in many cancer cells and is correlated with tumor progression (96). CTL specific for MAGE-1 and MAGE-3 have been isolated from the blood of melanoma patients (97–100). Although MAGE genes are expressed in spermatogonia and trophoblasts in the placenta, the absence of MHC class I on these cells ensures no problems regarding auto-immunity (101). Other antigens with similar expression are BAGE (102), GAGE (103) and NY ESO 1 (104). An antigen expressed in renal cell carcinoma, sarcoma, melanoma, and bladder carcinoma, but also in retina cells, is RAGE (105). With the use of tumor infiltrating lymphocytes or lymph node cells still new tumor-associated antigens are being discovered (106–109).

3.3. Tissue-Specific Antigens

Most tissues express unique proteins, that are also expressed on cancerous cells originating from that tissue. CTL can recognize the retina- and melanocyte-specific antigens gp100/pmel17 (110,111), MART-1/Melan-A (112,113), TRP1 (114) and TRP2 (115), and tyrosinase (116–118). Moreover, the presence of tyrosinase- and MelanA/MART-1-

Table 2. Human MHC class I-restricted peptide sequences of tissue-specific antigens

Gene/protein	Peptide	Restriction element	Tissue expression	Reference
Tyrosinase	MLLAVLYCL	HLA-A2	melanocytes	129
	YMNGTMSQV	HLA-A2	melanocytes	129
	SEIWRDIDF	HLA-B44	melanocytes	117
	AFLPWHRLF	HLA-A24	melanocytes	130
Melan-A/MART-1	AAGIGILTV	HLA-A2	melanocytes	131
	ILTVILGVL	HLA-A2	melanocytes	132
gp100/pmel17	ITDQVPFSY	HLA-A2	melanocytes	133
	YLEPGPVTA	HLA-A2	melanocytes	133,134
	KTWGQYWQV	HLA-A2	melanocytes	133,135
	LLDGTATLRL	HLA-A2	melanocytes	111,133
	VLYRYGSFSV	HLA-A2	melanocytes	133
	ALLAVGATK	HLA-A3	melanocytes	136
TRP-1/gp75	MSLQRQFLR	HLA-A31	melanocytes	137
TRP2	LLPGGRPYR	HLA-A31	melanocytes	115
HA-1	VLHDDLLEA	HLA-A2	haematopoietic cells	138

specific CTL in the blood of blood donors suggests common occurrence of such autoreactive CTL (Chapter 4,119,120). Indeed some melanoma patients that respond to immunotherapy show local depigmentation of the skin, as a token of destroyed melanocytes (121,122). Moreover, in melanoma patients evidence for *in vivo* activation of MART-1 specific CTL was found (123). It remains to be resolved whether or not the tissue-specific antigens in melanoma cells are expressed at a higher level than in normal melanocytes. *In vitro* normal melanocytes are recognized by tissue-specific CTL, but those melanocytes are unphysiologically stimulated. On the contrary, in a mouse model evidence was found for differential expression of TRP-1 in melanoma cells and melanocytes (124). Such differential expression might lead to differential recognition by tissue-specific CTL.

Minor histocompatibility antigens (mH antigens) with restricted tissue distribution, such as HA-1 and HA-2 that are expressed on human hematopoietic cells (125) and HB-1 that is expressed by B-lymphocyte blasts (126), can be regarded as tissue-specific antigens. The differences in mH antigens can be employed to generate mH-specific CTL that subsequently lyse leukemic cells (127).

The prostate-specific antigen is expressed by most adenocarcinomas of the prostate. An HLA-A2-binding peptide was discovered that was able to induce peptide-specific CTL. However, lysis of tumorous prostate cells, indicative of processing and presentation of this peptide, still needs to be proven (128).

3.4. Overexpressed Normal Antigens

The protein p53 is involved in the regulation of normal cell division. In tumor cells p53 is often overexpressed as a result of a p53 point mutation. This might lead to a higher amount of p53-derived peptides presented in MHC molecules on the cell surface. Perhaps this enhanced level of presentation lies at the root of the immune response against p53 observed in many cancer patients (139). Moreover, human-p53-specific CTL raised in HLA-transgenic mice can discriminate between tumor cells and normal cells, killing only the first (140). Similar results were obtained with the overexpressed mdm 2 protein (141). Based on these observations, we might conclude that the difference in expression levels leads to discrimination between normal and tumor cells.

The therapeutic potential of p53-specific CTL was recently illustrated in mice. The p53-specific CTL, raised in a p53 gene deficient mouse that is not tolerant to p53, were capable of preventing the outgrowth of p53-expressing tumor cells and of eradicating established tumors (142).

The HER-2/neu protein is expressed at low levels in epithelial cells, but is overexpressed in carcinoma of the breast (143), ovary (144), uterus (145), stomach (146), and adenocarcinoma of the lung (147). Recently an HLA-A2-resticted CTL epitope, spanning positions 654–662 of the HER-2/neu protein, was discovered that is naturally processed and presented in pancreatic cancer cells (148).

In addition, in a mouse model the mdm2 protein, which is overexpressed in tumor cells, was shown to encompass a naturally processed and presented CTL epitope (149).

Carcinoembryonic antigen (CEA) is expressed at high levels on tumors of epithelial origin, and on normal colon epithelium (150). One peptide was shown to bind stably to HLA-A*0201 and to elicitate peptide-specific CTL in HLA-A2-transgenic mice (151). However, recognition of CEA-expressing tumor cells still has to be proven.

3.5. Unconventional Antigens

Not only the normal open reading frame can give rise to CTL epitopes. A peptide resulting from a shifted reading frame proved to encompass a CTL epitope (152). Even intronic sequences, that are aberrantly expressed in tumor cells, can give rise to CTL epitopes (153). However, general expression of such antigens in more individuals is unlikely, and therefore they are less preferred for anti-tumor immunotherapy.

Both groups IIIc and IIId consist of auto-antigens. The common occurrence of CTL specific for auto-antigens (154) raises questions about negative selection in the thymus and clonal exhaustion or unresponsiveness in the periphery. It is speculated that only the CTL specific for high affinity self-peptides are deleted, and CTL for low affinity self-peptides are allowed to leave the thymus (155). However, we tested the affinity of those peptides and the stability of the complexes they form with MHC. They all form stable complexes, a feature that was shown to be better correlated with immunogenicity than binding affinity (156, Chapter 6). Moreover, the affinity, fine-specificity and T cell receptor composition of *in vitro*-induced tyrosinase-specific CTL was studied. No evidence for a poor performance, regarding these aspects, of these tissue-specific autoreactive CTL was found. Furthermore, in transgenic mice expressing a viral oncogene under the keratin-14 promotor, resulting in oncogene expression in epithelial cells and cortical thymocytes, no evidence for clonal deletion of oncogene-specific CTL was found (157). Oncogene-specific CTL could be induced in these transgenic mice upon peptide vaccination. *In vitro* these CTL were able to lyse cultured transgenic epithelial cells, but *in vivo* there was no sign of skin destruction. Nevertheless, tumor cells expressing the viral oncogene were efficiently eliminated by these vaccinated transgenic mice. Similar results were obtained in transgenic mice expressing a viral oncogene in lymphoid cells (158). Most likely autoreactive CTL can be put to good use in anti-tumor therapies. Indeed, melanoma patients undergoing immunotherapy who show destruction of melanocytes (vitiligo) have a better prognosis (159).

4. NATURALLY OCCURRING ANTI-TUMOR RESPONSES IN PATIENTS

Although antigens are present on tumor cells, the patient often does not generate CTL against his/her tumor. In melanoma patients tumor-specific CTL are more commonly found than in patients carrying lung-, colon-, or kidneycancers (160). Perhaps melanoma cells are more immunogenic due to their frequent expression of MHC class II molecules (161), adhesion molecules (162), CD40, and/or release of cytokines (163). Nevertheless, no signs of selective immune pressure on 39 malignant melanoma cell lines were found (164), although other reports show that immunoselection occurs *in vivo* (165). Some tumors actively prevent induction of CTL by mediating apoptosis via expression of DF3/MUC1 (166) or Fas-L (167), by downregulating the ζ-chain of the TCR /CD3 complex (168,169) and p56lck in T cells (170), or secreting cytokines that either hamper activation of CTL, such as TGFβ (171) and other factors (172), or hamper the maturation of antigen presenting cells (173). It is speculated that TGFβ is a growth factor for Th3 cells that suppress Th1 cells, subsequently preventing them from helping CTL (174). Indeed, in tumors that secreted TGFβ less TILs were found (175). Moreover, tumor cells can prevent presentation of their antigens by downregulating their MHC class I molecules (176), or by losing expression of the antigen (177). *In vitro* tumor specific tumor infiltrating lymphocytes can be easily expanded, though *in vivo* they were not able to eradicate the tumor (178). The level of antigen expression on melanoma cells was enough to activate CTL to lysis, but not enough for proliferation and detectable IL-2 secretion (179). These examples illustrate that often the anti-tumor response is not sufficient for tumor eradication.

Furthermore, if the immune system succeeds in mounting a tumor-specific CTL response, it is limited to a few antigens. Although many potential antigens are present on a tumor cell, the immune system only generates CTL against a few. These are called immuno-dominant antigens. The other antigens are not perceived. However, when the immune system is forced to generate a response against them, for instance by offering them in high amounts in an immunogenic form, CTL will react against these subdominant antigens presented on the original tumor cell (180,181). Thus, many CTL epitopes are available, but only a few will be used in the normal situation. Eliciting CTL specific for those subdominant antigens significantly enhances the anti-tumor CTL repertoire, and thereby the therapeutic potential of such CTL (182). Notably, in melanoma patients the majority of blood-derived HLA-A2-restricted CTL were directed against tumor-associated antigens, and not against tissue-specific antigens (183). A study on antigen-specificity of TIL revealed that the majority of HLA-A2-restricted CTL recognized the MART-1/Melan-A tissue-specific antigen (184), but CTL against the other tissue-specific antigens were rare.

Moreover, expression of melanocyte tissue-specific antigens in melanoma cells determined in biopsies was found to be heterogeneous. In 10 of 25 biopsies MART-1 was expressed in less than 50% of the tumor cells (185). A therapy focused only on inducing MART-1 specific CTL would thus result in outgrowth of MART-1 negative tumor cells. Generating CTL against a large panel of tissue-specific antigens would broaden the existing immune response and reduce the chance of tumor escape from immune recognition. When the generation of CTL against individual antigens is carried out at different sites, the problem of immunodominance can be overcome (186). It might be preferable to first optimize immunization protocols against such self-epitopes in animal models. For melanocyte tissue-specific antigens those are now available (187,188).

5. PREDICTING NOVEL CTL-EPITOPES/REVERSE IMMUNOLOGY

Owing to the laborious work of identifying the tumor antigens recognized by TIL, we can now start to predict which peptides can serve as CTL epitopes. Applying the

MHC-binding motifs (189,190), known tumor antigens can be screened for peptides that are likely to bind to other MHC class I molecules. Likewise, members of a gene family with a tumor antigen-bearing relative can be screened for other putative MHC class I binding peptides. Actual MHC-binding can be determined *in vitro* (191,192). Immunogenicity and natural processing of selected peptides can be studied *in vivo* in (transgenic) mice expressing the appropriate MHC molecule (193), or with *in vitro* CTL-induction methods. *In vitro* cytotoxicity assays on tumor cells reveal whether or not the chosen peptides are naturally processed and presented in the MHC molecule. We and others successfully employed this strategy to find novel CTL epitopes (194–196).

Screening of large proteins for peptides that fit the binding motif for a certain MHC molecule yields a list of peptides. The importance of aminoacids at non-anchor positions for binding is underscored, but the interplay between aminoacids at different positions is too complicated to be able to accurately predict actual binding (197). As a consequence, many peptides have to be synthesized and tested for genuine binding. Whereas MHC class I binding is assessed relatively easily on larger numbers of peptides, determining their immunogenicity is very laborious. A new assay was developed to narrow down the number of MHC-binding peptides that have to be tested for immunogenicity. Most peptide-MHC-binding assays measure peptide binding in the presence of solubilized peptide, and in many instances at temperatures lower than 37°C. Under these circumstances mainly on-rates are measured. *In vivo* it is likely that the off-rate is equally important. We developed an assay to measure the half-life of peptides bound to MHC at 37°C. Peptides that remained firmly bound to MHC molecules at 37°C were often immunogenic, whereas peptides that were released from the complex at this temperature were not.

Moreover, with reverse immunology strategies new tumor antigens can be identified. Analogous to the melanocyte-lineage tissue-specific antigens (i.e. MART-1, tyrosinase etc.), we studied the B lymphocyte-lineage tissue-specific antigen CD19. Since processing and presentation in MHC class I molecules of transmembrane molecules has been proven (198), they too can serve as tumor antigens for CTL-based immunotherapy. Several murine MHC-binding and immunogenic peptides were identified. After repetitive vaccinations CTL were isolated that specifically recognized both a mouse CD19-derived peptide and murine B-cell lymphoma cells (Chapter 3). We identified human CD19-derived HLA-A2-binding peptides (unpublished observations), but their immunogenicity and natural processing still need to be studied.

Several factors have to be taken into account in prediction of CTL epitopes starting from the protein sequence. Of course the expression levels of a putative tumor antigen have to be high enough for recognition by CTL. When the natural expression level of a peptide epitope is below threshold for CTL recognition, the usefulness of such an epitope in immunotherapy is less likely (199). Moreover, some proteins undergo post-translational modifications that alter the original aminoacid sequence (200) or add structures to the protein (201,202), that would not be predicted from the genomic information.

The human TAP heterodimer, in contrast to the murine TAP heterodimer, only selects peptides on length and hardly on amino acid composition (203). In theory all peptides that bind to MHC molecules will be transported by the human TAP heterodimer, thus no extra selection takes place at this point. However, single point mutations in one of the TAP proteins (TAP2) significantly alter the specificity of the transporter complex, opening the possibility that in some individuals the TAP proteins could select for specific peptides (204). Although such allelic differences have not been found so far, they might be of importance in some individual cases. Another factor that has to be taken into account in prediction of CTL epitopes concerns the processing of proteins. Preferential cutting of a protein could destroy a putative epitope. Such cleavage-motifs are now being studied (205). When in the near future motifs for protein processing are fully revealed, prediction of CTL epitopes becomes even more accurate.

REFERENCES

1. Yokoyama, W. M. 1995. Natural killer cell receptors. *Curr. Opin. Immunol. 7*:110–120.
2. Moretta, A., R. Biassoni, C. Bottino, D. Pende, M. Vitale, A. Poggi, M. C. Mingari, and L. Moretta. 1997. Major histocompatibility complex class I-specific receptors on human natural killer and T lymphocytes. *Immunol. Rev. 155*:105–117.
3. Held, W., J. Roland, and D. H. Raulet. 1995. Allelic exclusion of Ly49-family genes encoding class I MHC-specific receptors on NK cells. *Nature 376*:355–358.
4. Yu, Y. Y. L., T. George, J. R. Dortman, J. Roland, V. Kumar, and M. Bennett. 1996. The role of Ly49A and 5E6(Ly49C) molecules in hybrid resistance mediated by murine natural killer cells against normal T cell blasts. *Immunity 4*: 67–76.
5. Burshtyn, D. N., A. M. Scharenberg, N. Wagtmann, S. Rajagopalan, K. Berrada, T. Yi, J.-P. Kinet, and E. O. Long. 1996. Recruitment of tyrosine phosphatase HCP by the killer cell inhibitory receptor. *Immunity 4*:77–85.
6. Kärre, K., H. G. Ljunggren, G. Piontek, and R. Kiessling. 1986. Selective rejection of H-2-deficient lymphoma variants suggests alternative immune defense strategy. *Nature 319*:675–678.
7. Quillet, A., F. Presse, C. Marchiol-Fournigault, A. Harel-Bellan, M. Benbunan, H. L. Ploegh, and D. Fradelizi. 1988. Increased resistance to non-MHC-restricted cytotoxicity related to HLA A, B expression. Direct demonstration using β_2-microglobulin-transfected Daudi cells. *J. Immunol. 141*:17–20.
8. Storkus, W. J., J. Alexander, J. A. Payne, J. R. Dawson, and P. Cresswell. 1989. Reversal of natural killing susceptibility in target cells expressing transfected class I HLA genes. *Proc. Natl. Acad. Sci. U. S. A. 86*:2361–2364.
9. Chadwick, B. S., and R. G. Miller. 1992. Hybrid resistance *in vitro* . Possible role of both class-I MHC and self peptides in determining the level of target cell sensitivity. *J. Immunol. 148*:2307–2313.
10. Storkus, W. J., R. D. Salter, P. Cresswell, and J. R. Dawson. 1992. Peptide-induced modulation of target cell sensitivity to natural killing. *J. Immunol. 149*:1185–1190.
11. Townsend, A. R. M., J. Rothbard, F. M. Gotch, G. Bahadur, D. Wraith, and A. J. McMichael. 1986. The epitopes of influenza nucleoprotein recognized by cytotoxic T lymphocytes can be defined with short synthetic peptides. *Cell 44*:959–968.
12. . Kronenberg, M., G. Siu, L. Hood, and N. Shastri. 1986. The molecular genetics of the T-cell antigen receptor and T-cell antigen recognition. *Annu. Rev. Immunol. 4*:529–591.
13. . Seboun, E., M. A. Robinson, T. J. Kindt, and S. L. Hauser. 1989. Insertion/deletion-related polymorphisms in human T-cell receptor β gene complex. *J. Exp. Med. 170*:1263–1270.
14. Nikolic-Zugic, J., M. J. Bevan. 1990. Role of self-peptides in positively selecting the T-cell repertoire. *Nature 344*:65–67.
15. Kappler, J. W., U. D. Staerz, J. White, P. C. Marrack. 1988. Self-tolerance eliminates T cells specific for M1s-modified products of the major histocompatibility complex. *Nature 322*:35–40.
16. Kisielow, P., H. Bluthmann, U. D. Staerz, M. Steinmetz, H. Von Boehmer. 1988. Tolerance in T cell receptor transgenic mice involves deletion of immature CD4+CD8+ thymocytes. *Nature 333*:742–746.
17. Van Meerwijk, J. P. M., S. Marguerat, R. K. Lees, R. N. Germain, B. J. Fowlkes, and H. Robson MacDonald. 1997. Quantitative impact of thymic clonal deletion on the T cell repertoire. *J. Exp. Med. 185*:377–383.
18. Chien, Y. H., and M. M. Davis. 1993. How alfa beta T-cell receptors 'see' peptide/MHC complexes. *Immunol. Today 14*:597–602.
19. Garboczi, D. N., P. Ghosh, U. Utz, Q. R. Fan, W. E. Biddison, and D. C. Wiley. 1996. Structure of the complex between human T-cell receptor, viral peptide and HLA-A2. *Nature 384*:134–141.
20. Mueller, D. L., M. K. Jenkins, and R. H. Schwartz. 1989. Clonal expansion versus functinal clonal inactivation: a costimulatory signalling pathway determines the outcome of T cell antigen receptor occupancy. *Annu. Rev. Immunol. 7*:445–480.
21. Gajewski, T. F., F. Fallarino, C. Uyttenhove, and T. Boon. 1996. Tumor rejection requires a CTLA4 ligand provided by the host or expressed on the tumor. *J. Immunol. 156*:2909–2917.
22. Lanier, L. L. 1997. Natural killer cells: from no receptors to too many. *Immunity 6*:371–378.

23. Ikeda, H., B. Lethé, F. Lehmann, N. Van Baren, J.-F. Baurain, C. De Smet, H. Chambost, M. Vitale, A. Moretta, T. Boon, and P. Coulie. 1997. Characterization of an antigen that is recognized on a melanoma showing partial HLA loss by CTL expressing an NK inhibitory receptor. *Immunity 6*:199–208.
24. Kim, C. J., J. K. Taubenberger, T. B. Simonis, D. E. White, S. A. Rosenberg, and F. M. Marincola. 1996. Combination therapy with interferon gamma and interleukin-2 for the treatment of metastatic melanoma. *J. Immunother. 19*:50–58.
25. Karp, S. E., A. Faraber, J. C. Salo, P. Hwu, G. Jaffee, A. Asher, E. Shiloni, N. P. Restifo, J. J. Mulé, and S. A. Rosenberg. 1993. Cytokine secretion by genetically modified non-immunogenic murine fibrosarcoma. *J. Immunol. 150*:896–908.
26. Fearon, E., D. Pardoll, T. Itaya, P. Golumbek, H. Levitsky, J. Simons, H. Karasuyama, B. Vogelstein, and P. Frost. 1990. Interleukin-2 production by tumor cells bypasses T helper function in the generation of an anti-tumor response. *Cell 60*:397–403.
27. Gansbacher, B., K. Zier, B. Daniels, K. Cronin, R. Bannerji, and E. Gilboa. 1990. Interleukin-2 gene transfer into tumor cells abrogates tumorigenicity and induces protective immunity. *J. Exp. Med. 172*:1217–1224.
28. Handa, K., R. Suzuki, H. Matsui, Y. Shimizu, and K. Kumagai. 1983. Natural Killer (NK) cells as a responder to Interleukin 2 (IL 2). II. IL 2-induced Interferon γ production. *J. Immunol. 130*:988–992.
29. Colombo, M. P., and G. Forni. 1994. Cytokine gene transfer in tumor inhibition and tumor therapy: where are we now? *Immunol. Today 15*:48–51.
30. Toes, R. E. M., R. J. J. Blom, E. I. H. van der Voort, R. Offringa, C. J. M. Melief, and W. M. Kast. 1996. Protective anti-tumor immunity induced by immunization with completely allogeneic tumor cells. *Cancer Res. 56*:3782–3787.
31. Van Elsas, A., C. Aarnoudse, C. E. van der Minne, C. W. van der Spek, N. Brouwenstijn, S. Osanto, and P. I. Schrier. 1997. Transfection of IL-2 augments CTL response to human melanoma cells *in vitro*: immunological characterization of a melanoma vaccine. *J. Immunother. 20*:343–353.
32. Arienti, F., J. Sule-Suso, F. Belli, L. Mascheroni, L. Rivoltini, C. Melani, M. Maio, N. Cascinelli, M. P. Colombo, and G. Parmiani. 1996. Limited anti-tumor T cell response in melanoma patients vaccinated with interleukin-2 gene-transduced allogeneic melanoma cells. *Hum. Gene Ther. 7*:1955–1963.
33. Veelken, H., A. Mackensen, M. Lahn, G. Kohler, D. Becker, B. Franke, U. Brennscheidt, P. Kulmburg, R. M. Rosenthal, H. Keller, J. Hasse, W. Schultze-Seemann, E. H. Farthmann, R. Mertelsmann, and A. Lindemann. 1997. A phase-I clinical study of autologous tumor cells plus interleukin-2-gene-transfected allogeneic fibroblasts as a vaccine in patients with cancer. *Int. J. Cancer. 70*:269–277.
34. Golumbek, P. T., R. Azhari, E. M. Jaffe, H. Levitsky, A. Lazenby, K. Leong, and D. M. Pardoll. 1993. Controlled release, biodegradable cytokine depots: a new approach in cancer vaccine design. *Cancer Res. 53*:5841–5844.
35. Salvadori, S., B. Gansbacher, A. M. Pizzimenti, and K. S. Zier. 1994. Abnormal signal transduction by T cells of mice with parental tumors is not seen in mice bearing IL-2-secreting tumors. *J. Immunol. 153*:5176–5182.
36. Zier, K., B. Gansbacher, and S. Salvadori. 1996. Preventing abnormalities in signal transduction of T cells in cancer: the promise of cytokine gene therapy. *Immunol. Today 17*.39–45.
37. Pardoll, D. M. 1995. Paracrine cytokine adjuvants in cancer immunotherapy. *Annu. Rev. Immunol. 13*:399–415.
38. Huang, A. Y. C., P. Golumbek, M. Ahmadzadeh, E. Jaffee, D. M. Pardoll, and H. Levitsky. 1994. The role of bone marrow derived cells in presenting MHC class I-restricted tumor antigens. *Science 264*:961–965.
39. Kast, W. M., L. Roux, J. Curren, H. J. Blom, |A. C. Voordouw, R. H. Meloen, K. Kolakofsky, and C. J. M. Melief. 1991. Protection against lethal Sendai virus infection by *in vivo* priming of virus-specific cytotoxic T lymphocytes with a free synthetic peptide. *Proc. Nat. Acad. Sci. U. S. A. 88*:2283–2287.
40. Feltkamp, M. C. W., H. L. Smits, M. P. M. Vierboom, R. P. Minnaar, B. M. de Jongh, J. W. Drijfhout, J. ter Schegget, C. J. M. Melief, and W. M. Kast. 1993. Vaccination with cytotoxic T lymphocyte epitope-containing peptide protects against a tumor induced by human papillomavirus type 16-transformed cells. *Eur. J. Immunol. 23*:2242–2249.
41. Celis, E., V. Tsai, C. Crimi, R. DeMars, P. A. Wentworth, R. W. Chesnut, H. M. Grey, A. Sette, and H. M. Serra. 1994. Induction of anti-tumor cytotoxic T lymphocytes in normal humans using primary cultures and synthetic peptide epitopes. *Proc. Natl. Acad. Sci. U. S. A. 91*:2105–2109.
42. Salgaller, M. L., F. M. Marincola, J. N. Cormier, and S. A. Rosenberg. 1996. Immunization against epitopes in the human melanoma antigen gp100 following patient immunization with synthetic peptides. *Cancer Res. 56*:4749–4757.

43. Cormier, J. N., M. L. Salgaller, T. Prevette, K. C. Barracchini, L. Rivoltini, N. P. Restifo, S. A. Rosenberg, and F. M. Marincola. 1997. Enhancement of cellular immunity in melanoma patients immunized with a peptide from MART-1/Melan A. *Cancer J. Sci. Am. 3*:37–44.
44. Hu, X., N. G. Chakraborty, J. R. Sporn, S. H. Kurtzman, M. T. Ergin, and B. Mukherji. 1996. Enhancement of cytolytic T lymphocyte precursor frequency in melanoma patients following immunization with the MAGE-1 peptide loaded antigen presenting cell-based vaccine. *Cancer Res. 56*:2479–2483.
45. Toes, R. E. M., R. J. J. Blom, R. Offringa, W. M. Kast, and C. J. M. Melief. 1996. Enhanced tumor outgrowth after peptide vaccination. *J. Immunol. 156*:3911–3918.
46. Toes, R. E. M., R. Offringa, R. J. J. Blom, C. J. M. Melief, and W. M. Kast. 1996. Peptide vaccination can lead to enhanced tumor growth through specific T cell tolerance induction. *Proc. Natl. Acad. Sci. U. S. A. 93*:7855–7860.
47. Irvine, K. R., J. B. Rao, S. A. Rosenberg, and N. P. Restifo. 1996. Cytokine enhancement of DNA immunization leads to effective treatment of established pulmonary metastases. *J. Immunol. 156*:238–245.
48. Warnier, G., M. T. Duffour, C. Uyttenhove, T. F. Gajewski, C. Lurquin, H. Haddada, M. Perricaudet, and T. Boon. 1996. Induction of a cytolytic T cell response in mice with a recombinant adenovirus coding for tumor antigen P815A. *Int. J. Cancer. 67*:303–310.
49. Chen, P. W., M. Wang, V. Bronte, Y. F. Zhai, S. A. Rosenberg, and N. P. Restifo. 1996. Therapeutic anti-tumor response after immunization with a recombinant adenovirus encoding a model tumor associated antigen. *J. Immunol. 156*:224–231.
50. Zajac, P., D. Oertli, G. C. Spagnoli, C. Noppen, C. Schaefer, M. Heberer, and W. R. Marti. 1997. Generation of tumoricidal cytotoxic T lymphocytes from healthy donors after in vitro stimulation with a replication-incompetent vaccinia virus encoding MART-1/Melan-A 27–35 epitope. *Int. J. Cancer 71*:491–496.
51. Rao, J. B., R. S. Chamberlain, V. Bronte, M. W. Carroll, K. R. Irvine, B. Moss, S. A. Rosenberg, and N. P. Restifo. 1996. IL-12 is an effective adjuvant to recombinant vaccinia virus based tumor vaccines: enhancement by simultaneous B7.1 expression. *J. Immunol. 156*:3357–3365.
52. Jager, E., M. Ringhoffer, H. P. Dienes, M. Arand, J. Karbach, D. Jager, C. Ilsemann, M. Hagedorn, F. Oesch, and A. Knuth. 1996. Granulocyte macrophage colony stimulating factor enhances immune responses to melanoma associated peptides *in vivo*. *Int. J. Cancer 67*:54–62.
53. Chamberlain, R. S., M. W. Carroll, V. Bronte, P. Hwu, S. Warren, J. C. Yang, M. Nishimura, B. Moss, S. A. Rosenberg, and N. P. Restifo. 1996. Costimulation enhances the active immunotherapy effect of recombinant anti-cancer vaccines. *Cancer Res. 56*:2832–2836.
54. Reeves, M. E., R. E. Royal, J. S. Lam, S. A. Rosenberg, and P. Hwu. 1996. Retroviral transduction of human dendritic cells with a tumor associated antigen gene. *Cancer Res. 56*:5672–5677.
55. Eggers, M., B. Boes-Fabian, T. Ruppert, P. M. Kloetzel, and U. H. Koszinowski. 1995. The cleavage preference of the proteasome governs the yield of antigenic peptides. *J. Exp. Med. 182*:1865–1870.
56. Stock, D., P. M. Nederlof, E. Seemüller, W. Baumeister, R. Huber, and J. Löwe. 1996. Proteasome: from structure to function. *Curr. Opin. Biotechnol. 7*:376–385.
57. Groettrup, M., A. Soza, U. Kuckerkorn, and P.-M. Kloetzel. 1996. Peptide antigen production by the proteasome: complexity provides efficiency. *Immunol. Today 17*:429–435.
58. Cerundolo, V., A. Benham, V. Braud, S. Mukherjee, K. Gould, B. Macino, J. Neefjes, and A. Townsend. 1997. The proteasome-specific inhibitor lactacystin blocks presentation of cytotoxic T lymphocyte epitopes in human and murine cells. *Eur. J. Immunol. 27*:336–341.
59. Kelly, A., S. H. Powis, L. A. Kerr, I. Mockridge, T. Elliott, J. Bastin, B. Uchanska-Ziegler, A. Ziegler, J. Trowsdale, and A. Townsend. 1992. Assembly and function of the two ABC transporter proteins encoded in the human major histocompatibility complex. *Nature 355*:641–644.
60. Androlewicz, M. J., and P. Cresswell. 1994. Human transporters associated with antigen processing possess a promiscuous peptide-binding site. *Immunity 1*:7–14.
61. Momburg, F., J. Roelse, G. J. Hämmerling, and J. J. Neefjes. 1994. Peptide size selection by the major histocompatibility complex-encoded peptide transporter. *J. Exp. Med. 179*:1613–1623.
62. Sadasivan, B., P. J. Lehner, B. Ortmann, T. Spies, and P. Cresswell. 1996. Roles for calreticulin and a novel glycoprotein, tapasin, in the interaction of MHC Class I molecules with TAP. *Immunity 5*:103–114.
63. Madden, D. R. 1995. The three-dimensional structure of peptide-MHC complexes. *Annu. Rev. Immunol. 13*:587–622.
64. Lammert, E., D. Arnold, M. Nijenhuis, F. Momburg, G. J. Hammerling, J. Brunner, S. Stevanovic, H. G. Rammensee, and H. Schild. 1997. The endoplasmic reticulum-resident stress protein gp96 binds peptides translocated by TAP. *Eur. J. Immunol. 27*:923–927.
65. Anderson, K., P. Cresswell, M. Gammon, J. Hermes, A. Williamson, and H. Zweerink. 1991. Endogenously synthesized peptide with an endoplasmic reticulum signal sequence sensitizes antigen processing mutant cells to class I-restricted cell-mediated lysis. *J. Exp. Med. 174*:489–492.

66. J. Roelse, M. Gromme, F. Momburg, G. Hammerling, and J. Neefjes. 1994. Trimming of TAP-translocated peptides in the endoplasmic reticulum and in the cytosol during recycling. *J. Exp. Med. 180*:1591–1597.
67. Elliott, T., A. Willis, V. Cerundolo, and A. Townsend. 1995. Processing of major histocompatibility class I-restricted antigens in the endoplasmic reticulum. *J. Exp. Med. 181*:1481–1491.
68. Coulie, P. G., F. Lehmann, B. Lethé, J. Herman, C. A. Lurquin, and T. Boon. 1995. A mutated intron sequence codes for an antigenic peptide recognized by cytolytic T lymphocytes on a human melanoma. *Proc. Natl. Acad. Sci. U. S. A. 92*:7976–7980.
69. Robbins, P. F., M. El-Gamil, Y. F. Li, Y. Kawakami, D. Loftus, E. Appella, and S. A. Rosenberg. 1996. A mutated β-catenin gene encodes a melanoma-specific antigen recognized by tumor infiltrating lymphocytes. *J. Exp. Med. 183*:1185–1192.
70. Brandle, D., F. Brasseur, P. Weynants, T. Boon, and B. J. Van den Eynde. 1996. A mutated HLA A2 molecule recognized by autologous cytotoxic T lymphocytes on human renal cell carcinoma. *J. Exp Med. 183*:2501–2508.
71. Sijts, E. J. A. M., M. L. de Bruin, M. E. Ressing, J. D. Nieland, E. A. M. Mengede, C. J. Boog, F. Ossendorp, W. M. Kast, and C. J. M. Melief. 1994. Identification of an H-2 Kb-presented Moloney murine leukemia virus cytotoxic T lymphocyte epitope that displays enhanced recognition in H-2 Db mutant bm13 mice. *J. Virol. 68*:6038–6046.
72. Sijts, E. J. A. M., F. Ossendorp, E. A. M. Mengede, P. J. van den Elsen, and C. J. M. Melief. 1994. An immunodominant MCF Murine Leukemia Virus encoded CTL epitope identified by its MHC class I-binding motif, explains MuLV type specificity of MCF-directed CTL. *J. Immunol. 152*:106–116.
73. Kast, W. M., R. Offringa, P. J. Peters, A. C. Voordouw, R. H. Meloen, A. J. van der Eb, and C. J. M. Melief. 1989. Eradication of adenovirus E1-induced tumors by E1a-specific cytotoxic T lymphocytes. *Cell 59*:603–614.
74. Toes, R. E. M., R. Offringa, H. J. Blom, R. M. P. Brandt, A. J. van der Eb, C. J. M. Melief, and W. M. Kast. 1995. An Adenovirus type 5 early region 1B-encoded CTL epitope-mediating tumor eradication by CTL clones is down-modulated by an activated ras oncogene. *J. Immunol. 154*:3396–3405.
75. Tanaka, Y., M. J. Tevethia, D. Kalderon, A. E. Smith, and S. S. Tevethia. 1988. Clustering of antigenic sites recognized by cytotoxic T lymphocyte clones in the amino terminal half of SV40 T antigen. *Virology 162*:427–436.
76. Bertoletti, A., C. Ferrari, F. Fiaccadori, A. Penna, R. Margolskee, H.-J. Schlicht, P. Fowler, S. Guilhot, and F. V. Chisari. 1991. HLA class-I human cytotoxic T cells recognize endogenously synthesized hepatitis B virus nucleocapsid antigen. *Proc. Natl. Acad. Sci. USA 88*:10445–10449.
77. Penna, A., F. V. Chisari, a. Bertoletti, G. Missale, P. Fowler, T., Giuberti, F. Fiaccadori, and C. Ferrari. 1991. Cytotoxic T lymphocytes recognize an HLA-A2-restricted epitope within the hepatitis B virus nucleocapsid antigen. *J. Exp. Med. 174*:1565–1569.
78. Jacobson, S., H. Shida, D. McFarlin, A. Fauci, and S. Koenig. 1990. Circulating HTLV-1 pX-specific, $CD8^+$ cytotoxic T lymphocytes in patients with HTLV-1 associated neurologic disease. *Nature 348*:245–248.
79. Ressing, M. E., W. J. van Driel, E. Celis, A. Sette, R. M. P. Brandt, M. Hartman, J. D. Anholts, G. M. Schreuder, W. B. ter Harmsel, G. J. Fleuren, B. J. Trimbos, W. M. Kast, and C. J. M. Melief. 1996. Occasional memory cytotoxic T-cell responses of patients with human papillomavirus type 16-positive cervical lesions against a human leukocyte antigen-A*0201-restricted E7-encoded epitope. *Cancer Res. 56*:582–588.
80. Alexander, M., M. L. Salgaller, E. Celis, A. Sette, W. A. Barnes, S. A. Rosenberg, and M. A. Steller. 1996. Generation of tumor-specific cytolytic T lymphocytes from peripheral blood of cervical cancer patients by in vitro stimulation with a synthetic human papillomavirus type 16 E7 epitope. *Am. J. Obstet. Gynecol. 175*:1586–1593.
81. Skipper, J., and H. J. Stauss. 1993. Identification of two cytotoxic T lymphocyte-recognized epitopes in the Ras protein. *J. Exp. Med. 177*:1493–1498.
82. Peace, D. J., J. W. Smith, W. Chen, S. G. You, W. L. Cosand, J. Blake, and M. A. Cheever. 1994. Lysis of ras oncogene-transformed cells by specific cytotoxic T lymphocytes elicited by primary *in vitro* immunization with mutated ras peptide. *J. Exp. Med. 179*:473–479.
83. Fossum, B., T. Gedde-Dahl, J. Breivik, J. A. Eriksen, A. Spurkland, E. Thorsby, and G. Gaudemack. 1994. p21-ras-peptide-specific T-cell responses in a patient with colorectal cancer. $CD4^+$ and $CD8^+$ T cells recognize a peptide corresponding to a common mutation (13Gly->Asp). *Int J. Cancer 56*:40–45.
84. Yanuck, M., D. P. Carbone, C. D. Pendleton, T. Tsukui, S. F. Winter, J. D. Minna, and J. A. Berzofsky. 1993. A mutant p53 tumor suppressor protein is a target for peptide-induced $CD8^+$ cytotoxic T-cells. *Cancer Res. 53*:3257–3261.

85. Noguchi, Y., E. C. Richards, Y. T. Chen, L. J. Old. 1995. Influence of interleukin 12 on p53 peptide vaccination against established Meth A sarcoma. *Proc. Natl. Acad. Sci. U. S. A. 92*:2219–2223.
86. Ten Bosch, G. J. A., A. C. Toornvliet, T. Friede, C. J. M. Melief, and O. C. Leeksma. 1995. Recognition of peptides corresponding to the joining region of p210$^{BCR\text{-}ABL}$ protein by human T cells. *Leukemia 9*:1344–1348.
87. Cheever, M. A., M. L. Disis, H. Bernhard, J. R. Gralow, S. L. Hand, E. S. Huseby, H. L. Qin, M. Takahashi, and W. Chen. 1995. Immunity to oncogenic proteins. *Immunol. Rev. 145*:33–59.
88. De Plaen, E., K. Arden, C. Traversari, J. J. Gaforio, J.-P. Szikora, C. De Smet, F. Brasseur, P. van der Bruggen, B. Lethé, C. Lurquin, R. Brasseur, P. Chomez, O. De Backer, W. Cavenee, and T. Boon. 1994. Structure, chromosomal localization, and expression of twelve genes of the MAGE family. *Immunogenetics 40*:360–369.
89. Shichijo, S., K. Sagawa, F. Brasseur, T. Boon, and K. Itoh. 1996. MAGE 1 gene is expressed in T cell leukemia. *Int. J. Cancer 65*:709–710.
90. Ishida, H., T. Matsumura, M. L. Salgaller, Y. Ohmizono, Y. Kadono, and T Sawada. 1996. MAGE 1 and MAGE 3 or 6 expression in neuroblastoma related pediatric solid tumors. *Int. J. Cancer 69*:375–380.
91. Corrias, M. V., P. Scaruffi, M. Occhino, B. Debernardi, G. P. Tonini, and V. Pistoia. 1996. Expression of MAGE 1, MAGE 3 and MART 1 genes in neuroblastoma. *Int. J. Cancer 69*:403–407.
92. Sudo, T., T. Kuramoto, S. Komiya, A. Inoue, and K. Itoh. 1997. Expression of MAGE genes in osteosarcoma. *J. Orthopaedic Res. 15*:128–132.
93. Quillien, V., J. L. Raoul, D. Heresbach, B. Collet, L. Toujas, and F. Brasseur. 1997. Expression of MAGE genes in esophageal squamous cell carcinoma. *Anticancer Res. 17*:387–391.
94. Mori, M., H. Inoue, K. Mimori, K. Shibuta, K. Baba, H. Nakashima, M. Haraguchi, K. Tsuji, H. Ueo, G. F. Barnard, and T Akiyoshi. 1996 Expression of MAGE genes in human colorectal carcinoma. *Ann. Surg. 224*:183–188.
95. Russo, V., P. Dalerba, A. Ricci, C. Bonazzi, B. E. Leone, C. Mangioni, P. Allavena, C. Bordignon, and C. Traversari. 1996. MAGE, BAGE and GAGE genes expression in fresh epithelial ovarian carcinomas. *Int. J. Cancer 67*:457–460.
96. Desmet, C., O. Debacker, I. Faraoni, C. Lurquin, F. Brasseur, and T. Boon. 1996. The activation of human gene MAGE 1 in tumor cells is correlated with genome wide demethylation. *Proc. Natl. Acad. Sci. U. S. A. 93*:7149–7153.
97. Traversari, C., P. van der Bruggen, I. F. Luescher, C. Lurquin, P. Chomez, A. van Pel, E. De Plaen, A. Amar-Costesec, and T. Boon. 1992. A nonapeptide encoded by human gene MAGE-1 is recognized on HLA-A1 by cytolytic T lymphocytes directed against tumor antigen MZ2-E. *J. Exp. Med. 176*:1453–1457.
98. Gaugler, B., B. Van den Eynden, P. van der Bruggen, P. Romero, J. J. Gaforio, E. de Plaen, B. Lethé, F. Brasseur, and T. Boon. 1994. Human gene MAGE-3 codes for an antigen recognized on a melanoma by autologous cytolytic T lymphocytes. *J. Exp. Med. 179*:921–930.
99. Van der Bruggen, P., J.-P. Szikora, P. Boël, C. Wildmann, M. Somville, M.-L. Sensi, and T. Boon. 1994. Autologous cytolytic T lymphocytes recognize a MAGE-1 nonapeptide on melanomas expressing HLA-Cw*1601. *Eur. J. Immunol. 24*:2134–2140.
100. Herman, J., P. van der Bruggen, I. F. Luesscher, S. Mandruzzato, P. Romero, J. Thonnard, K. Fleischhauer, T. Boon, and P. G. Coulie. 1996. A peptide encoded by the human MAGE3 gene and presented by HLA B44 induces cytolytic T lymphocytes that recognize tumor cells expressing MAGE3. *Immunogenetics 43*:377–383.
101. Uytttenhove, C., C. Godfraind, B. Lethé, A. Amarcostesec, J. C. Renauld, T. F. Gajewski, M. T. Duffour, G. Warnier, T. Boon, and B. J. Van den Eynde. 1997. The expression of mouse gene P1A in testis does not prevent safe induction of cytolytic T cells against a P1A encoded tumor antigen. *Int. J. Cancer 70*:349–356.
102. Boël, P., C. Wildmann, M.-L. Sensi, R. Brasseur, J.-C. Renauld, P. Coulie, T. Boon, and P. van der Bruggen. 1995. BAGE: a new gene encoding an antigen recognized on human melanomas by cytolytic T lymphocytes. *Immunity 2*:167–175.
103. Van den Eynde, B., O. Peeters, O. de Backer, B. Gaugler, S. Lucas, and T. Boon. 1995. A new family of genes coding for an antigen recognized by autologous cytolytic T lymphocytes on a human melanoma. *J. Exp. Med. 182*:689–698.
104. Chen, Y. T., M. J. Scanlan, U, Sahin, O. Tureci, A. O. Gure, S. L. Tsang, B. Williamson, E. Stockert, M. Pfreundschuh, and L. J. Old. 1997. A testicular antigen aberrantly expressed in human cancers detected by autologous antibody screening. *Proc. Natl. Acad. Sci. U. S. A. 94*:1914–1918.
105. Gaugler, B., N. Brouwenstijn, V. Vantomme, J.-P. Szikora, C. W. van der Spek, J.-J. Patard, T. Boon, P. I. Schrier, and B. J. van den Eynde. 1996. A new gene coding for an antigen recognized by autologous cytolytic T lymphocytes on a human renal carcinoma. *Immunogenetics 44*:323–330.

106. Mazzocchi, A., W. J. Storkus, C. Traversari, P. Tarsini, M. J. Maeurer, L. Rivoltini, C. Vegetti, F. Belli, A. Anichini, G. Parmiani, and C. Castelli. 1996. Multiple melanoma associated epitopes recognized by HLA-A3 restricted CTLs and shared by melanomas but not melanocytes. *J. Immunol. 157*:3030–3038.
107. Bernhard, H., E. Jager, M. J. Maeurer, K.-H. Meyer zum Buschenfelde, and A. Knuth. 1996. Tumor associated antigens in human renal cell carcinoma: MHC restricted recognition by cytotoxic T lymphocytes. *Tissue Antigens 48*:22–31.
108. Takahashi, T., R. F. Irie, D. L. Morton, and D. S. Hoon. 1997. Recognition of gp43 tumor-associated antigen peptide by both HLA-A2 restricted CTL lines and antibodies from melanoma patients. *Cell. Immunol. 178*:162–171.
109. Hoshino, T., N. Seki, M. Kikuchi, T. Kuramoto, O. Iwamoto, I. Kodma, K. Koufuji, J. Takeda, and K Itoh. 1997. HLA class-I-restricted and tumor-specific CTL in tumor-infiltrating lymphocytes of patients with gastric cancer. *Int. J. Cancer 70*:631–638.
110. Bakker, A. B. H., M. W. J. Schreurs, A. J. De Boer, Y. Kawakami, S. A. Rosenberg, G. J. Adema, and C. G. Figdor. 1994. Melanocyte lineage-specific antigen gp100 is recognized by melanoma-derived tumor-infiltrating lymphocytes. *J. Exp. Med. 179*:1005–1009.
111. Kawakami, Y., S. Eliyahu, C. H. Delgado, P. F. Robbins, K. Sakaguchi, E. Appella, J. R. Yannelli, G. J. Adema, T. Miki, and S. A. Rosenberg. 1994. Identification of a human melanoma antigen recognized by tumor-infiltrating lymphocytes associated with *in vivo* tumor rejection. *Proc. Natl. Acad. Sci. U. S. A. 91*:6458–6462.
112. Coulie, P. G., V. Brichard, A. Van Pel, T. Wölfel, J. Schneider, C. Traversari, S. Mattei, E. De Plaen, C. Lurquin, J.-P. Szikora, and T. Boon. 1994. A new gene coding for a differentiation antigen recognized by autologous cytolytic lymphocytes on HLA-A2 melanomas. *J. Exp. Med. 180*:35–42.
113. Kawakami, Y., S. Eliyahu, C. H. Delgado, P. F. Robbins, L. Rivoltini, S. L. Topalian, T. Miki, and S. A. Rosenberg. 1994. Cloning of the gene for a shared human melanoma antigen recognized by autologous T cells infiltrating into tumor. *Proc. Natl. Acad. Sci. U. S. A. 91*:3515–3519.
114. Wang, R.-F., P. F. Robbins, Y. Kawakami, X.-Q. Kang, and S. A. Rosenberg. 1995. Identification of a gene encoding a melanoma tumor antigen recognized by HLA-A31-restricted tumor-infiltrating lymphocytes. *J. Exp. Med. 181*:799–804.
115. Wang, R. F., E. Appella, Y. Kawakami, X. Q. Kang, and S. A. Rosenberg. 1996. Identification of TRP 2 as a human tumor antigen recognized by cytotoxic T lymphocytes. *J. Exp. Med. 184*:2207–2216.
116. Brichard, V., A. Van Pel, T. Wölfel, C. Wölfel, E. De Plaen, B. Lethé, P. Coulie, and T. Boon. 1993. The tyrosinase gene codes for an antigen recognized by autologous cytolytic T lymphocytes on HLA-A2 melanomas. *J. Exp. Med. 178*:489–495.
117. Brichard, V. G., J. Herman, A. van Pel, C. Wildmann, B. Gaugler, T. Wölfel, T. Boon, and B. Lethé. 1996. A tyrosinase nonapeptide presented by HLA B44 is recognized on a human melanoma by autologous cytolytic T lymphocytes. *Eur. J. Immunol. 26*:224–230.
118. Yee, C., M. J. Gilbert, S. R. Riddell, V. G. Brichard, A. Fefer, J. A. Thompson, T. Boon, and P. D. Greenberg. 1996. Isolation of tyrosinase specific CD8(+) and CD4(+) T cell clones from the peripheral blood of melanoma patients following *in vitro* stimulation with recombinant vaccinia virus. *J. Immunol 157*:4079–4086.
119. Herr, W., J. Schneider, A. W. Lohse, K.-H. Meyer-Zumbuschenfelde, and T. Wölfel. 1996. Detection and quantification of blood derived CD8(+) T lymphocytes secreting tumor necrosis factor alpha in response to HLA A2.1 binding melanoma and viral peptide antigens. *J. Immunol. Meth. 191*:131–142.
120. Jager, E., M. Ringhoffer, M. Arand, J. Karbach, D. Jager, C. Ilsemann, M. Hagedorn, F. Oesch, and A. Knuth. 1996. Cytolytic T cell reactivity against melanoma associated differentiation antigens in peripheral blood of melanoma patients and healthy individuals. *Mel. Res. 6*:419–425.
121. Bystryn, J. C., D Rigel, R. J. Friedman, and A. Kopf. 1987. Prognostic significance of hypopigmentation in malignant melanoma. *Arch. Dermatol. 123*:1053–1055.
122. Rosenberg, S. A., and D. E. White. 1996. Vitiligo in patients with melanoma: normal tissue antigens can be targets for cancer immunotherapy. *J. Immunother. 19*:81–84.
123. Marincola, F. M., L. Rivoltini, M. L. Salgaller, T. Player, and S. A. Rosenberg. 1996. Differential anti MART-1/Melan A CTL activity in peripheral blood of HLA A2 melanoma patients in comparison to healthy donors: evidence of *in vivo* priming by tumor cells. *J. Immunother. 19*:266–277.
124. Hara, I., Y. Takechi, and A. N. Houghton. 1995. Implicating a role for immune recognition of self in tumor rejection: passive immunization against the Brown locus protein. *J. Exp. Med. 182*:1609–1614.
125. De Bueger, M., A. Bakker, J. J. van Rood, F. van der Woude, and E. Goulmy. 1992. Tissue distribution indicates heterogeneity among human cytotoxic T lymphocyte-defined non-MHC antigens. *J. Immunol. 149*:1788–1794.

126. Dolstra, H., H. Fredrix, F. Preijers, E. Goulmy, C. G. Figdor, T. M. de Witte, and E. van de Wiel-van Kemenade. 1997. Recognition of a B cell leukemia-associated minor histocompatibility antigen by CTL. *J. Immunol. 158*:560–565.
127. Falkenburg, J. H., W. M. Smit, and R. Willemze. 1997. Cytotoxic T-lymphocyte (CTL) responses against acute or chronic myeloid leukemia. *Immunol. Rev. 157*:223–230.
128. Xue, B. H., Y. Zhang, J. A. Sosman, and D. J. Peace. 1997. Induction of human cytotoxic T lymphocytes specific for prostate-specific antigen. *Prostate 30*:73–78.
129. Wölfel, T., A. Van Pel, V. Brichard, J. Schneider, B. Seliger, K.-H. Meyer zum Büschenfelde, and T. Boon. 1994. Two tyrosinase nonapeptides recognized on HLA-A2 melanomas by autologous cytolytic T lymphocytes. *Eur. J. Immunol. 24*:759–764.
130. Kang, W.-Q., Y. Kawakami, I. K. Sakaguchi, M. El-Gamil, R.-F. Wang, J. R. Yannelli, E. Appella, S. A. Rosenberg, and P. F. Robbins. 1995. Identification of a tyrosinase epitope recognized by HLA-A24 restricted tumor-infiltrating lymphocytes. *J. Immunol. 155*:1343–1348.
131. Kawakami, Y., R. Zakut, S. L. Topalian, H. Stötter, and S. A. Rosenberg. 1994. Identification of the immunodominant peptides of the MART-1 human melanoma antigen recognized by the majority of HLA-A2-restricted tumor infiltrating lymphocytes. *J. Exp. Med. 180*:347–352.
132. Castelli, C., W. J. Storkus, M. J. Maeurer, D. M. Martin, E. C. Huang, B. N. Pramanik, T. L. Nagabhushan, G. Parmiani, and M. T. Lotze. 1995. Mass spectrometric identification of a naturally processed melanoma peptide recognized by $CD8^+$ cytotoxic T lymphocytes. *J. Exp. Med. 181*:363–368.
133. Kawakami, Y., S. Eliyahu, C. Jennings, K. Sakaguchi, X.-Q. Kang, S. Southwood, P. F. Robbins, A. Sette, E. Appella, and S. A. Rosenberg. 1995. Recognition of multiple epitopes in the human melanoma antigen gp100 associated with *in vivo* tumor regression. *J. Immunol. 154*:3961–3968.
134. Cox, A. L., J. Skipper, Y. Cehn, R. A. Henderson, T. L. Darrow, J. Shabanowitz, V. H. Engelhard, D. F. Hunt, and C. L. Slingluff. 1994. Identification of a peptide recognized by five melanoma-specific human cytotoxic T cell lines. *Science 164*:716–719.
135. Bakker, A., M. Schreurs, G. Tafazzul, A. De Boer, Y. Kawakami, G. Adema, and C. Figdor. 1995. Identification of a novel peptide derived from the melanocyte-specific gp100 antigen as the dominant epitope recognized by an HLA-A2.1-restricted anti-melanoma CTL line. *Int. J. Cancer 62*:97–102.
136. Skipper, J. C., D. J. Kittlesen, R. C. Hendrickson, D. D. Deacon, N. L. Harthun, S. N. Wagner, D. F. Hunt, V. H. Engelhard, and C. L. Slingluff. 1996. Shared epitopes for HLA-A3-restricted melanoma-reactive human CTL include a naturally processed epitope from Pmel-17/gp100. *J. Immunol. 157*:5027–5033.
137. Wang, R.-F., M. Parkhurst, Y. Kawakami, P. F. Robbins, and S. A. Rosenberg. 1996. Utilization of an alternative open reading frame of a normal gene in generating a novel human cancer antigen. *J. Exp. Med.183*:1131–1140.
138. Den Haan, J. M. M., Meadows, L. M., W. Wang, J. Pool, E. Blokland, T. L. Bishop, C. Reinhardus, J. Shabanowitz, R. Offringa, D. F. Hunt, V. H. Engelhard, and E. Goulmy. 1997. The human immunodominant minor histocompatibility antigen HA-1 represents a diallelic gene with a single amino acid polymorphism. *Science, in press.*
139. Labrecque, S., N. Naor, D. Thomson, and G. Matlashewski. 1993. Analysis of the anti-p53 response in cancer patients. *Cancer Res. 53*:3468–3471.
140. Theobald, M., J. Biggs, D. Dittmer, A. J. Levine, and L. A. Sherman. 1995. Targeting p53 as a general tumor antigen. *Proc. Natl. Acad. Sci. U. S. A. 92*:11993–11997.
141. Dahl, A. M., P. C. Beverley, and H. J. Stauss. 1996. A synthetic peptide derived from the tumor-associated protein mdm2 can stimulate autoreactive, high avidity cytotoxic T lymphocytes that recognize naturally processed protein. *J. Immunol. 157*:239–246.
142. Vierboom, M. P. M., H. W. Nijman, R. Offringa, E. I. H. van der Voort, T. van Hall, L. van den Broek, G. J. Fleuren, P. Kenemans, W. M. Kast, and C. J. M. Melief. 1997. Tumor eradication by wild-type p53-specific cytotoxic T lymphocytes. *J. Exp. Med. 186*:695–704.
143. Slamon, D., W. Godolphin, L. Jones, J. Holt, S. Wong, D. Keith, w. Levine, S. Stuart, J. Udove, A. Ullrich, and M. Press. 1989. Studies of the HER-2/neu proto-oncogene in human breast and ovarian cancer. *Science 244*:707–712.
144. Berchuck, A., A. Kamel, R. Witaker, B. Kerns, G. Olt, R. Kinney, J. Soper, R. Dodge, D. Clarke-Pearson, and P. Marks. 1990. Overexpression of HER-2/neu is associated with poor survival in advanced epithelial ovarian cancer. *Cancer Res. 50*:4087–4091.
145. Berchuck, A., G. Rodriguez, R. B. Kinney, J. T. Soper, R. K. Dodge, D. L. Clarke-Pearson, and R. C. Bast. 1991. Overexpression of HER-2/neu in endometrial cancer is associated with advanced stage disease. *Am. J. Obstet. Gynecol. 164*:15–21.

146. Yonemura, Y., I. Ninomiya, A. Yamaguchi, S. Fushida, H. Kimua, S. Ohoyama, I. Miyazkil, Y. Endou, M. Tanaka, and T. Susaki. 1991. Evaluation of immunoreactivity for erbB-2 protein as a marker of poor short term prognosis in gastric cancer. *Cancer Res. 51*:1034–1038.
147. Weiner, D. B., J. Nordberg, R. Robinson, P. C. Nowell, A. Gazdar, M. I. Greene, W. V. Williams, J. A. Cohen, and J. A. Kern. 1990. Expression of the neu gene-encoded protein (P185neu) in human non-small cell carcinomas of the lung. *Cancer Res. 50*:421–425.
148. Peiper, M., P. S. Goedegebuure, D. C. Linehan, E. Ganguly, C. C. Douville, and T. J. Eberlein. 1997. The HER2/neu-derived peptide p654–662 is a tumor-associated antigen in human pancreatic cancer recognized by cytotoxic T lymphocytes. *Eur. J. Immunol. 27*:1115–1123.
149. Dahl, A. M., P. C. Beverley, and H. J. Stauss. 1996. A synthetic peptide derived from the tumor-associated protein mdm2 can stimulate autoreactive, high avidity cytotoxic T lymphocytes that recognize naturally processed protein. *J. Immunol. 157*:239–246.
150. Thompson, J. A., F. Grunert, and W. Zimmerman. 1991. Carcinoembryonic antigen gene family: molecular biology and clinical perspectives. *J. Clin. Lab. Anal. 5*:344–366.
151. Ras. E., S. H. van der Burg, S. T. Zegveld, R. M. P. Brandt, P. J. K. Kuppen, R. Offringa, S. O. Warnarr, C. J. H. van der Velde, and C. J. M. Melief. 1997. Identification of potential HLA-A*0201 restricted CTL epitopes derived from the epithelial cell adhesion molecule (Ep-CAM) and the carcinoembryonic antigen (CEA). *Human Immunol. 53*:81–89.
152. Wang, R. F., M. R. Parkhurst, Y. Kawakami, P. F. Robbins, and S. A. Rosenberg. 1996. Utilization of an alternative open reading frame of a normal gene in generating a novel human cancer antigen. *J. Exp. Med. 183*:1131–1140.
153. Guilloux, Y., S. Lucas, V. G. Brichard, A. van Pel, C. Viret, E. De Plaen, F. Brasseur, B. Lethé, F. Jotereau, and T. Boon. 1996. A peptide recognized by human cytolytic T lymphocytes on HLA A2 melanomas is encoded by an intron sequence of the N acetylglucosaminyltransferase V gene. *J. Exp. Med. 183*:1173–1183.
154. Lohse, A. W., M. Dinkelmann, M. Kimmig, J. Herkel, and K.-H. Meyer zum Buschenfelde. 1996. Estimation of the frequency of self-reactive T cells in health and inflammatory diseases by limiting dilution analysis and single cell cloning. *J. Autoimm. 9*:667–675.
155. Fairchild, P. J., and D. C. Wraith. 1996. Lowering the tone: mechanisms of immunodominance among epitopes with low affinity for MHC. *Immunol. Today 17*:80–85.
156. Burg, S. H. van der, M. J. W. Visseren, R. Offringa, and C. J. M. Melief. 1996. Do epitopes derived from auto-antigens display low affinity for MHC class I? *Immunol. Today 18*:97–98.
157. Melero, I., M. c. Singhal, P. McGowan, H. S. Haugen, J. Blake, K. E. Hellström, G. Yang, C. H. Clegg, and L. Chen. 1997. Immunological ignorance of an E7-encoded cytolytic T-lymphocyte epitope in transgenic mice expressing the E7 and E6 oncogenes of human papillomavirus type 16. *J. Virol. 71*:3998–4004.
158. Hu, J., W. Kindsvogel, S. Busby, M. C. Bailey, Y. Y. Shi, and P. D. Greenberg. 1993. An evaluation of the potential to use tumor-associated Ags as targets for anti-tumor T cell therapy using transgenic mice expressing a retroviral tumor Ag in normal lymphoid tissues. *J. Exp. Med. 177*:1681–1690.
159. Rosenberg, S. A., and D. E. White. 1996. Vitiligo in patients with melanoma: normal tissue antigens can be targets for cancer immunotherapy. *J. Immunother. 19*:81–84.
160. Parmiani, G., A. Anichini, and G. Fossati. 1990. Cellular immune response against autologous human malignant melanoma: are *in vitro* studies providing a framework for a more effective immunotherapy? *J. Natl. Cancer Inst. 82*:361–370.
161. Winchester, R. J., C. Y. Wang, A. Gibofsky, H. G. Kunker, and L. J. Old. 1978. Expression of Ia-like antigens on cultured human malignant melanoma cell lines. *Proc. Natl. Acad. Sci. U. S. A. 75*:6235–6239.
162. Anichini, A., R. Mortarini, and G. Parmiani. 1992. Beta 1-integrins on melanoma clones regulate the interaction with autologous cytolytic T-cell clones. *J. Immunother. 12*:183–186.
163. Guerder, S., and P. Matzinger. 1992. A fail-safe mechanism for maintaining self-tolerance. *J. Exp. Med. 176*:553–564.
164. Straten, P. T., A. F. Kirkin, T. Seremet, and J. Zeuthen. 1997. Expression of transporter associated with antigen processing 1 and 2 (TAP1/2) in malignant melanoma cell lines. *Int. J. Cancer 70*:582–586.
165. Jäger, E., M. Ringhoffer, J. Karbach, M. Arand, F. Oesch, and A. Knuth. 1996. Inverse relationship of melanocyte differentiation antigen expression in melanoma tissue and CD8+ cytotoxic-T-cell responses: evidence for immunoselection of antigen-loss variants in vivo. *Int. J. Cancer 66*:470–476.
166. Gimmi, C. D., B. W. Morrison, B. A. Mainprice, J. G. Gribben, V. A. Boussiotis, G. J. Freeman, S. Y. L. Park, M. Watanabe, J. L. Gong, D. F. Hayes, D. W. Kufe, and L. M. Nadler. 1996. Breast cancer associated antigen, DF3/MUC1, induces apoptosis of activated human T cells. *Nature-Med. 2*:1367–1370.
167. . Hahne, M., D. Rimoldi, M. Schroter, P. Romero, M. Schreier, L. E. French, P. Schneider, T. Bornand, A. Fontana, D. Lienard, J. Cerottini, and J. Tschopp. 1996. Melanoma cell expression of Fas(Apo-1/CD95) ligand: implications for tumor immune escape. *Science 274*:1363–1366.

168. Nakagomi, H., M. Petersson, I. Magnusson, C. Juhlin, M. Matsuda, H. Mellstedt, J. L. Taupin, E. Vivier, P. Anderson, and R. Kiessling. 1993. Decreased expression of the signal-transducing zeta chains in tumor-infiltrating T-cells and NK cells of patients with colorectal carcinoma. Cancer Res. 53:5610–5612.
169. Farace, F., E. Angevin, J. Vanderplancke, B. Escudier, and F. Triebel. 1994. The decreased expression of CD3 zeta chains in cancer patients is not reversed by IL-2 administration. *Int. J. Cancer 59*:752–755.
170. Finke, J. H., A. H. Zea, J. Stanley, D. L. Longo, H. Mizoguchi, R. R. Tubs, R. H. Wiltrout, J. J., O'Shea, S. Kudoh, E. Klein, R. M. Bukowski, and A. C. Ochoa. 1993. Loss of T-cell receptor zeta chain and p56lck in T-cells infiltrating human renal cell carcinoma. *Cancer Res. 53*:5613–5616.
171. Lopez, D. M., M. E. Handel-Fernandez, X. Cheng, V. Charyulu, L. M. Herbert, M. R. Dinapoli, and C. L. Calderon. 1996. Cytokine production by lymphoreticular cells from mammary tumor bearing mice: the role of tumor-derived factors. *Anticancer Res. 16*:2923–2929.
172. Sulitzeanu, D. 1993. Immunosuppressive factors in human cancer. *Adv. Cancer Res. 60*:247–267.
173. Gabrilovich, D. I., H. L. Chen, K. R. Girgis, H. T. Cunningham, G. M. Meny, S. Nadaf, D. Kavanaugh, and D. P. Carbone. 1996. Production of vascular endothelial growth factor by human tumors inhibits the functional maturation of dendritic cells. *Nat. Med. 2*:1096–1103.
174. Weiner, H. L. 1997. Oral tolerance: immune mechanisms and treatment of autoimmune diseases. *Immunol. Today 18*:335–343.
175. Merogi, A. J., A. J. Marrogi, R. Ramesh, W. R. Robinson, C. D. Fermin, and S. M. Freeman. 1997. Tumor-host interaction: analysis of cytokines, growth factors, and tumor-infiltrating lymphocytes in ovarian carcinomas. *Hum. Pathol. 28*:321–331.
176. Garrido, F., F. Ruiz-Cabello, T. Cabrera, J. J. Pérez-Villar, M. López-Botet, M. Duggan-Keen, and P. L. Stern. 1997. Implications for immunosurveillance of altered HLA class I phenotypes in human tumors. *Immunol. Today. 18*:89–94.
177. Maeurer, M. J., S. M. Gollin, D. Martin, W. Swaney, J. Bryant, C. Castelli, P. Robbins, G. Parmiani, W. J. Storkus, and M. T. Lotze. 1996. Tumor escape from immune recognition: lethal recurrent melanoma in a patient associated with downregulation of the peptide transporter protein TAP-1 and loss of expression of the immunodominant MART-1/Melan-A antigen. *J. Clin. Invest. 98*:1633–1641.
178. Toso, J. F., C. Oei, F. Oshidari, J. Tartaglia, E. Paoletti, H. K. Lyerly, S. Talib, and K. J. Weinhold. 1996. MAGE 1 specific precursor cytotoxic T lymphocytes present among tumor infiltrating lymphocytes from a patient with breast cancer: characterization and antigen specific activation. *Cancer Res. 56*:16–20.
179. Gervois, N., Y. Guilloux, E. Diez, and F. Jotereau. 1996. Suboptimal activation of melanoma infiltrating lymphocytes (TIL) due to low avidity of TCR/MHC-tumor peptide interactions. *J. Exp. Med. 183*:2403–2407.
180. Tanaka, Y., R. W. Anderson, W. L. Maloy, and S. S. Tevethia. 1989. Localization of an immunorecessive epitope on SV40 T antigen by H-2Db-restricted cytotoxic T-lymphocyte clones and a synthetic peptide. *Virology 171*:205–213.
181. Feltkamp, M. C. W., G. R. Vreugdenhil, M. P. M. Vierboom, E. Ras, S. H. Van der Burg, J. ter Schegget, C. J. M. Melief, and W. M. Kast. 1995. CTL raised against a subdominant epitope offered as a synthetic peptide eradicate human papillomavirus type 16-induced tumors. *Eur. J. Immunol. 25*:2638–2642.
182. Tsai, V., S. Southwood, J. Sidney, K. Sakaguchi, Y. Kawakami, E. Appella, A. Sette, and E. Celis. 1997. Identification of subdominant CTL epitopes of the GP100 melanoma-associated tumor antigen by primary in vitro immunization with peptide-pulsed dendritic cells. *J. Immunol. 158*:1796–1802.
183. Anichini, A., R. Mortarini, C. Maccalli, P. Squarcina, K. Fleischhauer, L. Mascheroni, and G. Parmiani. 1996. Cytotoxic T cells directed to tumor antigens not expressed on normal melanocytes dominate HLA A2.1 restricted immune repertoire to melanoma. *J. Immunol. 156*:208–217.
184. Kawakami, Y., S. Eliyahu, K. Sakaguchi, P. F. Robbins, L. Rivoltini, J. R. Yannelli, E. Appella, and S. A. Rosenberg. 1994. Identification of the immunodominant peptides of the MART-1 human melanoma antigen recognized by the majority of HLA-A2-restricted tumor infiltrating lymphocytes. *J. Exp. Med. 180*:347–352.
185. Marincola, F. M., Y. M. Hijazi, P. Fetsch, M. L. Salgaller, L. Rivoltini, J. Cormier, T. B. Simonis, P. H. Duray, M. Herlyn, Y. Kawakami, and S. A. Rosenberg. 1996. Analysis of expression of the melanoma associated antigens MART1 and gp100 in metastatic melanoma cell lines and in in situ lesions. *J. Immunother. 19*:192–205.
186. Van Waes, C., P. A. Monach, J. L. Urban, R. D. Wortzel, and H. Schreiber. 1996. Immunodominance deters the response to other tumor antigens thereby favoring escape: prevention by vaccination with tumor variants selected with cloned cytolytic T cells in vitro. *Tissue Antigens 47*:399–407.
187. Zhai, Y. F., J. C. Yang, P. Spiess, M. I. Nishimura, W. W. Overwijk, B. Roberts, N. P. Restifo, and S. A. Rosenberg. 1997. Cloning and characterization of the genes encoding the murine homologues of the human melanoma antigens MART-1 and gp100. *J. Immunother. 20*:15–25.

188. Bloom, M. B., D. Perrylalley, P. F. Robbins, Y. Li, M. Elgamil, S. A. Rosenberg, and J. C. Yang. 1997. Identification of tyrosinase related protein 2 as a tumor rejection antigen for the B16 melanoma. *J. Exp. Med. 185*:453–459.
189. Falk, K., O. Rötschke, S. Stevanovic, G. Jung, and H. G. Rammensee. 1991. Allele-specific motifs revealed by sequencing of self-peptides eluted from MHC molecules. *Nature 351*:290–296.
190. Engelhard, V. H. 1994. Structure of peptides associated with class I and class II MHC molecules. *Annu. Rev. Immunol. 12*:181–207.
191. Burg, S. H. van der, E. Ras, J. W. Drijfhout, W. E. Benckhuijsen, A. J. A. Bremers, C. J. M. Melief, and W. M. Kast. 1995. An HLA class I peptide-binding assay based on competition for binding to class I molecules on intact human B-cells: identification of conserved HIV-1 polymerase peptides binding to HLA-A*0301. *Human Immunol. 44*:189–198.
192. Sette, A., J. Sidney, M.-F. del Guerco, S. Southwood, J. Ruppert, C. Dahlberg, H. M. Grey, and R. T. Kubo. 1994. Peptide binding to the most frequent HLA-A class I alleles measured by quantitative binding assays. *Mol. Immunol. 31*:813–822.
193. Vitiello, A., D. Marchesini, J. Furze, L. A. Sherman, and R. W. Chesnut. 1991. Analysis of the HLA-restricted Influenza-specific cytotoxic T lymphocyte response in transgenic mice carrying a chimeric human-mouse class I major histocompatibility complex. *J. Exp. Med. 173:*1007–1015.
194. Fleischhauer, K., D. Fruci, P. Vanendert, J. Herman, S. Tanzarella, H. J. Wallny, P. Coulie, C. Bordignon, and C. Traversari. 1996. Characterization of antigenic peptides presented by HLA B44 molecules on tumor cells expressing the gene MAGE 3. *Int. J. Cancer 68*:622–628.
195. McIntyre, C. A., R. C. Rees, K. E. Platts, C. J. Cooke, M. O. Smith, K. A. Mulcahy, and A. K. Murray. 1996. Identification of peptide epitopes of MAGE 1, 2, 3 that demonstrate HLA A3 specific binding. *Cancer Immunol. Immunoth. 42*:246–250.
196. Kónya, J., C. Eklund, V. af Geijersstam, F. Yuan, G. Stuber, and J. Dillner. 1997. Identification of a cytotoxic T-lymphocyte epitope in the human papillomavirus type 16 E2 protein. *J. Gen. Virol. 78*:2615–2620.
197. Chen, W., S. Khilko, J. Fecondo, D. H. Margulies, and J. McCluskey. 1994. Determinant selection of Major Histocompatibility Complex Class I-restricted antigenic peptides is explained by Class I-peptide affinity and is strongly influenced by nondominant anchor residues. *J. Exp. Med. 180*:1471–1483.
198. Siliciano, R. F., and M. J. Soloski. 1995. MHC class-I restricted processing of transmembrane proteins: mechanism and biologic significance. *J. Immunol. 155*:2–5.
199. Valmori, D., D. Lienard, G. Waanders, D. Rimoldi, J. C. Cerottini, and P. Romero. 1997. Analysis of MAGE 3 specific cytolytic T lymphocytes in human leukocyte antigen A2 melanoma patients. *Cancer Res. 57*:735–741.
200. Skipper, J. C. A., R. C. Hendrickson, P. H. Gulden, V. Brichard, A. Van Pel, Y. Chen, J. Shabanowitz, T. Wolfel, C. L. Slingluff, T. Boon, D. F. Hunt, and V. H. Engelhard. 1996. An HLA-A2-restricted tyrosinase antigen on melanoma cells results from posttranslational modification and suggests a novel pathway for processing of membrane proteins. *J. Exp. Med. 183*:527–534.
201. Meadows, L., W. Wang, J. M. M. den Haan, E. Blokland, C. Reinhardus, J. W. Drijfhout, J. Shabanowitz, R. Pierce, A. I. Agulnik, C. E. Bishop, D. F. Hunt, E. Goulmy, and V. H. Engelhard. 1997. The HLA-A*0201-restricted H-Y antigen contains a posttranslationally modified cysteine that significantly affects T cell recognition. *Immunity 6*:273–281.
202. di Marzo Veronese, F., D. Arnott, V. Barnaba, D. J. Loftus, K. Sakaguchi, C. B. Thompson, S. Salemi, C. Mastroianni, A. Sette, J. Shabanowitz, D. F. Hunt, and E. Appella. 1996. Autoreactive cytotoxic T lymphocytes in human immunodeficiency virus type 1-infected subjects. *J. Exp. Med. 183*:2509–2516.
203. Androlewicz, M. J., and P. Cresswell. 1996. How selective is the transporter associated with antigen processing? *Immunity 5*:1–5.
204. Armandola, E. A., F. Momburg, M. Nijenhuis, N. Bulbuc, K. Fru, and G. J. Hammerling. 1996. A point mutation in the human transporter associated with antigen processing (TAP2) alters the peptide transport specificity. *Eur. J. Immunol. 26*:1748–1755.
205. Ossendorp, F., M. Eggers, A. Neisig, T. Ruppert, M. Groettrup, A. Sijts, E. Mengedé, P. M. Kloetzel, J. Neefjes, U. Koszinowski, and C. Melief. 1996. A single residue exchange within a viral CTL epitope alters proteasome-mediated degradation resulting in lack of antigen presentation. *Immunity 5*:115–124

DISCUSSION

Pierotti: I was wondering, if you have transgenic mice with the same p53 mutation that you have used to raise this response against the wild type, what do you predict that would happen?

Melief: Well, that would depend on how that mutation would affect the level of antigen in its tissues. Even a normal black-6 mouse might, theoretically, respond to these peptides and, in fact, there are studies in biopsy mice done by Michael Lotze where he showed that one and the same, wild type p53 peptide, can overcome tolerance and generate a CTL response that eradicates tumors.

Pierotti: Do you agree that it is just an amount, an issue of the amount of wild type expression of the protein?

Melief: Yes, we checked this. I did not have time to show this but the level of expression, for example, in normal thymocytes is such that CTL's do not recognize normal thymocytes unless they are activated by mitogens. If you look, by immunohistology you see very little expression of p53 in normal tissues. So the other experiment we did is to UV irradiate the skin of these nude mice, this leads to tremendous overexpression of p53, and then ask the question whether now that skin would be attacked by the CTL's. There was no sign of that but then again, there is also evidence that UV radiated skin is no longer rejected as an allograft so that may not be the right experiment. We have to check under which conditions, perhaps chemotherapeutic agents, could induce overexpression of p53 and cause it to induce autoimmunity.

Zanker: Did I correctly see your slides on C3 tumors and the E7 peptide that you used the amino acid sequences between 44 and 62? That means 8 amino acids. Are these processed like proteins as you have shown in the second slide? Do the proteasomes really recognize such small molecules?

Melief: It does not really matter whether you use the exact MHC binding ninemer peptide or a slightly longer sequence, or even the whole protein, as I showed you. For that very reason our strategy now for the next HPV vaccination trial would be to use, in fact, large overlapping peptides. These will have all predicted class 1 and class 2 epitopes of E6 and E7 and if you take 30-mers then you only end up with about 12 peptides that span the entire length of E6 and E7. We will first check whether vaccination with these long peptides work in the mouse model and, if it does, then I think it would be a good thing to try in the next clinical vaccine trial. Because also previously in a Sendai virus model we showed that peptide length is not critical for protection, as long as it contains the CTL epitope.

Hanahan: Yes, it was a nice strategy to use the p53 knockout mice to raise antibodies, I think this is going to be a big opportunity with all these knockout mice in general. But I am wondering with regard to your idea about proteins and what you have done with E-7. Do you think you could do the same thing now with p53 protein? I mean, do you think that is possible or is the tolerance such that you are not going to be able to break it?

Melief: We do not know this, but there was one report in which normal mice were vaccinated with an alvae canary pox virus with wild type p53 in it. That construct published in a PNAS paper was able to protect mice against a p53 mutant containing tumor. Now, unfortunately in that paper there were no T-helper or CTL responses demonstrable or not really investigated in detail but there might be protection with such a strategy. We do not know, therefore, for sure whether tolerance can be broken.

Parmiani: About your tolerizing peptide, the mechanism is a crucial issue because the peptide can induce tolerance or a cure.

Melief: The tolerance is very unlikely to be antibody mediated because there is no E1A at the cell surface, or E1B. This sequence is not at the cell surface except as an MHC bound peptide. That is number one. Secondly, we did not see any evidence of antibody formation against these peptides upon injection in IFA. So, I think it is similar to the phenomenon described by Zingernagel et al., as clonal exhaustion except that with this peptide it occurs at a very low subcutaneous dose, and so the message to you is that with some of these highly powerful peptides the balance between stimulation and down-regulation of the immune response is very finely tuned with these peptides and maybe we could try to shift that balance with the type of reagents that Jim Allison has discussed. So, we are going to do that experiment together, hopefully soon.

Zanker: Do you have any idea about the mechanism, what kind of motifs the proteosomes cleave out from a protein to show this as a peptide fraction for immune response induction on the cell surface? Because this is the key question of immunity, of cellular immunity, at least, because cellular immunity controls the inner of a cell for self or non-self.

Melief: As we were discussing this morning and as Thierry Boon was saying, there is at the moment no motif that predicts proteasome cleavage. In addition, the issue is quite complex because upon interferon stimulation the LMP-2 and LMP-7 components of the proteosome modulate the cleavage specificity. On top of that, there is the p28 activator which, if it binds, is also interferon induced and if it binds to the proteasome complex, this also affects cleavage specificity. So, the type of peptides generated under various conditions might in fact differ and I think we have to learn much more about it in order to know how exactly to manipulate it for tumor immunity. But, certainly it would be a major leap forward if some sort of prediction from a random protein sequence of processed peptides would become available.

Parmiani: You have shown that spleen derived or bone marrow derived dendritic cells work with more or less the same efficiency. Are those spleen derived dendritic cells the equivalent of blood dendritic cells in human?

Melief: We believe they are. They are not the natural resting dendritic cell which does not need any cytokine stimulation to differentiate. It only needs to become activated but this happens by shear adherence to plastic or glass and then it already up-regulates tremendously, all of the molecules involved in effective antigen presentation.

Bankert: Is therapeutic vaccination with dendritic cells only possible in experimental mouse models?

Melief: No, in fact we were able to isolate similar cells from human peripheral blood, and there is a firm associated with Stanford University which uses those particular DC's without any further cytokine stimulation for immunotherapy and have used them already with some degree of success for the treatment of B cell lymphoma with idiotypic immunoglobulin-loaded dendritic cells.

Bankert: You suggested the tolerance may be a clonal deletion, but in the low dose that would seem unlikely. Is it more likely that you are seeing possibly presentations of the antigen in the absence of co-stimulation which would also lead to the anergy that you see? As a follow up, I just want to ask you whether you are planning another phase-I trial with this vaccine? Are you going to be using dendritic cells or entertaining that possibility so that you would definitely not be seeing antigen in the absence of the co-stimulation?

Melief: Well, based on these observations we do not want to run the risk of this type of anergy by any of the potentially stimulatory peptides, so, indeed, we are planning to use the dendritic cells loaded with long overlapping peptides to avoid this theoretical problem. We have not seen it with any human peptides, so far, but it is a theoretical issue that needs to be addressed. And with regard to your first question, we do believe that, indeed, presentation in the absence of co-stimulation could be an explanation because these peptides very rapidly disseminate through the body, even from the IFA depot and could be loaded on non-professional APC's. And one reason to believe this is that if we co-inject this peptide shortly after we have intravenously injected the CTL clone recognizing that peptide, and we put the peptide subcutaneously in IFA then the animals die with a massive lung infiltration with inflammatory cells because of apparently some exhaustive anergising event in the CTL clone in those lungs and the cells die subsequent to that.

Zanker: Just as a comment, we think that there is really a big difference between the DC's at different locations because all these cells, if you stimulate them, have never seen glass and plastic in their life. So, if you stimulate them on glass and plastic you do get a tremendous misinformation and you have to look for them on a natural matrix and so you have to construct the natural matrix like collagen and to supplement collagen with laminin, fibrin, fibronectin and so on. All the data which we got on DC's stimulated on glass and plastic, in my opinion, are very misleading.

Melief: Well, I agree with you that it is an artifact, but that the artifact is useful for vaccination which in itself is an artifact, then I do not care. What we find is that these dendritic cells, under these conditions have extremely high MHC class 1 and class 2 expression and they process antigens very effectively and they have very high levels of co-stimulatory molecules. They have all the functional things that we believe are good to stimulate these responses. So, although I agree with you that we could do it in a nicer way, we also have to think about doing it in a GMP compatible way, so if you provide me with a GMP compatible collagen fibers, I would be willing to consider it.

REGULATION OF LEUKOCYTE-ENDOTHELIAL CELL INTERACTIONS IN TUMOR IMMUNITY

Sharon S. Evans,* Margaret Frey, David M. Schleider, Robert A. Bruce, Wan-chao Wang, Elizabeth A. Repasky, and Michelle M. Appenheimer

Department of Immunology
Roswell Park Cancer Institute
Buffalo, New York 14263

INTRODUCTION

Mechanisms of Immune Effector Cell Migration into Tumors *via* the Tumor Microvasculature

Successful immunotherapy ultimately depends on the ability of immune effector cells to infiltrate tumor tissues. A major site of extravasation of immune effector cells out of the blood and into tissues occurs across specialized post-capillary high endothelial venules (HEV).[1] Lymphocyte emigration into tissues involves a complex multistep adhesion cascade.[2,3] The L-selectin leukocyte adhesion molecule is responsible for mediating the initial attachment and slow rolling of lymphocytes along the luminal surface of HEV under hemodynamic shear conditions, a crucial first step in the extravasation of immune effector cells into lymph nodes and Peyer's patches as well as at extralymphoid sites. Firm adhesion of lymphocytes to HEV and transendothelial migration are dependent on the interaction of the leukocyte integrin, leukocyte function associated antigen-1 (LFA-1), with intercellular adhesion molecule-1 (ICAM-1) and ICAM-2 on endothelial cells. The $\alpha_4\beta_1$ and $\alpha_4\beta_7$ integrins have also been implicated in recruitment of lymphocytes to tissues through binding to their cognate receptors on endothelium including vascular cell adhesion molecule-1 (VCAM-1) and mucosal addressin cell adhesion molecule-1 (MAdCAM-1). Blockade of these leukocyte-endothelial cell adhesive interactions severely compromises anti-tumor immunity.[4–7] The immune effector cells that are recruited to tissues via L-selectin-, LFA-1-, $\alpha_4\beta_1$- and/or $\alpha_4\beta_7$-dependent mechanisms include: naive and memory CD4 and CD8 T cells, monocytes, neutrophils, and CD56^{bright}/CD56^{dim} natural killer (NK) cell subsets.[2,3,8]

* Corresponding author

The Biology of Tumors, edited by Mihich and Croce
Plenum Press, New York, 1998.

Emerging data from several recent studies suggest that tumor cells actively interfere with the mechanisms underlying effector cell extravasation, thereby subverting the immune response.[4–7,9–15] These observations provide an explanation for the finding that despite extensive vascularization of many tumors, there is frequently only limited or non-uniform lymphocyte infiltration in various regions within a given tumor. Leukocyte-endothelium interactions in a number of tumor types have been shown to be markedly diminished, compared to normal tissue microvessels, in a manner that cannot be accounted for by differences in blood flow shear rates.[10,16,17] Moreover, the ability of immune effector cells to interact with individual microvessels within a given tumor is highly heterogeneous.[16–19] At a mechanistic level, tumor-derived factors including basic fibroblast growth factor (bFGF) and transforming growth factor-β (TGF-β) have been implicated in suppression of endothelial adhesion molecules including L-selectin ligands, VCAM-1, ICAM-1, MAdCAM-1, and E-selectin.[9,11–15,20] Thus, downregulation of vascular adhesion molecules appears to be one mechanism by which tumors escape invasion by cytotoxic immune effector cells.

Potential for Enhanced Anti-Tumor Immunity through the Regulation of Leukocyte-Endothelium Adhesion

A major challenge is to develop strategies to overcome the suppressive effects of tumors on leukocyte-endothelium adhesion within tumor microenvironments in order to promote anti-tumor immune responses. Active recruitment of leukocytes to tumors via the modulation of cell-to-cell adhesion events is likely to culminate in significant anti-tumor immune activity as a result of an increase in the total number of immune effector cells recruited to tumors as well as a change in the effector population that is capable of migrating into tumor tissues. In addition, potent anti-tumor responses could also occur as a consequence of increased trafficking of lymphocytes to regional lymph nodes since these tissues are frequent sites of tumor metastasis as well as important sites of lymphocyte priming by tumor antigens. The significance of L-selectin-dependent homing of lymphocytes to lymph nodes for antigen priming is supported by evidence that mAb blockade of L-selectin adhesion effectively abrogates anti-tumor immune responses in animal models.[4–7,21]

This report discusses recent studies from our laboratory examining the underlying mechanisms that control L-selectin-mediated leukocyte-endothelial cell adhesion. It is becoming increasingly clear that L-selectin adhesion involves a delicate balance between opposing mechanisms which positively or negatively regulate the synthesis, shedding, and affinity/avidity of L-selectin and its cognate endothelial ligands. In particular, the immunomodulatory cytokine interferon-α (IFN-α) and IFN-α-inducible proteins appear to play a central role in controlling the L-selectin adhesion potential of lymphocytes. Based on new information regarding the complexity of the mechanisms that control L-selectin adhesion, novel strategies can be designed to enhance the delivery of immune effector cells to tumors and regional lymph nodes, potentially improving cancer treatment.

RESULTS AND DISCUSSION

IFN-α Regulation of L-selectin Gene Expression and Cell Surface Levels

Although the expression of L-selectin is known to be tightly regulated during lymphocyte maturation and differentiation,[2,3] information has only recently become available regarding the molecular mechanisms that control the synthesis of this adhesion molecule.

Data from our laboratory demonstrate for the first time the role of a specific cytokine, IFN-α, in the control of L-selectin gene expression and cell surface levels.[22–24] We have shown that recombinant human IFN-α markedly increases the cell surface density of L-selectin in lymphocyte populations isolated from human tonsil, spleen, and bone marrow.[22] IFN-α has further been shown to increase the cell surface density of L-selectin on clonal populations of malignant B cells from chronic lymphocytic leukemia patients.[25,26] Using paired IFN-α-sensitive and -resistant human B lymphoblastoid Daudi cell lines as model systems,[22–24,27,28] we have examined the mechanism by which IFN-α enhances L-selectin expression. These studies established that IFN-α induces an increase in L-selectin transcription, mRNA steady state levels, and surface expression in sensitive cells.[22,24] A potential physiological role for IFN-α in the control of L-selectin expression is suggested by evidence that this cytokine is constitutively produced in lymphoid tissues, in the absence of viral infection,[29,30] and by studies demonstrating that IFN-α influences the regional trafficking of lymphocytes *in vivo*.[31,32]

Our analysis has revealed striking similarities in the characteristics of IFN-α regulation of L-selectin and classical IFN-α-stimulated genes (ISG) such as ISG15, ISG54, and 9–27/Leu-13 in Daudi B lymphoid cells,[22,24] suggesting the involvement of a common signal transduction pathway. Elegant studies by Darnell, Kerr, and Stark, and their colleagues have elucidated the signal transduction pathway by which IFN-α effects immediate transcriptional changes in cells,[33,34] providing insight into the mechanisms potentially controlling L-selectin gene expression. Transcriptional stimulation of ISGs results from Janus-family tyrosine kinase (JAK)-mediated activation of latent cytoplasmic proteins termed signal transducers and activators of transcription (STAT). Transcriptional activation of IFN-α-induced primary response genes such as ISG15, ISG54, and 9–27/Leu-13 is mediated by STAT1 and STAT2 heterodimers which, in conjunction with a 48 kDa DNA-binding protein, form a complex termed IFN-stimulated gene factor 3 (ISGF3) which binds to IFN-stimulated DNA response elements (ISRE) in the 5′ region of target genes.

Multiple lines of evidence indicate that IFN-α directly activates L-selectin transcription in IFN-sensitive but not -resistant Daudi B cells through a signaling pathway that closely parallels JAK/STAT-mediated regulation of ISG15, ISG54, and 9–27/Leu-13. (1) Using flow cytometric analysis, IFN-α was shown to markedly increase the cell surface expression of L-selectin (Figure 1) as well as the ISG, 9–27/Leu-13, in IFN-α-sensitive Daudi cells but not in an IFN-α-resistant Daudi subclone that is defective in JAK/STAT-mediated activation of ISG.[22–24,27] (2) Northern analysis revealed that IFN-α also increases L-selectin, ISG15, ISG54, and 9–27/Leu-13 mRNA steady state levels in IFN-sensitive but not -resistant Daudi cells (Figure 2).[22,24] (3) Nuclear run-on analysis further demon-

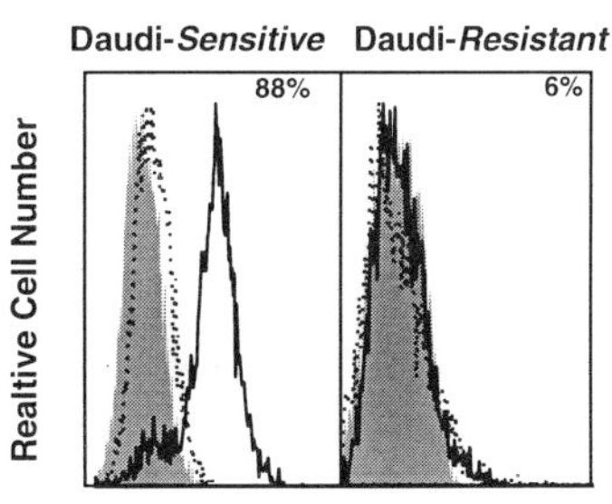

Figure 1. IFN-α upregulates L-selectin cell surface expression in IFN-sensitive, but not IFN-resistant Daudi B cells. Daudi cells were incubated 24 h in the absence (gray filled histograms) or presence of 500 IU/ml of IFN-α (black, unfilled histograms) and then stained with L-selectin-specific mAb (anti-Leu-8-FITC) and analyzed by flow cytometry.[22,23] The dotted histograms represent the background fluorescence of cells stained with a FITC-labeled isotype matched negative control antibody. In the absence of IFN-α, L-selectin levels were low, to non-detectable, in both IFN-sensitive and resistant cells. Following culture with IFN-α, a marked increase in L-selectin expression was detected exclusively in IFN-sensitive cells.

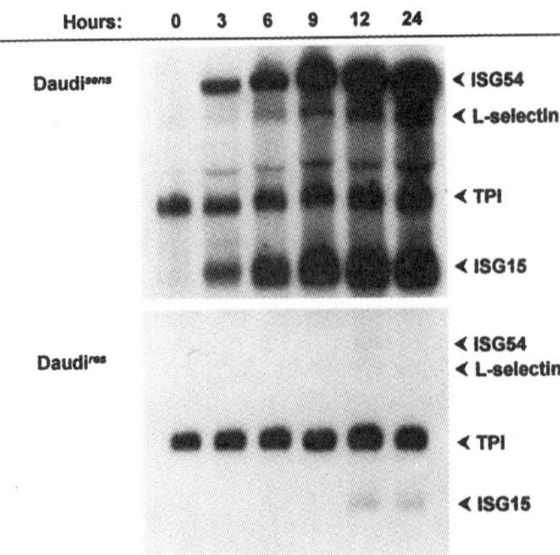

Figure 2. Northern analysis of IFN-α induction of L-selectin and ISG mRNA levels. Daudi-sensitive and -resistant cells were incubated for the indicated time intervals in the absence or presence of 500 IU/ml of IFN-α. Total cellular RNA was hybridized to [^{32}P]-labeled cDNA probes for L-selectin, ISG15, and ISG54, and analyzed by electrophoresis as described.[22] A cDNA probe for triose phosphate isomerase (TPI) was used as an internal loading control.

strated that IFN-α strongly activates the transcription of L-selectin and classical ISGs in IFN-α-sensitive Daudi B cells.[24,35] In sharp contrast, transcriptional activation of L-selectin by IFN-α does not occur in the IFN-α-resistant Daudi subclone. (4) The IFN-α signal transduction pathway that upregulates L-selectin gene expression operates through the activation of pre-existing cellular factors and is dependent on tyrosine kinase activity,[22,24] two hallmarks of the well characterized JAK/STAT signal transduction pathways used by IFN-α to transcriptionally activate ISG.[33,34] (5) Finally, we have recently identified a region of the L-selectin gene upstream of the transcript initiation site that confers IFN-α responsiveness onto a heterologous reporter gene.[24,35] This region of the L-selectin gene contains a complex array of potential regulatory elements including IFN-α-responsive consensus sequences (i.e., ISRE) as well as numerous potential binding sites for non-STAT transcription factors. These data support the hypothesis currently being tested by our laboratory that transcriptional activation of the L-selectin gene in hematopoietic cells is accomplished through combinatorial interactions of IFN-α-inducible STAT proteins and non-STAT transcription factors with multiple *cis*-acting elements in the L-selectin promoter.

The observed effect of IFN-α on L-selectin expression may have important implications for clinical studies. It is well established that lymphocyte activation in response to mitogens, monoclonal antibody-induced T cell receptor ligation, or stimulatory cytokines (IL-2, IL-15) *in vitro*, leads to loss of L-selectin from the surface of T cells and NK cells via a shedding mechanism.[8,23,36,37] Moreover, antigen-specific activation of cytolytic T cells *in vivo* is frequently accompanied by downregulation of L-selectin.[5] Thus, strategies that use stimulatory cytokines such as IL-2 or IL-15 to expand immune effector populations either *in vitro* or *in vivo* would likely exclude these cells from tissues requiring L-selectin in the extravasation process (i.e., regional lymph nodes, and potentially, tumor tissues). Studies are currently in progress in our laboratory to determine if IFN-α, perhaps in combination with other cytokines, can sustain L-selectin expression during lymphocyte activation. In this regard, we have recently shown that either IL-12 or IFN-α, which activate common JAK/STAT signal transduction molecules, act independently to increase L-selectin expression by CD56^{bright} and CD56^{dim} NK cells *in vitro*.[8] Moreover, McRae *et al.*[38] have shown that the combination of IL-12 and IFN-β (which functions similarly to IFN-α and binds the same cell surface receptor) are effective at maintaining L-selectin expression during lymphocyte activation *in vitro*.

Regulation of L-Selectin Shedding by the IFN-Inducible Leu-13 Protein

We have recently demonstrated a novel mechanism controlling L-selectin shedding from lymphocyte surface membranes that involves the 16 kDa IFN-inducible Leu-13 protein.[23] Leu-13 is constitutively expressed at high levels by B and T lymphocytes and is a member of a multimeric cell surface complex that controls lymphocyte proliferation and adhesion properties.[27,39–42] The lymphocyte cell surface molecules Leu-13 and CD81 (TAPA-1, target of an antiproliferative antibody) are expressed together in a complex with lineage-specific molecules (i.e., B cell-restricted molecules, CD19, CD21; T cell-restricted molecules, CD4, CD8).[40–43] Antibody binding to Leu-13 as well as to other members of this multimeric complex has been shown to inhibit growth factor-driven proliferation of normal and malignant lymphocytes,[39,40–44] supporting the notion that these molecules function as integral components of an important signal transduction complex in lymphocytes. In addition, ligation of these individual molecules with specific monoclonal antibodies (mAb) triggers a lymphocyte homotypic aggregation response through an undefined adhesion pathway.[27,39–44] Recent cloning of Leu-13[40] determined that this cell surface protein is encoded by the IFN-α-responsive 9–27 gene.[27,39,45] As discussed above for L-selectin, the IFN-α-inducible transcription factor, ISGF3, has been implicated in control of Leu-13 transcription.[45]

Recent findings in our laboratory[23] demonstrate that signaling via the Leu-13 surface protein by mAb-induced cross-linking triggers partial, but significant downregulation of L-selectin from the surface of B and T lymphocytes (Figure 3A). Leu-13 initiates L-selectin release from lymphocyte cell surfaces through a shedding mechanism, as indicated by an increase in soluble L-selectin levels in supernatants from anti-Leu-13 mAb treated cultures.[23] Significantly, anti-Leu-13-mAb-induced L-selectin downregulation was associated with a dramatic decrease in L-selectin-dependent adhesion of lymphocytes to vascular endothelium *in vitro* (Figure 3B).[23]

Our studies further demonstrate that Leu-13 initiates a novel signal transduction pathway regulating L-selectin shedding that is distinct from the previously well described protein kinase C (PKC)-dependent L-selectin shedding mechanism known to be activated in response to PMA (Figure 3A) or CD3 cross-linking.[2,3,23,36] Specifically, Leu-13 stimulates L-selectin shedding via a tyrosine kinase-dependent/PKC-independent signal transduction pathway. Notably, direct L-selectin ligation using L-selectin-specific monoclonal antibodies, modeling cross-linking interactions with endothelial cell ligands, similarly downregulates L-selectin surface expression through a tyrosine kinase-dependent/PKC-independent mechanism.[23]

Studies are currently in progress to determine whether the newly identified tyrosine kinase-dependent pathways (initiated by Leu-13 or L-selectin) or the PKC-dependent signaling pathways (triggered through cellular activators such as PMA and OKT3) converge on a common mechanism to initiate L-selectin shedding from lymphocyte surface membranes. Previous studies have shown that PMA activates PKC-dependent L-selectin shedding through zinc-dependent metalloproteases.[36,46,47] Although the endogenous metalloprotease(s) responsible for L-selectin shedding has not been identified, PKC or tyrosine kinase-dependent pathways may lead to downstream activation of membrane-associated metalloprotease activity, thereby increasing the rate of L-selectin shedding from the plasma membrane. Recent data from our laboratory, using selected hydroxamic acid-based metalloprotease inhibitors,[36] support the conclusion that Leu-13, like PMA, initiates L-selectin shedding through a zinc-dependent metalloprotease (manuscript in preparation).

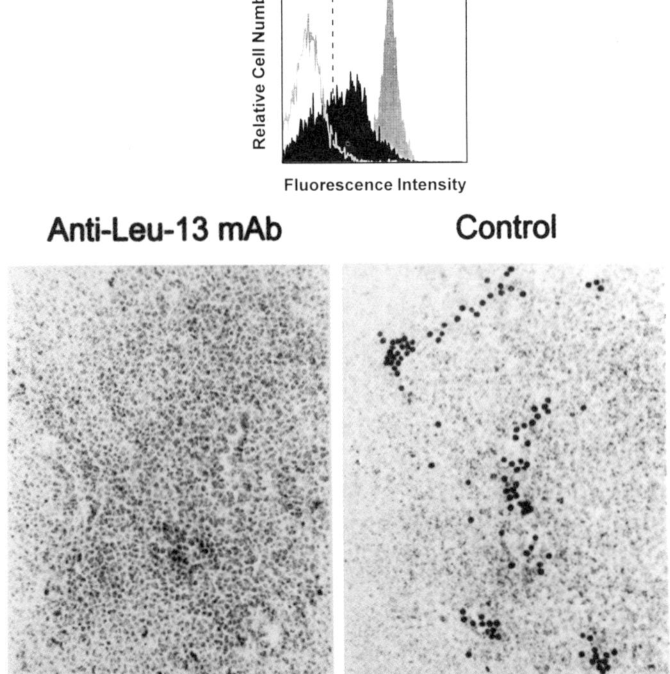

Figure 3. Leu-13 cross-linking downregulates L-selectin expression by human lymphocytes and inhibits L-selectin-mediated binding of lymphocytes to lymph node HEV. *(A).* Human peripheral blood lymphocytes (PBL) were incubated 24 h in medium alone (gray filled histogram) or in the presence of 10 μg/ml of anti-Leu-13 mAb (black filled histogram) or 50 ng/ ml of PMA (gray unfilled histogram) and L-selectin expression was determined by flow cytometry.[23] The dotted line indicates the fluorescence intensity of cells stained with a FITC-labeled isotype-matched negative control mAb. Following anti-Leu-13 treatment, L-selectin was partially downregulated (note the decrease in both L-selectin density and the frequency of L-selectin-positive cells). *(B).* Human PBL were preincubated 24 h in medium alone (control) or in the presence of anti-Leu-13 mAb (10 μg/ml) for 24 h and their ability to bind to HEV within lymph node cryosections was determined in an *in vitro* adhesion assay as described.[23] Note the numerous darkly stained, round lymphocytes bound to HEV within tissue cryosections in the control experiment whereas L-selectin-dependent adhesion of lymphocytes to HEV was significantly inhibited by anti-Leu-13 mAb treatment.

Based on these studies, it is evident that the mechanisms controlling L-selectin steady state levels on the lymphocyte surface are more complex than previously appreciated. Strategies which specifically target L-selectin shedding, either through blockade of lymphocyte signal transduction pathways or by specific metalloprotease inhibitors, may represent novel approaches toward stabilizing the L-selectin-dependent homing potential of immune effector cells including cytolytic T lymphocytes (CTL) and NK cells.

Hyperthermia and Proinflammatory Cytokines (i.e., IFN-α) Regulate L-selectin-Mediated Leukocyte-Endothelial Cell Adhesion

The use of hyperthermia as an adjuvant in the treatment of cancer has recently received renewed interest.[48–50] Although the anti-tumor mechanism of action of hyperthermia is largely undefined, there is increasing evidence suggesting that hyperthermia inhibits tumor progression through direct actions on the tumor vasculature as well as by indirect effects on immune effector cell activity. In this regard, high temperature hyperthermia (≥ 42°C) has been shown to damage the tumor microvasculature, inducing ischemic injury and death in surrounding neoplastic tissue.[51] Temperatures within the febrile range (i.e., 38–40°C) do not appear to disrupt the integrity of the tumor vasculature,[51,52] but have been shown to positively enhance various immune functions including (1) T cell proliferation and cytotoxicity,[53–55] (2) the antiviral activity of IFNs,[53,56] (3) the neutralizing capacity of Ab,[53] and (4) NK cell-mediated antitumor activity.[52] Recent studies, detailed below, revealed a previously unrecognized mechanism whereby hyperthermia acts at multiple, independent levels to enhance L-selectin-dependent lymphocyte-endothelial cell adhesion.

We have demonstrated that direct exposure of lymphocytes to fever-range temperatures (38–41°C) *in vitro* results in a ≥ 100% increase in L-selectin-dependent adhesion of lymphocytes to lymph node HEV (Table 1).[57] Exposure of human, mouse, or rat lymphocytes to hyperthermia similarly stimulates L-selectin-dependent lymphocyte adhesion to vascular endothelium *in vitro*, indicating that this response is conserved across species. Moreover, culture of lymphocytes under hyperthermia conditions markedly enhances the ability of these cells to traffic in an L-selectin-dependent manner to peripheral lymph nodes, mesenteric lymph nodes, and Peyer's patches.[57] In sharp contrast, febrile temperatures do not increase LFA-1 activity, suggesting that hyperthermia plays a relatively restricted role in controlling the initial attachment and rolling of lymphocytes on vascular endothelium during extravasation. Studies examining the mechanisms underlying hyperthermia control of L-selectin adhesion revealed that febrile temperatures do not increase lymphocyte L-selectin surface density or L-selectin-dependent recognition of soluble carbohydrate substrates,[57] suggesting that elevated temperatures enhance the avidity of preexisting L-selectin molecules for physiologic ligands.

Table 1. Hyperthermia and IFN-α enhance L-selectin-dependent adhesion of PBL to lymph node HEV and L-selectin association with the detergent-insoluble cytoskeleton[a]

	Average no. cells bound/HEV (fold-increase)	% Detergent-insoluble L-selectin
Cell incubation		
37°C	1.35 ± 0.20	0.0%
40°C	3.02 ± 0.35 (2.3)	71.7%
Conditioned medium		
37°C	1.39 ± 0.23	0.0%
40°C	3.25 ± 0.36 (2.3)	64.2%
Recombinant cytokine		
IFN-α (500 IU/ml)	2.69 ± 0.17 (1.9)	69.3%

[a]Human PBL were either (1) cultured continuously for 12 h at 37°C or 40°C, (2) maintained for 12 hours at 37°C in the presence of conditioned medium derived from PBL treated at 37°C or 40°C, or (3) cultured at 37°C (12 h) in the presence of 500 IU/ml of recombinant IFN-α. L-selectin-dependent adhesion of PBL to lymph node HEV was evaluated in an in vitro adhesion assay.[57] The mean ± the standard deviation of triplicate samples is shown; values for adhesion in the presence of the L-selectin-function blocking mAb, DREG-56, have been subtracted. The fold-increase of adhesion in hyperthermia- or IFN-α-treated samples, relative to control cells maintained at 37°C is indicated in parentheses. The amount of cell surface L-selectin that is associated with the detergent-insoluble (0.5% NP-40) cytoskeletal matrix was assessed by flow cytometry.[71] Data are representative of 6 independent experiments.

The avidity of several transmembrane adhesion molecules, including L-selectin, is thought to be controlled by dynamic interactions between the cytoplasmic domains of adhesion proteins and the lymphocyte cortical cytoskeleton.[58,59] Thus, it is noteworthy that a marked increase in the amount of L-selectin that associates with the detergent-insoluble cytoskeletal matrix is consistently observed in lymphocytes in response to hyperthermia (Table 1).[60] In addition, soluble factors in the conditioned medium of heat-treated cells mediate thermal stimulation of L-selectin adhesion (Table 1).[57] Studies are in progress to identify these soluble factors. Candidate factors include several proinflammatory cytokines such as TNF-α, IL-1β, IL-6, IFN- γ, and IFN-α that are produced in response to hyperthermia *in vitro* and *in vivo*.[53,61–64] In support of this notion, we have demonstrated that incubation of PBL with recombinant IFN-α causes a marked increase in L-selectin-mediated adhesion of lymphocytes to lymph node HEV (Table 1), raising the possibility that IFN-α and/or other cytokines participate in control of L-selectin avidity.

In related studies, fever-range whole body hyperthermia (WBH) has been shown to act at the level of the endothelium to increase L-selectin-mediated lymphocyte-endothelial cell adhesion.[65,66] In this regard, fever-range, WBH (40°C, 8 h) markedly stimulates the ability of lymph node HEV to support L-selectin-dependent adhesion of lymphocytes *in vitro* and homing *in vivo*. The increase in L-selectin binding of lymph node HEV observed in hyperthermia-treated animals is not due to an increase in the level of expression of L-selectin ligands on HEV, suggesting that the avidity of these endothelial adhesion molecules is enhanced.

We have examined the effect of fever-range WBH on the expression of adhesion molecules on microvessels[66] within human breast cancer xenografts in a novel chimeric animal model developed by Repasky et al.[52,67] In this model, highly vascularized patient primary tumors are grown in the gonadal fat pad of B- and T-cell-deficient SCID (severe combined immunodeficient) mice. Following fever-range WBH (40°C, 9 h),[66] a marked increase in the expression of endothelial L-selectin ligands (identified by reactivity with the MECA-79 mAb) was detected on blood vessels within human tumors engrafted in SCID mice. In the absence of hyperthermia treatment, MECA-79-reactivity in the tumor vasculature could not be detected. Notably, hyperthermia does not appear to increase the expression of the L-selectin endothelial ligands, VCAM-1, or MAdCAM-1, in normal tissue vascular beds including lung, liver, or kidney, suggesting that in a clinical setting, hyperthermia effects on vascular adhesion would be relatively restricted to tumor tissues and lymph nodes (as described in the preceding section).

These results provide a possible explanation for the observation of Burd *et al.*[52] that fe ver-range WBH causes dense accumulations of fluorescence-labeled xenogeniec lymphocytes in tumor blood vessels and throughout the tumor stroma, whereas relatively few lymphocytes are localized in tumors of unheated animals. Furthermore, fever-range hyperthermia treatment has been associated with delayed growth of human breast tumor xenografts in SCID mice that is dependent on endogenous NK cell activity.[52] Increased anti-tumor responses following fever-range WBH have also been observed in tumor-bearing animals with an intact immune system.[52,68–70] Notably, in several of these animal models, a decrease in lymph node metastases is also observed in response to fever-range hyperthermia.[68,69] Collectively, these results raise the intriguing possibility that hyperthermia treatment could result in enhanced recruitment of immune effector cells to highly vascularized tumor sites and regional lymph nodes through the regulation of L-selectin-dependent adhesion.

SUMMARY

These data support the concept that cytokines (i.e., IFN-α and/or other proinflammatory cytokines) in combination with fever-range hyperthermia can overcome the immunosuppressive effects of tumor microenvironments by stimulating L-selectin-dependent adhesive interactions between circulating immune effector cells and endothelium lining blood vessels in tumor tissues and regional lymph nodes. Studies are in progress to elucidate the precise role of the L-selectin adhesion molecule in antitumor immune responses and to determine the molecular mechanisms underlying cytokine and/or hyperthermia control of L-selectin adhesion. Further studies are indicated to determine the efficacy of combined hyperthermia and cytokine modalities as an adjuvant to conventional cancer chemotherapy regimens, immunotherapy involving adoptive transfer of immune effector cells (expressing appropriate adhesion molecules), vaccine therapy, or anti-angiogenic therapy.

REFERENCES

1. Girard, J. P., and T. A. Springer. 1995. High endothelial venules (HEVs): specialized endothelium for lymphocyte migration. *Immunol. Today 16:449–457.*
2. Butcher, E. C., and L. J. Picker. 1996. Lymphocyte homing and homeostasis. *Science 272:60–66.*
3. Springer, T. A. 1995. Traffic signals on endothelium for lymphocyte recirculation and leukocyte emigration. *Annu. Rev. Physiol. 57:827–872.*
4. Schweighoffer, T., W. Schmidt, M. Buschle, and M. L. Birnstiel. 1996. Depletion of naive T cells of the peripheral lymph nodes abrogates systemic antitumor protection conferred by IL-2 secreting cancer vaccines. *Gene Therapy 3:819–824.*
5. Kagamu, H., J. E. Touhalisky, G. E. Plautz, J. C. Krauss, and S. Shu. 1996. Isolation based on L-selectin expression of immune effector T cells derived from tumor-draining lymph nodes. *Cancer Res. 56:4338–4342.*
6. Rosato, A., A. Zambon, B. Macino, S. Mandruzzato, V. Bronte, G. Milan, P. Zanovello, and D. Collavo. 1996. Anti-L-selectin monoclonal antibody treatment in mice enhances tumor growth by preventing CTL sensitization in peripheral lymph nodes draining the tumor area. *Int. J. Cancer 65:847–851.*
7. Rosato, A., V. Bronte, S. Mandruzzato, A. Zambon, F. Calderazzo, G. Biasi, P. Zanovello, and D. Collavo. 1992. Role of adhesion molecules in the immune reaction to M-MSV induced tumors. *Int. J. Cancer Suppl. 7:24–27.*
8. Frey, M., N. Packianathan, T. A. Fehniger, M. E. Ross, W. C. Wang, C. C. Stewart, M. A. Caligiuri, and S. S. Evans. 1998. Differential expression and function of L-selectin on $CD56^{bright}$ and $CD56^{dim}$ natural killer cell subsets. *J. Immunol. 161:400–408.*
9. Piali, L., A. Fichtel, H. J. Terpe, B. A. Imhof, and R. H. Gisler. 1995. Endothelial vascular cell adhesion molecule 1 expression is suppressed by melanoma and carcinoma. *J. Exp. Med. 181:811–816.*
10. Wu, N. Z., B. Klitzman, R. Dodge, and M. W. Dewhirst. 1992. Diminished leukocyte-endothelial interaction in tumor microvessels. *Cancer Res. 52:4265–4268.*
11. Melder, R. J., G. C. Koenig, B. P. Witwer, N. Safabakhsh, L. L. Munn, and R. K. Jain. 1996. During angiogenesis, vascular endothelial growth factor and basic fibroblast growth factor regulate natural killer cell adhesion to tumor endothelium. *Nature Med. 2: 992–997.*
12. Onrust, S. V., P. M. Hartl, S. D. Rosen, and D. Hanahan. 1996. Modulation of L-selectin ligand expression during an immune response accompanying tumorigenesis in transgenic mice. *J. Clin. Invest. 97:54–64.*
13. Chin, Y. H., M. W. Ye, J. P. Cai, and X. M. Xu. 1996. Differential regulation of tissue-specific lymph node high endothelial venule cell adhesion molecules by tumor necrosis factor and transforming growth factor-β1 *Immunology 87:559–565.*
14. Bereta, J., M. Bereta, S. Cohen, and M. C. Cohen. 1993. Regulation of VCAM-1 expression and involvement in cell adhesion to murine microvascular endothelium. *Cell. Immunol. 147:313–330.*
15. Ferrara, N. 1996. Natural killer cells, adhesion and tumor angiogenesis. *Nature Med. 2:971–972.*
16. Jain, R. K., G. C. Koenig, M. Dellian, J. D. Fukumura, L. L. Mun, and R. J. Melder. 1996. Leukocyte-endothelial adhesion and angiogenesis in tumors. *Cancer & Metastasis Rev. 15:195–204.*
17. Sasaki, A., R. J. Melder, T. L. Whiteside, R. B. Herberman, and R. K. Jain. 1991. Preferential localization of human adherent lymphokine-activated killer cells in tumor microcirculation. *J. Nat. Ca. Inst. 83:433–437.*
18. Fukumura, D., H. A. Salehi, B. Witwer, R. F. Tuma, R. J. Melder, and R. K. Jain. 1995. Tumor necrosis factor alpha-induced leukocyte adhesion in normal and tumor vessels: effects of tumor type, transplantation site, and host strain. *Cancer Res. 55:4824–4829.*

19. Ohkubo, C., D. Bigos, and R. K. Jain. 1991. Interleukin-2-induced leukocyte adhesion to the normal and tumor microvascular endothelium in vivo and its inhibition by dextran sulfate: Implications for vascular-leak syndrome. *Cancer Res. 51:1561–1563.*
20. Gamble, J. R., and M. A. Vadas. 1988. Endothelial adhesiveness for blood neutrophils is inhibited by transforming growth factor-β. *Science 242:97–99.*
21. Yang, G., M. T. Mizuno, K. E. Hellstrom, and L. Chen. 1997. B7-negative versus B7-positive P815 tumor: differential requirements for priming of an antitumor immune response in lymph nodes. *J. Immunol. 158:851–858.*
22. Evans, S. S., R. P. Collea, M. M. Appenheimer, and S. O. Gollnick. 1993. Interferon-α induces the expression of the L-selectin homing receptor in human B lymphoid cells. *J. Cell Biol. 123:1889–1998.*
23. Frey, M., M. M. Appenheimer, and S. S. Evans. 1997. Tyrosine kinase-dependent regulation of L-selectin expression through the Leu-13 signal transduction molecule. *J. Immunol.158:5424–5434.*
24. Appenheimer, M. M., P. H. Silva, R. A. Bruce, S. O. Gollnick, P. D. Aplan, and S. S. Evans. Identification of the 5′ transcriptional regulatory region of the interferon-α-inducible human L-selectin gene. (Manuscript submitted)
25. Jewell, A .P., and K. L. Yong. 1997. Regulation and function of adhesion molecules in B-cell chronic lymphocytic leukemia. *Acta Hematologica 97:67–72.*
26. Evans, S. S., M. M. Appenheimer, and S. O. Gollnick. 1994. Interferon-α induces L-selectin expression by human B lymphoid cells. *J. Cell. Biochem. 18D:270.*
27. Evans, S. S., R. Collea, J. A. Leasure, and D. B. Lee. 1993. IFN-α induces homotypic adhesion and Leu-13 expression in human B lymphoid cells. *J. Immunol. 150:736–747.*
28. Scarozza, A. M., J. T. Collins, and S. S. Evans. 1992. DNA synthesis in nuclei isolated from Daudi B cells: A model to study the antiproliferative mechanisms of interferon-α. *J. Interferon Res. 12:35–42.*
29. Tovey, M.G., M. Streuli, I. Gresser, J. Gugenheim, B. Blanchard, J. Guymarho, F. Vignaux, and M. Gigou. 1987. Interferon messenger RNA is produced constitutively in the organs of normal individuals. *Proc. Natl. Acad. Sci. 84:5038–5042.*
30. Proietti, E., C. Vanden Broecke, P. Di Marzio, S. Gessani, I. Gresser, and M. G. Tovey. 1992. Specific interferon genes are expressed in individual cells in the peritoneum and bone 1992. Specific interferon genes are expressed in individual cells in the peritoneum and bone marrow of normal mice. *J. Interferon Res.. 12:27–34.*
31. Gresser, I., D. Guy-Grand, C. Maury, and M. T. Maunoury. 1981. Interferon induces peripheral lymphadenopathy in mice. *J. Immunol. 127:1569–1575.*
32. Hein, W.R. and A. Supersaxo. 1988. Effect of interferon-alpha-2a on the output of recirculating lymphocytes from single lymph nodes. *Immunology 64:469–474.*
33. Darnell, J.E., Jr. 1997. STATs and gene regulation. *Science 277:1630–1635.*
34. Leaman, D.W., S. Leung, X. Li, and G.R. Stark. 1996. Regulation of STAT-dependent pathways by growth factors and cytokines. *FASEB J. 10:1578–1588.*
35. Appenheimer, M. M., P. H. Silva, R. A. Bruce, S. O. Gollnick, P. D. Aplan, and S. S. Evans. Identification of interferon responsive elements within the L-selectin gene regulatory region. Presented at the joint AAAAI/AAI/CIS meeting, Feb. 21–26, 1997, San Francisco, CA.
36. Walcheck, B., J. Kahn, J. M. Fisher, B. B. Wang, R. S. Fisk, D. G. Payan, C. Feehan, R. Betageri, K. Darlak, A. F. Spatola, and T. K. Kishimoto. 1996. Neutrophil rolling altered by inhibition of L-selectin shedding *in vitro. Nature 380:720–723.*
37. Steen, P. D., J. R. McGregor, C. M. Lehman, and W. E. Samlowski. 1989. Changes in homing receptor expression on murine lymphokine-activated killer cells during IL-2 exposure. *J. Immunol. 143:4324–4330.*
38. McRae, B. L., L. J. Picker, and G. A. van Seventer. 1997. Human recombinant interferon-β influences T helper subset differentiation by regulating cytokine secretion pattern and expression of homing receptors. *Eur. J. Immunol. 27:2650–2656.*
39. Evans, S. S., D. B. Lee, T. Han, T. B. Tomasi, and R. L. Evans. 1990. Monoclonal antibody to the interferon-inducible protein Leu-13 triggers aggregation and inhibits proliferation of leukemic B cells. *Blood 76:2583–2593.*
40. Deblandre, G. A., O. P. Marinx, S. S. Evans, S. Majjaj, O. Leo, D. Caput, G. A. Huez, and M. G. Wathelet. 1995. Expression cloning of an interferon-inducible 17-kDa membrane protein implicated in the control of cell growth. *J. Biol. Chem. 270:23860–23866.*
41. Tedder, TF, L. J. Zhou, and P. Engel. 1994. The CD19/CD21 signal transduction complex of B lymphocytes. *Immunol. Today 15:437–442.*
42. Fearon, D. T., and R. H. Carter. 1995. The CD19/CR2/TAPA-1 complex of B lymphocytes: linking natural to acquired immunity. *Annu. Rev. Immunol. 13:127–149.*

43. Levy, S., S. C. Todd, and H. T. Maecker. 1998. CD81 (TAPA-1), a molecule involved in signal transduction and cell adhesion in the immune system. *Annu. Rev. Immunol. 16:89–109.*
44. Chen, Y. X., K. Welte, D. H Gebhard, and R. L. Evans. 1984. Induction of T cell aggregation by antibody to a 16kd human leukocyte surface antigen. *J. Immunol. 133:2496–2501.*
45. Reid, L. E., A. H. Brasnett, C. S. Gilbert, A. C. Porter, D. R. Gewert, G. R. Stark, and I. M. Kerr. 1989. A single DNA response element can confer inducibility by both α- and γ-interferons. *Proc. Natl. Acad. Sci. USA 86:840–844.*
46. Preece, G., G. Murphy, and A. Ager. 1996. Metalloproteinase-mediated regulation of L-selectin levels on leukocytes. *J. Biol. Chem. 271:11634–11640.*
47. Bennett, T. A., E. B. Lynam, L. A. Sklar, and S. Rogelj. 1996. Hydroxamate-based metalloprotease inhibitor blocks shedding of L-selectin adhesion molecule from leukocytes: functional consequences for neutrophil aggregation. *J. Immunol. 156:3093–3097.*
48. Shecterle, L. M., and J. A. St. Cyr. 1995. Whole body hyperthermia as a potential therapeutic option. *Cancer Biotherapy 10:253–256.*
49. Kapp, D. S. 1996. Thermal dose response, systemic hyperthermia, and metastases: old friends revisited. *Int. J. Rad. Oncol. Biol. Phys. 35:189–194.*
50. Alpard, S. K., R. A. Vertees, W. Tao, D. J. Deyo, R. L. Brunston, Jr., and J. B. Zwischenberger. 1996. Therapeutic hyperthermia. *Perfusion 11:425–435.*
51. Fajardo, L. F., and S. D. Prionas. 1994. Endothelial cells and hyperthermia. *Int. J. Hyperthermia 10:347–353.*
52. Burd, R., S. Dziedzic, Y. Xu, M. A. Caligiuri, J. R. Subjeck, and E. A. Repasky. Tumor cell apoptosis, lymphocyte recruitment and tumor vascular changes are induced by low temperature, long duration whole body hyperthermia. *J. Cell. Physiol., in press.*
53. Roberts, N. J., Jr. 1991. Impact of temperature elevation on immunologic defenses. *Reviews Infect. Dis. 13:462–472.*
54. Smith, J. B., R. P. Knowlton, and S. S. Agarwal. 1978. Human lymphocyte responses are enhanced by culture at 40°C. *J. Immunol. 121:691–694.*
55. Roberts, N. J., Jr., and R. T. Steigbigel. 1977. Hyperthermia and human leukocyte functions: Effects on response of lymphocytes to mitogen and antigen and bactericidal capacity of monocytes and neutrophils. *Infect. Immunity 18:673–679.*
56. Heron, I., and K. Berg. 1978. The actions of interferon are potentiated at elevated temperature. *Nature 274:508–510.*
57. Wang, W.C., L. M. Goldman, D. M. Schleider, M. M. Appenheimer, J. R. Subjeck, E. A. Repasky, and S. S. Evans. 1998. Fever-range hyperthermia enhances L-selectin-dependent adhesion of lymphocytes to vascular endothelium. *J. Immunol. 160:961–969.*
58. Pavalko, F. M., and C. A. Otey. 1994. Role of adhesion molecule cytoplasmic domains in mediating interactions with the cytoskeleton. *Proc. Soc. Exp. Biol. Med. 205:282–293.*
59. Pavalko, F. M., D. M. Walker, L. Graham, M. Goheen, C. M. Doerschuk, and G. S. Kansas. 1995. The cytoplasmic domain of L-selectin interacts with cytoskeletal proteins via α-actinin: receptor positioning in microvilli does not require interaction with α-actinin. *J. Cell Biol. 129:1155–1164.*
60. Evans, S. S., D. M. Schleider, L. Bowman, G. S. Kansas, and J. D. Black. Dynamic association of L-selectin with the lymphocyte cytoskeletal matrix. (Manuscript submitted)
61. D'Oleire, F., C. L. Schmitt, H. I. Robins, J. D. Cohen, and D. Spriggs. 1993. Cytokine induction in humans by 41.8°C whole-body hyperthermia. *J. Natl. Can. Inst. 85:833–834.*
62. Robins, H. I., M. Kutz, G. J. Wiedemann, D. M. Katschinski, D. Paul, E. Grosen, C. L. Tiggelaar, D. Spriggs, W. Gillis, and F. d'Oleire. 1995. Cytokine induction by 41.8°C whole body hyperthermia. *Cancer Letters 97:195–201.*
63. Roberts, N. J., Jr. 1986. Differential effects of hyperthermia on human leukocyte production of interferon-α and interferon-γ. *Proc. Soc. Exper. Biol. Med. 183:42–47.*
64. Downing, J. F., M. W. Taylor, K. M. Wei, and R. S. Elizondo. 1987. In vivo hyperthermia enhances plasma antiviral activity and stimulates peripheral lymphocytes for increased synthesis of interferon-γ. *J. Interferon Res. 7:185–193.*
65. Evans, S. S., W. C. Wang, R. Burd, D. M. Schleider, J. R. Subjeck, and E. A. Repasky. Whole body hyperthermia augments L-selectin-dependent adhesion of lymphocytes to lymph node high endothelial venules. (Manuscript in preparation)
66. Evans, S. S., W. C. Wang, L. M. Goldman, J. Subjeck, and E. A. Repasky. Hyperthermia-induced enhancement of lymphocyte adhesion to vascular endothelium. Presented at the Radiation Research Society Meeting, April 1996.

67. Sakakibara, T., H. L. Bumpers, F. Chen, R. B. Bankert, M. A. Arredondo, S. B. Edge, and E. A. Repasky. 1996. Growth and metastasis of surgical specimens of human breast carcinomas in SCID mice. *Cancer J. Sci. Am. 2:291–300.*
68. Matsuda, H., F. R. Strebel, T. Kaneko, L. L. Danhauser, G. N. Jenkins, N. Toyota, J. M. Bull. 1997. Long duration-mild whole body hyperthermia of up to 12 hours in rats: feasibility, and efficacy on primary tumor and auxillary lymph node metastases of a mammary adenocarcinoma: implications for adjuvant therapy. *Int. J. Hyperthermia. 13:89–98.*
69. Hachiya, H., K. Okada, A. Sakurai, N. Satomi, and K. Haranaka. 1992. Antitumor activity of recombinant human tumor necrosis factor in combination with hyperthermia against heterotransplanted human prostatic carcinoma and its lymph node metastasis in nude mice. *Mol. Biother. 4:34–39.*
70. Sakaguchi, Y., M. Makino, T. Kaneko, L. C. Stephens, F. R. Strebel, L. L. Danhauser, G. N. Jenkins, and J. M. Bull. 1994. Therapeutic efficacy of long duration-low temperature whole body hyperthermia when combined with tumor necrosis factor and carboplatin in rats. *Cancer Res. 54:2223–2227.*
71. Geppert, T. D., and P. E. Lipsky. 1991. Association of various T cell-surface molecules with the cytoskeleton. Effect of cross-linking and activation. *J. Immunol. 146:3298–3305.*

DISCUSSION

Bankert: With cytokines therapy, one of the natural sequels is fever and one of the first things a physician reaches for with a fever is aspirin. So maybe we ought to do the cytokine trials over again without aspirin.

Evans: Well, there is an interesting aspect with a lot of the non-steroidal inflammatory agents which induce selectin shedding very nicely so there are several reasons to consider maybe not taking aspirin as your first attack.

Bankert: Do you know whether the increased adhesion that is associated with the hyperthermia is due to one particular heat shock protein?

Evans: We really do not know. We have looked and we know that HSP70 and HSP110 are both induced over the same time frame, but whether it is functionally related to adhesion is not clear yet. We are very interested in looking into that.

Melief: Has anybody tried to combine hyperthermia with either vaccination or adoptive transfer?

Evans: Well, that is, in fact, exactly where the field of hyperthermia is beginning to go and very few people though are considering using hyperthermia at fever range temperatures, which I think is going to be the critical issue. So, there are unpublished data, from Betsy Repasky's lab for example, using hyperthermia in combination with gamma interferon where you see a much greater antitumor activity which is what you would expect. One of the things we, of course, are really interested is applying the same kind of story to interferon-alpha because it has such great effects and I think, overall, we do not view this as an end, but it may be an interesting combinatorial type of activity in terms of using this first to take advantage of the tumor vasculature, get the cells in and then come in, for example, with anti-angiogenic approaches later to really break down the vascular beds. So we have multiple mechanisms in play. I think that is where things will go but interferon-alpha could be interesting in that same setting.

Zanker: First, I would like to congratulate you for this clear-cut work on migration. Second, I would like to ask you, operationally or functionally speaking, do you think that shedding a molecule like L-selectin has something to do with the possibility that the lymphocyte can detach and to roll on and exit from the vascular?

Evans: Yes, exactly, that is definitely where our thinking is, and many people in the field are really considering. Because if L-selectin just bound firmly, we would all be clogged up in about one minute so really the attachment and rolling has to incorporate an adhesion process. The adhesion may involve the shedding, the low level shedding, but it also may involve interactions, transient interactions with the cytoskeleton. A collaborator with us on these studies has actually published studies where they have used hydroxamic inhibitors in the matrix metal proteonases and they have shown that if they apply these to the cells, they are no longer able to roll, so they firmly bind through L-selectin dependent mechanisms and they do not move. Probably even under the cases of sheer you really need to have this very low level shedding that would occur along the surface.

Zanker: Do you think that one can extend this paradigm that shedding of other antigens of tumor cells has something to do with detachment and with migration? For instance, integrins on other cells, not only to lymphocytes?

Evans: You mean in terms of malignant cells, for example? It is unlikely that the integrins are really shedded at very high levels, but in terms of any metastatic cells that is using L-selectin as a migratory pathway, L-selectin is shed at very high levels and, in fact, in itself is angiogenic, so V-CAM is shed as well. So I think the story is still out really as far as these shed molecules are playing a role functionally, no one is really quite sure how to really deal with that, but even the L-selectin ligands are shed at high levels in the peripheral blood.

Boon: What would be the acceptable way to induce hyperthermia in patients who have been vaccinated, for instance?

Evans: Yes, I think that is a very good question. We are using whole body hyperthermia where the animals are put in a very warm room, but one of the concepts that we really want to address is whether you need to have the whole body hyperthermia or could you do something like in breast tumor, where it could be much more localized; you could have regional hyperthermia, take out the lymphocytes and heat them independently in the presence of, for example, pro-inflammatory cytokines that could increase L-selectin shedding and then put the whole system back together and we have to really see. Joan Bull at M.D. Anderson has done whole body hyperthermia treatment of patients and, in fact, unlike the more conventional therapies which can be very stressful and many patients cannot really withstand the high temperatures that they are exposed to, the whole body hyperthermia at fever range temperatures is tolerated.

Boon: There are no acceptable agents that can produce a mild fever?

Evans: Well, as it was mentioned before, a whole lot of these pro-inflammatory cytokines in fact do give you mild fever--interferon-alpha is famous in its clinical trials for that property. One thing we have not done yet that we are anxious to do is look to see if cytokines exactly duplicate the effect of hyperthermia. You heat the cells and add in the pro-inflammatory cytokines and that may help us to decide if that would be a possible approach.

Bankert: It is interesting, it is almost 103 years, I think, to the day that Coley first reported his phenomenon. Maybe we now have an explanation for the observed effects with Coley's toxin.

White: There is increasing reason to believe that there is a homing fingerprint on the end of endothelial cells of specific tissues. Do you know whether that fingerprint is maintained in the tumor cells? Do the endothelial cells of the tumor reflect the origin of that tumor?

Evans: That is a very good question and I actually do not know the answer to that. I know that there can be differential expression of certain adhesion molecules in tumors that grow out, but in terms of some of the specific things you are talking about, I do not know how carefully that has been addressed.

White: So it could be another interesting targeting approach?

Evans: Yes, I agree.

DePinho: I guess there is a lot known about how lymphocytes negotiate tight junctions and how they penetrate in. Is there any inkling as to whether that mode, that process is disregulated in some manner in tumors versus normal organs? Are there mechanistic differences in the means by which lymphocytes negotiate tight junctions of tumor versus normal endothelium?

Evans: That is a very good question. To my knowledge, some of the molecules that are very critical in terms of migration are maintained in tumors. For example, P-CAM is expressed normally, I-CAM is expressed, and I-CAM is not thought to be subject to the levels of regulation that the adhesion molecules are at the lymphocytes. But I do not think this has been really addressed. What is known though, is that if you take model systems where you grow tumors in close proximity to endothelial cells, they release factors that feed back on the endothelium to prevent V-CAM expression in that system. So, there are factors that feedback, in terms of certain adhesion properties, but in terms of migration through the actual tight junctions between the endothelial cells, I do not know of any studies that have addressed that.

Bankert: You alluded to the fact that you found some increase in binding and you suggested maybe of NK cells. Have you looked at the phenotype of the lymphocyte? Is there a skewed binding of NK cell over CD-4 or CD-4 over CD-8? After the hyperthermia you showed binding, you showed the increase in binding, but you just called them lymphocytes, and in one case you showed an enriched population of NK cells where there was also binding. But if you do not enrich, if you just take peripheral blood, can you speculate or have you looked at the phenotype of the cells?

Evans: We have not looked at it in terms of the *in vivo* homing studies. We have looked at that in terms of adhesion *in vitro* and we know that the CD-56 bright cells are enhanced in terms of their ability to bind as well as T cells. We have not broken it down yet to CD-4 and CD-8. We just got those assays really worked out so we can really delineate which cells are doing the effective adhesion. Interestingly though CD-56 dim and NK cells really do not bind very well in that system.

Bankert: Clinical trials must be supplemented with viable animal models to evaluate patients' immune response to their tumors and to evaluate a patient's augmented response to tumor vaccines. One is now able to do this using the severe combined immunodeficient mouse model in which patients' lymphocytes are co-engrafted with the patients' tumor cells.

21

ROLE OF ENDOTHELIAL VERSUS SMOOTH MUSCLE CELLS IN BLOOD VESSEL FORMATION

Peter Carmeliet* and Désiré Collen

The Center for Transgene Technology and Gene Therapy
Flanders Interuniversity Institute for Biotechnology
Leuven, B-3000, Belgium

ABSTRACT

The process of how blood vessel form has long been studied by descriptive analysis. Recently, however, substantial progress in our understanding of the molecular mechanisms mediating new blood vessel formation has been achieved. Initially, blood vessels form as endothelium-lined channels by *in situ* differentiation of endothelial cells. Subsequently, they sprout and remodel into a highly organized and interconnected vascular network. During further maturation of the blood vessels, a sheet of primitive vascular smooth muscle cells surrounds the endothelium-lined channels, which controls endothelial cell function and provides structural support. The *in vivo* role of the candidate molecules responsible for these processes has been further established by targeted gene manipulation in transgenic mice. This review highlights recent developments in the genetic analysis of blood vessel formation, as deduced from analysis of gene-inactivated mice.

1. DEVELOPMENT OF ENDOTHELIAL-LINED CHANNELS

Distinct cellular processes mediate blood vessel formation[1]. Initially, mesenchymal cells differentitate *in situ* into early hemangioblasts which form cellular aggregates (blood islands), in which the inner cell population differentiates into hematopoietic pre-

* Address for correspondence: Peter Carmeliet, M.D., Ph.D., Center for Transgene Technology & Gene Therapy, Campus Gasthuisberg, Herestraat 49, University of Leuven, Leuven, B-3000, Belgium; tel: 32-16-34.57.72; fax: 32-16-34.59.90; e-mail: peter.carmeliet@med.kuleuven.ac.be

The Biology of Tumors, edited by Mihich and Croce
Plenum Press, New York, 1998.

cursors and the outer cell population gives rise to the primitive endothelial cells (vasculogenesis). Basic fibroblast growth factor (bFGF) may participate in this process[2]. The larger vessels of the embryo arise by this process. The primitive vascular plexus subsequently develops into a complex organized and interconnecting network either via intussusception of preexisting vessels (whereby a preexisting vessel is split into two daughter vessels by invagination of surrounding pericytes/smooth muscle cells and extracellular matrix)[3], or via a process called "angiogenesis", whereby new blood vessels sprout from a preexisting blood vessel[1]. Angiogenesis mediates vascularization of organs such as the brain and kidney. Remodeling of this vascular plexus into large and small vessels by metabolic, hydrodynamic, hypoxic or rheologic factors may further reshape this network into a functional vasculature.

1.1. The Vascular Endothelial Growth Factor (VEGF) System

Vascular endothelial growth factor (VEGF-A), unlike other known angiogenic factors, has a unique combination of properties: (i) it is produced by cells in close vicinity of endothelial cells, suggesting paracrine regulation of blood vessel formation; (ii) it is secreted and exerts a direct effect via interaction with endothelial receptors Flk-1 and Flt-1; (iii) its expression is highly regulated by hypoxia, providing a physiological feedback mechanism to accomodate insufficient tissue oxygenation by promoting blood vessel formation; and (iv) it is a potent growth factor since its over- or under-expression significantly affects blood vessel formation *in vivo*[4–6]. Several isoforms of VEGF-A exist with different mitogenic potency, affinity for extracellular matrix proteins or cellular receptors, and tissue-specific and temporal pattern of expression. More recently, other VEGF-related factors have been identified and homo- or heterodimerization of these ligands may determine their biological specificity: placenta growth factor (PlGF)[7], VEGF-B[8], VEGF-C (also called VEGF-related factor or VRP)[9,10] and FIGF, a c-*fos* induced growth factor[11]. The differential role of the VEGF isoforms and homologues in governing vascular development during embryonic development or adult pathology is largely unkown. Three receptor tyrosine kinases bind the VEGF family members with different specifity and affinity: VEGF receptor-1 (Flt-1) binds VEGF-A and PlGF[12,13], VEGF receptor-2 (Flk-1) binds VEGF-A and VEGF-C (possibly with a lower affinity)[9,14], and VEGF receptor-3 (Flt-4) binds VEGF-C[9,10].

Targeted inactivation of a single *VEGF-A* allele resulted in haploinsufficiency with embryonic lethality due to abnormal blood vessel development[15–17]. The dorsal aorta had a much smaller lumen, sprouting of new vessels was reduced, and connections of the large blood vessels with the heart appeared abnormal. In the yolk sac, large vitello-embryonic blood vessels (that connect the yolk sac with the embryo) were absent and only an irregular plexus of enlarged capillaries was present. Since proliferation of vitelline endothelial cells was normal, abnormal fusion of endothelial cells or intussusception of blood vessels may have caused the impaired vitelline vascular development. Blood vessel development was more affected in homozygous VEGF-A deficient embryos, generated by aggregation of homozygous VEGF-A deficient embryonic stem cells with tetraploid embryos, than in heterozygous VEGF-A deficient embryos, suggesting a tight gene dosage-dependent relationship and production of only minimally required VEGF levels during embryonic development. Indeed, at 8.5 days post coitum, there were only some scattered endothelial cells within the homozygous VEGF-A deficient embryo that failed to organize into a pair of dorsal aortas. Expression of early endothelial cell markers was also lower in homozygous than in heterozygous VEGF-A deficient embryos, suggesting that endothelial cell development was delayed but not aborted. Taken together, these studies indicate that VEGF-A is essential for embryonic vascular development, not via early endothelial cell differentiation but via endothelial cell maturation, expansive growth, lumen formation, vascular patterning and possibly via induction of vascular connections.

More recently, we have generated other transgenic mice such as mice that only expressed the VEGF-A^{120} isoform. Initial analysis indicates that homozygous mutant VEGF-A^{120} embryos die *in utero*, indicating that VEGF-A^{120} is able to rescue the defective embryonic vascular development of heterogygous VEGF-A deficient embryos but insufficient to sustain normal vascular development in a homozygous form (unpublished observations in collaboration with P. d'Amore et al, Harvard Medical School, Boston, USA). Deficiency of PlGF did not compromize development, feritility or placentation (unpublished observations in collaboration with G. Persico et al, Naples, Italy). This was not anticipated in view of the presumed role of PlGF in establishing vascular connections in the placenta. However, initial analysis suggests that healing of skin wounds was retarded possibly due to impaired formation of granulation tissue, neovascularization and reduced vascular permeability.

Deficiency of the VEGF receptor Flk-1 resulted in embryonic lethality around 10 days of gestation due to abnormal vascular development[18]. Histological examination revealed complete absence of organized blood vessels and necrosis in the mutant embryo proper. LacZ expression (by the targeted *flk-1* promoter) was only detected in some putative endothelial cell precursors. Thus, the developing mesoderm cannot complete differentiation into blood islands in the absence of a functional receptor. Interestingly, the observation that VEGF-A deficiency did not result in a similar phenotype as the Flk-1 deficiency, strongly suggests the presence other VEGF-related ligands for Flk-1. Targeting of the *flt-1* gene also resulted in embryonic lethality around 10 days of gestation[19]. Staining for LacZ (expressed by the targeted *flt-1* promoter) revealed the presence of numerous differentiated endothelial cells, which, however, failed to form an organized vascular network. These vascular structures were abnormally large and fused, and contained enclosed LacZ positive endothelial cells. Thus, these findings suggest a possible role of Flt-1 in contact inhibition of endothelial cell growth, or in endothelial cell assembly via controlling adhesion between endothelial cell precursors.

1.2. The Endothelial Cell Receptor Tyrosine Kinase TIE-1

TIE-1 (tyrosine kinase with immunoglobulin and epidermal growth factor homology domains) and TIE-2/TEK (tunica interna endothelial cell kinase) comprise another family of endothelial cell specific receptor tyrosine kinases[20,21]. A ligand has been only identified for the TIE-2/TEK receptor (angiopoietin-1)[22]. Although both TIE-1 and TIE-2/TEK are expressed by endothelial cells during vascular development, current evidence suggests that TEK affects blood vessel formation indirectly via an effect on periendothelial cell function (and is therefore discussed in the next section), whereas thus far only an effect on endothelial cell function has been ascribed for TIE-1.

Targeted inactivation of the *tie-1* gene resulted in normal differentiation of the endothelium, formation and patterning of early blood vessels by 13 days of gestation[23]. However, whithin one day, homozygous TIE-1 deficient embryos manifested hemorrhages distributed throughout the body surface, became oedematous and died by 14 days of gestation. Upon microscopic analysis, most large blood vessels appeared normal but staining for LacZ (expressed by the targeted *tie-1* promoter) was reduced in the microvasculature, often associated with breakdown of the vessel integrity and hemorrhage. In another *tie-1* gene inactivation study, only a fraction (15%) of homozygous TIE-1 mutant embryos died at 13 days of gestation due to localized hemorrhage and edema[24]. The remainder fraction of mutant embryos displayed severe edema and hemorrhage at multiple locations throughout the body one day prior to birth, and died immediately after birth due to respiratory problems resulting from pulmonary congestion. Despite a normal vessel pattern, the structural integrity of the endothelial cells in the mutant embryos was defective, as revealed by the extravasation of erythrocytes through, but not between endothelial cells. Taken together, the "leaky syndrome" of the TIE-1 deficient embryos suggests that TIE-1 is not required for early endothelial cell differentiation, vasculogenesis and the initial steps of angiogenesis. TIE-1 appears, however, to be required for the structural integrity of microvascular endothelial cells during maturation of blood vessels. Whether this is due to an endothelial cell-specific defect, or related to a defect in peri-endothelial cells (such as in the angiopoietin-1 deficient embryos; cfr infra) remains to be determined.

2. DEVELOPMENT OF THE VASCULAR SMOOTH MUSCLE LAYER AROUND THE ENDOTHELIUM

Once endothelial cells are assembled into vascular channels, they become surrounded by smooth muscle cells/pericytes that may affect maturation of the blood vessels not only by providing the fragile primitive blood vessels structural support, required to accomodate the increased blood pressure, but also by controlling endothelial cell proliferation, vascular permeabilty and tone[25]. Although a role for vascular smooth muscle cells/pericytes in the pathogenesis of vasculopathies during adulthood including atherosclerosis, restenosis and diabetic retinopathy has been documented, their role during vascular development has been poorly characterized.

2.1. The Coagulation System

Initiation of the plasma coagulation system is triggered by tissue factor, which functions as a cellular receptor and cofactor for activation of the serine proteinase factor VII to factor VIIa[26]. This complex activates factor X either directly or indirectly via activation of factor IX, resulting in the generation of thrombin and in the conversion of fibrinogen to fibrin[27]. Thrombin and factor Xa produce a positive feedback stimulation of coagulation by activating factor VIII and factor V which serve as membrane-bound receptors/cofactors for the proteolytic enzymes factor IXa and factor Xa, respectively. Evidence has been provided that the coagulation system may also be involved in other functions beyond hemostasis including cellular migration and proliferation, immune response, angiogenesis, embryonic development, metastasis and brain function[28–30].

2.1.1. The Cellular Tissue Factor Receptor and Its Ligand Factor VII. Targeted inactivation of the tissue factor gene resulted in increased fragility of endothelial cell-lined channels in the yolk sac, which are essential for transferring maternally derived nutrients from the yolk sac to the rapidly growing embryo[31]. At a time when the blood pressure increased during embryogenesis (day 9 of gestation), the immature tissue factor deficient blood vessels ruptured, formed micro-aneurysms and 'blood lakes', and failed to sustain proper circulation between the yolk sac and embryo. Secondarily, the embryo became wasted and died due to generalized necrosis. Only in advanced stages of deterioration became the immature blood vessels leaky, resulting in bleeding into the extracoelomic cavity. Similar observations were made when tissue factor deficient embryos were cultured *in vitro*, suggesting that the observed vascular defects in the yolk sac were not merely due to a possible defect in feto-maternal exchange. A role for tissue factor in embryonic development is also suggested by the observations of Toomey et al[32,33].

Visceral endoderm and endothelial cell function appeared normal, suggesting that these cell types were not responsible for the vascular defects. In contrast, defective development and/or maturation of mesenchymal cells (smooth muscle cell/pericyte-like cells) appeared to be the likely reason for the vascular defects. These cells surround the endothelium in yolk sac vessels, form a primitive "muscular" wall and provide structural support by their close physical association and increasing production of extracellular matrix proteins. Microscopic and ultrastructural analysis revealed that deficiency of tissue factor resulted in a 75% reduction of the number of mesenchymal cells, and a diminished amount of extracellular matrix[31]. Immunocytochemical analysis further revealed a reduced level of smooth muscle α-actin staining in these cells, suggesting impaired differentiation and/or accumulation. Because these primitive smooth muscle cells provide structural support for the endothelium, the vessels in the mutant embryos are too fragile and break open. Pericytes have also been implicated in adult diabetic retinopathy where pericyte 'drop out' results in the formation and rupture of micro-aneurysms and blindness[25].

Unresolved questions are how tissue factor exerts this morphogenic action, i.e., via intracellular signaling as suggested previously[30], and/or whether fibrin formation occurs and is essential during early vascular development, as suggested by others[34]. Thusfar, there is no conclusive evidence that coagulation factors pass the feto-maternal barrier and that the early stage embryo has a functional clotting system. Indeed, factor VII levels in the plasma from 11.5 day old wild type embryos are $< 0.5\%$ of those during adulthood[35]. There is also no evidence for fibrin deposits in the visceral yolk sac from ultrastructural analysis of wild type embryos[35], as would be expected if TF deficient embryos died due to a hemostatic defect. Moreover, intracardial injection of high doses of thrombin in 9.5 day old embryos fails to induce fibrin thrombi[35]. Thus, hemostasis during early embryonic development (9 days of gestation) appears to be minimally dependent on fibrin formation. Notably, embryos lacking platelets also develop normally without signs of bleeding[36], suggesting that fibrin- and platelet-dependent hemostasis during early development is significantly compromized, if at all operational.

Since the yolk sac contains tissue factor-mediated procoagulant activity, which is immunocytochemically localized in the visceral endoderm cells and possibly in the mesenchymal periendothelial cells[31], it is possible that tissue factor provides spatial migratory cues for recruitment of mesenchymal cells to envelope the endothelial cells, that it provides a signal for mesenchymal cells to differentiate into pericytes/smooth muscle cells, and/or that it affects their growth. Although it could do so by generation of downstream coagulation molecules (including factor Xa and thrombin, which both can affect migration/proliferation of vascular cells), our recent analysis that the factor VIIa inhibitor rNAPc2 fails to induce the typical TF deficient-like vascular fragility in cultured embryos[35] might suggest that TF could act via alternative pathways, for example as a cellular adhesion molecule, or by affecting novel cellular mechanisms (for example the production of a mesenchymal recruitment signal), or via interaction with another ligand/receptor. Although analysis revealed no endothelial cell abnormalities and normal expression of VEGF, a possible (in)direct effect of TF on endothelial cells cannot be excluded in view of the close structural and functional relationship between endothelial cells and pericytes.

In contrast to the severe embryonic TF deficient lethality, factor VII deficient mice develop normally untill birth and die early postnatally due to massive hemorrhaging[35]. Interestingly, factor VII levels in the plasma from 11.5 day old factor VII deficient embryos are undetectable (<0.05%) during the entire development. Moreover, intravenous injection in pregnant mice of human recombinant factor VIIa induces supraphysiological plasma levels in the mother but only background levels in early stage embryos, indicating that transfer of maternal factor VII is, if any, vanishingly low. Taken together, the TF and factor VII targeting studies suggest that hemostasis during early embryonic development (around 9.5–10.5 days of gestation) may be less dependent on TF-dependent fibrin formation than anticipated but that TF-mediated vascular integrity may relate to other morphogenic processes, the nature of which remain to be defined. That hemostasis during ealry embryogenesis may differ from that during adulthood is further supported by previous gene-inactivation studies revealing that embryos deficient in platelets develop normally without bleeding[36] but that inactivation of genes involved in function or assembly of (peri)endothelial cells is the most frequent cause of bleeding and defective hemostasis during development[16,24,37–42].

2.1.2. Other Coagulation Factors. Thrombin has been implicated in processes beyond hemostasis[29]. Indeed, it is mitogenic for fibroblasts and vascular smooth muscle cells, chemotactic for monocytes and it activates endothelial cells. Many of the cell signaling activities of thrombin appear to be mediated by the thrombin receptor. Expression studies have suggested that the thrombin receptor participates in inflammatory, proliferative or reparative responses such as restenosis, atherosclerosis, neovascularization and tumorigenesis. In addition, *in situ* analysis indicated that this receptor is expressed during early embryogenesis in the developing heart and blood vessels, in the brain and in several epithelial tissues[43]. Targeting of the trombin receptor resulted in blocking of embryonic development in approximately 50% of the homozygous deficient embryos around a similar developmental stage as in tissue factor deficient embryos, presumably resulting from abnormal yolk sac vascular development[44,45]. Although the cellular defect was not characterized in detail, vitelline vessels appeared to be defective, resulting in increased fragility of blood vessels with secondary rupture, blood leakage and pallor of the embryo. Similarly enlarged pericardial cavities in the thrombin receptor deficient embryos suggested compromized vitello-embryonic blood circulation.

Other coagulation factors also appear to be involved in morphogenic processes during early embryogenesis, possibly in blood vessel formation. Similar to the thrombin receptor deficient phenotype, deficiency of factor V resulted in embryonic lethality in approximately half of the homozygously deficient embryos, possibly due to vascular defects in the yolk sac[46]. The leakage of blood from the defective blood vessels in tissue factor, thrombin receptor and factor V deficient embryos ("vascular" bleeding) contrasts with the postnatal bleeding in mice deficient of factor VII[35], factor VIII[47] and fibrinogen[48] and in the surviving fraction of factor V deficient mice, which occurs due to defective clot formation following trauma of normally developed blood vessels ("hemostatic" bleeding). Bleeding in the latter mice occurred shortly after birth (factor V, factor VII and fibrinogen) or was associated with injury (factor VIII). Thus, it appears that several coagulation factors (tissue factor, factor V, thrombin receptor) participate in morphogenic processes beyond control of hemostasis, whereas other coagulation factors (factor VII, factor VIII, fibrinogen) play a predominant role in hemostasis via clot formation. This raises the question whether tissue factor and the thrombin receptor, which are expressed during restenosis, atherosclerosis and neovascularization, also play a similar (non-hemostatic) role in these processes. If so, this could open an attractive therapeutic avenue to selectively inhibit the hemostatic or morphogenic properties of these molecules.

2.2. The Tyrosine Receptor Kinase TIE-2/TEK and Its Ligand Angiopoietin-1

Recently, a ligand for the TIE-2/TEK receptor (angiopoietin-1) was identified, which specifically induced tyrosine phosphorylation of TIE-2/TEK in endothelial cells, but, surprisingly, failed to induce endothelial cell proliferation, migration or tube formation in collagen gels[22]. In the early embryo (9.0 days of gestation), angiopoietin-1 is most prominently expressed in the myocardium, and becomes subsequently (>13 days of gestation) more widely expressed throughout the embryo, most often in the mesenchyme surrounding developing vessels (intersegmental vessels, eye vessels etc), in close association with endothelial cells. Targeted disruption of the *angiopoietin-1* gene resulted in embryonic lethality around 11.0 to 12.5 days of gestation due to abnormal cardiovascular development[38]. Vascular defects included a less complex and immature vascular network, characterized by more dilated, and almost syncitial appearance of the vessels, much less distinction in size between large and small vessels, and fewer and straighter branches. Yet, there was no difference in endothelial cell accumulation. These abnormal vessels contrast with the highly branched vascular network with clear distinction in size between major vessels and smaller sprouts. In addition, certain vessels, such as the intersomitic vessels, appeared to regress beyond 11 days of gestation. Vascular defects were more prominent in the yolk sac than within the embryo proper. Together, these findings suggest that remodeling of the initially homogeneous capillary network into both large and small vessels is defective. Ultrastructural analysis revealed that, in angiopoietin deficient embryos, the endothelial cells were more rounded and failed to make close contact with the subendothelial extracellular matrix, and that there were fewer periendothelial mesenchymal cells (most likely immature smooth muscle cell/pericyte-like cells). In addition, the intervening space contained fewer and scattered collagen-like fibers. These findings can be explained by a failure in recruitment of and association with peri-endothelial mesenchymal cells, possibly because endothelial cells fail to transmit a recruiting signal (PDGF-BB?) for mesenchymal cells in the absence of angiopoietin-1. This results in impaired branching and vessel segmentation (intussuception), leading to more homogeneously dilated vessels. Stabilization and maturation of the immature vessels is compromized, resulting in increased fragility, rupture and "vascular" bleeding, and in other cases, in vessel regression. Whether angiopoietin-1 directly affects endothelial cell function *in vivo* remains to be elucidated.

Interference with *tek* gene function was accomplished by generation of transgenic mice with a dominant negative TEK mutant due to lysine-to-alanine mutation of codon 853 (tek^{L853A}) in the intracellular domain, which abolished (auto)phosphorylation[49]. In addition, *tek* gene function was disrupted via homologous recombination in embryonic stem cells ($tek^{-/-}$). tek^{L853A} and $tek^{-/-}$ embryos displayed an enlarged pericardial cavity and died *in utero* around 9.5 days of gestation. Up until 9 days of gestation, mutant embryos developed normally, containing a normally patterned vasculature in both extra- and intra-embryonic tissues. However, there were fewer large vessels and lack of vascular sprouts for example in the head mesenchyme. Overall, the immature blood vessels were dilated without distinction between large and small vessels, they were more fragile and ruptured, inducing hemorrhaging in diverse cavities of the body. In the yolk sac, there were 75% less endothelial cells in the mutant blood islands. Taken together, TEK does not appear to be required for the initial appearance and differentiation of the endothelial cell lineage but is necessary for remodeling of the primary capillary network into large and small vessels. In light of the phenotypic analysis of angiopoietin-1 deficient embryos, it is possible (but not demonstrated) that the fragility and reduced branching/remodeling of blood vessels in TEK mutant embryos is also due to similar periendothelial cell defects.

Recently, an activating arginine-to-tryptophan mutation in the kinase domain at position 849 of TIE-2/TEK receptor has been related to the development of venous malformations, characterized by significantly enlarged lumina of the vessels, with disproportionally large number of endothelial versus smooth muscle cells[50]. In fact, the smooth muscle cell layer was frequently underdeveloped and disorganized, appeared patchy or was locally absent. Somewhat paradoxically, deficiency of the angiopoietin-1 ligand and superactivation of the TIE-2/TEK receptor resulted in impaired smooth muscle cell recruitment, accumulation or differentiation. These observations may be reconciled by postulating that angiopoietin-1 activation of the TIE-2/TEK receptor is functionally different from activation by the R849W mutation, the precise molecular mechanisms of which, however, need to be further unraveled.

2.3. Transforming Growth Factor-β1

Transforming growth factor-β1 (TGF-β1) is a multifunctional cytokine capable of affecting blood vessel formation[51,52]. It can bind to several cellular receptors on endothelial cells of which the receptor serine/threonine kinases type I and II have been identified as signal transducing TGF-β1 receptors. In addition, TGF-β1 interacts with other molecules such as betaglycan and endoglin and combinatorial interactions between these TGF-β binding molecules may modulate its biological role. Although TGF-β1 is angiogenic *in vivo*, it inhibits macrovascular endothelial cell proliferation, induces apoptotic endothelial cell death and impairs endothelial cell migration via reducing their proteolytic balance and expression of adhesion molecules *in vitro*[52–54]. Based on these findings, it has been assumed that TGF-β1 induces blood vessel formation *in vivo* indirectly via affecting inflammatory or connective tissue cells which in turn can control angiogenesis. However, the pleiotropic effect of TGF-β1 *in vitro* on endothelial cells depends on its concentration, the type and density of endothelial cells and the presence of extracellular matrix and other growth factors such as bFGF and VEGF[52,55]. In fact, TGF-β1 has been shown to promote proliferation of angiogenic endothelial cells, to induce formation of capillary-like tubes and to increase expression of endothelial PECAM, fibronectin and tight cell junctions *in vitro*[52,56]. Besides a possible direct endothelial cell effect, TGF-β1 may control blood vessel formation via an effect on peri-endothelial mesenchymal cells by inhibiting their growth and migration, by promoting their differentiation and by inducing their production of extracellular matrix[57–60]. Endothelial and smooth muscle cells secrete TGF-β1 in a biologically inactive form which can be activated by plasmin but only when both cell types make close contact with each other[54]. TGFβ-1 has been implicated in pathological vessel development, restenosis, hypertensive remodeling and atherosclerosis.

Targeted inactivation of the *TGF-β1* gene resulted in embryonic lethality in approximately 50% of homozygous TGFβ1 deficient embryos around 9.5 days post coitum[37]. Analysis of TGF-β1 deficient embryos revealed vascular defects in the yolk sac ranging from delay in vasculogenesis, development of small, disorganized and fragile vessels to regional absence of yolk sac vessel formation. These vascular defects in the yolk sac appeared to compromize vitello-embryonic circulation and induce wasting of the embryos, which revealed no specific defects or abnormalities *per se*. In some embryos, the allantois failed to fuse with the chorion compromizing the chorio-allantoic placental circulation. Deficiency of TGF-β1 did not appear to affect early differentiation of endothelial cells, nor to reduce endothelial cell proliferation. However, there were frequently no contacts between the endothelial cell layers in the yolk sac and the apposed visceral endoderm and mesothelial cells, respectively, suggesting that these contacts had either not formed or were disrupted. As a result, the immature vitelline vessels were fragile, lacking the structural support, and ruptured with leakage of blood in the yolk sac ("vascular bleeding"). These results suggest that TGF-β1 is essential for vascular integrity and maturation of the vascular bed. Although the precise cellular mechanism was not elucidated, it is possible that deficiency of TGF-β1 impaired differentiation of mesenchymal cells to pericytes and/or reduced their production of extracellular matrix. Alternatively, or additionally, it is possible that loss of TGF-β1 affected the interaction between endothelial and periendothelial mesenchymal cells and/or compromized vascular tube formation or terminal differentiation of endothelial cells as suggested by the reduced number of endothelial cells expressing Flk-1. The impaired terminal differentiation of endothelial cells may, however, also be secondarily due to a defective interaction between endothelial and periendothelial pericytes, which are known to affect each other's function[25]. Patients suffering from hereditary hemorrhagic teleangiectasia (Rendu-Osler-Weber syndrome) have been shown to carry mutations in two TGF-β1 binding proteins, endoglin and activin receptor-like kinase, further underscoring the role of TGF-β1 in development and maturation of the vascular bed[61].

2.4. Platelet-Derived Growth Factor (PDGF) and Its Receptor

The platetelet-derived growth factor (PDGF) family includes three dimeric ligands, PDGF-AA, PDGF-AB and PDGF-BB, encoded by different genes[52,62–64]. The PDGF receptor type alpha (PDGFR-α) binds the A- or the B-chain of the PDGF dimers with high affinities, whereas the PDGF receptor type-β (PDGFR-β) only binds the B-chain. PDGF-BB is angiogenic *in vivo*. However, it is not clear whether PDGF-BB elicits these effects by recruitment of inflammatory cells and/or mesenchymal cells, or acts directly on endothelial cells. *In vivo*, angiogenic microcapillary endothelial cells coexpress PDGF-BB and PDGFR-β[65] whereas regenerating macrovascular endothelial cells coexpress PDGF-A and PDGFR-α[66], suggesting autocrine endothelial growth control by PDGF. *In vitro*, PDGF-BB induces proliferation and tube formation via interaction with PDGF-β receptors but only of developing angiogenic endothelial cell cords[52,67]. In contrast, macrovascular endothelial cells maintain expression of PDGF-BB but lack expression of PDGFR-β and fail to proliferate in response to PDGF-BB[52]. Apart from a possible direct endothelial effect, PDGF-BB may affect blood vessel formation via inducing migration or proliferation of PDGFR-β expressing periendothelial mesenchymal cells or smooth muscle cells in a paracrine loop, whereas PDGF-AA stimulates proliferation of smooth muscle cells in an autocrine loop[62,64–68].

Targeted disruption of the *PDGF-B* gene resulted in defective development of the vascular smooth muscle cell lineage[69]. This was particularly striking during kidney glomerular development beyond 17 days of gestation, revealing abnormal development of the smooth muscle cells around the glomerular vessels. In contrast to the detailed capillary network of a normal glomerular tuft, the capillary tuft of the mutant glomeruli was missing, and instead, the Bowman's capsular space contained one or several capillary aneurysm-like structuress filled with blood. Ultrastructurally, these aneurysm-like structures were delineated by basement membranes, with podocytes on their outer surface and endothelial cells lining their inside[69]. However, mesangial cells, which are pericyte-like cells of the kidney vasculature, could not be identified in mutant glomeruli. Taken together, it is possible that PDGF-BB, initially expressed by the differentiating glomerular epithelium, recruits mesangial progenitor cells and subsequently, when PDGF-BB and PDGF-β receptor are expressed by mesangial cells, stimulates proliferation of mesangial cells in an autocine loop[70]. Lack of pericytes throughout the entire microvascular bed, possibly due to defects in pericyte recruitment or proliferation may also explain why PDGF-B deficient embryos died perinatally due to fatal hemorrhage.

Loss of *PDGF-B* gene function also caused other defects of the smooth muscle lineage development. Indeed, the large arteries of the mutant embryos were markedly dilated. However, since the number of smooth muscle cell layers appeared normal, deficiency of PDGF-B resulted in dysfunction rather than hypoplasia or hypotrophy of the smooth muscle cells, possibly related to the vasoconstrictor properties of PDGF (38). Similar defects in kidney development and hematopoiesis were observed in mice lacking the PDGFR-β, suggesting that these processes result from signaling of PDGF-BB through the PDGFR-β[71]. Taken together, PDGF-BB appears to play a central role in providing the required structural integrity of the maturing vasculature possibly via an effect on recruitment, proliferation and/or other functions of pericytes.

3. INTEGRATED VIEW OF EARLY BLOOD VESSEL DEVELOPMENT

These targeting studies suggest the following hypothetical sequence of events during blood vessel development[72]. bFGF participates in the differentiation of the early mesenchymal cells into hemangioblasts. Subsequent development of the angioblasts into endothelial cells and haematopoietic precursors is in part mediated by the action of VEGF-A interacting with the Flk-1 receptor, as suggested by the defective development of both cell lineages in the Flk-1 deficient embryos. However, since VEGF-A deficient embryos display normal hematopoiesis and only retarded, but not aborted endothelial cell development, other factors, possibly related to VEGF-A, might interact with Flk-1. VEGF-A does play, however, a dominant and strict gene dosage-dependent role in the organization of these early endothelial cells into functional blood vessels via affecting the formation of their lumen, the sprouting of new blood vessels and their organization into a network. Some of the morphogenic properties of VEGF-A, e.g. the assembly of a vascular network, are mediated by Flt-1. The VEGF-A homologue PlGF is redundant for embryonic vascular development and placentation, but particpates in the neovascularization response in adult mice after wounding.

Once endothelial cell lined channels are formed, vascular integrity is maintained by organization of a primitive muscular wall, providing the required structural support against increased blood pressure and modulating endothelial cell function. The recently cloned angiopoietin-1, via interaction with its TIE-2/TEK receptor, may activate endothelial cells to produce a recruitment factor for mesenchymal cells, possibly PDGF-BB. Once the mesenchymal cells contact the endothelium, TGF-β1 may be activated and induce differentiation of the mesenchymal cells into pericytes and smooth muscle cells, inhibit endothelial cell proliferation and stimulate matrix deposition. Activation of latent TGF-β1 may occur via plasmin, generated from plasminogen by the action of plasminogen activators, which can be produced by endothelial, mesenchymal and, in the yolk sac, by visceral endoderm cells. Smooth muscle cells may further control their own growth via autocrine production of

PDGF-AA. The cellular initiator of blood coagulation, tissue factor, and possibly other coagulation factors (including factor V and the thrombin receptor) may also participate in recruitment, differentiation and/or proliferation of mesenchymal cells, but the precise molecular mechanisms remain to be determined. A mutual interrelationship between coagulation and fibrinolytic factors and vascular cytokines such as VEGF, PDGF and TGF-β1 has been previously established (28–31,37,38,50,52). TIE-1 appears to play a similar role in the maintenance of vascular integrity or remodeling of blood vessels, but it remains to be defined whether this is due to an effect on endothelial or peri-endothelial cells.

CONCLUSIONS

Targeted gene manipulation has resulted in a better understanding of the molecular mechanisms governing endothelial cell function, the formation of a muscular wall around the endothelial cell-lined channels and the assembly of these endothelial/smooth muscle cells into a connecting vasculature. Quite surprisingly, the coagulation system appears to participate in vascular development, revealing a role beyond hemostasis. Future studies will be required to elucidate whether tissue factor operates independently of its currently only known ligand factor VII, or whether it might interact with another (novel) ligand. Such insights might allow to design more rational therapeutic strategies aimed at inhibiting the hemostatic versus morphogenic properties of this pathway, which could be relevant for treatment of restenosis, atherosclerosis or related processes..

ACKNOWLEDGMENTS

The authors thank the members of the Center for Transgene Technology and Gene Therapy and the external collaborators who contributed to these studies, and M. Deprez for the artwork illustrations.

REFERENCES

1. Risau, W. 1995. Differentiation of endothelium. *Faseb J.* 9 (10):926–33.
2. Flamme, I., and W. Risau. 1992. Induction of vasculogenesis and hematopoiesis in vitro. *Development.* 116 (2):435–9.
3. Patan, S., B. Haenni, and P. H. Burri. 1996. Implementation of intussusceptive microvascular growth in the chicken chorioallantoic membrane (CAM): 1. pillar formation by folding of the capillary wall. *Microvasc Res.* 51 (1):80–98.
4. Dvorak, H. F., L. F. Brown, M. Detmar, and A. M. Dvorak. 1995. Vascular permeability factor/vascular endothelial growth factor, microvascular hyperpermeability, and angiogenesis. *Am J Pathol.* 146 (5):1029–39.
5. Terman, B. I., and M. Dougher Vermazen. 1996. Biological properties of VEGF/VPF receptors. *Cancer Metastasis Rev.* 15 (2):159–63.
6. Ferrara, N. 1995. Vascular endothelial growth factor. The trigger for neovascularization in the eye. *Lab Invest.* 72 (6):615–8.
7. Maglione, D., V. Guerriero, G. Viglietto, P. Delli Bovi, and M. G. Persico. 1991. Isolation of a human placenta cDNA coding for a protein related to the vascular permeability factor. *Proc Natl Acad Sci U S A.* 88 (20):9267–71.
8. Oloffson, B., K. Pajusola, A. Kaipainen, G. von Eulen, V. Joukow, O. Saksela, A. Orpona, R. F. Pettersoson, K. Alitalo, and U. Eriksson. 1996. Vascular endothelial growth factor B, a novel growth factor for endothelial cells. *Proc Natl Acad Sci USA.* 93:2576–81.

9. Joukov, V., K. Pajusola, A. Kaipainen, D. Chilov, I. Lahtinen, E. Kukk, O. Saksela, N. Kalkkinen, and K. Alitalo. 1996. A novel vascular endothelial growth factor, VEGF-C, is a ligand for the Flt4 (VEGFR-3) and KDR (VEGFR-2) receptor tyrosine kinases. *Embo J.* 15 (2):290–8.
10. Lee, J., A. Gray, J. Yuan, S. M. Luoh, H. Avraham, and W. I. Wood. 1996. Vascular endothelial growth factor-related protein: a ligand and specific activator of the tyrosine kinase receptor Flt4. *Proc Natl Acad Sci U S A.* 93 (5):1988–92.
11. Orlandini, M., L. Marconcini, R. Ferruzzi, and S. Oliviero. 1996. Identification of a c-fos-induced gene that is related to the platelet-derived growth factor/vascular endothelial growth factor family [corrected; erratum to be published]. *Proc Natl Acad Sci U S A.* 93 (21):11675–80.
12. de Vries, C., J. A. Escobedo, H. Ueno, K. Houck, N. Ferrara, and L. T. Williams. 1992. The fms-like tyrosine kinase, a receptor for vascular endothelial growth factor. *Science.* 255 (5047):989–91.
13. Sawano, A., T. Takahashi, S. Yamaguchi, M. Aonumura, and M. Shibuya. 1996. Flt-1 but not KDR/Flk-1 tyrosine kinase is a receptor for placenta growth factor, which is related to vascular endothelial growth factor. *Cell Growth Differ.* 7:213–21.
14. Quinn, T. P., K. G. Peters, C. De Vries, N. Ferrara, and L. T. Williams. 1993. Fetal liver kinase 1 is a receptor for vascular endothelial growth factor and is selectively expressed in vascular endothelium. *Proc Natl Acad Sci U S A.* 90 (16):7533–7.
15. Ferrara, N., K. Carver Moore, H. Chen, M. Dowd, L. Lu, K. S. O'Shea, L. Powell Braxton, K. J. Hillan, and M. W. Moore. 1996. Heterozygous embryonic lethality induced by targeted inactivation of the VEGF gene. *Nature.* 380 (6573):439–42.
16. Carmeliet, P., V. Ferreira, G. Breier, S. Pollefeyt, L. Kieckens, M. Gertsenstein, M. Fahrig, A. Vandenhoeck, H. Kendraprasad, C. Eberhardt, C. Declercq, J. Pawling, L. Moons, D. Collen, W. Risau, and A. Nagy. 1996. Abnormal blood vessel development and lethality in embryos lacking a single vascular endothelial growth factor allele. *Nature.* 380:435–9.
17. Carmeliet, P., and D. Collen. 1997. Insights into vascular biology via targeted gene inactivation and adenovirus-mediated gene transfer of the plasminogen system. *In* Coronary Restenosis. From Genetics to Therapeutics. G. Z. Feuerstein, editor. Marcel Dekker, Inc, New York. 225–40.
18. Shalaby, F., J. Rossant, T. P. Yamaguchi, M. Gertsenstein, X. F. Wu, M. L. Breitman, and A. C. Schuh. 1995. Failure of blood-island formation and vasculogenesis in Flk-1-deficient mice. *Nature.* 376 (6535):62–6.
19. Fong, G. H., J. Rossant, M. Gertsenstein, and M. L. Breitman. 1995. Role of the Flt-1 receptor tyrosine kinase in regulating the assembly of vascular endothelium. *Nature.* 376 (6535):66–70.
20. Mustonen, T., and K. Alitalo. 1995. Endothelial receptor tyrosine kinases involved in angiogenesis. *J Cell Biol.* 129 (4):895–8.
21. Sato, T. N., Y. Qin, C. A. Kozak, and K. L. Audus. 1993. Tie-1 and tie-2 define another class of putative receptor tyrosine kinase genes expressed in early embryonic vascular system [published erratum appears in Proc Natl Acad Sci U S A 1993 Dec 15;90(24):12056]. *Proc Natl Acad Sci U S A.* 90 (20):9355–8.
22. Davis, S., T. H. Aldrich, P. F. Jones, A. Acheson, D. L. Compton, V. Jain, T. E. Ryan, J. Bruno, C. Radziejewski, P. C. Maisonpierre, and G. D. Yancopoulos. 1996. Isolation of angiopoietin-1, a ligand for the TIE2 receptor, by secretion-trap expression cloning. *Cell.* 87 (7):1161–9.
23. Puri, M. C., J. Rossant, K. Alitalo, A. Bernstein, and J. Partanen. 1995. The receptor tyrosine kinase TIE is required for integrity and survival of vascular endothelial cells. *Embo J.* 14 (23):5884–91.
24. Sato, T. N., Y. Tozawa, U. Deutsch, K. Wolburg Buchholz, Y. Fujiwara, M. Gendron Maguire, T. Gridley, H. Wolburg, W. Risau, and Y. Qin. 1995. Distinct roles of the receptor tyrosine kinases Tie-1 and Tie-2 in blood vessel formation. *Nature.* 376 (6535):70–4.
25. Nehls, V., and D. Drenckhahn. 1993. The versatility of microvascular pericytes: from mesenchyme to smooth muscle? *Histochemistry.* 99 (1):1–12.
26. Edgington, T. S., N. Mackman, K. Brand, and W. Ruf. 1991. The structural biology of expression and function of tissue factor. *Thromb Haemost.* 66 (1):67–79.
27. Davie, E. W. 1995. Biochemical and molecular aspects of the coagulation cascade. *Thromb Haemost.* 74 (1):1–6.
28. Altieri, D. C. 1995. Xa receptor EPR-1. *Faseb J.* 9 (10):860–5.
29. Coughlin, S. R. 1994. Molecular mechanisms of thrombin signaling. *Semin Hematol.* 31 (4):270–7.
30. Camerer, E., A. B. Kosto, and H. Prydz. 1996. Cell biology of tissue factor, the principal initiator of blood coagulation. *Thromb Res.* 81 (1):1–41.
31. Carmeliet, P., N. Mackman, L. Moons, T. Luther, P. Gressens, I. Van Vlaenderen, H. Demunck, M. Kasper, G. Breier, P. Evrard, M. Müller, W. Risau, T. Edgington, and D. Collen. 1996. Role of tissue factor in embryonic blood vessel development. *Nature.* 383:73–5.

32. Toomey, J. R., K. E. Kratzer, N. M. Lasky, J. J. Stanton, and G. J. Broze, Jr. 1996. Targeted disruption of the murine tissue factor gene results in embryonic lethality. *Blood.* 88 (5):1583–7.
33. Toomey, J. R., K. E. Kratzer, N. M. Lasky, and G. J. J. Broze. 1997. Effect of tissue factor deficiency on mouse and tumor development. *Proc Natl Acad Sci USA.* 94:(in press).
34. Bugge, T. H., Q. Xiao, K. W. Kombrinck, M. J. Flick, K. Holmback, M. J. Danton, M. C. Colbert, D. P. Witte, K. Fujikawa, E. W. Davie, and J. L. Degen. 1996. Fatal embryonic bleeding events in mice lacking tissue factor, the cell-associated initiator of blood coagulation. *Proc Natl Acad Sci U S A.* 93 (13):6258–63.
35. Rosen, E., J. Y. Chan, I. Esohe, F. Clotman, G. Vlasuk, S. Albrecht, A. Lissens, T. Luther, L. Jalbert, L. Schoonjans, L. Moons, D. Collen, F. J. Castellino, and P. Carmeliet. 1997. Factor VII deficient mice develop normally but suffer fatal perinatal bleeding. *Nature (in press).*
36. Shivdasani, R. A., M. F. Rosenblatt, D. Zucker Franklin, C. W. Jackson, P. Hunt, C. J. Saris, and S. H. Orkin. 1995. Transcription factor NF-E2 is required for platelet formation independent of the actions of thrombopoietin/MGDF in megakaryocyte development. *Cell.* 81 (5):695–704.
37. Dickson, M. C., J. S. Martin, F. M. Cousins, A. B. Kulkarni, S. Karlsson, and R. J. Akhurst. 1995. Defective haematopoiesis and vasculogenesis in transforming growth factor-beta 1 knock out mice. *Development.* 121 (6):1845–54.
38. Suri, C., P. F. Jones, S. Patan, S. Bartunkova, P. C. Maisonpierre, S. Davis, T. N. Sato, and G. D. Yancopoulos. 1996. Requisite role of angiopoietin-1, a ligand for the TIE2 receptor, during embryonic angiogenesis. *Cell.* 87 (7):1171–80.
39. Oshima, M., H. Oshima, and M. M. Taketo. 1996. TGF-beta receptor type II deficiency results in defects of yolk sac hematopoiesis and vasculogenesis. *Dev Biol.* 179 (1):297–302 *LHM: This title is not owned by this library.
40. George, E. L., E. N. Georges Labouesse, R. S. Patel King, H. Rayburn, and R. O. Hynes. 1993. Defects in mesoderm, neural tube and vascular development in mouse embryos lacking fibronectin. *Development.* 119 (4):1079–91 *LHM: This title is not owned by this library.
41. Yang, J. T., H. Rayburn, and R. O. Hynes. 1995. Cell adhesion events mediated by alpha 4 integrins are essential in placental and cardiac development. *Development.* 121 (2):549–60 *LHM: This title is not owned by this library.
42. Offermans, S., V. Mancino, J.-P. Revel, and M. I. Simon. 1997. Vascular system defects and impaired cell chemokinesis as a result of G alpha 13 deficiency. *Science.* 275:533–36.
43. Soifer, S. J., K. G. Peters, J. O'Keefe, and S. R. Coughlin. 1994. Disparate temporal expression of the prothrombin and thrombin receptor genes during mouse development. *Am J Pathol.* 144 (1):60–9.
44. Connolly, A. J., H. Ishihara, M. L. Kahn, R. V. Farese, Jr., and S. R. Coughlin. 1996. Role of the thrombin receptor in development and evidence for a second receptor. *Nature.* 381 (6582):516–9.
45. Darrow, A. L., W. P. Fung-Leung, R. D. Ye, R. J. Santulli, W. M. Cheung, C. K. Derian, C. L. Burns, B. P. Damiano, L. Zhou, C. M. Keenan, P. A. Peterson, and P. Andrade-Gordon. 1996. Biological consequences of thrombin receptor deficiency in mice. *Thromb Haemost.* 76:860–6.
46. Cui, J., K. S. O'Shea, A. Purkayastha, T. L. Saunders, and D. Ginsburg. 1996. Fatal haemorrhage and incomplete block to embryogenesis in mice lacking coagulation factor V. *Nature.* 384 (6604):66–8.
47. Bi, L., A. M. Lawler, S. E. Antonarakis, K. A. High, J. D. Gearhart, and H. H. Kazazian, Jr. 1995. Targeted disruption of the mouse factor VIII gene produces a model of haemophilia A. *Nat Genet.* 10 (1):119–21.
48. Suh, T. T., K. Holmback, N. J. Jensen, C. C. Daugherty, K. Small, D. I. Simon, S. Potter, and J. L. Degen. 1995. Resolution of spontaneous bleeding events but failure of pregnancy in fibrinogen-deficient mice. *Genes Dev.* 9 (16):2020–33.
49. Dumont, D. J., G. Gradwohl, G. H. Fong, M. C. Puri, M. Gertsenstein, A. Auerbach, and M. L. Breitman. 1994. Dominant-negative and targeted null mutations in the endothelial receptor tyrosine kinase, tek, reveal a critical role in vasculogenesis of the embryo. *Genes Dev.* 8 (16):1897–909.
50. Vikkula, M., L. M. Boon, K. L. r. Carraway, J. T. Calvert, A. J. Diamonti, B. Goumnerov, K. A. Pasyk, D. A. Marchuk, M. L. Warman, L. C. Cantley, J. B. Mulliken, and B. R. Olsen. 1996. Vascular dysmorphogenesis caused by an activating mutation in the receptor tyrosine kinase TIE2. *Cell.* 87 (7):1181–90.
51. Kingsley, D. M. 1994. The TGF-beta superfamily: new members, new receptors, and new genetic tests of function in different organisms. *Genes Dev.* 8 (2):133–46.
52. Battegay, E. J. 1995. Angiogenesis: mechanistic insights, neovascular diseases, and therapeutic prospects. *J Mol Med.* 73 (7):333–46.
53. Frank, S., G. Hubner, G. Breier, M. T. Longaker, D. G. Greenhalgh, and S. Werner. 1995. Regulation of vascular endothelial growth factor expression in cultured keratinocytes. Implications for normal and impaired wound healing. *J Biol Chem.* 270 (21):12607–13.

54. Nunes, I., J. S. Munger, J. G. Harpel, Y. Nagano, R. L. Shapiro, P. E. Gleizes, and D. B. Rifkin. 1996. Structure and activation of the large latent transforming growth factor-beta complex. *Int J Obes Relat Metab Disord*. 20 Suppl 3:S4–8.
55. Pepper, M. S., J. D. Vassalli, L. Orci, and R. Montesano. 1993. Biphasic effects of transforming growth factor-beta 1 on in vitro angiogenesis. *Exp Cell Res*. 204:356–63.
56. Iruela-Arispe, M. L., and E. H. Sage. 1993. Endothelial cells exhibiting angiogenesis in vitro proliferate in response to TFG-beta1. *J Cell Biochem*. 52:414–30.
57. Antonelli Orlidge, A., K. B. Saunders, S. R. Smith, and P. A. D'Amore. 1989. An activated form of transforming growth factor beta is produced by cocultures of endothelial cells and pericytes. *Proc Natl Acad Sci U S A*. 86 (12):4544–8.
58. Rohovsky, S. A., K. K. Hirschi, and P. A. D'Amore. 1996. Growth factor effects on a model of vessel formation. *Surg. Forum*. 47:390–391.
59. Metcalfe, J. C., and D. J. Grainger. 1995. TGF-β: implications for vascular disease. *J Hum Hypertens*. 9:679–83.
60. D'Amore, P. A., and S. R. Smith. 1993. Growth factor effects on cells of the vascular wall: a survey. *Growth Factors*. 8:61–75.
61. McAllister, K. A., K. M. Grogg, D. W. Johnson, C. J. Gallione, M. A. Baldwin, C. E. Jackson, E. A. Helmbold, D. S. Markel, W. C. McKinnon, J. Murrell, and et al. 1994. Endoglin, a TGF-beta binding protein of endothelial cells, is the gene for hereditary haemorrhagic telangiectasia type 1. *Nat Genet*. 8 (4):345–51.
62. Ross, R., E. W. Raines, and D. F. Bowen Pope. 1986. The biology of platelet-derived growth factor. *Cell*. 46 (2):155–69.
63. Heldin, C. H., and B. Westermark. 1989. Platelet-derived growth factor: three isoforms and two receptor types. *Trends Genet*. 5 (4):108–11.
64. Klagsbrun, M., and S. Dluz. 1993. Smooth muscle cell and endothelial cell growth factors. *Trends Cardiovasc Med*. 3:213–7.
65. Holmgren, L., A. Glaser, S. Pfeifer-Ohlsson, and R. Ohlsson. 1991. Angiogenesis during human extraembryonic development involves the spatiotemporal control of the PDGF ligand and receptor gene expression. *Development*. 113:749–54.
66. Lindner, V., and M. A. Reidy. 1995. Platelet-derived growth factor ligand and receptor expression by large vessel endothelium in vivo. *Am J Pathol*. 146:1488–97.
67. Battegay, E. J., J. Rupp, L. Iruela Arispe, E. H. Sage, and M. Pech. 1994. PDGF-BB modulates endothelial proliferation and angiogenesis in vitro via PDGF beta-receptors. *J Cell Biol*. 125 (4):917–28.
68. Grotendorst, G. R., T. Chang, H. E. Seppa, H. K. Kleinman, and G. R. Martin. 1982. Platelet-derived growth factor is a chemoattractant for vascular smooth muscle cells. *J Cell Physiol*. 113 (2):261–6.
69. Levéen, P., M. Pekny, S. Gebre Medhin, B. Swolin, E. Larsson, and C. Betsholtz. 1994. Mice deficient for PDGF B show renal, cardiovascular, and hematological abnormalities. *Genes Dev*. 8 (16):1875–87.
70. Alpers, C. E., R. A. Seifert, K. L. Hudkins, R. J. Johnson, and D. F. Bowen-Pope. 1992. Developmental patterns of PDGFB-chain, PDGF receptor and alpha-acti expression in human glomerulonogenesis. *Kidney Int*. 42:390–9.
71. Soriano, P. 1994. Abnormal kidney development and hematological disorders in PDGF beta-receptor mutant mice. *Genes Dev*. 8 (16):1888–96.
72. Folkman, J., and P. A. D'Amore. 1996. Blood vessel formation: what is its molecular basis? [comment]. *Cell*. 87 (7):1153–5.

DISCUSSION

Jain: George Palade and Greg Roberts have shown that VEGF leads to the formation of fenestrations. Now you are saying that the placenta growth factor is involved in the formation of VVO's. Harold Dvorak proposed that the VVO's are a result of VEGF. George Palade has proposed that these fenestrations are collapsed VVO's. So, how do you put all this together, VEGF and PlGF?

Carmeliet: Well, I am not implying that PlGF is directly interacting or causing the VVO's to open. PlGF modulates the action of VEGF in this respect, and may actually cooperate in keeping them open or inducing their permeability and if PlGF is not there then this might cause a problem. There are lots of things that still need to be tested. For example, we do not know at this moment whether PlGF is expressed there.

Jain: Have you looked at endothelial junctions in VEGF versus PlGF mice? As you know, VEGF also opens up the endothelial junctions and does PlGF facilitate that? When we look at tumors, the pathway for transport does not appear to be VVO's as much as the endothelial junctions. They are wide in tumors along with the abundance of VEGF.

Carmeliet: The pathologist has told me on occasion that there might be also lesions common with endothelial cell junctions, but there is nothing really that I can make a statement about at this point.

Jain: So you have not made an attempt to quantify these three different structures?

Carmeliet: No, not yet.

Hanahan: Given the differences between the VEGF receptor knockouts and VEGF itself, I was just wondering whether you want to comment or speculate a bit about which the compensatory molecules are, do you have any speculations about VEGF B and C as to what they might be doing?

Carmeliet: No, I think the critical experiments for that are the knockouts and Dr. Alitalo has told me that he has a phenotype but I am not aware of it. These molecules may interact with flk, as you know, at least some of them. I can only speculate here at this moment, maybe Dr. Jain can comment on that.

Jain: We did examine VEGF-C transgenics in collaboration with Dr. Alitalo. I was going to show that data later on today. *In vitro* VEGF-C is mitogenic for both endothelial cells as well as lymphatic endothelial cells, but *in vivo,* the blood vessels of a VEGF-C transgenic are normal, there is no difference, no proliferation, no abnormality. The only place it affects is the lymphatic endothelial cells.

Hanahan: These are over-expressors, not knockouts though.

Jain: Not knockouts, only overexpressors.

Azizkhan: So there is no functional redundancy in this system?

Carmeliet: Well that is the surprising thing to me; given that there are more and more ligands being identified, it is surprising that VEGF-A even in a single allele deficiency already causes that much of a problem with the vascular development. We did a series of experiments with a presumed dominant negative mutant form created by the knockout strategy. We believe that this is not a case but, again, I am very much surprised that, for example, the difference between a haplo insufficiency of a VEGF versus a VEGF-121 isoform is only causing problems in the homozygous, whereas the PEGF knockout is surviving and does not seem to have a major phenotype change. I was told that VEGF-B knockout also causes a change in phenotype. But I am not aware of further details.

Azizkhan: So would you hypothesize that the different VEGF's have different roles in embryonic development each one, being critical but clearly not overlapping?

Carmeliet: That is certainly a possible explanation.

Jain: What about in the adult mouse?

Carmeliet: You mean transgenics? Well, we are generating these mice, we do not have any. The only information available at this moment was obtained in a system where VEGF expression can be switched: will VEGF stable tumor cell lines xenograft vascularization of tumors is massively induced and then when VEGF is switched off, by reducing the amount of tetracyclin, then the tumors collapse. What we are doing now is actually making these mice *in vivo* so that we should be able to test that strategy. We are going to adapt this specific expression of the switch. also in tumor tissue.

Mihich: The function of these isoforms is the same in normal tissues and in tumor tissues or are there selective differences in function among these isoforms?

Carmeliet: These differences are only started to be understood at this moment. The VEGF-164 seems to have the biological activity and that is used in most experiments also for gene therapy trials. Now, it seems, with more recent data that these different isoforms have a different tissue specific expression and also different function inasmuch the smaller isoforms, for example, have a higher proliferation capacity than the longer ones. So the idea was what was represented on the slide, namely, that the shorter one migrates the furthest away to the site where endothelial cells have to start proliferating and then can attract maybe the longer forms to form a patterning information. But the *in vivo* situation is still not proven.

Jain: That is a very good hypothesis about the shorter molecules traveling further. But the fact is that the endothelial cells also produce VEGF, *in vivo* and *in vitro. In vivo* we cannot see it because the single is much lower than the perivascular cells. If we put endothelial cells in culture under hypoxia gradients we find that they form a vascular network and then if we probe for VEGF mRNA or protein we find them in abundant quantity *in vitro*. Further, if we use an anti-VEGF antibody from Napoleon Ferrara, we find out that they do not form these tubes *in vitro*. Given that scenario why do you think patterning is influenced by the longer forms.

Carmeliet: I am not an expert; maybe Georg Breier can comment on that. But, I thought, peri and endothelial usually produce VEGF only when there is a high degree of stress, for example, of hypoxic stress. Then the endothelial cells also start to produce VEGF *in vivo. In vitro,* there may be a whole different structure.

Breier: But, as you said, VEGF might be upregulated *in vitro,* however, both in embryos and in tumors, we see expression of VEGF in the neighboring tissue whereas the receptors are mostly detected in endothelium. Of course, this does not exclude that endothelial cells *in vivo*, might, to some extent, also express VEGF, however, at least from this, one would conclude that the paracrine mechanism is, in most cases, the more relevant mechanism.

Jain: But in an emergency you need the endothelials to compensate for that, that is what I am trying to say. If the endothelial cell has VEGF, then as you said, in a stressed condition it should be able to produce VEGF and hopefully compensate for that. And if it can, then I am just wondering would this patterning hypothesis make sense?

Carmeliet: This is just a hypothesis that we would like to test but again I am already surprised by the data that the VEGF-121 homozygous mutant mice actually do show a normal patterning to a certain extent. I do not know how far this hypothesis will go, we will have to prove it or disprove it.

Anderson: I was curious on your slide of the structure of the VEGF gene, you had an upstream hypoxia responsive site and also a downstream site. I was wondering if there is cooperativity between these two sites? Is there biphasic induction? Are they both used? What goes on there?

Carmeliet: Yes, the five prime is the one that seems to induce transcription, whereas the three prime end is more involved in stability and it seems there to be a cooperation between both. The three prime end seems to be quite important also for the sustained level. We are making transgenics in which these hypoxic responsive elements, at least in the five prime end, are being deleted. Soon, we should be able to tell what are the effects, if they survive, and whether we can have a model of reduced responsiveness to hypoxia.

22

THE ROLE OF VASCULAR ENDOTHELIAL GROWTH FACTOR IN TUMOR ANGIOGENESIS

Georg Breier,[1,*] Annette Damert,[1] Sabine Blum,[1] Ernst Reichmann,[2] Karl H. Plate,[3] and Werner Risau[1]

[1]Max-Planck-Institute for Physiological and Clinical Research
Bad Nauheim, Germany
[2]Intitut Suisse de Recherches Experimentales sur le Cancer
Epalinges, Switzerland
[3]Neurocenter
University of Freiburg
Germany

INTRODUCTION

Angiogenesis, the sprouting of capillaries from pre-existing vessels, is of fundamental importance during embryonic vascular development and during certain physiological processes in the adult organism, for example menstruation, pregnancy, or wound healing. Moreover, angiogenesis plays a major role in the pathogenesis of several diseases, such as proliferative retinopathy, psoriasis, or solid tumor growth (Folkman, 1995). The growth of solid tumors is largely dependent on the supply of oxygen and nutrients from the blood stream. Small tumors of less than 1–2 mm in diameter are not vascularized, and they can be nourished by simple diffusion. Such avascular tumors do not induce neovascularization, and they rarely metastasize. However, further tumor growth is dependent on a vascular network that is able to fulfill the demands of the growing tumor for oxygen and nutrients. Vascularized tumors induce host vessels to extend vascular sprouts. They have the potential to expand their cell population, and they may eventually metastasize (for review, see Folkman, 1995).

One of the central questions of tumor biology is how a solid tumor progresses from the prevascular phase to the vascular phase. It has been proposed by Judah Folkmann

* Address for Correspondence: Georg Breier, Ph.D., Max-Planck-Institute for Physiological and Clinical Research, Parkstr. 1, D-61231 Bad Nauheim, Germany. Phone: +49-6032-705293; Fax: +49-6032-72259; E-mail: GBreier@kerckhoff.mpg.de

The Biology of Tumors, edited by Mihich and Croce
Plenum Press, New York, 1998.

more than twenty years ago that the angiogenic switch may be accomplished through the release of angiogenic growth factors from tumor cells that are able to attract blood vessels from the periphery of tumors. The concept of tumor angiogenesis has greatly stimulated research for stimulators of angiogenesis. These molecules are able to stimulate the proliferation and migration of endothelial cells (which represent main events of blood vessel growth) and angiogenesis in in vivo test systems. Angiogenic activity may also arise from host cells that are infiltrating the tumor (e.g. macrophages), or be mobilized from the extracellular matrix. More recently, endogenous negative regulators of angiogenesis that are associated with vascularized tumors have been discovered. It has therefore been proposed that angiogenesis in tumors requires the concomitant loss of physiological inhibition of endothelial cell proliferation. Therefore, the switch to the angiogenic phenotype may depend upon the outcome of a net balance between endogenous stimulators and inhibitors of angiogenesis (Folkman, 1995; Hanahan and Folkman, 1996).

VASCULAR ENDOTHELIAL GROWTH FACTOR (VEGF) AND VEGF RECEPTORS

While the regulation of angiogenesis requires the coordinate interaction of a great variety of molecules, it it is reasonable to assume that soluble angiogenic growth factors and their endothelial receptors play a central role in these processes by functioning as signaling molecules. A great number of angiogenic growth factors have been identified, including the fibroblast growth factors, acidic FGF and basic FGF, and vascular endothelial growth factor (VEGF) (for reviews, see Klagsbrun and D´Amore, 1991; Plate et al., 1994). Following its first description in 1989, a body of evidence has accumulated demonstrating that VEGF is a major regulator of both physiologcal and pathological blood vessel growth (for recent reviews, see Ferrara, 1993; Breier and Risau, 1996). VEGF has a variety of biological functions that are required for the process of angiogenesis: VEGF stimulates the proliferation and the migration of endothelial cells, it is secreted by a variety of normal and—at particularly high levels—tumor cells; and it stimulates the production of plasminogen activators that are involved in the proteolytical degradation of the basement membrane of blood vessels. VEGF is also a potent vascular permeability factor; this activity may also be relevant during the process of capillary sprouting. In addition, VEGF acts as a survival factor for newly formed retinal blood vessels (Alon et al., 1995).

The high affinity cellular VEGF receptors, VEGF receptor-1 (Flt-1) and VEGF receptor-2 (Flk-1), are receptor tyrosine kinases characterized by an extracellular domain with seven immunoglobulin-like motifs, a single transmembrane domain, and a split intracellular tyrosine kinase domain (Figure. 1). Both VEGF receptors are expressed predominantly in endothelial cells in vitro and in vivo (seee below). The intracellular signaling events evoked by VEGF binding to VEGF receptors are currently under investigation. The binding of VEGF to endothelial cells leads to the activation of the MAP kinase, phospholipase C-γ, phosphoinositide 3-kinase and protein kinase C pathways (D'Angelo et al., 1995; Guo et al., 1995; Xia et al., 1996). However, the specific pathways utilized by the individual VEGF receptors are only poorly characterized. This is in part due to the fact that at least some of these pathways are apparently specific for endothelial cells, which makes their analysis in heterologous cell systems difficult (Seetharam et al., 1995). It is, however, generally assumed that VEGF receptor-2 (Flk-1) acts as a signaling receptor because ligand binding induces receptor autophosphorylation (Millauer et al., 1994; Waltenberger et al., 1994). The function of the VEGF receptor-1 (Flt-1) is at present unclear

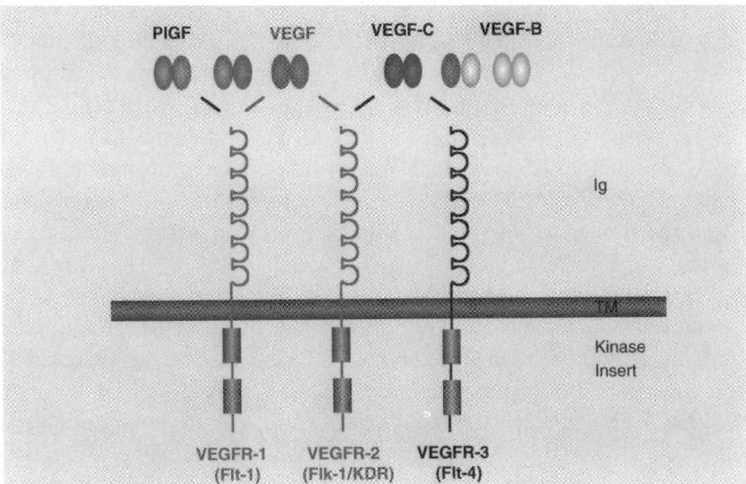

Figure 1. The vascular endothelial growth factor (VEGF) and the VEGF receptor families. VEGF, the related placenta growth factor (PlGF), and VEGF-C are the ligands of the VEGF receptor family members (VEGFR-1, VEGFR-2, VEGFR-3; for review, see Mustonen and Alitalo, 1995). VEGF receptors are transmembrane receptor tyrosine kinases with an extracellular domain, a single transmembrane domain and a split intracellular tyrosine kinase domain. VEGF binds to VEGF receptor-1 (Flt-1) with higher affinity (10–100 pM) than to VEGF receptor-2 (Flk-1; 400–800 pM). VEGFR-1 is also a receptor for PlGF, an angiogenic factor that is expressed abundantly in the placenta. VEGFR-3 (Flt-4) binds VEGF-C, a chemotactic factor for endothelial cells. VEGFR-3 is expressed in early vascular development of the mouse, but its expression becomes restricted to the lymphatic endothelium later. The receptor for VEGF-B, an endothelial cell mitogen which is expressed during mouse development, has not yet been reported. The promiscuity of receptors and of ligands, as well as heterodimerization of the ligands, may result in a complex pattern of receptor-ligand interactions, but the specific functions of the new VEGF family members is less well characterized. Abbreviations: Ig, immunoglobulin-like domain; TM, transmembrane region; Kinase, tyrosine kinase domain.

because receptor autophosphorylation is only very weak (Waltenberger *et al.*, 1994). However, the influx of Ca^{2+} ions is stimulated in Xenopus laevis embryos that express VEGF receptor-1 (Flt-1)(De Vries et al., 1992), and direct binding of the p85 subunit of PI 3-kinase has been demonstrated (Cunningham et al., 1995). In monocytes, which do not express VEGF-receptor-2 (Flk-1), VEGF receptor-1 (Flt-1) activation has been shown to induce migration, and the production of tissue factor (Clauss et al., 1996), indicating that VEGF receptor-1 (Flt-1) mediates biological functions.

VEGF IS A MAJOR REGULATOR OF EMBRYONIC VASCULAR DEVELOPMENT

The first evidence for an important role of VEGF in vascular growth and development came from studies correlating VEGF and VEGF receptor expression and phases of embryonic vascular development. VEGF and the VEGF receptors are expressed throughout embryonic vascular development (Breier et al., 1992; Millauer et al., 1993; Yamaguchi et al., 1993; Breier et al., 1995). Both VEGF receptors are expressed in endothelial cells of apparently all developing blood vessels and capillaries. The expression of the VEGF receptors is relatively specific for endothelium. Exceptions include VEGF receptor-2 (Flk-1) expression in early haematopoietic cells (Kabrun et al., 1997) and in retinal progenitor cells (Yang and Cepko,

1996), and VEGF receptor-1 (Flt-1) expression in monocytes (Clauss *et al.*, 1996), and in spongiotrophoblast cells of the placenta (Breier *et al.*, 1995). Mice lacking functional genes encoding either VEGF or the VEGF receptors fail to develop a functional vascular system and, as a consequence, die in utero before day 10.5. Mice defective for VEGF-receptor-2 show the most severe phenotype because differentiation of endothelial cells is aborted (Shalaby et al., 1995). This phenotype demonstrates that the Flk-1 receptor is required for the differentiation of endothelial cells in the embryo, consistent with the observation that VEGF receptor-2 (Flk-1) is the first endothelial receptor to be expressed in angioblast precursors in the primitive mesoderm of vertebrate embryos (Yamaguchi *et al.*, 1993). Mouse embryos lacking a single VEGF allele show a severe haploid-insufficient phenotype (Carmeliet et al., 1996; Ferrara et al., 1996). This indicates that tightly controlled VEGF protein dose is required for maintaining endothelial cell differentiation during early stage vascular development. Based upon their temporal expression patterns, it is evident that VEGF and the VEGF receptors have a function also at later embryonic stages. In most organs of the adult organism, VEGF receptor-2 (Flk-1) expression in the endothelium is down-regulated—with the exception of a few structures, such as the choroid plexus or the kidney glomeruli—indicating that this receptor has a function in endothelial cell proliferation, and, perhaps, differentiation (Millauer *et al.*, 1993). In contrast, VEGF receptor-1 (Flt-1) expression persists in adult organs, although at somewhat lower levels than in embryonic endothelium (Peters et al., 1993; Breier *et al.*, 1995). The function of VEGF receptor-1 (Flt-1) may thus be related to continous turnover or permeability of endothelium.

VEGF IS A TUMOR ANGIOGENESIS FACTOR IN VIVO

VEGF has also been implicated in the regulation of pathological angiogenesis in certain diseases, such as proliferative retinopathy, or solid tumor growth (Ferrara, 1993; Plate *et al.*, 1994). Our studies on the expression of VEGF and VEGF receptors in human glioma indicated for the first time that VEGF may function as tumor angiogenesis factor in vivo (Plate et al., 1992). Gliomas are tumors of astrocytic origin; they represent the most common brain tumors in men. Based upon histological and clinical criteria, low grade and high grade glioma can be distinguished. High grade glioma (in particular glioblastoma) is the most malignant form of the disease; the mean survival time following diagnosis is less than a year. High grade glioma may arise de novo (primary glioblastoma), but there is good evidence that they can develop by progression from low grade glioma (secondary glioblastoma). Tumor progression involves a variety of genetic alterations, including loss of putative tumor suppressor genes and the activation of oncogenes (for review, see Plate and Risau, 1995). Low-grade gliomas are moderately vascularized tumors whereas high-grade gliomas show prominent microvascular proliferations and areas of high vascular density, indicating that tumor progression is accompanied by the onset of angiogenesis. In contrast to the normal brain, where VEGF and VEGF receptor-2 (Flk-1) expression is down-regulated, the switch to the angiogenic phenotype is accompanied by the upregulation of VEGF mRNA in certain tumor areas, and by the induction of VEGF receptor expression in the tumor vasculature (Plate *et al.*, 1992). This indicates that VEGF secreted by tumor cells stimulates tumor angiogenesis in a paracrine fashion upon binding to its receptors on the tumor vasculature.

We have investigated the function of VEGF during tumor angiogenesis in a rat C6 glioma model. These cells form rapidly growing tumors following intracerebral implantation in pseudosyngeneic rats, or following subcutaneous inoculation in nude mice. The pattern of VEGF and VEGF receptor expression in C6 glioma closely resembles the pattern observed in human glioma (Plate et al., 1993). Dominant-negative VEGF receptor-2 (Flk-1) mutants were generated that inhibited signaling via the endogenous functional VEGF receptors (Millauer *et al.*, 1994). Retrovirus-mediated gene transfer of the VEGF receptor -2 (Flk-1) mutants significantly inhibited tumor growth and angiogenesis in C6 tumors in nude mice. This demonstrates that the VEGF signal transduction system is essential for tumor angiogenesis. A similar inhibition of tumor growth was achieved by the application of neutralizing antibodies directed against VEGF (Kim et al., 1993).

VEGF and the VEGF receptors are upregulated in a variety of human tumors, including von Hippel-Lindau-associated hemangioblastoma and breast carcinoma (for recent reviews, see Ferrara, 1995; Wizigmann-Voos and Plate, 1996). A correlation between blood vessel densitity and VEGF expression levels was observed in human breast tumors (Toi et al., 1995). These observations indicate that VEGF may function as a tumor angiogenesis factor in a variety of different tumor types. This hypothesis is supported by the observation that dominant-negative VEGF receptor-2 mutants inhibited the growth of various experimental tumors, including mammary, ovarian, and lung carcinoma (Millauer et al., 1996).

VEGF EXPRESSION IS UPREGULATED BY HYPOXIA

A striking observation was that glioma cells localized in the vicinity of necroses—the so-called palisading cells—expressed particularly high levels of VEGF mRNA (Plate *et al.*, 1992; Shweiki et al., 1992). These areas are thought to represent regions of low oxygen and glucose concentration. This suggested that VEGF expression is stimulated by hypoxia and/or hypoglycemia, and that this may represent a mechanism by which the tumor tries to escape the detrimental consequences of hypoxia in poorly supplied tumor areas (Figure 2). Both hypoxia and hypoglycemia are potent inducers of VEGF mRNA expression in vitro (Plate *et al.*, 1993; Shweiki et al., 1995). We and others have shown that the upregulation of VEGF mRNA expression in cultured C6 glioma cells is mediated by two distinct mechanisms, first, transcriptional activation and second, increased mRNA stability (Ikeda et al., 1995; Stein et al., 1995). Mutational analysis demonstrated that transcriptional activation of the VEGF gene proceeds through the binding of hypoxia-inducible factor-1 (HIF-1), a basic helix-loop-helix/PAS domain transcription factor—or a HIF-1 related factor—to regulatory cis-acting sequences in the 5′ flanking region of the VEGF gene, whereas mRNA stabilization is mediated by 3′ untranslated sequences.

Although these observations strongly suggest that hypoxia upregulates VEGF in vivo, direct evidence was still lacking. We have therefore used reporter gene assays to identify the regulatory elements of the VEGF gene that are involved in the upregulation of VEGF mRNA expression in vivo (Damert et al., 1997). Rat GS-9L glioma cells were stably transfected with expression vectors that contained regulatory cis-acting elements of the mouse VEGF gene, fused to the β-galactosidase reporter gene. The elements analysed were the 5′ flanking region, including or excluding the HIF-1 binding site, and the 3′ flanking region of the human VEGF gene. Strong and complete β-galactosidase staining of perinecrotic palisading cells was observed only when the VEGF promoter including the HIF-1 binding site was used in combination with the 3′ flanking region. This indicates that both transcriptional activation via HIF-1 and an increased mRNA stability are required for VEGF expression in perinecrotic palisading cells in vivo. Immunohistochemical staining with the hypoxia marker, EF5, show localization of β-galactosidase expressing palisading cells in hypoxic tumor areas. These data demonstrate that the regulatory elements that are

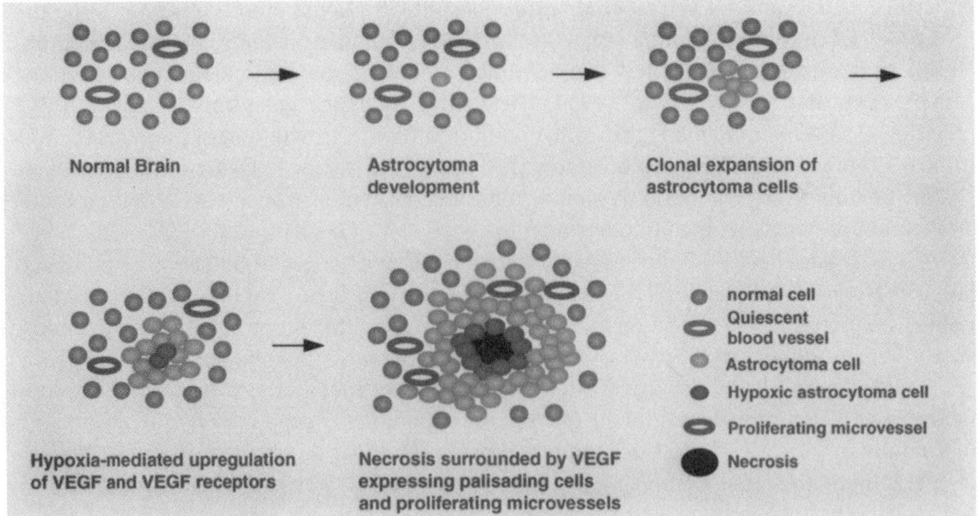

Figure 2. Regulation of glioma angiogenesis. A model. Astrocytoma development and clonal expansion of astrocytoma cells leads to local hypoxia in the growing tumor tissue. VEGF expression is upregulated in response to hypoxia, accompanied by the induction of VEGF receptors in the tumor endothelium. If the newly formed tumor vasculature is not able to supply the tumor tissue sufficiently with oxygen and nutrients, local necrosis occurs. These areas are lined by palisading cells that express high levels of VEGF.

involved in the upregulation of VEGF gene expression in response to hypoxia in vitro, are also relevant in experimental glioma in vivo.

ONCOGENIC TRANSFORMATION AND THE MICROENVIRONMENT COOPERATE TO STIMULATE VEGF EXPRESSION AND TUMOR ANGIOGENESIS

Neovascularization of a tumor requires that a certain number of its cells have switched to the angiogenic phenotype. The molecular mechanisms by which tumor cells become angiogenic are currently under investigation. The relatively high levels of VEGF produced by tumor cells suggests that, besides environmental factors, also genetic factors contribute to the regulation of VEGF expression. The role of oncogenes in the control of tumor cell proliferation is well documented; we have analysed whether oncogenic transformation might also have an effect on tumor angiogenesis.

We have used a mouse mammary carcinoma model to investigate the regulation of VEGF expression and tumor angiogenesis. Since the de-regulation of the signaling pathways involving the Ras protein is one of the most frequent events observed in human carcinoma, we have introduced the v-Ha-Ras oncogene into immortalised mammary epithelial cells and have studied the effect of Ras over-expression on tumor growth and angiogenesis. Ras transformed mammary epithelial cells form rapidly growing tumors following subcutaneous injection into nude mice, whereas the parental mammary epithelial cells are non-tumorigenic (Oft et al., 1996). During tumor growth in vivo, these cells undergo an epithelial to fibroblastoid transformation. This phenotypic conversion is induced by transforming growth factor-β (TGF-β) whose expression is induced in tumor cells during tumor progression. Tumor angiogenesis was accompanied by the expression of VEGF in tumor cells, and of VEGF receptors in the tumor vasculature. Particularly high VEGF mRNA levels were observed in perinecrotic areas. In vitro, Ras-transformed tumorigenic mammary epithelial cells secreted higher levels of VEGF protein than their non-tumorigenic parental cells. VEGF expression was upregulated in response to hypoxia, with particularly high VEGF levels produced by Ras-transformed cells. This indicates that Ras and hypoxia synergistically upregulate VEGF expression in mammary epithelial cells. Thus, Ras has not only a direct effect on the growth of tumor cells, but it may also stimulate tumor growth indirectly, by stimulating tumor angiogenesis. The increased proliferation rate of Ras transformed cells may stimulate tumor angiogenesis also indirectly, through the generation of hypoxia in areas of rapidly proliferating tumor cells, and the subsequent upregulation of VEGF expression. The observation that Ras stimulates VEGF expression was also made by Rak et al. (1995). Furthermore, introduction of activated H-ras into immortalized endothelial cells is capable of activating the angiogenic switch (Arbiser et al., 1997).

Since TGF-β has been shown to stimulate VEGF expression in the Ha-CaT keratinocyte cell line (Frank et al., 1995) and in smooth muscle cells (Brogi et al., 1994), we have analysed whether TGF-β stimulates VEGF expression in mammary epithelial cells. TGF-β caused an upregulation of VEGF expression in both non-tumorigenic and in Ras-transformed tumorigenic epithelial cells in vitro. This effect was cooperative with the effect of hypoxia in these cells. Thus, TGF-β and hypoxia appear to act in concert to stimulate VEGF expression and tumor angiogenesis in vivo.

To investigate whether the over-expression of VEGF in immortalised mammary epithelial cells alone is sufficient to confer tumorigenicity, we generated stable mammary epithelial cell lines that over-expressed VEGF, and inoculated them into nude mice. No tumor growth was observed, indicating that VEGF does not act as an oncogene. Thus, high levels of VEGF expression support but are not sufficient for tumor growth.

VEGF EXPRESSION IS CONTROLLED BY TUMOR SUPPRESSOR GENES

A first indication for a role of tumor suppressor genes in the regulation of VEGF expression came from studies of hemangioblastomas of patients suffering from von Hippel Lindau (VHL) disease. These tumors are characterized by a low proliferation rate and by a dense network of blood vessels. It is therefore unlikely that hypoxia occurs in these tumors. Despite the apparent absence of hypoxia, a strong VEGF expression in stromal cells was observed (Wizigmann-Voos et al., 1995). This indicated that, in hemangioblastoma, a mechanism other than hypoxia leads to VEGF upregulation, with the VHL tumor suppressor gene acting as a negative regulator of VEGF expression (Wizigmann-Voos and Plate, 1996). Direct evidence came from studies of cells lacking functional VHL genes which show elevated VEGF mRNA and protein levels that can be reduced if a functional VHL gene is re-introduced (Siemeister et al., 1996). The VHL gene product controls VEGF mRNA levels by post-transcriptional mechanisms (Gnarra et al., 1996).

VEGF RECEPTOR-2 (Flk-1) EXPRESSION IS INDUCED BY VEGF

In contrast to the mechanisms that are involved in the upregulation of VEGF mRNA expression, much less is known about the factors that are involved in the induction of VEGF receptors in tumor endothelium. It has been poposed that hypoxia may upregulate also the expression of VEGF receptors because this occurs in the hypoxic lung (Tuder et al., 1995). However, we and others have found that the exposure of endothelial cells to hypoxia did not result in increased VEGF receptor-2 mRNA expression (Detmar et al., 1997). Thus, the observed upregulation of VEGF receptors may occur by an indirect mechanism (Brogi et al., 1996). This hypothesis was supported by the observation that treatment of endothelial cells with conditioned media from hypoxic smooth muscle cells upregulated VEGF-receptor-2 (KDR) expression. An attractive candidate would be VEGF itself. We have therefore investigated whether the receptor ligand is capable to induce VEGF receptor-2 (Flk-1) expression in slices of 30-day postnatal mouse brain in which VEGF receptor-2 (Flk-1) expression is downregulated. First, a significant up-regulation of VEGF receptor-2 (Flk-1) protein expression was observed in response to hypoxia. The addition of recombinant VEGF to brain slices led to an up-regulation of VEGF receptor -2 (Flk-1) expression, which could be inhibited by neutralizing antibodies directed against VEGF. In contrast, PDGF-BB, placenta growth factor or erythropoietin had no effect. These results indicate that hypoxia has a differential but synergistic effect on VEGF and VEGF receptor expression. The concerted effect of hypoxia on VEGF and VEGF receptor expression may represent a mechanism to stimulate tumor angiogenesis. It seems likely, however, that additional (hypoxia-inducible?) factors are also involved in the up-regulation of VEGF receptor expression in tumors.

CONCLUSIONS

The neovascularization of tumors is a complex process that requires the coordinate interaction of molecules that are involved in cell-cell or cell-extracellular matrix interactions, such as integrins, endothelial growth factors and their endothelial receptors. Taking in account the multitude of factors that have been involved in these processes, it is remarkable to see that the inhibition of a single signal transduction system resulted in the inhibition of tumor growth and angiogenesis. Inhibition of VEGF binding or signaling has also been successfully used to inhibit ischemia-induced retinal neovascularization (Aiello et al., 1995; Adamis et al., 1996).

Thus, the VEGF signal transduction represents a useful target for anti-angiogenic therapy in angiogenesis-dependent diseases. Strategies designed to inhibit VEGF function, however, have to consider that VEGF is expressed in certain organs of the adult organism, and interfering with VEGF function in the adult organism might affect normal organ function. Further research will demonstrate whether the signaling pathways that are utilized under pathological conditions might represent specific targets for inhibiting VEGF or VEGF receptor expression in tumors or other angiogenesis-dependent diseases.

ACKNOWLEDGMENTS

We wish to thank Christine Kremer, Simone Erhardt, Michaela Fahrig, Richard Haas, Marion Krieg, Marcia Machein, Janos Peli, Masato Sasaki , Astrid Stratmann, Rembert Stratmann, Carola Wild and Susanne Wizigmann-Voos. Our work was supported by the Max Planck society, by grants from the Dr. Mildred-Scheel Stiftung and the Bundesministerium für Bildung und Forschung (to K.H.P. and W.R.), by the Center for Clinical Research I, Freiburg University Medical School and the Deutsche Forschungsgemeinschaft (to K.H.P.), and by the Schweizerische Krebsliga (E.R.).

REFERENCES

Adamis, A. P., Shima, D. T., Tolentino, M. J., Gragoudas, E. S., Ferrara, N., Folkman, J., D'Amore, P. A., and Miller, J. W. (1996). Inhibition of vascular endothelial growth factor prevents retinal ischemia-associated iris neovascularization in a nonhuman primate. *Arch. Ophthalmol.* 114, 66–71.

Aiello, L. P., Pierce, E. A., Foley, E. D., Takagi, H., Chen, H., Riddle, L., Ferrara, N., King, G. L., Smith, L. E., Funatsu, H., Hori, S., Yamashita, H., and Kitano, S. (1995). Suppression of retinal neovascularization in vivo by inhibition of vascular endothelial growth factor (VEGF) using soluble VEGF-receptor chimeric proteins. *Proc. Natl. Acad. Sci. USA* 92, 10457–10461.

Alon, T., Hemo, I., Itin, A., Peer, J., Stone, J., and Keshet, E. (1995). Vascular endothelial growth factor acts as a survival factor for newly formed retinal vessels and has implications for retinopathy of prematurity. *Nat. Med.* 1, 1024–1028.

Arbiser, J. L., Moses, M. A., Fernandez, C. A., Ghiso, N., Cao, Y., Klauber, N., Frank, D., Brownlee, M., Flynn, E., Parangi, S., Byers, H. R., and Folkman, J. (1997). Oncogenic H-ras stimulates tumor angiogenesis by two distinct pathways. *Proc. Natl. Acad. Sci. USA* 94, 861–866.

Breier, G., Albrecht, U., Sterrer, S., and Risau, W. (1992). Expression of vascular endothelial growth factor during embryonic angiogenesis and endothelial cell differentiation. *Development* 114, 521–532.

Breier, G., Clauss, M., and Risau, W. (1995). Coordinate expression of vascular endothelial growth factor receptor-1 (flt-1) and its ligand suggests a paracrine regulation of murine vascular development. *Dev. Dyn.* 204, 228–239.

Breier, G., and Risau, W. (1996). The role of vascular endothelial growth factor in blood vessel formation. *Trends Cell Biol.* 6, 454–456.

Brogi, E., Schatteman, G., Wu, T., Kim, E. A., Varticovski, L., Keyt, B., and Isner, J. M. (1996). Hypoxia-induced paracrine regulation of vascular endothelial growth factor receptor expression. *J. Clin. Invest.* 97, 469–476.

Brogi, E., Wu, T., Namiki, A., and Isner, J. M. (1994). Indirect angiogenic cytokines upregulate VEGF and bFGF gene expression in vascular smooth muscle cells, whereas hypoxia upregulates VEGF expression only. *Circulation* 90, 649–652.

Carmeliet, P., Ferreira, V., Breier, G., Pollefeyt, S., Kieckens, L., Gertsenstein, M., Fahrig, M., Vandenhoeck, A., Harpal, K., Eberhardt, C., Declerq, C., Pawling, J., Moons, L., Collen, D., Risau, W., and Nagy, A. (1996). Abnormal blood vessel development and lethality in embryos lacking a single VEGF allele. *Nature* 380, 435–439.

Clauss, M., Weich, H., Breier, G., Knies, U., Röckl, W., Waltenberger, J., and Risau, W. (1996). The vascular endothelial growth factor receptor Flt-1 mediates biological activities: implications for a functional role of placenta growth factor in monocyte activation and chemotaxis. *J. Biol. Chem.* 271, 17629–17634.

Cunningham, S. A., Waxham, M. N., Arrate, P. M., and Brock, T. A. (1995). Interaction of the Flt-1 tyrosine kinase receptor with the P85 subunit of phosphatidylinositol 3-kinase—mapping of a novel site involved in binding. *J. Biol. Chem.* 270, 20254–20257.

D'Angelo, G., Struman, I., Martial, J., and Weiner, R. I. (1995). Activation of mitogen-activated protein kinases by vascular endothelial growth factor and basic fibroblast growth factor in capillary endothelial cells is inhibited by the antiangiogenic factor 16-kDa N-terminal fragment of prolactin. *Proc. Natl. Acad. Sci. USA.* 92, 6374–6378.

Damert, A., Machein, M., Breier, G., Fujita, M. Q., Hanahan, D., Risau, W., and Plate, K. H. (1997). Upregulation of vascular endothelial growth factor expression in a rat glioma is conferred by two distinct hypoxia-driven mechanisms. *Cancer Res.* in press.

De Vries, C., Escobedo, J. A., Ueno, H., Houck, K., Ferrara, N., and Williams, L. T. (1992). The fms-like tyrosine kinase, a receptor for vascular endothelial growth factor. *Science* 255, 989–991.

Detmar, M., Brown, L. F., Berse, B., Jackman, R. W., Elicker, B. M., Dvorak, H. F., and Claffey, K. P. (1997). Hypoxia regulates the expression of vascular permeability factor/vascular endothelial growth factor (VPF/VEGF) and its receptors in human skin. *J. Invest. Dermatol.* 108, 263–268.

Ferrara, N. (1993). Vascular endothelial growth factor. *Trends Cardiovasc. Med.* 3, 244–250.

Ferrara, N. (1995). The role of vascular endothelial growth factor in pathological angiogenesis. *Breast Cancer Res. Treatm.* 36, 127–137.

Ferrara, N., Carver-Moore, K., Chen, H., Dowd, M., Lu, L., O'Shea, K. S., Powell-Braxton, L., Hillan, K., and Moore, M. W. (1996). Heterozygous embryonic lethality induced by targeted inactivation of the VEGF gene. *Nature* 380, 439–442.

Folkman, J. (1995). Angiogenesis in cancer, vascular, rheumatoid and other disease. *Nat. Med.* 1, 27–31.

Frank, S., Hübner, G., Breier, G., Longaker, M. T., Greenhalgh, D. G., and Werner, S. (1995). Regulation of vascular endothelial growth factor expression in cultured keratinocytes: implications for normal and impaired wound healing. *J. Biol. Chem.* 270, 12607–12613.

Gnarra, J. R., Zhou, S., Merrill, M. J., Wagner, J. R., Krumm, A., Papavassiliou, E., Oldfield, E. H., Klausner, R. D., and Linehan, W. M. (1996). Post-transcriptional regulation of vascular endothelial growth factor mRNA by the product of the VHL tumor suppressor gene. *Proc. Natl. Acad. Sci. USA* 93, 10589–10594.

Guo, D., Jia, Q., Song, H. Y., Warren, R. S., and Donner, D. B. (1995). Vascular endothelial cell growth factor promotes tyrosine phosphorylation of mediators of signal transduction that contain SH2 domains. Association with endothelial cell proliferation. *J. Biol. Chem.* 270, 6729–6733.

Hanahan, D., and Folkman, J. (1996). Patterns and emerging mechanisms of the angiogenic switch during tumorigenesis. *Cell* 86, 353–364.

Ikeda, I., Achen, M. G., Breier, G., and Risau, W. (1995). Hypoxia-induced transcriptional activation and increased mRNA stability of vascular endothelial growth factor in C6 glioma cells. *J. Biol. Chem.* 270, 19761–19766.

Kabrun, N., Bühring, H.-J., Choi, K., Ullrich, A., Risau, W., and Keller, G. (1997). Flk-1 expression defines a population of early embryonic hematopoietic precursors. *Development* in press.

Kim, K. J., Li, B., Winer, J., Armanini, M., Gillett, N., Phillips, H. S., and Ferrara, N. (1993). Inhibition of vascular endothelial growth factor-induced angiogenesis suppresses tumour growth in vivo. *Nature* 362, 841–844.

Klagsbrun, M., and D'Amore, P. A. (1991). Regulators of angiogenesis. *Annu. Rev. Physiol.* 53, 217–239.

Millauer, B., Longhi, M. P., Plate, K. H., Shawver, L. K., Risau, W., Ullrich, A., and Strawn, L. M. (1996). Dominant-negative inhibition of Flk-1 suppresses the growth of many tumor types in vivo. *Cancer Res.* 56, 1615–1620.

Millauer, B., Shawver, L. K., Plate, K. H., Risau, W., and Ullrich, A. (1994). Glioblastoma growth inhibited in vivo by a dominant-negative Flk-1 mutant. *Nature* 367, 576–579.

Millauer, B., Wizigmann-Voos, S., Schnürch, H., Martinez, R., Moller, N. P., Risau, W., and Ullrich, A. (1993). High affinity VEGF binding and developmental expression suggest Flk-1 as a major regulator of vasculogenesis and angiogenesis. *Cell* 72, 835–846.

Mustonen, T., and Alitalo, K. (1995). Endothelial receptor tyrosine kinases involved in angiogenesis. *J. Cell Biol.* 129, 895–898.

Oft, M., Peli, J., Rudaz, C., Schwarz, H., Beug, H., and Reichmann, E. (1996). TGFb1 and Ha-Ras collaborate in modulating the phenotypic plasticity and invasiveness of epithelial tumor cells. *Genes Dev.* 10, 2462–2477.

Peters, K. G., De Vries, C., and Williams, L. T. (1993). Vascular endothelial growth factor receptor expression during embryogenesis and tissue repair suggests a role in endothelial differentiation and blood vessel growth. *Proc. Natl. Acad. Sci. U S A* 90, 8915–8919.

Plate, K. H., Breier, G., Millauer, B., Ullrich, A., and Risau, W. (1993). Up-regulation of vascular endothelial growth factor and its cognate receptors in a rat glioma model of tumor angiogenesis. *Cancer Res.* 53, 5822–5827.

Plate, K. H., Breier, G., and Risau, W. (1994). Molecular mechanisms of developmental and tumor angiogenesis. *Brain Pathol.* 4, 207–218.

Plate, K. H., Breier, G., Weich, H. A., and Risau, W. (1992). Vascular endothelial growth factor is a potential tumour angiogenesis factor in human gliomas in vivo. *Nature* 359, 845–848.

Plate, K. H., and Risau, W. (1995). Angiogenesis in malignant gliomas. *Glia* 15, 339–347.

Rak, J., Mitsuhashi, Y., Bayko, L., Filmus, J., Shirasawa, S., Sasazuki, T., and Kerbel, R. S. (1995). Mutant ras oncogenes upregulate VEGF/VPF expression—implications for induction and inhibition of tumor angiogenesis. *Cancer Res.* 55, 4575–4580.

Seetharam, L., Gotoh, N., Maru, Y., Neufeld, G., Yamaguchi, S., and Shibuya, M. (1995). A unique signal transduction from FLT tyrosine kinase, a receptor for vascular endothelial growth factor VEGF. *Oncogene* 10, 135–147.

Shalaby, F., Rossant, J., Yamaguchi, T. P., Gertsenstein, M., Wu, X. F., Breitman, M. L., and Schuh, A. C. (1995). Failure of blood-island formation and vasculogenesis in Flk-1-deficient mice. *Nature* 376, 62–66.

Shweiki, D., Itin, A., Soffer, D., and Keshet, E. (1992). Vascular endothelial growth factor induced by hypoxia may mediate hypoxia-initiated angiogenesis. *Nature* 359, 843–845.

Shweiki, D., Neeman, M., Itin, A., and Keshet, E. (1995). Induction of vascular endothelial growth factor expression by hypoxia and by glucose deficiency in multicell spheroids: implications for tumor angiogenesis. *Proc. Natl. Acad. Sci. U.S.A.* 92, 768–772.

Siemeister, G., Weindel, K., Mohrs, K., Barleon, B., Martiny-Baron, G., and Marme, D. (1996). Reversion of deregulated expression of vascular endothelial growth factor in human renal carcinoma cells by von Hippel-Lindau tumor suppressor protein. *Cancer Res.* 56, 2299–2301.

Stein, I., Neeman, M., Shweiki, D., Itin, A., and Keshet, E. (1995). Stabilization of vascular endothelial growth factor mRNA by hypoxia and hypoglycemia and coregulation with other ischemia-induced genes. *Mol. Cell. Biol.* 15, 5363–5368.

Toi, M., Inada, K., Suzuki, H., and Tominaga, T. (1995). Tumor angiogenesis in breast cancer—its importance as a prognostic indicator and the association with vascular endothelial growth factor expression. *Breast Cancer Res. Treatm.* 36, 193–204.

Tuder, R. M., Flook, B. E., and Voelkel, N. F. (1995). Increased gene expression for VEGF and the VEGF receptors KDR/flk and flt in lungs exposed to acute or to chronic hypoxia—modulation of gene expression by nitric oxide. *J. Clin. Invest.* 95, 1798–1807.

Waltenberger, J., Claesson-Welsh, L., Siegbahn, A., Shibuya, M., and Heldin, C. H. (1994). Different signal transduction properties of KDR and Flt1, two receptors for vascular endothelial growth factor. *J. Biol. Chem.* 269, 26988–26995.

Wizigmann-Voos, S., Breier, G., Risau, W., and Plate, K. H. (1995). Up-regulation of vascular endothelial growth factor and its receptors in von Hippel-Lindau disease-associated and in sporadic hemangioblastomas. *Cancer Res.* 55, 1358–1364.

Wizigmann-Voos, S., and Plate, K. H. (1996). Pathology, genetics and cell biology of hemangioblastomas. *Histol. Histopathol.* 11, 1049–1061.

Xia, P., Aiello, L. P., Ishii, H., Jiang, Z. Y., Park, D. J., Robinson, G. S., Takagi, H., Newsome, W. P., Jirousek, M. R., and King, G. L. (1996). Characterization of vascular endothelial growth factor's effect on the activation of protein kinase C, its isoforms, and endothelial cell growth. *J. Clin. Invest.* 98, 2018–26.

Yamaguchi, T. P., Dumont, D., Conlon, R. A., Breitman, M. L., and Rossant, J. (1993). flk-1, an flt-related receptor tyrosine kinase is an early marker for endothelial cell precursors. *Development* 118, 489–498.

Yang, K., and Cepko, C. L. (1996). Flk-1, a receptor for vascular endothelial growth factor (VEGF), is expressed by retinal progenitor cells. *J. Neurosc.* 16, 6089–6099.

DISCUSSION

Carmeliet: How do you explain that Flt-1 dominant negative receptor also suppresses the tumor growth?

Breier: There of course are several possible explanations. First, there are indications that Flt-1 and Flk-1 may form heterodimers, and if this were the case under these conditions, the most likely explanation would be that the dominant-negative Flt-1 receptor mutant inhibits Flk-1 signaling and tumor angiogenesis. Another possibility would be that the VEGF receptors utilize independent signaling mechanisms; at least the knockouts point in that direction since the Flk-1 and the Flt-1 knockouts have a different phenotype indicating that at least some of the signaling events are diverging. If both of the receptors serve an essential function in tumor angiogenesis, the inhibition of any of the receptors would result in the inhibition of angiogenesis and tumor growth.

Carmeliet: But you have no experimental proof that that occurs in the tumors? The dimerization of the two receptors?

Breier: No, but we are working on this.

Carmeliet: And then in one *in situ* hybridization, I presume that was HIF-1 α. What about HIF-1β?

Breier: HIF-1β is expressed quite ubiquitously. HIF-1β is also known as ARNT, the arylhydrocarbon receptor nuclear translocator which is known to form heterodimers with a variety of related molecules, not only HIF-1α, so, we do not think that this would be very telling about the mechanism of VEGF regulation.

Carmeliet: But does it overlap, have you looked at the expression? The topographic expression pattern of both the alpha and the beta?

Breier: No, we have not looked at the expression of HIF-1β.

Carmeliet: Because the data from Gledel in England, indicates that deficiency of that and also the knockout of HIF-1 actually do suggest that it does play a significant role.

Breier: Maybe I should add here that, recently, a novel molecule related to HIF-1α has been identified which is called HLF (HLF-1α-like factor) or, HRF (HIf-1α related factor). This factor has also been shown to stimulate VEGF expression, and might also be involved in the regulation of VEGF expression in vivo.

Livingston: Yes, I think you may remember a report about a year ago from Zoltan Arany at Dana-Farber who reported that the co-activator for HIF is a member of the p300 CBP family of a very large, nuclear, integrating transcription factor-like molecules. One prediction of his results was that E1A should interfere with the expression of certain HIF-activated genes. So the question I would ask is whether E1A interferes with the hypoxia-induction effects you have so nicely demonstrated in gliomas?

Breier: We have not done this experiment.

Livingston: It is an important experiment because it offers a therapeutic strategy I think.

Breier: It would be interesting to look for this, I agree.

Zanker: Concerning the Ras transfection experiments, what was the speculation behind that the epithelial cells now show a phenotype of fibroblast morphology? Because it is really astonishing that RAS is in each cell and it should be a mutated Ras that is transfected and is the site increased or is the change of CTP through GMP or CDP changed?

Breier: Right. It was the v-Ha-ras oncogene, in a retroviral vector, that was used for the infection. You should consider also that the parental cells were already immortalized. This explains why a single oncogene is capable to transform the cells. Concerning the mechanism, Ernst Reichmann has shown that TGF-β and HA-Ras collaborate in modulating the phenotype of these cells.

Zanker: Is there an increase on p21?

Breier: We have not looked for this. As I said, our interest was in the regulation of VEGF expression there.

Livingston: And so, if you implant your cells in the mammary fatpad, do you get what looks like breast cancer?

Breier: We have not done this experiment. One can look at the behavior of the cells in the three dimensional serum-free collagen gels and this tells us about the ability to behave like organotypic cells. In contrast to the immortalized non-transformed cells, which form nice tubes and that you can even induce to produce milk proteins if you add some lactating hormones, the Ras transformed cells do not, rather they form large cysts. And if you add serum to the collagen shells, they even do not form a lumen but only cord like structures, which can be explained by the TGF-β that is present in the serum.

Livingston: Do the RAS tumors, when analyzed pathologically, resemble breast cancer? Are there any breast-like structures present?

Breier: Not that I know of.

Livingston: Are they sarcomas?

Breier: It is very difficult to observe any morphological correlates of human cancer in the animal model because tumor progression is very fast.

Anderson: On your cells that were transformed by v-Harvey Ras where you said it was a retroviral construct, were you actually using the Harvey virus itself? In other words, are there any of the VL30 sequences that flank the v-Harvey Ras in the Harvey virus still present?

Breier: I am not sure whether these sequences were present because the cell line was generated in another lab.

Anderson: The reason I am asking you is because you showed some slides showing how VL30 itself was strongly oxygen responsive.

Breier: What we observe if we analyze different individual cell clones is that there seems to be a correlation between Ras expression levels and VEGF expression levels.

Carmeliet: There has been some confusion on whether the expression of HIF's are regulated at the translation level. Your data are quite surprising in that respect. Can you comment on this and what is known in general in other models that you have, for instance, all types of ischemia in different models.

Breier: It has been published that HIF1 mRNA is induced by hypoxia itself, however, as I understand, a variety of laboratories could not reproduce these results. So the current thinking of many colleagues is that HIF1α is indeed not regulated at the transcriptional level, at least *in vitro*. This has, to my knowledge, not been thoroughly investigated with reference to the thesis you are alluding to and the current thinking is that the HIF1α protein is stabilized by hypoxia. Although this is difficult to prove since no antibodies are available for HIF1α.

Livingston: That is not true. There are multiple clonal antibodies they are published and widely available and have been disseminated around the world.

Breier: Against mouse?

Livingston: Against the human, it worked quite well. I do not know if they crossed the mouse, but it is easy enough to try.

Breier: O.K. We will be happy to use them.

Azizkhan: Does anybody know anything about the 3 prime regulatory sequences of HIF1 and VEGF that are involved in stabilization? Are they the same?

Breier: The 3 prime UTR sequence is not very well characterized; it is still a very large region that has not been narrowed down yet.

Jain: A couple of questions: One is how does TGFβ stabilize or up-regulate VEGF expression, is that known?

Breier: No, that is not known, but this has also been observed in other cell types, for example, in keratinocytes.

Jain: Right. In keratinocytes it is known.

Breier: The mechanism is unknown.

Jain: Another question: Is Ras the only oncogene which up-regulates VEGF, or almost any oncogene or suppressor gene as mutation would do that?

Breier: The VHL tumor suppressor gene has been implicated, p53 also.

Jain: There is a hypothesis recently put forward by Bob Kerbel that he calls angiogenesis progression hypothesis, where he follows the Bert Vogelstein model of colon carcinogenesis and puts a similar sequence for angiogenesis related genes coming up as different oncogenes turn on or off or are mutated. So, I am just wondering if Ras alone or a multiple multistep process is involved?

Breier: Yes, that is a difficult question that we cannot answer at present.

Jain: Another question: You say that there is a synergism between Ras, hypoxia and TGFβ. Synergism has a very specific meaning, that means one plus one is greater than two. Has that been tested, or are you using this word loosely?

Breier: If you look at the expression levels you see that the combination of two of these factors is stronger than the saturation levels of one factor alone. I would like to use the term synergism in that sense.

Jain: There was a presentation at the 1997 AACR meeting from Adrian Harris—Ron Bicknell's group, which shows that part of the reason VEGF goes up is because there is a better translation and they have looked at the ELF4 expression and that goes up more and that is maybe part of the explanation for VEGF up-regulation by hypoxia. Have you looked at that or any other correlation of that with HIF1, anybody in the audience or you yourself?

Breier: No, we have not looked at this. I agree with you, we only know part of the processes that lead to up-regulation to VEGF in hypoxia. It seems from what we know, that this occurs at a variety of different levels.

TUMOR MICROCIRCULATION

Role in Delivery of Molecular and Cellular Medicine

Rakesh K. Jain*

Department of Radiation Oncology
Massachusetts General Hospital
Harvard Medical School
Boston, Massachusetts 02114

ABSTRACT

To reach cancer cells in a tumor, a blood-borne therapeutic molecule or cell must make its way into the tumor microcirculation and across the blood vessel wall into the interstitium, and finally migrate through the interstitium. Unfortunately, tumors often develop in ways that hinder each of these steps. Our research goals are to analyze each of these steps experimentally and theoretically, and then integrate the resulting information in a unified theoretical framework. This paradigm of analysis and synthesis has allowed us to obtain a better understanding of physiological barriers in solid tumors, and to develop novel strategies to exploit and/or to overcome these barriers for improved cancer detection and treatment.

INTRODUCTION

Cancer is the second leading cause of death in the United States and in many industrialized countries (1). After the primary tumor has been surgically removed and/or sterilized by radiation, the residual disease is usually managed with a variety of systemic therapies (Table 1). For these therapies to be successful, they must satisfy two requirements: (a) the relevant agent must be effective in the *in vivo* orthotopic microenvironment of tumors, and (b) this agent must reach the target cells *in vivo* in optimal quantities. The

* TEL: (617) 726-4083; FAX: (617) 726-4172; EMAIL: jain@steele.mgh.harvard.edu; WWW: http://steele.mgh.harvard.edu/

The Biology of Tumors, edited by Mihich and Croce
Plenum Press, New York, 1998.

Table 1. Agents used in various conventional and novel therapies can be divided in three categories: molecules, particles, and cells

Therapy \ Agent	Molecules	Particles	Cells
Radiotherapy	✓	✓	
Chemotherapy	✓	✓	.
Immunotherapy	✓	✓	✓
Gene therapy	✓	✓	✓
Hyperthermia	✓		
Phototherapy	✓	✓	

goal of our research is to examine the latter issue—the delivery of diagnostic and therapeutic agents to solid tumors and normal host tissues.

All conventional and novel therapeutic agents can be divided into three categories—molecules, particles and cells [Table 1]. A blood-borne molecule or particle that enters the tumor vasculature reaches cancer cells via distribution through the vascular compartment, transport across the microvascular wall, and transport through the interstitial compartment. For a molecule of given size, charge, and configuration, each of these transport processes may involve diffusion and convection. In addition, during the journey the molecule may bind nonspecifically to proteins or other tissue components, bind specifically to the target(s), or be metabolized (2). Although lymphokine-activated killer (LAK) cells (lymphocytes activated by the lymphokine interleukin-2) or tumor-infiltrating lymphocytes (TIL) are capable of deformation, adhesion, and migration, they encounter the same barriers that restrict their movement in tumors. Some of these physiological parameters are also important for heat transfer in normal and tumor tissues during hyperthermic treatment of cancer (3).

The overall aim of our research is to develop a quantitative understanding of each of the above mentioned steps involved in the delivery of various agents. More specifically, our goals are to understand 1) how angiogenesis takes place and what determines blood flow heterogeneities in tumors, 2) how blood flow influences the metabolic microenvironment in tumors, and how microenvironment affects the biological properties of tumors (e.g., vascular permeability; cell adhesion), 3) how material moves across the microvascular wall, and 4) how it moves through the interstitial compartment and the lymphatics. In addition, we are examining the role of cell deformation and adhesion in the delivery of cells. Following analysis of these processes for molecules, particles and cells, we integrate this information in a unified framework for scale-up from mice to men (Figure 1). In this article, I will briefly describe various experimental and theoretical approaches used in our lab, our recent findings in these six areas, and finally, how we have taken some of these concepts from bench to bedside for potential improvement in cancer detection and treatment.

EXPERIMENTAL AND THEORETICAL APPROACHES

We have utilized five approaches to gain insight into the pathophysiology of solid tumors:

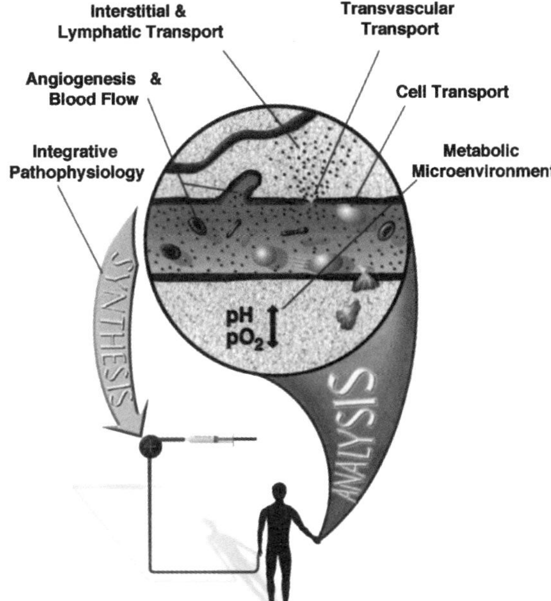

Figure 1. Quantitative understanding of various steps involved in the delivery of therapeutic agents is studied by analyzing the underlying processes and then integrating the resulting information in a unified framework. More specifically, our goal is to develop a quantitative understanding of (a) angiogenesis and blood flow; (b) metabolic microenvironment; (c) transvascular transport; (d) interstitial and lymphatic transport; (e) cell transport, and (f) systemic distribution and interspecies scale-up.

a. A tissue-isolated tumor which is connected to the host's circulation by a single artery and a single vein (4, 5). This technique was originally developed by P.M. Gullino at the National Cancer Institute in 1961 for rats (6); we have recently adapted it to mice (7, 8) and humans (9).
b. A modified Sandison rabbit ear chamber (10, 11), a modified Algire mouse dorsal chamber (12, 13), and a cranial window in mice and rats (14). The ear chamber has the advantage of superior optical quality and the mice of working with immunodeficient and genetically engineered animals (15–17). Recently we have developed a quantitative angiogenesis assay using these windows to study the physiology of vessels induced by individual growth factors (18, 19) (Figure 2). We also perfuse single vessels of tumors in these windows (20). We also utilize two acute preparations: liver and mesentery (21).
c. *In vitro* methods to assess the deformability, adhesion and permeability of normal and neoplastic cells (22–25), as well as measurements of adhesion molecules' expression in intact monolayers (26, 27) (Figure 3).
d. Routine molecular biology techniques (e.g., *in situ* hybridization, Southern, Northern and Western blotting).
e. Mathematical models to describe and integrate the data obtained from the above four approaches, to scale up biodistribution data from mice to men, and to design future experiments (28–42).

While each of these approaches has its limitations, it is their combination that has permitted us to develop the framework for tumor microcirculation and drug delivery described in this article.

Figure 2. Various microcirculatory preparations used to study delivery of therapeutic agents in solid tumors: (a) Sandison window in the rabbit ear (11); (b) Algire window in the dorsal skin of rodents (13); (c) cranial window in rodents (14); and (d) collagen I gel, containing angiogenic factors, sandwiched between nylon mesh (3 mm x 3 mm) to permit the growth of blood vessels (18). These preparations allow noninvasive, continuous measurement of angiogenesis and blood flow; metabolites, such as pH, pO2; transport of molecules and particles; and cell-cell interactions *in vivo*.

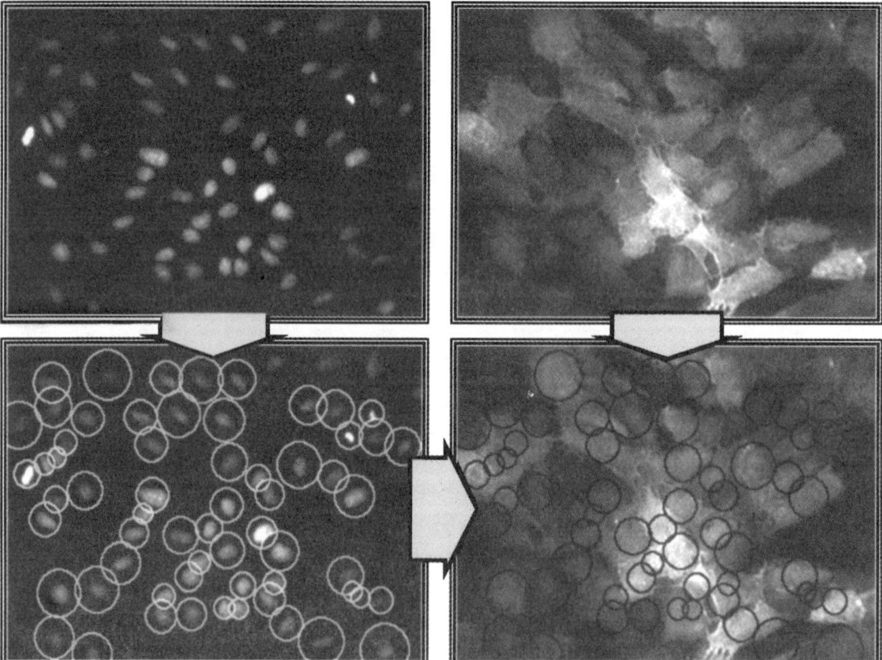

Figure 3. Targeted Sampling Fluorometry (TSF) allows the quantification of adhesion molecule expression over an intact cell monolayer on a cell-by-cell basis. At top are the two images acquired for analysis: the nuclei are stained with propidium iodide (red) and the adhesion molecule is labeled with fluorescein (green) using double immunostaining. The nuclei are first located in the propidium iodided channel, and regions of interest (ROIs) formed around each nucleus (bottom left); these ROIs are then applied to the immunostain image to find the fluorescence intensity in each region, corresponding to one cell. The procedure yields a histogram of intensities for the monolayer (Adapted from 26).

DISTRIBUTION THROUGH VASCULAR SPACE

The tumor vasculature consists of both vessels recruited from the preexisting network of the host vasculature, and vessels resulting from the angiogenic response of host vessels to cancer cells (43, 44). Movement of molecules through the vasculature is governed by the vascular morphology (*i.e.*, the number, length, diameter, and geometric arrangement of various blood vessels) and the blood flow rate (29, 45–47).

Although the tumor vasculature originates from the host vasculature and the mechanisms of angiogenesis are similar (43, 48), its organization may be completely different depending on the tumor type, its growth rate, and its location (47). The fractal dimensions and minimum path lengths of tumor vasculature are different from those of the normal host vessels (45, 46). The architecture and blood flow are different not only among various tumor types but also between a spontaneous tumor and its transplants (44, 49). For example, unlike normal tissue, where RBC velocity is dependent on vessel diameter, there is no such dependence in tumors (13, 14, 21). Furthermore, the RBC velocity may be an order of magnitude lower in some tumors compared to the host vessels (Figure 4). The temporal and spatial heterogeneity in tumor blood flow may, in part, be a result of elevated geometric and viscous resistance in tumor vessels (5, 9, 50, 51), coupling between high vascular permeability and elevated interstitial fluid pressure (38), and vascular remodeling by intussusception (48) and solid stress generated by proliferating cancer cells (52).

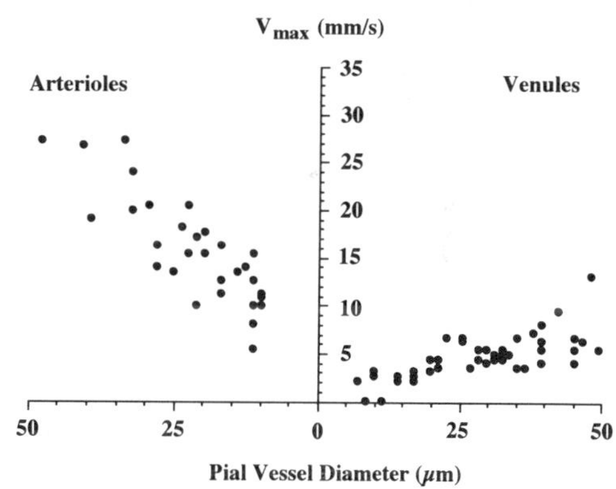

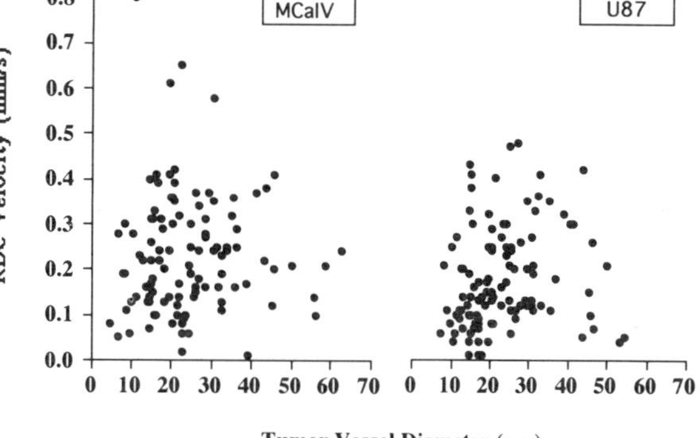

Figure 4. Blood velocity as a function of vessel diameter in (a) normal pial vessels, and (b) a human glioma (U87) xenograft on the pial surface. Note that in normal microcirculation, blood velocity is dependent on vessel diameter, whereas in tumors there is no such dependence. Furthermore, the blood velocity in tumor vessels is about an order of magnitude lower than in host vessels. (Adapted from 14).

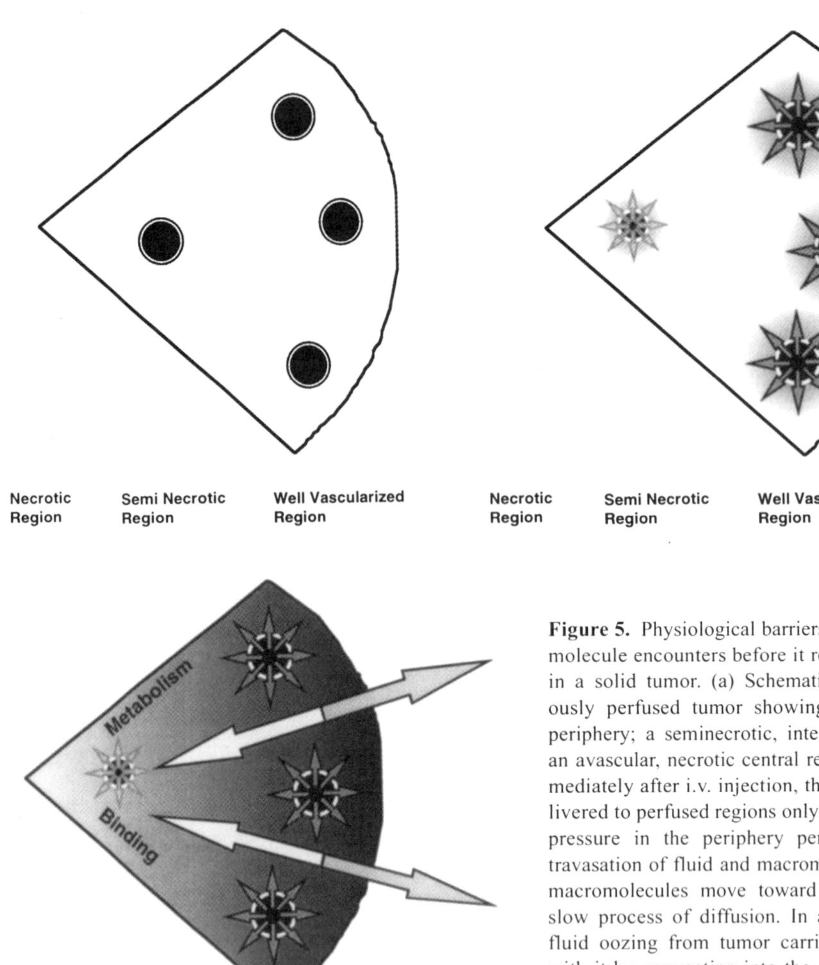

Figure 5. Physiological barriers that a blood-borne molecule encounters before it reaches a cancer cell in a solid tumor. (a) Schematic of a heterogeneously perfused tumor showing well-vascularized periphery; a seminecrotic, intermediate zone; and an avascular, necrotic central region. Note that immediately after i.v. injection, the molecules are delivered to perfused regions only. (b) Low interstitial pressure in the periphery permits adequate extravasation of fluid and macromolecules. (c) These macromolecules move toward the center by the slow process of diffusion. In addition, interstitial fluid oozing from tumor carries macromolecules with it by convection into the normal tissue. Note that the interstitial movement may be further retarded by binding. Products of metabolism may be cleared rapidly by blood. (Reproduced from 109).

Based on perfusion rates, four regions can be recognized in a tumor: an avascular, necrotic region, a seminecrotic region, a stabilized microcirculation region, and an advancing front (53) (Figure 5-a). Intratumor blood flow distributions in spontaneous animal and human tumors are now being investigated using nuclear magnetic resonance, positron emission tomography, and functional computed tomography (34, 54, 55). While limited, these results are in concert with the transplanted tumor studies: blood flow rates in necrotic and seminecrotic regions of tumors are low, while those in non-necrotic regions are variable and can be substantially higher than in surrounding (contralateral) host normal tissues (56). Considering these spatial and temporal heterogeneities in blood supply coupled with variations in the vascular morphology at both microscopic and macroscopic levels, it is not surprising that the spatial distribution of therapeutic agents in tumors is heterogeneous and that the average uptake decreases, in general, with an increase in tumor

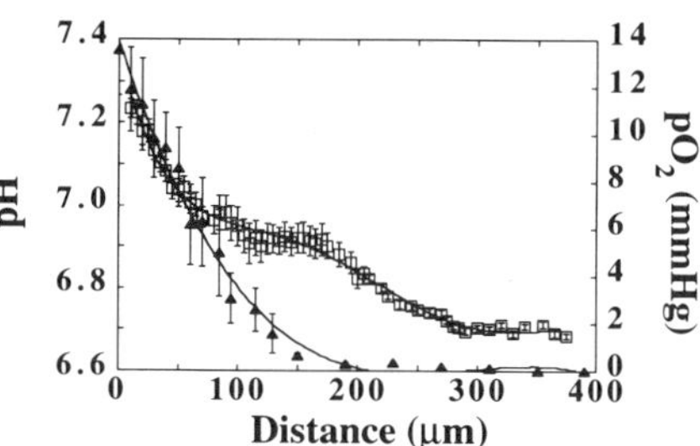

Figure 6. Spatial gradients of metabolites in tumors. pH gradients measured using fluorescence ratio imaging microscopy. pO_2 gradients measured using phosphorescence quenching. Distance from the vessel wall, in microns, is shown on the x-axis, with zero being the vessel wall. (Adapted from 58).

weight. This perfusion heterogeneity also makes it difficult to heat the tumor periphery during hyperthermia (3).

METABOLIC MICROENVIRONMENT

The temporal and spatial heterogeneities in blood flow are expected to lead to a compromised metabolic microenvironment in tumors. To quantify the spatial gradients of key metabolites, we have recently adapted two optical techniques: fluorescence ratio-imaging microscopy (FRIM) and phosphorescence quenching microscopy (PQM) (57–61). As shown in Figure 6, both pH and pO_2 decrease as one moves away from tumor vessels leading to acidic and hypoxic regions in tumors. While low pO_2 and pH are detrimental to some therapies (e.g., radiation), they might enhance the effect of certain drugs, if the drug could be delivered in adequate quantities in those regions (62–64).

To gain further insight into tumor metabolism, we have combined two powerful approaches: magnetic resonance spectroscopy and tissue isolated tumors. The former allows us to measure the energy level in tumors while the latter allows us to control the supply of individual substrates (e.g., glucose, oxygen) to the tumor. Using this approach, we have recently shown that solid tumors depend more on glucose than oxygen to maintain their ATP level (65).

TRANSPORT ACROSS THE MICROVASCULAR WALL

Once a blood-borne molecule has reached an exchange vessel, its extravasation, Js (g/s), occurs by diffusion and convection and, to some extent, presumably by transcytosis (66). Diffusive flux is proportional to the exchange vessel's surface area, S (cm^2), and the difference between the plasma and interstitial concentrations, $C_p - C_i$ (g/m). Convection is proportional to the rate of fluid leakage, J_f (m/s), from the vessel. J_f, in turn, is proportional to S and the difference between the vascular and interstitial hydrostatic pressures, $p_v - p_i$ (mm Hg), minus the osmotic reflection coefficient (σ) times the difference between the vascular and interstitial osmotic pressures $\pi_v - \pi_i$ (mm Hg). The proportionality constant that relates transluminal diffusion flux to concentration gradients, ($C_p - C_i$), is referred to as the vascular permeability coefficient, P (cm/s), and the constant that relates fluid leakage to pressure gradients is referred to as the hydraulic conductivity, L_p (cm/mm Hg·s). The effectiveness of the transluminal osmotic pressure difference in producing fluid movement across a vessel wall is characterized by σ, which is close to 1 for a macromolecule and close to zero for a small molecule. Thus, the transport of a molecule across nor-

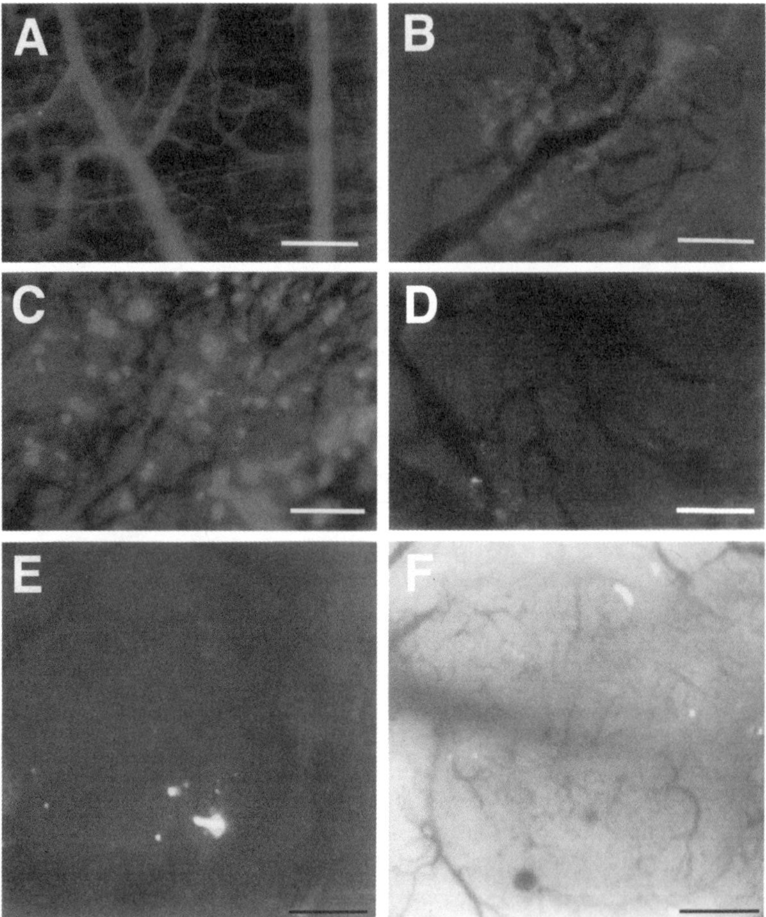

Figure 7. Transvascular transport in dorsal skin and tumors. (A) There is hardly any extravasation of 90 nm diameter liposomes from normal vessels; (B) Heterogeneous extravasation of 90 nm diameter liposomes from LS174T tumor vessels, 48 hours after injection. Note that some vessels are leaky as indicated by the yellow fluorescence for Rhodamine, while others are not. Extravasated liposomes do not diffuse far from blood vessels. (Adapted from (14)). (C) Liposomes of 400 nm diameter (yellow fluorescent spots) extravasate adequately from LS174T tumor. (D) Liposomes of 600 nm diameter do not extravasate, suggesting that LS174T vessels have pore-size cut-off of about 500 nm. (Adapted from (72)). (E) The human glioma (HGL21) xenograft is permeable to Lissamin green (i.e., tumor tissue becomes green) when grown subcutaneously (Yuan and Jain, unpublished results). (F) The same glioma develops blood-brain barrier properties (i.e., impermeable to Lissamin green) when grown in the cranial window (Adapted from 14).

mal or tumor vessels is governed by three transport parameters (P, L_p, and σ), the surface area for exchange, and the transvascular concentration and pressure gradients.

Vascular permeability and hydraulic conductivity of tumors in general is significantly higher than that of various normal tissues (14, 66–71), and hence, these vessels may lack permselectivity (72) (Figure 7-a,b). Positively changed molecules have a higher permeability (73). Despite increased overall permeability, not all blood vessels of a tumor are leaky (Figure 7-b). Even the leaky vessels have a finite pore size, which we have been able to measure in a variety of human and rodent tumors (74) including a human colon carcinoma (LS174T) xenografted in the dorsal window (Figure 7-c,d). Our hypothesis is that the large pore size in tumors represents wide inter-endothelial junctions (75). Not only does the vascular permeability vary from one tumor to the next, but within the same tumor it varies both spatially and temporally (66). The local microenvironment plays an important role in controlling vascular permeability. For example, a human glioma (HGL21) is fairly leaky when grown subcutaneously in immunodeficient mice, but it exhibits blood-brain barrier properties in the cranial window (Figure 7-e,f). We have seen such site-dependent differences for other tumors in other orthotopic sites (21). Our working hypothesis is that the host-tumor interactions control the production and secretion of cytokines associated with permeability changes (e.g., VPF/VEGF and its inhibitors). A better understanding of the molecular mechanisms of permeability-regulation in tumors is likely to yield strategies for improved drug delivery (76).

If tumor vessels are indeed "leaky" to fluid and macromolecules, then what leads to the poor extravasation of these agents in various regions of tumors? As shown by us and others (77–90), experimental and human tumors exhibit high interstitial fluid pressure. Furthermore, the uniformly high pressure drops precipitously to normal values in the tumor's periphery or in the peritumor region (28, 35, 78). This may lower fluid extravasation in the high pressure regions, especially because the oncotic and hydrostatic pressures are also equal between the intravascular and extravascular space (79, 91, 92). Because the transvascular transport of macromolecules in normal tissues occurs primarily by convection (66, 93), convective transport of macromolecules in the center of tumors may be less than in the tumor periphery (20, 28, 35). Additionally, the average vascular surface area per unit tissue weight decreases with tumor growth, hence reduced transvascular exchange would be expected in large tumors compared with small tumors (28, 29).

TRANSPORT THROUGH INTERSTITIAL SPACE AND LYMPHATICS

Once a molecule has extravasated, its movement through the interstitial space occurs by diffusion and convection (85). Diffusion is proportional to the concentration gradient in the interstitium, and convection is proportional to the interstitial fluid velocity, u_i (cm/s). The latter, in turn, is proportional to the pressure gradient in the interstitium. Just as the interstitial diffusion coefficient, D (cm^2/s), relates the diffusive flux to the concentration gradient, the interstitial hydraulic conductivity, K (cm^2/mm Hg·s), relates the interstitial velocity to the pressure gradient (85). Values of these transport coefficients are determined by the structure and composition of the interstitial compartment as well as the physicochemical properties of the solute molecule (94–100).

Using fluorescence recovery after photobleaching (FRAP) we have found D of various molecules to be about $^1/_3$ that in water (101) and similar to that in the host tissue (95). Similarly, the value of K for a human colon carcinoma xenograft (LS174T) measured using two different methods (102, 103) was found to be higher than that of a hepatoma (100), which in turn was higher than that of the liver. Given these relatively high values of D and K, why do exogenously injected macromolecules not distribute uniformly in tumors? As discussed next, there are two reasons for this apparent paradox.

The time constant for a molecule with diffusion coefficient D to diffuse across distance L is approximately $L^2/4D$. For diffusion of IgG in tumors, this time constant is on the order of one hour for a 100-μm distance, days for a 1-mm distance, and months for a 1-cm distance. So for a 1-mm tumor, diffusional transport would take days and for a 1-cm

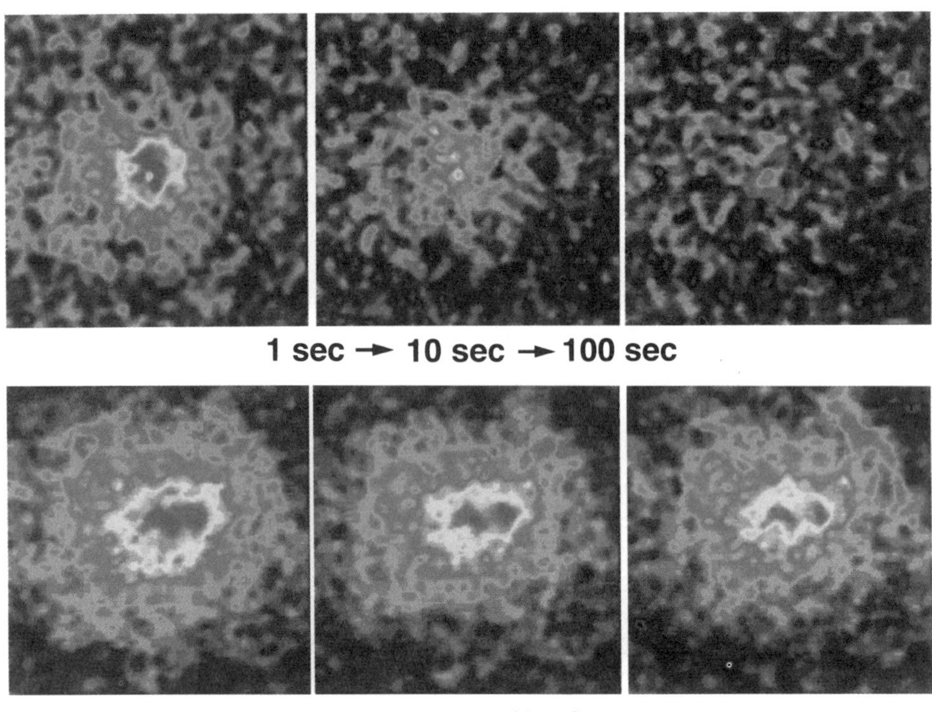

Figure 8. Role of binding in the interstitial transport in tumors, measured using fluorescence recovery after photobleaching. (a) Recovery of a photobleached spot is complete in about 100 seconds for a non-specific monoclonal antibody. (b) Recovery is incomplete for an antibody against carcino-embryonic antigen, present on the surface of many carcinoma cells. (Adapted from 101).

tumor, it would take months. If the central vessels have collapsed completely due to cellular proliferation (52) and interstitial matrix rearrangement there would be no delivery of macromolecules by blood flow to this necrotic center. Binding may further retard the transport in tumors (30, 31, 101, 104–108). The role of binding is clearly illustrated in Figure 8, which compares the rate of fluorescence recovery of a photobleached spot in tumor tissue injected with a non-specific *vs.* specific IgG. In addition to the heterogeneity in D in tumors, the most unexpected result of these photobleaching studies was the large extent (30–40%) of non-specific binding (101).

As mentioned earlier, interstitial fluid pressure is high in the center of tumors and low in the periphery and surrounding tissue (28, 35, 78). Therefore, one would expect interstitial fluid motion from the tumor's periphery into the surrounding normal tissue (Figure 5-b,c). In various animal and human (xenograft) tumors studied to date, 6–14% of plasma entering the tumor has been found to leave from the tumor's periphery (66, 109). This fluid leakage leads to a radially outward interstitial fluid velocity of 0.1 to 0.2 μm/s at the periphery of 1 cm "tissue-isolated" tumor (66). (The radially outward velocity is likely to be an order of magnitude lower in a tumor grown in the subcutaneous tissue or muscle (28)). A macromolecule at the tumor periphery has to overcome this outward convection to diffuse into the tumor. The relative contribution of this mechanism of heterogeneous distribution of antibodies in tumors may be smaller than the contribution of heterogeneous extravasation due to elevated pressure and necrosis (28).

In most normal tissues, extravasated macromolecules are taken up by the lymphatics and brought back to the central circulation. Because of the lack of functional lymphatics within the tumor, the fluid and macromolecules oozing from the tumor surface must be picked by the peri-tumor host lymphatics (29). To characterize the transport into and within the lymphatic capillaries, we have recently developed a mouse tail model (110). We have measured uptake and transport in this model using a macroscopic approach (RTD analysis) and a microscopic approach (FRAP) (111, 112). Our current efforts are directed towards uncovering mechanisms of lymphangiogenesis (113), and understanding changes in lymphatic transport in the presence of a tumor (114).

TRANSPORT OF CELLS

So far we have discussed the parameters that govern the transport of molecules and particles (e.g., liposomes) in tumors. When a leukocyte enters a blood vessel, it may continue to move with flowing blood, collide with the vessel wall, adhere transiently or stably, and finally extravasate. These interactions are governed by both local hydrodynamic forces and adhesive forces. The former are determined by the vessel diameter and fluid velocity, and the latter by the expression, strength and kinetics of bond formation between adhesion molecules and by surface area of contact (115, 116). Deformability of cells affects both types of forces. Despite their importance in immunotherapy and gene therapy, the determinants of cell transport in tumors have not been examined.

Using intravital microscopy, we have recently shown that rolling of endogenous leukocytes is generally low in tumor vessels, whereas stable adhesion ($\geq$ 30 sec) is comparable between normal and tumor vessels (Figure 9-a,b) (117). On the other hand, both rolling and stable adhesion are nearly zero in angiogenic vessels induced in collagen gels by bFGF or VEGF/VPF, two of the most potent angiogenic factors (18). Whether the latter is due to a low flux of leukocytes into angiogenic vessels and/or downregulation of adhesion molecules in these immature vessels is currently under investigation. The age of the animal also plays an important role in leukocyte-endothelial interactions (118).

To gain further insight into the type of cells that adhere to tumor vessels, we examined the localization of IL-2 activated natural killer (A-NK) cells in normal and tumor tissues in mice using positron emission tomography (22, 119). Following systemic injection, we found that these cells localized primarily in the lungs immediately after injection and a non-detectable number of cells arrived in the tumor (22). These findings were consistent with our previous work on the deformability of these cells using micropipet aspiration technique, in which we showed that IL-2 activation makes these cells rigid and predicted their mechanical entrapment in the lung microcirculation (24, 120). Constitutive expression of certain adhesion molecules in the lung vasculature also facilitates their localization in the lungs (121).

One approach to reduce lung entrapment is to reduce the rigidity of these cells (122). Instead, to circumvent the lung, we decided to inject A-NK cells into the blood supply of tumors, and we found that A-NK cells, both xenogenic and syngeneic, adhered to blood vessels in three different tumor models (119, 123, 124). These results also supported the hypothesis that the endogenous cells that adhere to tumor vessels after systemic IL-2 injection are mostly activated lymphocytes (125).

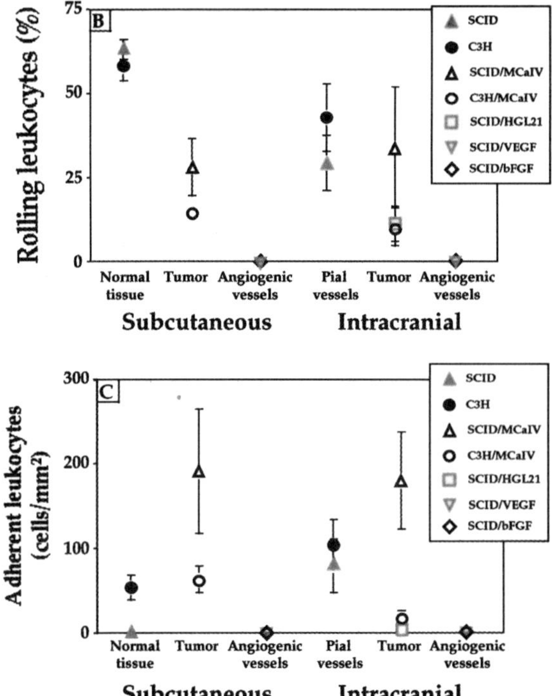

Figure 9. Leukocyte-endothelial interactions in normal and tumor (117) and angiogenic (18) vessels in the dorsal skin window and the cranial window: (a) Rolling, and (b) Adhesion. Note that rolling is significantly reduced in tumor vessels compared to host vessels, while stable adhesion is similar in both vessel types. Both rolling and adhesion are negligible in angiogenic vessels.

To find out the adhesion molecules involved in the A-NK cell adhesion to tumor vessels, we utilized two in vitro approaches. In the first approach, we simulated the tumor vasculature in vitro, by incubating the human umbilical vein endothelial cells (HUVECs) in the tumor interstitial fluid collected using a micropore chamber (6, 37, 62, 126). Using targeted sampling fluorometry (Figure 3), we were able to quantify the expression of relevant adhesion molecules on the HUVEC monolayers (26). To determine the relative contributions of these molecules in adhesion under physiological flow conditions, we utilized the flow chamber (23). Using appropriate antibodies, we found that molecules upregulated on the HUVECs include ICAM-1 and VCAM-1, which bind to CD18 and VLA-4 on the A-NK cells. We also observed sporadic upregulation of E-selectin. We were able to confirm the role of these molecules in vivo by treating A-NK cells with antibodies against CD18 and VLA-4 prior to injecting them into the arterial supply of tumors. As in our *in vitro* studies, blocking these adhesion molecules nearly eliminated the adhesion of A-NK cells to tumor vessels (126).

What leads to the upregulation of these molecules in the tumor vasculature? We already knew that these molecules can be upregulated by TNF-α and a protein of 90kD molecular weight (p90) secreted by some neoplastic cells (115, 127, 128), and downregulated by TGFβ (129–131). We wanted to find out if there are other molecules present in the tumor milieu that are also inducing this upregulation. Since tumor growth and metastasis are angiogenesis dependent, we decided to focus on the two most potent angiogenic molecules—bFGF and VEGF/VPF (43, 121, 132). We found that VEGF can mimic tumor interstitial fluid, and upregulate these molecules (17). bFGF, on the other hand, exhibited no effect when used alone, but abrogated the upregulation induced by VEGF or TNFα (126). These findings were in concert with earlier reports that bFGF retards the transmigration of lymphocytes across endothelial monolayer (133) and reduces adhesion of endothelial cells to collagen (134). They also offer a possible explanation for lower leukocyte-endothelial interactions in tumors; bFGF might have downregulated adhesion molecules in these tumors. Our current efforts are directed towards defining interactions between angiogenic and adhesion molecules using various *in vitro* and *in vivo* approaches, including genetically engineered mice (15, 17, 121).

PHARMACOKINETIC MODELING: SCALE UP FROM MOUSE TO HUMAN

So far we have analyzed each of the steps in the delivery of molecules and cells to and within solid tumors. Can we take this information and integrate it in a unified framework? We have been successful to some extent in this endeavor, using physiologically-based pharmacokinetic modeling. This approach, pioneered by two chemical engineers K. Bischoff and R.L. Dedrick in the 1960's, has been applied successfully to describe and scale up the biodistribution of low molecular weight agents (for a review, see (3, 135, 136)). We have extended this approach to macromolecules and cells (32, 33, 137–139).

In this approach, a mammalian body is represented by a number of physiological compartments interconnected anatomically (Figure 10). The volume and blood flow rate to each of these compartments/organs are known or can be measured. The parameters that characterize transport across the sub-compartments (i.e., vascular, interstitial and cellular) and the metabolism of various agents are not generally known and cannot be easily measured. Our philosophy has been to use as many measured parameters as possible and estimate the remaining parameters by fitting the model to the murine biodistribution data. By scaling-up the parameters using well-defined scale-up laws (135), we then predict the biodistribution in human patients and compare with clinical data. Discrepancies between predictions and actual data help us in identifying inter-species differences and force us to question our model assumptions. This is an evolutionary process—as our understanding of underlying physiology and biochemistry improves, the relevant parameters are modified and the model is refined further. The model is useful not only for designing murine experiments and/or clinical trials, but also in identifying the sensitive parameters that need careful measurement and analysis. If we need detailed spatial information about a tissue/organ, then we develop a distributed parameter model for that organ, e.g., tumor (28–33, 36, 140, 141). While simple in principle, this cyclic approach of analysis and synthesis has served

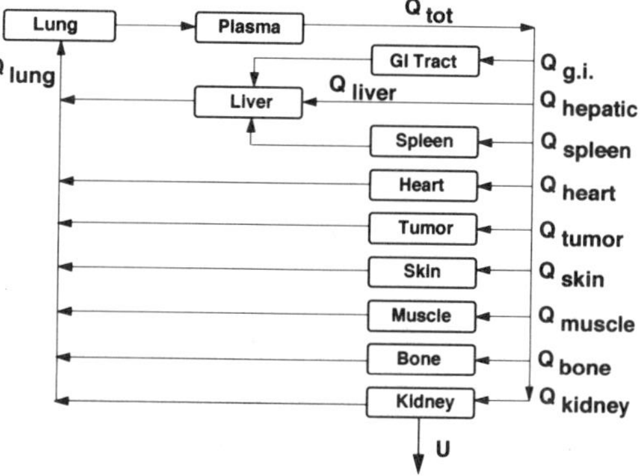

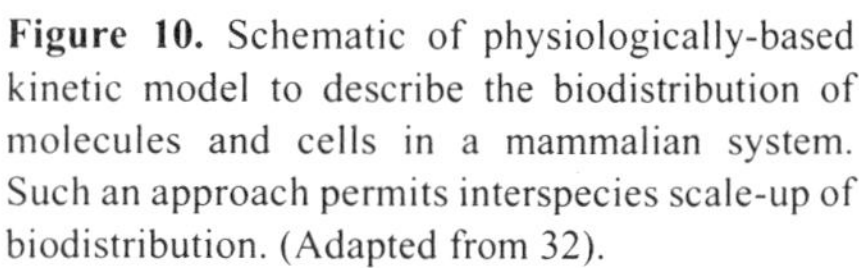
Figure 10. Schematic of physiologically-based kinetic model to describe the biodistribution of molecules and cells in a mammalian system. Such an approach permits interspecies scale-up of biodistribution. (Adapted from 32).

as a useful paradigm for developing a deeper understanding of drug and cell distribution in normal and malignant tissues. The level of sophistication of these models is likely to improve with our understanding of underlying principles (45).

BENCH TO BEDSIDE

The physiologic factors that contribute to the poor delivery of therapeutic agents to tumors include heterogeneous blood supply, interstitial hypertension, relatively long transport distances in the interstitium, and cellular heterogeneities (Figure 5). How can these physiologic barriers be exploited or overcome? Can we take our findings about these barriers from the bench to the bedside? Two recently developed strategies that have the potential to improve the detection and treatment of solid tumors in patients are described here.

As stated earlier, all solid tumors in patients exhibit interstitial hypertension (Table 2), provided the patient has not received any anti-edema treatment (82). We have also shown theoretically and confirmed experimentally that IFP rises quite steeply in the tumor boundary (35, 78). We have used this knowledge in improving the design of the needle used by radiologists to localize the tumor for surgical excision (142). We can facilitate the needle placement in a tumor by placing a pressure-sensor in the needle. Since tumors begin to exhibit interstitial hypertension almost from the onset of angiogenesis (92), this needle may be able to help in localizing early disease. The same concept may be useful in optimizing location and infusion pressure of needles employed in intratumor infusion of therapeutic agents (102), and for monitoring response to therapy (89).

Several physical (*e.g.*, radiation, heat) and chemical (*e.g.*, vasoactive drugs) agents may lead to an increase in tumor blood flow or vascular permeability (49, 66, 143–148), or lower pH (62, 64). Another approach may be based on increasing the interstitial transport rate of molecules by increasing K or D enzymatically (100, 102, 109) or using multistep approaches (33, 41, 149, 150). We have used several physical and chemical agents to lower IFP in tumors (13, 103, 151–156). Since microvascular and interstitial pressures in tumors are approximately equal, any change in one is followed rapidly by a similar change in the other, and thus the convective enhancement disappears rapidly (39, 79, 157, 158). By adapting a poroelastic model to solid tumors, we have calculated theoretically and confirmed experimentally that the time constant of pressure transmission across the tumor vasculature is on the order of 10 seconds (39). During such a short time, the convective enhancement is calculated to be very small (~1%). However, if the vascular pressure is in-

Table 2. Interstitial fluid pressure (mmHg) in normal and neoplastic tissues in patients

TISSUETYPE			
Normal skin	5	0.4	-1.0 - 3.0
Normal breast	8	0.0	-0.5 - 3.0
Head/neck carcinomas	27	19.0	1.5 - 79.0
Cervical carcinomas	26	23.0	6.0 - 94.0
Lung carcinomas	26	10.0	1.0 - 27.0
Metastatic melanomas	14	21.0	0.0 - 60.0
Metastatic melanomas	12	14.5	2.0 - 41.0
Breast carcinomas	13	29.0	5.0 - 53.0
Breast carcinomas	8	15.0	4.0 - 33.0
Brain tumors	17	7.0	2.0 - 15.0
Brain tumors	11	1.0	-0.5 - 8.0
Colorectal liver mets	8	21.0	6.0 - 45.0
Lymphomas	7	4.5	1.0 - 12.5
Renal cell carcinoma	1	38.0	------

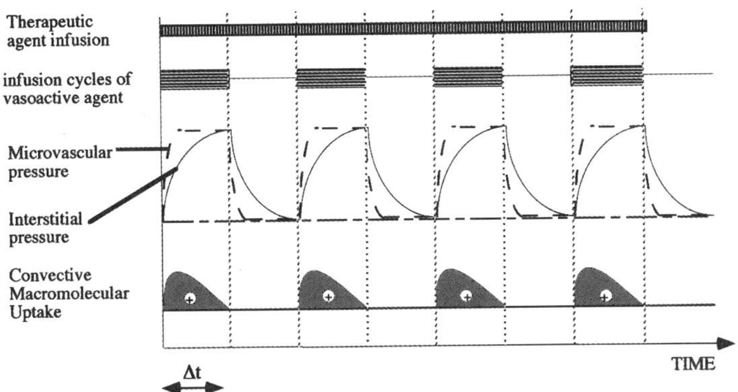

Figure 11. A novel approach to increase convective transport of molecules across tumor vessels based on the finding that there is a ~ 10 second delay in the transmission of intravascular pressure to the interstitial compartment. For this approach to work, the transvascular transport has to be uni-directional *or* the extravasated molecule must bind avidly so that it does not intravasate when the intravascular pressure is lower than interstitial pressure. (Adapted from 39).

creased repeatedly and if the transvascular transport is unidirectional or if the molecule binds avidly in the extravascular region, then we can , in principle, increase drug delivery to solid tumors significantly (Figure 11).

In contrast, the physiologic barriers discussed here may be less of a problem for a) radioimmunodetection; b) treating leukemias, lymphomas, and small tumors (*e.g.*, micrometastases) in which the physiological barriers are not yet fully established; c) treatment of adequately perfused, low-pressure regions of large tumors for debulking; and d) treatment with antibodies or other agents directed against the host cells (e.g., tumor endothelial cells, fibroblasts) or the sub-endothelial matrix. These physiologic barriers also may pose less problems for treatment with a molecule or cell that has nearly 100% specificity for cells in the tumor. Until such selective molecules or cells are developed, methods are urgently needed to overcome or exploit these physiologic barriers in tumors. It is hoped that an improved understanding of transport in tumors will help in developing these strategies (159).

ACKNOWLEDGMENTS

I am grateful to my former and current collaborators who have made working on this difficult and often frustrating problem a real joy. They and others working independently on this problem have contributed significantly to the accomplishments summarized in this article. I wish to thank Pietro M. Gullino, Marcos Intaglietta and Herman D. Suit for their encouragement, wise counsel and unconditional support. I also thank Carol Lyons for typing this manuscript, Larry Baxter for his help with the references, Lance Munn for his help with Figures 1–3, Fan Yuan with Figures 4 and 7 and Table 3, Gabriel Helmlinger with Figure 6, David Berk with Figure 8, Dai Fukumura with Figure 9, Larry Baxter with Figure 10, Paolo Netti with Figure 11, and Yves Boucher with Table 2. Research described here was primarily supported by grants from the National Cancer Institute, the National Science Foundation and the American Cancer Society.

Table 3. List of former and current collaborators

Post Doc Fellows and Junior Faculty		Tarbell J	1997-	**M.S. Students**				Griffith-Cima L	1992-
		Traykov T	1984-85			*DRACO, Lund, Sweeden*		Hynes R	1994-
		Ward K	1986-87	Chrysanthopoulos G	1977-79	Svensjo E	1987	Langer R	1996-
Baish J	1994-95	Willsey L	1996-	Brekken C	1994-95				
Baxter L	1991-	Yamada S	1993-94	Duggins E	1978-79	*Genentech*		*National Cancer Institute*	
Berk D	1992-	Yuan F	1990-96	Lai C M	1980-82	Ferrara N	1995-	Gullino P M	1974-85
Boucher Y	1988-	Zlotecki R	1992-93	Lindholm P	1986-88				
Davies C	1997-			Martin D	1983-85	*Hadassah Medical School*		*Sloan-Kettering*	
Dellian M	1993-95	**Ph.D. Students**		Mengato R	1978-80	Keshet E	1995 -	Sirotnak FM	1978
Dimitrov D	1983-84			Misiewicz M	1984-86				
Endo M	1994-	Baxter L	1985-90	Peloso R	1979-81	*Harvard Medical School*		*Temple University*	
Friedrich S	1996-	Chary S	1984-89	Rebar V	1982-84	Brisco D	1995-	Tuma R F	1993-
Fukumura D	1994-	Clauss M	1983-90	Sien H	1976-78	Brown D	1995-96		
Gohongi T	1997-	Dudar T	1978-82	Simpson S	1976-79	Brownell A L	1992-	*UC-San Diego*	
Hansen N	1997-	Eskey C	1988-92	Townsend J G	1978-80	Brownell G	1992-	Holger A	1994-
Hartford A	1997-	Gazit Y	1993-96	Volpe B	1979-81	Burnstein D	1996-	Intaglietta M	1993-
Heijn M	1997-	Gerlowski L	1979-84	Zawicki D	1976-79	Choi N	1991-	Skalak R	1994-
Helmlinger G	1994-	Hobbs S	1992-97			Dvorak H	1994-		
Ivanov I	1977-78	Kaufman E	1988-92	**Technicians**		Fishman M	1996-	*UC-San Francisco*	
Jiang P	1997-	Koenig G	1994-			Gimbrone M	1991-	Papahadjopoulos D	1994
Kristensen C	1995-	Maldarelli C	1977-81	Chen Y	1994-	Harsh G	1993-		
Kristjansen P	1992-94	Martin G	1984-94	Finney P	1980-84	Hamberg L	1994-	*University of Delaware*	
Lee I	1991-94	Nugent L	1978-82	Kahn J	1992-	Isselbacher K	1995-	Denn M M	1972-74
Less J	1989-91	Sevick E	1985-89	Roberge S	1989-	Jones R	1994-	Ruckenstein E	1972-75
Leu A	1993-94	Sha D	1996-	Schultz D	1984-91	Kopans D	1991-	Wei J	1974-75
Leunig M	1991-93	Swartz M	1994-			Mayadas T	1995		
Lichtenbeld H	1994-	Ward K	1980-86	**Collaborators**		Milstone D	1995-	*University of Munich*	
Melder R	1989-	Weissbrod J	1976-79			Seed B	1996-	Kastbenbauer E	1990-91
Monsky W	1996-	Zhu H	1992-96	*Allegheny General Hospital*		Slavin S	1995-	Messmer K	1990-91
Munn L	1993-	Znati C	1990-95	DiPette D	1986	Suit H D	1991-		
Netti P	1994-					Torchilin V P	1994-	*University of Pittsburgh*	
Nozue M	1993-95	**M.D. Students**		*Carnegie Mellon University*		Van den Abbeele A	1995-	Bloomer W	1989-91
Ohkubo C	1989			Domach M M	1988-92	Wagner D	1994-	Goldfarb R	1988-91
Patan S	1994-	Demhartner T	1992	Grossmann I E	1982	Wang C C	1991-	Herberman R	1988-91
Pluen A	1997-	Greif D	1995	Koretsky A P	1988-	Weissleder R	1996-	Kirkwood J	1989-91
Roh H	1989-91	Loskin B	1995-96	Tuma D T	1984	Whitman G	1993-96	Posner M	1989-91
Sasaki A	1987-89	Salehi H	1993-94			Wolf G	1994-	Whiteside T	1989-91
Sckell A	1996-97	Safabakhsh N	1995-96	*Columbia University*				Wolmark N	1989-91
Shah S	1979-82	Witwer B	1994-95	Chien S	1976-78	*Massachusetts Institute of*			
Stohrer M	1993-94			Pierson R N	1976-78	*Technology*		*University of Virginia*	
Stock R	1987-91			Schmid-Schönbein G	1976-78	Deen W	1993-	Skalak T	1989-91
Tanda S	1994-96			Somasundaran P	1976-78	Gray M	1996-		

An earlier version of this article was published as "1995 Whitaker Lecture: Delivery of Molecules, Particles and Cells to Solid Tumors," in the *Annals of Biomedical Engineering,* 24:457–473 (1996). The author thanks the Biomedical Engineering Society for allowing him to reproduce this article.

REFERENCES

1. Beardsley, T. Trends in cancer epidemiology: A war not won. Scientific American, 270: 118–126, 1994.
2. Jain, R.K. Barriers to drug delivery in solid tumors. Scientific American, 271: 58–65, 1994.
3. Jain, R.K. Transport phenomena in tumors. Advances in Chemical Engineering, 20: 129–200, 1994.
4. Sevick, E.M., and Jain, R.K. Blood flow and venous pH of tissue-isolated Walker 256 carcinoma during hyperglycemia. Cancer Research, 48: 1201–1207, 1988.
5. Sevick, E.M., and Jain, R.K. Geometric resistance to blood flow in solid tumors perfused ex vivo: effects of tumor size and perfusion pressure. Cancer Research, 49: 3506–3512, 1989.
6. Gullino, P.M. Techniques in tumor pathophysiology. *In:* Methods in Cancer Research. H. Busch (ed.), New York: Academic Press, pp. 45–92, 1970.
7. Kristjansen, P.E., Roberge, S., Lee, I., and Jain, R.K. Tissue-isolated human tumor xenografts in athymic nude mice. Microvascular Research, 48: 389–402, 1994.
8. Kristjansen, P.E.G., Brown, T.J., Shipley, L.A., and Jain, R.K. Intratumor pharmacokinetics, flow resistance, and metabolism during gemcitabine infusion in ex vivo perfused human small cell lung cancer. Clinical Cancer Research, 2: 359–367, 1996.
9. Less, J.R., Posner, M.C., Skalak, T., Wolmark, N., and Jain, R.K. Geometric resistance to blood flow and vascular network architecture in human colorectal carcinoma. Microcirculation, 4: 25–33, 1997.
10. Dudar, T.E., and Jain, R.K. Microcirculatory flow changes during tissue growth. Microvascular Research, 25: 1–21, 1983.
11. Zawicki, D.F., Jain, R.K., Schmid-Schoenbein, G.W., and Chien, S. Dynamics of neovascularization in normal tissue. Microvascular Research, 21: 27–47, 1981.
12. Leunig, M., Yuan, F., Berk, D.A., Gerweck, L.E., and Jain, R.K. Angiogenesis and growth of isografted bone: quantitative in vivo assay in nude mice. Laboratory Investigation, 71: 300–307, 1994.
13. Leunig, M., Yuan, F., Menger, M.D., Boucher, Y., Goetz, A.E., Messmer, K., and Jain, R.K. Angiogenesis, microvascular architecture, microhemodynamics, and interstitial fluid pressure during early growth of human adenocarcinoma LS174T in SCID mice. Cancer Research, 52: 6553–6560, 1992.
14. Yuan, F., Salehi, H.A., Boucher, Y., Vasthare, U.S., Tuma, R.F., and Jain, R.K. Vascular permeability and microcirculation of gliomas and mammary carcinomas transplanted in rat and mouse cranial windows. Cancer Research, 54: 4564–4568, 1994.
15. Yamada, S., Mayadas, T., Yuan, F., Wagner, D., Hynes, R., Melder, R., and Jain, R.K. Rolling in P-selectin deficient mice is reduced but not eliminated in the dorsal skin. Blood, 86: 3487–3492, 1995.
16. Milstone, D.S., Fukumura, D., Padget, R.C., O'Donnell, P.E., Davis, V.M., Benavidez, O.J., Melder, R.J., Jain, R.K., and Gimbrone, M.A. Mice lacking E-selectin show normal rolling but reduced arrest of leukocytes on cytokine-activated microvascular endothelium. Submitted, 1997.
17. Detmar, M., Brown, L.F., Schön, M.P., Elicker, B.M., Richard, L., Velasco, P., Fukumura, D., Monsky, W., Claffey, K.P., and Jain, R.K. Tortuous blood vessels and enhanced leukocyte adhension in VEGF transgenic mice. Submitted, 1997.
18. Dellian, M., Witwer, B.P., Salehi, H.A., Yuan, F., and Jain, R.K. Quantitation and physiological characterization of bFGF and VEGF/VPF induced vessels in mice: Effect of microenvironment on angiogenesis. American Journal of Pathology, 149: 59–72, 1996.
19. Jain, R.K., Schlenger, K., Höckel, M., and Yuan, F. Quantitative angiogenesis assays: Progress and problems. Submitted, 1997.
20. Lichtenbeld, H.C., Yuan, F., Michel, C.C., and Jain, R.K. Perfusion of single tumor microvessels: Application to vascular permeability measurement. Microcirculation, 3: 349–357, 1996.
21. Fukumura, D., Yuan, F., Monsky, W.L., Chen, Y., and Jain, R.K. Effect of host microenvironment on the microcirculation of human colon adenocarcinoma. Submitted, 1997.
22. Melder, R.J., Brownell, A.L., Shoup, T.M., Brownell, G.L., and Jain, R.K. Imaging of activated natural killer cells in mice by positron emission tomography: preferential uptake in tumors. Cancer Research, 53: 5867–5871, 1993.

23. Munn, L.L., Melder, R.J., and Jain, R.K. Analysis of cell flux in the parallel plate flow chamber: implications for cell capture studies. Biophysical Journal, 67: 889–895, 1994.
24. Sasaki, A., Jain, R.K., Maghazachi, A.A., Goldfarb, R.H., and Herberman, R.B. Low deformability of lymphokine-activated killer cells as a possible determinant of in vivo distribution. Cancer Research, 49: 3742–3746, 1989.
25. Traykov, T.T., and Jain, R.K. Effect of glucose and galactose on red blood cell membrane deformability. International Journal of Microcirculation: Clinical and Experimental, 6: 35–44, 1987.
26. Munn, L., Koenig, G.C., Jain, R.K., and Melder, R. Kinetics of adhesion molecule expression and spatial organization using targeted sampling fluorometry. BioTechniques, 19: 622–631, 1995.
27. Jain, R.K., Munn, L.L., Fukumura, D., and Melder, R.J. Methods for the *in vitro* and *in vivo* quantification of adhesion between leukocytes and vascular endothelium. *In:* Methods in Molecular Medicine—Methods in Tussie Engineering. New Jersey: Humana Press, pp. In Press, 1997.
28. Baxter, L.T., and Jain, R.K. Transport of fluid and macromolecules in tumors. I. Role of interstitial pressure and convection. Microvascular Research, 37: 77–104, 1989.
29. Baxter, L.T., and Jain, R.K. Transport of fluid and macromolecules in tumors. II. Role of heterogeneous perfusion and lymphatics. Microvascular Research, 40: 246–263, 1990.
30. Baxter, L.T., and Jain, R.K. Transport of fluid and macromolecules in tumors. III. Role of binding and metabolism. Microvascular Research, 41: 5–23, 1991.
31. Baxter, L.T., and Jain, R.K. Transport of fluid and macromolecules in tumors. IV. A microscopic model of the perivascular distribution. Microvascular Research, 41: 252–272, 1991.
32. Baxter, L.T., Zhu, H., Mackensen, D.G., Butler, W.F., and Jain, R.K. Biodistribution of monoclonal antibodies: scale-up from mouse to man using a physiologically based pharmacokinetic model. Cancer Research, 55: 4611–4622, 1995.
33. Baxter, L.T., Zhu, H., Mackensen, D.G., and Jain, R.K. Physiologically based pharmacokinetic model for specific and nonspecific monoclonal antibodies and fragments in normal tissues and human tumor xenografts in nude mice. Cancer Research, 54: 1517–1528, 1994.
34. Eskey, C.J., Wolmark, N., McDowell, C.L., Domach, M.M., and Jain, R.K. Residence time distributions of various tracers in tumors: implications for drug delivery and blood flow measurement. Journal of the National Cancer Institute, 86: 293–299, 1994.
35. Jain, R.K., and Baxter, L.T. Mechanisms of heterogeneous distribution of monoclonal antibodies and other macromolecules in tumors: significance of elevated interstitial pressure. Cancer Research, 48: 7022–7032, 1988.
36. Jain, R.K., and Wei, J. Dynamics of drug transport in solid tumors: Distributed parameter model. Journal of Bioengineering, 1: 313–329, 1977.
37. Jain, R.K., Wei, J., and Gullino, P.M. Pharmacokinetics of methotrexate in solid tumors. Journal of Pharmacokinetics and Biopharmaceutics, 7: 181–194, 1979.
38. Netti, P.A., Roberge, S., Boucher, Y., Baxter, L.T., and Jain, R.K. Effect of transvascular fluid exchange on arterio-venous pressure relationship: Implication for temporal and spatial heterogeneities in tumor blood flow. Microvascular Research, 52: 27–46, 1996.
39. Netti, P.A., Baxter, L.T., Boucher, Y., Skalak, R., and Jain, R.K. Time dependent behavior of interstitial pressure in solid tumors: Implications for drug delivery. Cancer Research, 55: 5451–5458, 1995.
40. Pierson, R.N., Price, D.C., Wang, J., and Jain, R.K. Extracellular water measurements: Organ tracer kinetics of bromide and sucrose in rats and man. American Journal of Physiology, 235: 254–264, 1978.
41. Yuan, F., Baxter, L.T., and Jain, R.K. Pharmacokinetic analysis of two-step approaches using bifunctional and enzyme-conjugated antibodies. Cancer Research, 51: 3119–3130, 1991.
42. Netti, P., Baxter, L.T., Boucher, Y., Skalak, R., and Jain, R.K. Analysis of macro and microscopic fluid transport mechanisms in living tissues. AICHE Journal, 43: 818–834, 1997.
43. Folkman, J. Tumor angiogenesis. *In:* The molecular basis of cancer. P.M. Mendelsohn, M.A.P. Howley and L.A. Liotta (ed.), Philadephia: W.B. Saunders, pp. 206–232, 1995.
44. Jain, R.K. Determinants of tumor blood flow: a review. Cancer Research, 48: 2641–2658, 1988.
45. Baish, J.W., Gazit, Y., Berk, D.A., Nozue, M., Baxter, L.T., and Jain, R.K. A novel approach to examine the role of vascular heterogeneity in nutrient and drug delivery for tumors: An invasion percolation model. Microvascular Research, 51: 327–346, 1996.
46. Gazit, Y., Berk, D.A., Leunig, M., Baxter, L.T., and Jain, R.K. Scale-invariant behavior and vascular network formation in normal and tumor tissue. Physical Review Letters, 75: 2428–2431, 1995.
47. Less, J.R., Skalak, T.C., Sevick, E.M., and Jain, R.K. Microvascular architecture in a mammary carcinoma: branching patterns and vessel dimensions. Cancer Research, 51: 265–273, 1991.
48. Patan, S., Munn, L., and Jain, R.K. Intussusceptive microvascular growth in solid tumors: A novel mechanism of tumor angiogenesis. Microvascular Research, 51: 260–272, 1996.

49. Jain, R.K., and Ward-Hartley, K.A. Tumor blood flow: Characterization, modifications and role in hyperthermia. IEEE Transactions in Sonics and Ultrasonics, 31: 504–526, 1984.
50. Sevick, E.M., and Jain, R.K. Viscous resistance to blood flow in solid tumors: effect of hematocrit on intratumor blood viscosity. Cancer Research, 49: 3513–3519, 1989.
51. Sevick, E.M., and Jain, R.K. Effect of red blood cell rigidity on tumor blood flow: increase in viscous resistance during hyperglycemia. Cancer Research, 51: 2727–2730, 1991.
52. Helmlinger, G., Netti, P.A., Lichtenbeld, H.C., Melder, R.J., and Jain, R.K. Solid stress inhibits the growth of multicellular tumor spheroids. Nature Biotechnology, In Press, 1997.
53. Endrich, B., Reinhold, H.S., Gross, J.F., and Intaglietta, M. Tissue perfusion inhomogeneity during early tumor growth in rats. Journal of the National Cancer Institute, 62: 387–395, 1979.
54. Eskey, C.J., Koretsky, A.P., Domach, M.M., and Jain, R.K. 2H-nuclear magnetic resonance imaging of tumor blood flow: spatial and temporal heterogeneity in a tissue-isolated mammary adenocarcinoma. Cancer Research, 52: 6010–6019, 1992.
55. Hamberg, L.M., Kristjansen, P.E., Hunter, G.J., Wolf, G.L., and Jain, R.K. Spatial heterogeneity in tumor perfusion measured with functional computed tomography at 0.05 microliter resolution. Cancer Research, 54: 6032–6036, 1994.
56. Vaupel, P., and Jain, R.K. Tumor blood supply and metabolic microenvironment: Characterization and therapeutic implications. Stuttgart: Gustav Fischer Publications. 1991.
57. Dellian, M., Helmlinger, G., Yuan, F., and Jain, R.K. Fluorescence ratio imaging and optical sectioning: effect of glucose on spatial and temporal gradients. British Journal of Cancer, 74: 1206–1215, 1996.
58. Helmlinger, G., Yuan, F., Dellian, M., and Jain, R.K. Interstitial pH and pO2 gradients in solid tumors in vivo: Simultaneous high-resolution measurements reveal a lack of correlation. Nature Medicine, 3: 177–182, 1997.
59. Martin, G.R., and Jain, R.K. Fluorescence ratio imaging measurement of pH gradients: calibration and application in normal and tumor tissues. Microvascular Research, 46: 216–230, 1993.
60. Martin, G.R., and Jain, R.K. Noninvasive measurement of interstitial pH profiles in normal and neoplastic tissue using fluorescence ratio imaging microscopy. Cancer Research, 54: 5670–5674, 1994.
61. Torres-Filho, I.P., Leunig, M., Yuan, F., Intaglietta, M., and Jain, R.K. Noninvasive measurement of microvascular and interstitial oxygen profiles in a human tumor in SCID mice. Proceedings of the National Academy of Sciences of the USA, 91: 2081–2085, 1994.
62. Jain, R.K., Shah, S.A., and Finney, P.L. Continuous noninvasive monitoring of pH and temperature in rat Walker 256 carcinoma during normoglycemia and hyperglycemia. Journal of the National Cancer Institute, 73: 429–436, 1984.
63. Nozue, M., Lee, I., Manning, J.M., Manning, L.R., and Jain, R.K. Oxygenation in tumors by modified hemoglobins. Journal of Surgical Oncology, 62: 109–114, 1996.
64. Ward, K.A., and Jain, R.K. Response of tumours to hyperglycaemia: characterization, significance and role in hyperthermia. International Journal of Hyperthermia, 4: 223–250, 1988.
65. Eskey, C.J., Koretsky, A.P., Domach, M.M., and Jain, R.K. Role of oxygen vs. glucose in energy metabolism in a mammary carcinoma perfused ex vivo: direct measurement by 31P NMR. Proceedings of the National Academy of Sciences of the USA, 90: 2646–2650, 1993.
66. Jain, R.K. Transport of molecules across tumor vasculature. Cancer & Metastasis Reviews, 6: 559–593, 1987.
67. Dvorak, H.F., Brown, L.F., Detmar, M., and Dvorak, A.M. Vascular permeability factor/vascular endothelial growth factor, microvascular hyperpermeability, and angiogenesis. American Journal of Pathology, 146: 1029–1039, 1995.
68. Gerlowski, L.E., and Jain, R.K. Microvascular permeability of normal and neoplastic tissues. Microvascular Research, 31: 288–305, 1986.
69. Sevick, E.M., and Jain, R.K. Measurement of capillary filtration coefficient in a solid tumor. Cancer Research, 51: 1352–1355, 1991.
70. Yuan, F., Leunig, M., Berk, D.A., and Jain, R.K. Microvascular permeability of albumin, vascular surface area, and vascular volume measured in human adenocarcinoma LS174T using dorsal chamber in SCID mice. Microvascular Research, 45: 269–289, 1993.
71. Yuan, F., Leunig, M., Huang, S.K., Berk, D.A., Papahadjopoulos, D., and Jain, R.K. Microvascular permeability and interstitial penetration of sterically stabilized (stealth) liposomes in a human tumor xenograft. Cancer Research, 54: 3352–3356, 1994.
72. Yuan, F., Dellian, M., Fukumura, D., Leunig, M., Berk, D.A., Torchiliin, V.P., and Jain, R.K. Vascular permeability in a human tumor xenograft: Molecular size-dependence and cut off size. Cancer Research, 55: 3752–3756, 1995.

73. Dellian, M., Yuan, F., Trubetskoy, V., Torchilin, V.P., and Jain, R.K. Dependence of Microvascular Permeability on Molecular Charge in a Human Tumor Xenograft. International Journal of Microcirculation: Clinical and Experimental, 16: 152, 1996.
74. Hobbs, S.K., Yuan, F., Cima, L.G., and Jain, R.K. Tumor Microvascular Pore Cutoff Size: Implications for Macromolecules and Particle Drug Delivery. International Journal of Microcirculation: Clinical and Experimental, 16: 218, 1996.
75. Roberts, W.G., and Palade, G. Neovasculature induced by vascular endothelial growth factor is fenestrated. Cancer Research, 57: 1207–1211, 1997.
76. Yuan, F., Chen, Y., Dellian, M., Safabakhsh, N., Ferrara, N., and Jain, R.K. Time-dependent changes in vascular permeability and morphology in established human tumor xenografts induced by an anti-VEGF/VPF antibody. Proceedings of the National Academy of Sciences of the USA, 93: 14765–14770, 1996.
77. Arbit, E., Lee, J., and DiResta, G. Interstitial hypertension in human brain tumors: possible role in peritumoral edema formulation. *In:* Intracranial Pressure. H. Nagai, K. Kamiya and S. Ishi (ed.), Tokyo: Springer-Verlag, pp. 609–614, 1994.
78. Boucher, Y., Baxter, L.T., and Jain, R.K. Interstitial pressure gradients in tissue-isolated and subcutaneous tumors: implications for therapy. Cancer Research, 50: 4478–4484, 1990.
79. Boucher, Y., and Jain, R.K. Microvascular pressure is the principal driving force for interstitial hypertension in solid tumors: implications for vascular collapse. Cancer Research, 52: 5110–5114, 1992.
80. Boucher, Y., Lee, I., and Jain, R.K. Lack of general correlation between interstitial fluid pressure and pO2 in tumors. Microvascular Research, 50: 175–182, 1995.
81. Boucher, Y., Kirkwood, J.M., Opacic, D., Desantis, M., and Jain, R.K. Interstitial hypertension in superficial metastatic melanomas in humans. Cancer Research, 51: 6691–6694, 1991.
82. Boucher, Y., Salehi, H., Witwer, B., Harsh, G.R., and Jain, R.K. Interstitial fluid pressure in intracranial tumors in patients and in rodents: Effect of anti-edema therapy. British Journal of Cancer, In Press, 1997.
83. Curti, B.D., Urba, W.J., Alvord, W.G., Janik, J.E., Smith, J.W., Madara, K., and Longo, D.L. Interstitial Pressure of subcutaneous nodules in melanoma and lymphoma patients: Changes during treatment. Cancer Research, 53: 2204–2207s, 1993.
84. Gutmann, R., Leunig, M., Feyh, J., Goetz, A.E., Messmer, K., Kastenbauer, E., and Jain, R.K. Interstitial hypertension in head and neck tumors in patients: correlation with tumor size. Cancer Research, 52: 1993–1995, 1992.
85. Jain, R.K. Transport of molecules in the tumor interstitium: A review. Cancer Research, 47: 3039–3051, 1987.
86. Less, J.R., Posner, M.C., Boucher, Y., Borochovitz, D., Wolmark, N., and Jain, R.K. Interstitial hypertension in human breast and colorectal tumors. Cancer Research, 52: 6371–6374, 1992.
87. Nathanson, S.D., and Nelson, L. Interstitial fluid pressure in breast cancer, benighn breast conditions, and breast parenchyma. Annals of Surgical Oncology, 1: 333–338, 1994.
88. Roh, H.D., Boucher, Y., Kalnicki, S., Buchsbaum, R., Bloomer, W.D., and Jain, R.K. Interstitial hypertension in carcinoma of uterine cervix in patients: possible correlation with tumor oxygenation and radiation response. Cancer Research, 51: 6695–6698, 1991.
89. Znati, C.A., Karasek, K., Faul, C., Roh, H.-D., Boucher, Y., Rosenstein, M., Kalnicki, S., Buchsbaum, R., Chen, A., Bloomer, W.D., and Jain, R.K. Interstitial fluid pressure changes in cervical carcinomas in patients undergoing radiation therapy: A potential prognostic factor. Submitted, 1997.
90. Znati, C.A., Rosenstein, M., Boucher, Y., Epperly, M.W., Bloomer, W.D., and Jain, R.K. Effect of radiation on interstitial fluid pressure and oxygenation in a human colon carcinoma xenograft. Cancer Research, 56: 964–968, 1996.
91. Stohrer, M., Boucher, Y., Stangassinger, M., and Jain, R.K. Oncotic pressure in human tumor xenografts. Proceedings of the American Association of Cancer Research, 1995.
92. Boucher, Y., Leunig, M., and Jain, R.K. Tumor angiogenesis and interstitial hypertension. Cancer Research, 56: 4264–4266, 1996.
93. Rippe, B., and Haraldsson. Fluid and protein fluxes across small and large pores in the microvasculature. Application of two-pore equations. Acta Physiologica Scandinavia, 131: 411–428, 1987.
94. Berk, D.A., Yuan, F., Leunig, M., and Jain, R.K. Fluorescence photobleaching with spatial Fourier analysis: measurement of diffusion in light-scattering media. Biophysical Journal, 65: 2428–2436, 1993.
95. Chary, S.R., and Jain, R.K. Direct measurement of interstitial convection and diffusion of albumin in normal and neoplastic tissues by fluorescence photobleaching. Proceedings of the National Academy of Sciences of the USA, 86: 5385–5389, 1989.
96. Johnson, E.M., Berk, D.A., Jain, R.K., and Deen, W.M. Diffusion and partitioning of proteins in charged agarose gels. Biophysical Journal, 68: 1561–1568, 1995.

97. Johnson, E.M., Berk, D.A., Jain, R.K., and Deen, W.M. Hindered diffusion in agarose gels: Test of effective medium model. Biophysical Journal, 70: 1017–1026, 1996.
98. Johnson, M., Berk, D.A., Blankschtein, D., Golan, D.E., Jain, R.K., and Langer, R. Lateral diffusion of small compounds in human stratum corneum and model lipid bilayer systems. Biophysical Journal, 71: 2656–2668, 1996.
99. Nugent, L.J., and Jain, R.K. Extravascular diffusion in normal and neoplastic tissues. Cancer Research, 44: 238–244, 1984.
100. Swabb, E.A., Wei, J., and Gullino, P.M. Diffusion and convection in normal and neoplastic tissues. Cancer Research, 34: 2814, 1974.
101. Berk, D.A., Yuan, F., Leunig, M., and Jain, R.K. Direct in vivo measurement of targeted binding in a human tumor xenograft. Proceedings of the National Academy of Sciences of the USA, 94: 1785–1790, 1997.
102. Boucher, Y., Brekken, C., Netti, P.A., Baxter, L.T., and Jain, R.K. Hydraulic conductivity of solid tumors: A novel in vivo measurement technique and implications for drug delivery. Submitted, 1997.
103. Znati, C.A., Boucher, Y., Rosenstein, M., Turner, D., Watkins, S., and Jain, R.K. Effect of radiation on the interstitial matrix and hydraulic conductivity of tumors. Submitted, 1997.
104. Juweid, M., Neumann, R., Paik, C., Perez-Bacete, M.J., Sato, J., Van Osdol, W., and Weinstein, J.N. Micropharmacology of monoclonal antibodies in solid tumor: Direct experimental evidence for a binding site barrier. Cancer Research, 52: 5144, 1992.
105. Kaufman, E.N., and Jain, R.K. Quantification of transport and binding parameters using fluorescence recovery after photobleaching. Potential for in vivo applications. Biophysical Journal, 58: 873–885, 1990.
106. Kaufman, E.N., and Jain, R.K. Measurement of mass transport and reaction parameters in bulk solution using photobleaching. Reaction limited binding regime. Biophysical Journal, 60: 596–610, 1991.
107. Kaufman, E.N., and Jain, R.K. Effect of bivalent interaction upon apparent antibody affinity: experimental confirmation of theory using fluorescence photobleaching and implications for antibody binding assays. Cancer Research, 52: 4157–4167, 1992.
108. Kaufman, E.N., and Jain, R.K. In vitro measurement and screening of monoclonal antibody affinity using fluorescence photobleaching. Journal of Immunological Methods, 155: 1–17, 1992.
109. Jain, R.K. Delivery of novel therapeutic agents in tumors: Physiological barriers and strategies. Journal of the National Cancer Institute, 81: 570–576, 1989.
110. Leu, A.J., Berk, D.A., Yuan, F., and Jain, R.K. Flow velocity in the superficial lymphatic network of the mouse tail. American Journal of Physiology, 267: H1507–1513, 1994.
111. Berk, D.A., Swartz, M.A., Leu, A.J., and Jain, R.K. Transport in lymphatic capillaries: II. Microscopic velocity measurement with fluorescence recovery after photobleaching. American Journal of Physiology, 270: H330-H337, 1996.
112. Swartz, M.A., Berk, D.A., and Jain, R.K. Transport in lymphatic capillaries: I. Macroscopic measurements Using residence time distribution theory. American Journal of Physiology, 270: H324-H329, 1996.
113. Jeltsch, M., Kaipainen, A., Joukov, V., Meng, X., Lakso, M., Rauvala, H., Swartz, M., Fukumura, D., and Jain, R.K. Hyperplasia of lymphatic vessels in VEGF-C transgenic mice. Science, 276: 1423–1425, 1997.
114. Leu, A., Berk, D., and Jain, R.K. Search for initial lymphatics in solid tumors. Submitted, 1997.
115. Melder, R.J., Munn, L.L., Yamada, S., Ohkubo, C., and Jain, R.K. Selectin and integrin mediated T lymphocyte rolling and arrest on TNFa-activated endothelium is augmented by erythrocytes. Biophysical Journal, 69: 2131–2138, 1995.
116. Munn, L.L., Melder, R.J., and Jain, R.K. Role of erythrocytes in leukocyte-endothelial interactions: Mathematical model and experimental validation. Biophysical Journal, 71: 466–478, 1996.
117. Fukumura, D., Salehi, H., Witwer, B., Tuma, R.F., Melder, R.J., and Jain, R.K. TNF-alpha-induced leukocyte-adhesion in normal and tumor vessels: Effect of tumor type, transplantation site and host. Cancer Research, 55: 4824–4829, 1995.
118. Yamada, S., Melder, R.J., Leunig, M., Ohkubo, C., and Jain, R.K. Leukocyte-rolling increases with age. Blood, 86: 4707–4708, 1995.
119. Melder, R.J., Elmaleh, D., Brownell, A.L., Brownell, G.L., and Jain, R.K. A method for labeling cells for positron emission tomography (PET) studies. Journal of Immunological Methods, 175: 79–87, 1994.
120. Melder, R.J., and Jain, R.K. Kinetics of interleukin-2 induced changes in rigidity of human natural killer cells. Cell Biophysics, 20: 161–176, 1992.
121. Jain, R.K., Koenig, G., Dellian, M., Fukumura, D., Munn, L.L., and Melder, R.J. Leukocyte-endothelial adhesion and angiogenesis in Tumors. Cancer and Metastasis Reviews, 15: 195–204, 1996.
122. Melder, R.J., and Jain, R.K. Reduction of rigidity in human activated natural killer cells by thioglycollate treatment. Journal of Immunological Methods, 175: 69–77, 1994.
123. Melder, R.J., Salehi, H.A., and Jain, R.K. Localization of activated natural killer cells in MCaIV mammary carcinoma grown in cranial windows in C3H mice. Microvascular Research, 50: 35–44, 1995.

124. Sasaki, A., Melder, R.J., Whiteside, T.L., Herberman, R.B., and Jain, R.K. Preferential localization of human adherent lymphokine-activated killer cells in tumor microcirculation. Journal of the National Cancer Institute, 83: 433–437, 1991.
125. Ohkubo, C., Bigos, D., and Jain, R.K. Interleukin 2 induced leukocyte adhesion to the normal and tumor microvascular endothelium in vivo and its inhibition by dextran sulfate: implications for vascular leak syndrome. Cancer Research, 51: 1561–1563, 1991.
126. Melder, R.J., Koenig, G., Witwer, B.P., Safabakhsh, N., Munn, L.L., and Jain, R.K. During angiogenesis, vascular endothelial growth factor and basic fibroblast growth factor regulate natural killer cell adhesion to tumor endothelium. Nature Medicine, 2: 992–997, 1996.
127. Jallal, B., Powell, F., Zachwieja, J., Brakebusch, C., Germain, L., Jacobs, J., Iacobelli, S., and Ullrich, A. Suppression of tumor growth *in vivo* by local and systemic 90K level increase. Cancer Research, 55: 3223–3227, 1995.
128. Melder, R.J., Koenig, G., Munn, L.L., and Jain, R.K. Adhesion of activated natural killer cells to TNF-alpha treated endothelium under physiological flow conditions. Natural Immunity, 15: 154–163, 1997.
129. Gamble, J.R., and Vadas, M.A. Endothelial adhesiveness for blood neutrophils is inhibited by transforming growth factor-beta. Science, 242: 97–99, 1988.
130. Gamble, J.R., and Vadas, M.A. Endothelial cell adhesiveness for human T lymphocytes is inhibited by transforming growth factor-beta. Journal of Immunology, 146: 1149–1154, 1991.
131. Gamble, J.R., and Khew-Goodall, Y. Transforming growth factor-beta inhibits E-selectin expression on human endothelial cells. Journal of Immunology, 150: 4494–4503, 1993.
132. Fidler, I.J. Modulation of the organ microenvironment for treatment of cancer metastasis. Journal of the National Cancer Institute, 87: 1588–1592, 1995.
133. Kitayama, J., Nagawa, J., Yasuhara, H., Tsuno, N., Kimura, W., Shibata, Y., and Muto, T. Suppressive effect of basic fibroblast growth factor on transendothelial emigration of CD4(+) T-lymphocyte. Cancer Research, 54: 4729–4733, 1994.
134. Haying, J.B., and Williams, S.K. Reduced adhesion of human microvascular endothelial cells to collagen in response to basic FGF is mediated by $\beta 1$ integrin. FASEB Journal, 8: Abstr. 263, 1994.
135. Dedrick, R.L. Animal scale-up. Journal of Pharmacokinetics and Biopharmaceutics, 1: 435–461, 1973.
136. Gerlowski, L.E., and Jain, R.K. Physiologically based pharmacokinetic modeling: principles and applications. Journal of Pharmaceutical Sciences, 72: 1103–1127, 1983.
137. Zhu, H., Melder, R., Baxter, L., and Jain, R.K. Physiologically based kinetic model of effector cell biodistribution in mammals: implications for adoptive immunotherapy. Cancer Research, 56: 3771–3781, 1996.
138. Zhu, H., Baxter, L., and Jain, R.K. Potential and limitations of radioimmunodetection and radioimmunotherapy with monoclonal antibodies: evaluation Using a physiologically-based pharmacokinetic model. Journal of Nuclear Medicine, 96: 256–265, 1997.
139. Zhu, H., Jain, R.K., and Baxter, L.T. Tumor pretargeting for radioimmunodetection and radioimmunotherapy: Evaluation using a physiologically-based pharmacokinetic model. Submitted, 1997.
140. Jain, R.K. Effect of inhomogeneities and finite boundaries on temperature distribution in a perfused medium with application to tumors. Transactions of the ASME Journal of Biomechanical Engineering, 100: 235–241, 1978.
141. Jain, R.K. Transient temperature distributions in an infinite perfused medium due to a time-dependent, spherical heat source. Transactions of the ASME Journal of Biomechanical Engineering, 101: 82–86, 1979.
142. Jain, R.K., Boucher, Y., Stacey-Clear, A., Moore, R., and Kopans, D. Method for locating tumors prior to needle biopsy, U.S. Patent Number 5,396,897, March 14. 1995.
143. Dudar, T.E., and Jain, R.K. Differential response of normal and tumor microcirculation to hyperthermia. Cancer Research, 44: 605–612, 1984.
144. Fukumura, D., Yuan, F., Endo, M., and Jain, R.K. Role of nitric oxide in tumor microcirculation: Blood flow, vascular permeability, and leukocyte-endothelial interactions. American Journal of Pathology, In Press, 1996.
145. Gerlowski, L.E., and Jain, R.K. Effect of hyperthermia on microvascular permeability to macromolecules in normal and tumor tissues. International Journal of Microcirculation: Clinical & Experimental, 4: 363–372, 1985.
146. Kristensen, C.A., Nozue, M., Boucher, Y., and Jain, R.K. Reduction of interstitial fluid pressure after TNF-alpha treatment of human melanoma xenografts. British Journal of Cancer, 74: 533–536, 1996.
147. Kristensen, C.A., Roberge, S., and Jain, R.K. Effect of tumor necrosis factor-alpha on vascular resistance, nitric oxide production, glucose and oxygen consumption in perfused, tissue-isolated human melanoma xenografts. Clinical Cancer Research, 3: 319–324, 1997.
148. Fukumura, D., and Jain, R.K. Role of nitric oxide in angiogenesis and microcirculation in tumors. Cancer and Metastasis Reviews, In Press, 1997.

149. Baxter, L.T., and Jain, R.K. Pharmacokinetic analysis of the microscopic distribution of enzyme-conjugated antibodies and prodrugs: comparison with experimental data. British Journal of Cancer, 73: 447–456, 1996.
150. Baxter, L.T., Yuan, F., and Jain, R.K. Pharmacokinetic analysis of the perivascular distribution of bifunctional antibodies and haptens: comparison with experimental data. Cancer Research, 52: 5838–5844, 1992.
151. Kristjansen, P.E., Boucher, Y., and Jain, R.K. Dexamethasone reduces the interstitial fluid pressure in a human colon adenocarcinoma xenograft. Cancer Research, 53: 4764–4766, 1993.
152. Lee, I., Boucher, Y., and Jain, R.K. Nicotinamide can lower tumor interstitial fluid pressure: mechanistic and therapeutic implications. Cancer Research, 52: 3237–3240, 1992.
153. Lee, I., Boucher, Y., Demhartner, T.J., and Jain, R.K. Changes in tumour blood flow, oxygenation and interstitial fluid pressure induced by pentoxifylline. British Journal of Cancer, 69: 492–496, 1994.
154. Lee, I., Demhartner, T.J., Boucher, Y., Jain, R.K., and Intaglietta, M. Effect of hemodilution and resuscitation on tumor interstitial fluid pressure, blood flow, and oxygenation. Microvascular Research, 48: 1–12, 1994.
155. Leunig, M., Goetz, A.E., Dellian, M., Zetterer, G., Gamarra, F., Jain, R.K., and Messmer, K. Interstitial fluid pressure in solid tumors following hyperthermia: possible correlation with therapeutic response. Cancer Research, 52: 487–490, 1992.
156. Leunig, M., Goetz, A.E., Gamarra, F., Zetterer, G., Messmer, K., and Jain, R.K. Photodynamic therapy-induced alterations in interstitial fluid pressure, volume and water content of an amelanotic melanoma in the hamster. British Journal of Cancer, 69: 101–103, 1994.
157. Zlotecki, R.A., Baxter, L.T., Boucher, Y., and Jain, R.K. Pharmacologic modification of tumor blood flow and interstitial fluid pressure in a human tumor xenograft: network analysis and mechanistic interpretation. Microvascular Research, 50: 429–443, 1995.
158. Zlotecki, R.A., Boucher, Y., Lee, I., Baxter, L.T., and Jain, R.K. Effect of angiotensin II induced hypertension on tumor blood flow and interstitial fluid pressure. Cancer Research, 53: 2466–2468, 1993.
159. Jain, R.K. Delivery of molecular medicine to solid tumors. Science, 271: 1079–1080, 1996.

DISCUSSION

Rauscher: Can you induce lymph-angiogenesis with VEGF-C in tumors?

Jain: Probably you cannot for many reasons. They may not have receptors for it. Alternatively, lymphatic vessels are formed but they get "crushed" similar to normal vessels; probably the latter is the likely hypothesis. If you stain a tumor for FLT4 probably you will be able to stain it, that means there are lymphatic capillaries, that means lymphatic endothelium is there, but it may not be functional. That implies somehow these vessels are "crushed". We have just published a paper in *Nature Biotechnology* which shows why vessels are crushed: When tumor cells proliferate they generate solid stress, not fluid pressure but solid pressure, similar to somebody punching you. The pressure generated by cancer cells when they proliferate is about 45 to 120 mm mercury. A vessel has to have pressure greater than 45–120 mm mercury to avoid being crushed by cancer cells. The same stress may be crushing the normal vessels and that is why you see blood flow heterogeneity in tumor. The same stress may be also crushing the lymphatic vessels.

Rauscher: So they are formed, but they are just not functional?

Jain: We do not know if they are formed or not, but they certainly do not function, that we do know. There are very limited studies. We are looking at both animal and human tumors in collaboration with Dr. Kari Alitalo in our own laboratory. But I think there is a bigger message here: In molecular biology or in cellular biology we have a very simplistic view of the world, that is, with one molecule we can explain a lot of things. That is not the case. I showed the data on VEGF. If you look in the literature, VEGF is the primary factor which leads to vascular hyperpermeability in tumors. I showed two tumors, with equal amount of VEGF, one is permeable the other is not. I can go on and on, and give a number of examples of why the simplistic view of cancer is going to get us into trouble, or has gotten us into trouble, and will continue to unless we enlarge the vision and begin to look at integrative pathophysiology of tumors.

Mihich: How is the interstitial fluid in the tumor exported?

Jain: How does it come out? Very nice question.

Mihich: I remember the old model of Dr. Gullino in the ovarian tumor where you had an efferent artery and efferent vein. How does it get out here?

Jain: Very good question. If you measure the amount of plasma or blood entering a solid tumor via artery and the amount coming out from the vein, in a normal organ the two are close or equal. In tumors, for every 100 ml which enters via the artery only approximately 90 ml comes out of the vein, so where is this 10 ml coming out from? It comes out from the periphery of the tumor. It oozes out. If you were to look at a tumor it oozes interstitial fluid from the surface. What happens to this fluid? It is the peri-tumor lymphatics, and that is why the peri-tumor lymphatics get enlarged. When a pathologist sees lymphatics invaded by tumor, he says, O.K. there is lymphangiogenesis. They are host lymphatics. They are getting VEGF-C and why wouldn't they? If the tumor is producing large amount of VEGF-C, where is it going to go with the fluid leaving the tumor. The peri-tumor lymphatics. So they become hyperplastic and get enlarged. This can also facilitate lymphatic metastasis. Now you have larger lymphatics in the peri-tumor region and this oozing fluid carries cancer cells and drugs with it, and delivers it to the peri-tumor lymphatics. These cells go to the nearest lymphnode and set up home there. If the right soil is there and the right growth factors are present, lymphatic metastasis would occur.

Croce: How do some anticancer drugs, like taxol affect the endothelial cell? The normal component in the tumor, and, in fact, the fact that we can see them just knocking out angiogenesis.

Jain: Absolutely. As a matter of fact, I believe that a majority of time when an anticancer drug works, that does not have to be just taxol it can be any cytotoxic agent, a significant component of response may be anti-vascular and why wouldn't it be? Well, tumor endothelial cells proliferate fairly rapidly, so if these drugs work against proliferating cells and that is the first cell in the line of fire, it should go first. I think the data of Dr. Steve Rosenberg in patients on LAK and TIL therapy can be explained also that way. You cannot explain that data by direct cytotoxicity to cancer cells alone, because these cells may not extravasate in sufficient quantity. Yet, they are killing the tumor. One endothelial cell feeds hundred to thousand cancer cells, so you kill one endothelial cell and you can get 100–1,000 fold amplification. So, your point is well taken. The anticancer agents may be also anti-vascular.

Arndt: What is the driving force for extravasation of a drug, a liposome or microcapsule and is increased pressure in the tumor directed against this diffusion?

Jain: The question being asked is what is driving extravasation of liposomes or, gene targeting vectors, or viral particles of the same size. It is mostly diffusion, because convection is less in tumors. Let me explain. Molecules move in our body or any place by two mechanisms: Diffusion and convection. Diffusion is a process we know from daily experience. It is driven by concentration gradients. In the morning when you make your coffee, you put a little sugar in your coffee. If you do not do anything, do not even touch it, let it sit on the table for a couple of hours, the whole coffee becomes sweet by diffusion. But none of us wait for a couple of hours, right? What do we do? We stir it. What is stirring doing? It is called convection. You just set fluid in motion there. And that is what our body does. It does not wait for diffusion to take care of things. To drive convection you need pressure gradients. In normal vessels, pressure is about 25 mm Hg, extravascular pressure zero, so there is convection. In tumors, as I showed to you, the intra, and extravascular pressures are approximately equal, so convection disappears. So primarily molecules extravasate in tumors by diffusion. Diffusion is inherently slow. So, going back to what I showed you earlier, the vessels where the liposomes were coming out from, the diffusive pathways are large.

Livingston: So, Rakesh, just two questions. First of all, is it fair to argue that the major effect of chemotherapy, when it works, likely cannot be only, or even primarily, on endothelial cells? Otherwise it would be easy to treat tumors.

Jain: I agree with you. That is why I said earlier it is not a major but a significant factor. I totally agree with you. And the reason for that is very simple. Only a small fraction of tumor endothelium is proliferating. If you listen to others you might get the impression they are all proliferating, but it is less, depending upon the tumor.

Livingston: Question two. Can you set up technology to follow diffusion of a small molecule drug, versus death in a tumor lump? If you follow diffusion of a small molecule drug, the size of the various blood vessels and apoptosis of tumor cells, what result do you get?

Jain: I have not done that experiment, unfortunately. We just began to ask the question of tumor treatment. We have done only two types of experiments: anti-VEGF antibody treatment and hormonal withdrawal. We have just started radiation studies to see what that does differentially to the endothelial ells versus cancer cells. We would like to look at the effect of taxol and a number of different toxic agents. Part of the reason we have not done these experiments is because of the lack of NIH funds for these types of studies. But, let us face it, these are important questions. I have been asked to go to the pharmaceutical companies to get the funding for this type of work and when I go to pharmaceutical companies, they say this is an academic research.

Livingston: One needs to change that.

Jain: Well that needs to be changed at NCI, David, and at the industrial level.

Livingston: One can work on that. But in theory it should be possible now, with say, the right intracellular reporters even in viable cells to measure apoptosis versus where the drug is? Where the vessels are? And if you can do that kind of experiment then you can begin to ask, it seems to me, interesting questions regarding therapeutic outcome, based on all these parameters.

Jain: We can follow, David, different tracers simultaneously in these windows. Not only can we look at apoptosis, but also drug and endothelial cells and other elements. If adequate funds are available, we can do this type of work.

Anderson: I am curious, what do you think the possibility is that in fact the absence of lymphatics is what drives necrosis in the cores of tumors? I have always been dissatisfied with the explanations that exist on necrosis, and it seems like if you put an artificial drain in the middle of a tumor it might change therapeutic responses to radiation.

Jain: Yes, you can put a drain in the middle of a tumor to deliver drugs.

Anderson: But can you keep necrosis from arising? In other words, can you create artificial lymphatics to reduce this pressure problem?

Jain: We have not done it that way. I do want to draw to your attention that necrosis does not occur just because of a lack of lymphatics. I think it occurs because the proliferating tumor cells crush the blood vessels, as the tumor is growing in a confined space. If the tumor could grow in a free space that would not happen, or at least it would happen at a much reduced extent.

Melief: I am wondering what the absence of lymphatics means in terms of the immunizing potential of the tumors? So, the tumor antigen now only oozes out, in a very inefficient way. Is it true that because of this, you would predict that both T-lymphocytes that get it cannot get out of the tumor anymore, as well as antigen presenting cells that normally would travel free of the lymphatics to the draining nodes, cannot get out efficiently? Has this been studied at all?

Jain: The lymphatic model was recently developed and we are asking these kinds of questions.

Evans: I have a specific question and a more general question. The specific one is derived from your earlier data that NK cells had a higher potential to bind in tumor vasculature than T cells. In your VEGF transgenic model where you saw increased binding, was that NK cells or did you also see binding of T cells?

Jain: We do not know. The only thing we have measured is the overall leukocyte population and as you saw the overall adhesion goes up. We need to do the differential count to see what is really adhering there.

Evans: The increased adhesion was really impressive. The other aspect I was curious about, is in these strategies that are proposing anti-angiogenic inhibition of tumor growth, some of the things that are surprising is that some tissues that are established actually are undergoing angiogenesis as well. I was wondering is there any evidence, for example, in lymphoid tissues that are quite dynamic, that there is angiogenesis going on in that let us say, anti-VEGF could effect the endothelium, for example, in lymphnodes or in Peyer's patches and so on?

Jain: Please ask Peter or Georg--they might have a better answer to that question than me. They have probably stained everything in the body with the anti-VEGF antibody, I have not, so I cannot answer that question. There was a paper last month in *Nature Medicine* where TNP470 was given for a long time and there were undesirable effects in the female mice. So anti-angiogenic therapy would definitely have some limitations. In addition to physiological angiogenesis, during menstrual cycle, there is wound healing. If you are on anti-angiogenic therapy and you get hit by a truck you are in trouble and so on and so forth. So there are going to be a lot of settings, but it is always cost to benefit ratio. This is a small cost to pay for a large benefit. For example, unlike females in child-bearing ages, males would be protected from anti-angiogenic therapy just to give you an example. That means novel strategies would have to be developed. For example, if you are undergoing surgery or if you have a wound, some antidote can be given just around that time to help you recover from wound healing or surgery. But we are looking five or ten years down the road for that, right now several drugs are under clinical trials.

Zanker: Did you ever observe that tumor cells are entering the tumor vessels and escaping from the tumor? You can show this on the video?

Jain: You can collect the tumor cells coming out. As a matter of fact, my mentor Dr. Gullino, in 1974, published a paper in *Cancer Research* showing that a 1 gram tumor releases about 1–2 million cells per day. We have just repeated those studies. The good news is that most of these cells are dead, otherwise we would be in trouble.

Zanker: Because it is known that the best modulator of vascular diameter is nitric oxide, do you have any idea about the nitric oxide production?

Jain: We published a paper in the February 1997 issue *of The American Journal of Pathology* where we looked at the role of nitric oxide in tumor microcirculation, and the answer is very long. Yes, nitric oxide does play a role in tumor angiogenesis, permeability and metastasis. The first author of the paper is Dr. Dai Fukumura.

INDEX